Student Solutions Manual and Study Guide

INTRODUCTORY LINEAR ALGEBRA with Applications

FIFTH EDITION

Bernard Kolman
Drexel University

Prepared by
David R. Hill
Temple University

MACMILLAN PUBLISHING COMPANY
New York

MAXWELL MACMILLAN CANADA, INC.
Toronto

MAXWELL MACMILLAN INTERNATIONAL
New York Oxford Singapore Sydney

Printed in the United States of America.

Macmillan Publishing Company
866 Third Avenue, New York, New York 10022

Macmillan Publishing Company is part of the
Maxwell Communications Group of Companies.

Maxwell Macmillan Canada, Inc.
1200 Eglinton Avenue East
Suite 200
Don Mills, Ontario M3C 3N1

ISBN 0-02-354955-6

Printing: 1 2 3 4 5 6 7 8 Year: 3 4 5 6 7 8 9 0 1 2

Contents

Chapter 5. Eigenvalues and Eigenvectors

Chapter 6. Linear Programming

Chapter 7. Applications

Chapter 8. Numerical Linear Algebra

Appendix A. Complex Numbers

PREFACE

This *Student Solutions Manual and Study Guide* is to accompany the Fifth Edition of Bernard Kolman's *Introductory Linear Algebra with Applications*. Detailed solutions to all odd numbered computational and theoretical exercises are included. The majority of steps are included in many of the exercises or the procedure used to obtain an intermediate result is stated. Many exercises specifically refer to an example given in the text which closely parallels the technique used. In the theoretical exercises reasons for many of the steps involved cite a particular theorem from the text. When a group of exercises are solved by a particular technique, this is indicated by briefly describing the strategy. Such occurrences are denoted by the heading **<<Strategy>>**.

New to this edition is Chapter 0, Introduction to Proofs. Chapter 0 presents a brief introduction to logic, techniques of proof, fundamental notions on sets, and an optional section on mathematical induction. The purpose of this chapter is to provide a foundation for and a bridge to the abstract mathematical concepts encountered by students in linear algebra. This material can be used for self-study and reviewed as needed. Complete solutions to all exercises in Chapter 0 are provided.

Also new to this edition are the MATLAB Exercises. Either full or partial solutions, often with MATLAB commands, are given for all these exercises.

This *Student Solutions Manual and Study Guide* was developed with the aid of students and from experience using the text. I am grateful to John Edenhofner Jr., a former student, for his valuable assistance in developing the first edition of this manual. For this second edition I was assisted by Andrea Stout who tested the MATLAB exercises and provided valuable comments on other new material. It was a pleasure working with her. In addition I appreciate the cooperation of Bernard Kolman throughout the development of this manual. I am grateful for permission to use portions of the *Answer Manual* from previous editions. Thanks also to Robert Pirtle, Mathematics Editor of Macmillan, for his interest and cooperation during this project.

Please direct any comments or suggestions concerning this book to :

David R. Hill
Department of Mathematics
Temple University
Philadelphia, Pa. 19122

Chapter 0. Introduction to Proofs

In many instances a first course in linear algebra is used to introduce students to active participation in abstract mathematics. That is, not only will new concepts and proofs of such concepts be given in the text, but the student is expected to create proofs for related concepts. In many cases a student is not prepared for this "jump" in mathematical sophistication. Most high school mathematics and calculus emphasize computation and manipulation based on the ideas introduced, rather than on building new ideas (proofs) from previous ideas. It is not that the "jump" is so hard, rather it is often the case that there has been very little foundation laid to understand the language of the statement of theorems and the nature of the techniques of proof.

Instead of blaming the system or previous instructors, what is developed here is a short summary and explanation of fundamental notions that are valuable on two fronts:

1. Reading and interpreting statements in the language of mathematics.

2. Developing proofs.

In order to use this material effectively, you must actively participate in it. One way to do this is to record your work in a notebook, both scratch work, which often contains initial ideas and attempts, and any refined versions. That way you can compare your work with others and with the solutions supplied here. Moreover, you (and your instructor) will be able to see that the strategies you have employed will vary and mature as you work throughout your linear algebra course. Remember that proofs presented in books are polished versions that have undergone numerous refinements. Such proofs are meant to be read by those who understand the language of mathematics. You should not expect your proofs to appear as slick or succinct as those in a book. With practice and experience your proofs will improve.

Throughout your previous mathematical experiences you have been introduced to some of the vocabulary of mathematics. Such terms as variable, equation, and real number should be quite familiar. In fact, there is a set of mathematical terminology that has become almost automatic due to your courses in algebra, trigonometry, and calculus. As you progress through your linear algebra course your mathematical vocabulary will expand. What is important in abstract mathematics is the way that the mathematical vocabulary is woven into the grammar used to describe concepts and make statements about concepts. We must communicate these in an unambiguous way; such precision is a crucial aspect of mathematics. The verification (proof) of a statement must be done so that there is no ambiguity present, for only then, will there be no doubt about its correctness.

In order to develop the precision of exposition required, we include a discussion of logic. Our focus will be to use the structure of logic as part of the method of mathematical proofs and their validity. We introduce a symbolic language, with symbols having precise meanings and uses. In this way known facts can be combined to prove new facts in a systematic fashion. We also avoid the imprecise overtones of everyday language that can cloud our reasoning. The use of connectives for building compound statements and the use of conditional statements is developed.

This Introduction to Proofs is meant to be for self-study. It has been the author's experience that small groups of students working together can greatly aid one another when using this material. It is recommended that this material be studied early in conjunction with the first few sections of Chapter 1. Thus the ideas will be available as needed for proofs as you progress through Chapter 1 and in later chapters which are more theoretically oriented.

Section 0.1 Logic

Logic is an analytical theory of the art of reasoning. We shall introduce symbols which facilitate the development of rules and techniques for determining the validity of an argument. Most of the things we will prove or that you are asked to prove in linear algebra appear in the form of a declarative sentence. That is, mathematical language that makes a statement. We adopt the following terminology.

Definition 1. A statement or proposition is a declarative sentence that is either true or false.

Example 1. Each of the following is a statement.

(a) Nero is dead.

(b) Every rectangle is a square.

(c) $4 + 3 = 7$

(d) $x^2 \geq 0$, for every real number x.

(e) $\sqrt{x^2} = x$, for every real number x.

Parts (a), (c) and (d) are true statements, while (b) and (e) are false. (In (e), $\sqrt{(-3)^2} = \sqrt{9} = 3 \neq -3$, so it is false.) ■

Example 2. Each of the following is not a statement because it is not a declarative sentence or its truth value is not known.

(a) $x^2 - 3x + 2 = 0$ {It is a declarative mathematical sentence, but its truth value depends on the value of x.}

(b) Can you read Russian? {This is question, not a statement.}

(c) Call me tomorrow. {This is a command, not a statement.}

■

In algebra we studied equations for which our focus was to determine a solution. The mathematical formulation involved using a variable, often represented by x, y, z,... , to represent a real number solution of the equation. In (mathematical) logic we represent statements using letters p, q, r, ... In this case the letters p, q, r, ... , are variables representing statements. Thus we can assign

p : Six is prime.

q : The earth is larger than the moon.

Hence in the context of the problem where these assignments are made we can use symbols p and q rather than the explicit verbal statements.

In algebra we manipulated numbers and variables representing numbers by using the arithmetic operations +, −, ×, ÷, and exponentiation. In logic we manipulate statements and variables representing statements using connective operations and conditional operations. We consider three connectives which are defined next.

Definition 2. Given statements p and q.

(a) The statement "p and q" is denoted $p \wedge q$ and is called the conjunction of p and q. Statement $p \wedge q$ is true when both p and q are true.

(b) The statement "p or q" is denoted $p \vee q$ and is called the disjunction of p and q. Statement $p \vee q$ is true when either p or q or both are true.

(c) The statement "not p" is denoted ~p and is called the negation of p. Statement ~p is true when p is false.

The use of connective operations builds compound statements from p and q, the original statements. The relationship between the statements p and q and compound statements $p \wedge q$, $p \vee q$, and ~p is conveniently displayed using a truth table. A truth table is a visual display of the truth value of the compound statement given the various possibilities for the truth of the original statements. Directly from Definition 2 we have the following truth tables.

p	q	$p \wedge q$
T	T	T
T	F	F
F	T	F
F	F	F

Table 1.

p	q	$p \vee q$
T	T	T
T	F	T
F	T	T
F	F	F

Table 2.

p	~p
T	F
F	T

Table 3.

The use of logic and connectives for aid in proofs requires that we recognize the verbal equivalents of the symbols $\wedge$, $\vee$, and ~. We present some common verbal forms next:

$\wedge$ and, together with, combined with, all of

$\vee$ or, one of, any

~ not, opposite

In most cases you will see the concise verbal equivalents "and", "or", and "not".

Several different compound statement forms often appear. Examples 3 through 6 briefly discuss these forms using truth tables.

Example 3. Given statements p and q determine the truth table of compound statement $\sim(p \wedge q)$. (In words, not (p and q).)

p	q	$p \wedge q$	$\sim(p \wedge q)$
T	T	T	F
T	F	F	T
F	T	F	T
F	F	F	T

Table 4.

We see that column 3 comes directly from Table 1 and then column 4 is obtained by negating (finding the opposite) of the truth values of column 3. As Table 4 shows $\sim(p \wedge q)$ is false only when both p and q are true.

■

Example 4. Given statements p and q determine the truth table of compound statement $\sim(p \vee q)$. (In words, not (p or q).)

p	q	p ∨ q	~(p ∨ q)
T	T	T	F
T	F	T	F
F	T	T	F
F	F	F	T

Table 5.

We see that column 3 comes directly from Table 2 and then column 4 is obtained by negating column 3. As Table 5 shows ~(p ∨ q) is false when either p or q is true.

■

Example 5. Given statements p and q determine the truth table of compound statement (~p) ∨ (~q). (In words, not p and not q.)

p	q	~p	~q	(~p) ∨ (~q)
T	T	F	F	F
T	F	F	T	T
F	T	T	F	T
F	F	T	T	T

Table 6.

■

Example 6. Given statements p and q determine the truth table of compound statement (~p) ∧ (~q). (In words, not p or not q.)

p	q	~p	~q	(~p) ∧ (~q)
T	T	F	F	F
T	F	F	T	F
F	T	T	F	F
F	F	T	T	T

Table 7.

■

From the truth tables in Tables 4, 5, 6, and 7 we make the following observations.

(a) The truth table for $\sim(p \wedge q)$ is identical to the truth table for $(\sim p) \vee (\sim q)$. (We mean that the final columns in Tables 4 and 6 are the same.)

(b) The truth table for $\sim(p \vee q)$ is identical to the truth table for $(\sim p) \wedge (\sim q)$. (We mean that the final columns in Tables 5 and 7 are the same.)

Definition 3. Two statements are <u>equivalent</u> provided they have the same truth table.

Thus we can rephrase (a) and (b) above as follows.

(a) $\sim(p \wedge q)$ is equivalent to $(\sim p) \vee (\sim q)$.

(b) $\sim(p \vee q)$ is equivalent to $(\sim p) \wedge (\sim q)$.

As far as proofs are concerned we can use an equivalent statement to replace a given statement without changing the problem. Hence **we can substitute an equivalent statement for a given statement to help simplify things or make things easier or just to change the point of view.** You did this type of maneuver many times in algebra as the following illustrates:

- Fraction 5/3 is equivalent to mixed number $1\frac{2}{3}$.
- Multiplying by 1 is equivalent to multiplying by 5/5, or by $\frac{x - 2}{x - 2}$, $x \neq 2$.
- Dividing by 2 is equivalent to multiplying by 1/2.
- Simplifying or putting in lowest terms.

We give further illustrations of the power and versatility of equivalent statements later. The use of equivalent statements is an important technique in constructing proofs.

Another way to construct compound statements is to use conditional operations. The simplest of these operations is defined as follows.

Definition 4.	Given statements p and q, the conditional statement or implication is if p then q which is denoted symbolically as $p \Longrightarrow q$ This is read "p implies q".

The conditional statement is the most important kind of statement in mathematics. You have used conditional statements many times as illustrated in the following example.

Example 7.

- If two lines are parallel (p), then the lines do not intersect (q).
- If it does not rain today (p), I will cut the grass (q).
  (Sometimes, "then" does not appear explicitly.)
- Quantity $\frac{-b + \sqrt{b^2 - 4ac}}{2a}$ is a complex number (q) if $b^2 - 4ac < 0$ (p). ∎

To appropriately use the compound statement $p \Longrightarrow q$ we must identify statements p and q correctly. The following terminology is used in this regard:

The **if-statement** p is called the antecedent or hypothesis.

The **then-statement** q is called the consequent or conclusion.

In this manual we use the terms hypothesis for the if-statement and conclusion for the then-statement.

Mathematicians have agreed that the truth value of a conditional statement will be given as follows.

Definition 4. (continued)	The conditional statement $p \Longrightarrow q$ is true whenever the hypothesis is false or the conclusion is true.

Thus we have the following truth table.

p	q	p ==> q
T	T	T
T	F	F
F	T	T
F	F	T

Table 8.

An unexpected result from Table 8 is that conditional statements may be true even when there is no connection between the hypothesis and conclusion. For example,

Eagles can fly ==> 1 + 1 = 2.

We will have no occasion to exploit this curiosity.

A primary objective in our work will be to construct a proof of p ==> q. This means we want to show

If p is true, then q is true.

In order to construct a **proof, a logical argument in the language of mathematics**, we must identify the hypothesis and conclusion. A conditional statement can appear disguised in a number of different ways. The following expressions are all equivalent to p ==> q.

if p then q	q whenever p
p implies q	p only if q
q if p	p is sufficient for q
q provided that p	q is necessary for p
	from p it follows that q

Example 7 gives several illustrations. Exercise 6 at the end of the section provides an opportunity to identify the hypothesis and conclusion.

We digress briefly to comment on two forms equivalent to p ==> q which occur frequently in mathematics. The form

p is sufficient for q

can be thought of in the following way. It is sufficient to know that statement p is true in order to conclude that statement q is true. We sometimes say <u>p is a sufficient condition for q</u>. The form

q is necessary for p

can be thought of in the following way. It necessarily follows that if statement p is true then statement q is true. We sometimes say q is a necessary consequence of p.

In constructing proofs for conditional statements, p ==> q, it is sometimes simpler to prove an equivalent statement called the **contrapositive**.

> **Definition 5.** Given statements p and q the contrapositive of p ==> q is (~q) ==> (~p).

Verification that p ==> q and (~q) ==> (~p) are equivalent is done by showing that their truth tables are identical. See Exercise 7.

Example 8. Form the contrapositive of each of the following conditional statements.

(a) Statement p ==> q: If two lines are parallel, then the lines do not intersect.
Contrapositive (~q) ==> (~p): If two lines intersect, then the lines are not parallel.

(b) Statement p ==> q: If a quadratic equation has two distinct real roots, then its graph crosses the x-axis.
Contrapositive (~q) ==> (~p): If the graph of a quadratic equation fails to cross the x-axis, then the equation does not have two distinct real roots.

An effort should be made to make the contrapositive statement grammatically correct.

■

Another type of construction that is related but not equivalent to a conditional statement is defined next.

> **Definition 6.** Given statements p and q, the converse of p ==> q is q ==> p.

Note that the converse merely interchanges the roles of the hypothesis and the conclusion. To show that the converse is not equivalent to the conditional statement we construct the truth tables in Table 9. We see that the third and fourth columns are not identical hence a conditional statement and its converse are not equivalent.

Warning: In proofs we can **not** substitute the converse for the original statement.

p	q	$p \Longrightarrow q$	$q \Longrightarrow p$
T	T	T	T
T	F	F	T
F	T	T	F
F	F	T	T

Table 9.

Example 9. Construct the converse of each of the following conditional statements.

(a) Statement $p \Longrightarrow q$: If numbers a and b are positive, then ab is positive.

Converse $q \Longrightarrow p$: If ab is positive, then a and b are positive.

Note that the original statement $p \Longrightarrow q$ is true while its converse $q \Longrightarrow p$ is false because $ab > 0$ for $a = -1$, $b = -2$.

(b) Statement $p \Longrightarrow q$: If $n + 1$ is odd, then n is even.

Converse $q \Longrightarrow p$: If n is even, then $n + 1$ is odd.

Note that in this case both $p \Longrightarrow q$ and its converse $q \Longrightarrow p$ are true.

■

There is another type of conditional operation that is closely related to a conditional statement $p \Longrightarrow q$.

Definition 7. Given statements p and q, the <u>biconditional statement</u> is

p if and only if

which is denoted symbolically as

$$p \Longleftrightarrow q$$

A biconditional is true when p and q have the same truth value.

It follows directly from Definition 7 that a biconditional has the following truth table.

p	q	p $\iff$ q
T	T	T
T	F	F
F	T	F
F	F	T

Table 10.

Because a biconditional statement p $\iff$ q is true precisely when the truth values of p and q are the same, it follows from Definition 3 that a biconditional can be used to test whether p and q are equivalent. Thus from Examples 4 and 6 we have

$$\sim(p \wedge q) \iff (\sim p) \vee (\sim q)$$

and from Examples 5 and 7 we have

$$\sim(p \vee q) \iff (\sim p) \wedge (\sim q).$$

Example 10. Each of the following is a biconditional statement.

(a) An algebraic equation represents a line if and only if it can be put into the form $ax + by = c$ for some constants a, b, c.

(b) $a > b$ if and only if $a - b > 0$.

(c) A function f(x) is a polynomial if and only if it can be put into the form $a_n x^n + \cdots + a_1 x + a_0$.

(d) An integer $n > 1$ is prime if and only if its only divisors are 1 and itself.

■

A convenient way to think of a biconditional is

p $\iff$ q is true exactly when p and q are equivalent.

Several additional comments concerning biconditionals are useful for learning to read abstract mathematics and learning to do proofs. As you have seen in this brief discussion of logic, definitions give us a set of (elementary) facts which can be used to build more complicated statements or relationships. Any definition supplies us with an equivalence or biconditional statement. Hence definitions give us items which can be used to substitute for one another. In the text you will see many definitions which have the following form:

Definition: A matrix is called 'some name' if

'algebraic type expression'.

Because a definition is an equivalence (a biconditional), any time the 'some name' appears we can substitute the corresponding 'algebraic type expression' if it more useful in a given circumstance. Thus we often can convert a verbal description to an equivalent algebraic form. This is quite valuable when justifying a particular step of a proof.

From Definition 7, $p \Longleftrightarrow q$ is read "p if and only if q". Some times "if and only if" is abbreviated "iff" and in such cases the biconditional would appear in the form p iff q. A biconditional is actually a conjunction of the two conditional statements $p \Longrightarrow q$ and $q \Longrightarrow p$. To verify this we show that the truth table for compound statement $(p \Longrightarrow q) \wedge (q \Longrightarrow p)$ is the same as Table 10. (See Exercise 13.) Thus it follows that

To prove $p \Longleftrightarrow q$ we must show <u>both</u> $p \Longrightarrow q$ and (its converse) $q \Longrightarrow p$ are true.

Keep in mind that to prove an if and only if statement there are two things that must be proved. We will return to this in the section on proof techniques.

In the language of mathematics a biconditional ($p \Longleftrightarrow q$) or an if and only if statement is often referred to as <u>a set of necessary and sufficient conditions</u>. This is just another way of saying statements p and q are equivalent. The search for alternative sets of necessary and sufficient conditions is very important in mathematics. With several such sets, we have flexibility in substituting equivalent expressions which in many cases permit us to change the point of view adopted in a proof. You will see that the text emphasizes this by presenting sets of equivalent properties in summaries of important concepts.

Exercises 0.1

1. Determine which of the following are statements. For those which are statements give its truth value.

   (a) $5 - 1 = 8$
   (b) The diagonals of a square are perpendicular.
   (c) $x^2 - 1 = 0$
   (d) Are we having fun yet?
   (e) $\sqrt{x^2} = |x|$, for any real number x.
   (f) If p(x) is a polynomial, then it is a function whose domain is the set of all real numbers.

2. Given statements p, q, and r. Write out each of the following compound statements.

   p: The door is open.
   q: The tables are set.
   r: The food is ready.

   (a) $p \wedge q$ (b) $q \vee (\sim r)$ (c) $p \wedge (\sim q)$ (d) $p \wedge q \wedge r$

3. Given statements p, q, and r. Express each of the following symbolically.

   p: The car won't start.
   q: It is freezing outside.
   r: I am going back to bed.

   (a) The car will start and its is freezing outside.
   (b) Either I am going back to bed or it is freezing outside.
   (c) I am not going back to bed and the car won't start.
   (d) I am staying up.

4. Construct a truth table for each of the following statements.

   (a) $p \wedge (\sim q)$ (b) $p \vee (q \wedge (\sim p))$
   (c) $p \wedge (q \vee (\sim p))$ (d) $p \vee (\sim q)$

5. Formulate an (algebraically) equivalent statement for each of the following.

   (a) Multiply both sides of the equation by 1/3.
   (b) Subtract 3 from both sides of the equation.
   (c) $5 - 3$
   (d) $\dfrac{x^2 + x - 2}{x^2 - 1}$

6. Identify the hypothesis p and conclusion q in each of the following conditional statements.

   (a) I will go to the movies if it rains.
   (b) If the diagonals of a rectangle are not perpendicular, it is not a square.

Exercises 0.1

(c) If $x > 0$ and $y > 0$, then $x + y > 0$.
(d) $f(x)$ is a function on $[a,b]$ implies that $f(x)$ is continuous on $[a,b]$.
(e) I will pay $5.00 for a cup of coffee only if elephants fly.
(f) All sides are equal provided that T is an equilateral triangle.
(g) Parallel opposite sides is sufficient for a quadrilateral to be a parallelogram.

7. Show that $p \Longrightarrow q$ and the contrapositive $(\sim q) \Longrightarrow (\sim p)$ are equivalent statements using truth tables.

8. Form the contrapositive of each of the following statements.

(a) If two lines are perpendicular, then they intersect at right angles.
(b) Function $f(x)$ is differentiable provided that $f(x)$ is a polynomial.

9. Form the converse of each of the statements in Exercise 8.

10. Show that $\sim(p \Longrightarrow q) \quad \Longleftrightarrow \quad p \wedge (\sim q)$.

11. Show that $\sim(p \wedge q) \quad \Longleftrightarrow \quad (p \Longrightarrow (\sim q))$.

12. Complete the following biconditional statements.

(a) Positive integer n is even $\Longleftrightarrow$ ________________.

(b) Angles a and b are complementary if and only if _________.

(c) $x^2 \leq 4$ iff ____________.

(d) $n + 1$ is even $\Longleftrightarrow$ ____________.

(e) Parabola $y = ax^2 + 1$ opens upward iff ____________.

13. Show that $(p \Longrightarrow q) \Longleftrightarrow ((p \Longrightarrow q) \wedge (q \Longrightarrow p))$.

Section 0.2 Techniques of Proof

In this section we discuss techniques for constructing proofs of conditional statements $p \Longrightarrow q$ and biconditionals $p \Longleftrightarrow q$. As you progress through your linear algebra course the material changes, hence the content of statements p and q will change. Once p and q have been determined you should identify the type of material related to p and q. In other words, the context or mathematical area that is relevant to the problem. This helps you focus on the area from which you may be able to use definitions, previous theorems, or even previous problems as justifications or reasons (replacements) for steps in a proof.

To prove $p \Longrightarrow q$ we must show that whenever p is true it follows that q is true using a logical argument in the language of mathematics. The construction of this logical argument may be quite elusive, while the logical argument itself is the thing we call the proof. Conceptually the proof that p implies q is a sequence of steps that logically connect p to q. Each step in the "connection" must be justified or have a reason for its validity which is usually a previous definition, a property or axiom that is known to be true, a previously proven theorem or problem, or even a previously verified step in the current proof. Thus we connect p and q by logically building blocks of known (or accepted) facts. Often it is not clear what building blocks (facts) to use, and especially how to get started on a fruitful path. In many cases the first step of the proof is crucial. Unfortunately we have no explicit guidelines in this area other than to recommend that you carefully read the hypothesis p and conclusion q in order to clearly understand them. Only in this way can you begin to seek relationships (connections) between them. Experience and practice are a must in developing proof skills.

The construction of a proof requires us to build a step-by-step connection (a logical bridge) between p and q. If we let b_1, b_2, ..., b_n represent logical building blocks then conceptually our proof appears as

$$p \Longrightarrow b_1 \Longrightarrow b_2 \Longrightarrow \cdots \Longrightarrow b_n \Longrightarrow q$$

where each conditional must be justified. This approach is known as a direct proof. We illustrate this in Examples 1 and 2.

Example 1. Prove: If m and n are even integers, then m + n is even.

Proof: Let p: m and n are even integers
q: m + n is even
Note that the mathematical area is numbers, particularly even numbers. Reasons will appear in braces $\{\cdots\}$.

To start we ask ourselves, assuming p is true, what facts do we know that can lead to q. Since both p and q include the use of even numbers, for building block b_1 we try for something involving even numbers. We know that an even number is a multiple of 2. Try the following.

$p \Longrightarrow \underbrace{m = 2k,\ n = 2j}_{b_1}$ for some integers k and j {property of even numbers}

Since q involves the sum m + n, we try involving the sum in b_2.

$b_1 \Longrightarrow \underbrace{m + n = 2k + 2j = 2(k + j)}_{b_2}$ {properties of arithmetic}

We observe that b_2 implies that the sum m + n is a multiple of 2. Hence m + n is even. This is just q, hence we have

$b_2 \Longrightarrow q$

In summary $p \Longrightarrow b_1 \Longrightarrow b_2 \Longrightarrow q$.

■

Example 2. Let x be a real number.
Prove: $|x + 2| < 1$ implies $-3 < x < -1$.

Proof: Let p: $|x + 2| < 1$
q: $-3 < x < -1$

The mathematical area involves inequalities.

To start we ask ourselves, assuming p is true, what facts do we know that can lead to q. The conclusion involves an expression that has x isolated in the middle. Thus we will try an algebraic rearrangement of p.

$p \Longrightarrow \underbrace{-1 < x + 2 < 1}_{b_1}$ {properties of absolute value and inequalities}

Next try an arithmetic operation to isolate x in the middle; subtract 2 from each piece of the string of inequalities.

$b_1 \Longrightarrow \underbrace{-3 < x < -1}_{q}$

In summary $p \Longrightarrow b_1 \Longrightarrow q$.

■

In both Examples 1 and 2 we began with

"What fact(s) concerning hypothesis p can be used to start a bridge to q?"

We obtained a building block b_1 such that $p \Longrightarrow b_1$. We proceeded by asking ourselves

"What fact(s) concerning hypothesis b_1 can be used to continue the bridge to q?"

This scenario was continued until q was obtained. Examples 1 and 2 proceeded forward from p to q, building logical connections. We call this forward building.

Alternatively we could ask ourselves

"What fact(s) must be known to conclude that q is true?"

Call this b_1. Continue backwards from b_1 by asking

"What fact(s) must be known to conclude that b_1 is true?"

Call this b_2. Continue in this fashion until we reach p. Such a logical bridge is called backward building. Proofs can sometimes be built either way, but we have no way to determine in a particular instance which technique may be more fruitful. As you might expect, the two techniques can be combined. Build forward a few steps, build backward a few steps and strive to logically join the two ends. Often it is an individual's choice based on intangibles that determines the approach used. Most proofs in books appear to use forward building, but remember such proofs are the result of many refinements. Any of these constructions is called a direct proof.

Often building at both ends is used in proving trigonometric identities since subsitutions that express both sides in terms of sines and cosines are easy to make. See Example 3. Another interesting combination of forward and backward building occurs in calculus limit proofs involving the definition of limits in terms of epsilon and delta. Students are often amazed at how it was known what δ (in terms of ϵ) should be chosen so that the proof works out perfectly. The answer is that what you are reading is a refined version of a proof in which the δ was not known but determined from the manipulation of the proof and then used to make a "clean" looking proof for publication. See Example 4.

Example 3. Prove the trigonometric identity $\cos^2\theta = \dfrac{1 + \cos 2\theta}{2}$.

Proof: Let p: $\cos^2\theta$ and q: $\dfrac{1 + \cos 2\theta}{2}$

Starting forward from p using the Pythagorean identity we have

$p \Longrightarrow 1 - \sin^2\theta$, that is $\cos^2\theta = 1 - \sin^2\theta$

Starting backward from q using the identity for the cosine of a sum of two angles we have

$$q \Longrightarrow \frac{1 + \cos(\theta + \theta)}{2} \Longrightarrow \frac{1 + \cos^2\theta - \sin^2\theta}{2}$$

Replacing $\cos^2\theta$ by $1 - \sin^2\theta$ as in the forward step we have

$$\frac{1 + \cos^2\theta - \sin^2\theta}{2} \Longrightarrow \frac{1 + (1 - \sin^2\theta) - \sin^2\theta}{2} \Longrightarrow \frac{2(1 - \sin^2\theta)}{2}$$
$$\Longrightarrow 1 - \sin^2\theta$$

Connecting the forward and backward building we conclude that $p \Longrightarrow q$. Actually an identity is a biconditional. Since algebraic operations and substitutions are really equivalences, each of the preceding $\Longrightarrow$ can be replaced by $\Longleftrightarrow$. Thus we have simultaneously proved $q \Longrightarrow p$. Note that this is a special case and in general we can not replace $\Longrightarrow$ by $\Longleftrightarrow$.

■

Warning: Example 4 is more difficult than any of the preceding examples and uses theory that appears in calculus. This example can be omitted without loss of continuity in the remaining discussions in this chapter. If you choose to omit this example, then Exercise 7 should also be omitted.

Example 4. Prove $\lim_{x \to 3} x^2 = 9$.

Proof: According to the definition of a limit we proceed as follows Let $\epsilon > 0$ be given. We must find $\delta > 0$ such that whenever x is chosen such that $0 < |x - 3| < \delta$, then we have $|x^2 - 9| < \epsilon$. Let p: $0 < |x - 3| < \delta$ for some appropriate choice of δ

$$q: |x^2 - 9| < \epsilon$$

In order to determine δ, which will depend on ϵ, we start backwards.

$$q: |x^2 - 9| < \epsilon \Longleftrightarrow \underbrace{|x + 3|\ |x - 3| < \epsilon}_{b_1}$$

Note that the term $|x - 3|$ which is important to p has appeared in this equivalent form of q. Near $x = 3$, $|x + 3|$ is near 6 and $|x - 3|$ is near 0. The next step is to move backwards from b_1 so that we leave the term $|x - 3|$ alone but replace the term $|x + 3|$ with a constant. Eventually as $x \to 3$ we can consider x in a small interval around 3. Suppose we take $2.9 < x < 3.1$ which is equivalent to saying $|x - 3| < .1$ and which is equivalent to $5.9 < x + 3 < 6.1$.

For $2.9 < x < 3.1$, $x + 3 = |x + 3|$ so $|x + 3| < 6.1$. Thus we can see that b_1 is implied by b_2 where

$$b_1 \Leftarrow \underbrace{6.1\,|x - 3| < \epsilon \text{ when } 0 < |x - 3| < .1}_{b_2}$$

We see that

$$b_2 \Longleftrightarrow \underbrace{|x - 3| < \frac{\epsilon}{6.1} \quad \text{when } 0 < |x - 3| < .1}_{b_3}$$

In b_3 we have two conditions that $|x - 3|$ is to satisfy in order to imply b_2 which in turn implies b_1 which in turn implies q. We combine these conditions into one as follows. Let

$$\delta = \min\{.1, \epsilon/6.1\},$$

then

$$b_3 \Leftarrow \underbrace{0 < |x - 3| < \delta}_{p}$$

In summary, if we choose δ as defined above (or smaller), then $p \Longrightarrow q$.

∎

The proof of a biconditional requires that we prove both $p \Longrightarrow q$ and $q \Longrightarrow p$. If in proving $p \Longrightarrow q$ each building block used is an equivalence then we have simultaneously proved $q \Longrightarrow p$. See Example 3. Warning: You must check to determine that each building block in the bridge from p to q is an equivalence. You can **not** assume that it is.

Another proof technique replaces the original statement $p \Longrightarrow q$ by an equivalent statement and then proves the new statement. Such a procedure is called an indirect method of proof. One indirect method uses the equivalence between $p \Longrightarrow q$ and its contrapositive $(\sim q) \Longrightarrow (\sim p)$. The proof of $(\sim q) \Longrightarrow (\sim p)$ is done directly. We call this proof by contrapositive. Unfortunately there is no specific indicator in a conditional that informs you that an indirect proof by contrapositive may be fruitful. Sometimes the appearance of the word not in the conclusion q is a suggestion to try this method. There are no guarantees that it will work. We illustrate the use of proof by contrapositive in Examples 5 and 6.

Example 5. Use proof by contrapositive to show:

If 3m is an odd number, then m is odd.

<u>Proof:</u> Let p: 3m is odd, and q: m is odd. The contrapositive, (~q) ⟹ (~p) is

If $\underbrace{\text{m is even}}_{(\sim q)}$, then $\underbrace{\text{3m is even}}_{(\sim p)}$.

Using forward building we have

$(\sim q) \Longrightarrow \underbrace{m = 2k, \text{ for some integer } k}_{b_1}$ {a meaning of even}

We see that (~p) involves 3m, so we arrange to get 3m into the act.

$b_1 \Longrightarrow \underbrace{3m = 3(2k) = 2(3k)}_{b_2}$ {by algebra}

Thus 3m is expressed as 2 times another number, hence 3m is even. That is,

$b_2 \Longrightarrow (\sim p)$

In summary, (~q) ⟹ (~p) which is equivalent to p ⟹ q. ■

Example 6. Use proof by contrapositive to show:

If a and b are positive real numbers such that 4ab is not equal to $(a + b)^2$ then a is not equal to b.

<u>Proof:</u> Let p: $4ab \neq (a + b)^2$, and q: $a \neq b$. The contrapositive (~q) ⟹ (~p) is

If a = b, then $4ab = (a + b)^2$.

Using backward building we have

$(\sim p) \Longleftarrow \underbrace{4ab = (a + b)^2 = a^2 + 2ab + b^2}_{b_1}$ {by algebra}

$b_1 \Longleftarrow \underbrace{0 = a^2 - 2ab + b^2}_{b_2}$ {by algebra}

$$b_2 \Longleftarrow \underbrace{0 = (a - b)^2}_{b_3} \quad \text{\{by algebra\}}$$

$$b_3 \Longleftarrow \underbrace{0 = |a - b|}_{b_4} \quad \text{\{take square root of both sides\}}$$

$$b_4 \Longleftarrow \underbrace{a = b}_{(\sim q)} \quad \text{\{by algebra\}}$$

In summary, $(\sim p) \Longleftarrow b_1 \Longleftarrow b_2 \Longleftarrow b_3 \Longleftarrow b_4 \Longleftarrow (\sim q)$ and the equivalence of conditional and contrapositive tells us $p \Longrightarrow q$. In this proof each $\Longleftarrow$ can be replaced by $\Longleftrightarrow$ because the rules of algebra employed are equivalences. Hence the original statement could have been phrased as an if and only if statement.

■

A second indirect method of proof , called <u>proof by contradiction</u>, uses the equivalence

$$(p \Longrightarrow q) \Longleftrightarrow (\,(p \wedge (\sim q)) \Longrightarrow c)$$

where c is a statement that is always false. (See Exercise 10.) The motivation for this procedure can be seen by referring to Table 8 in Section 1. The only way $p \Longrightarrow q$ can be false is if hypothesis p is true and the conclusion q is false. Proof by contradiction assumes "p is true and q is false" and then attempts to build a logical bridge to a statement that is known never to be true. When this is done we say we have reached a contradiction, hence our additional hypothesis "q is false" must be incorrect. The preceding equivalence then implies that $p \Longrightarrow q$ is true. If we are unable to build our bridge to some always false statement, then this technique of proof just fails to verify that $p \Longrightarrow q$ is true. We can not claim $p \Longrightarrow q$ is false. Possibly we were not clever enough to build the bridge. In such cases an alternate proof technique may yield the result. As with proof by contrapositive, unfortunately, there is no specific indicator within $p \Longrightarrow q$ that informs you that proof by contradiction may be fruitful. Sometimes the appearance of the word not in the conclusion q is a suggestion to try this method. Another possible indicator is if q is one of two possible alternatives like even/odd, rational/irrational, or prime/composite (having factors other than itself and 1). (One indicator of this type in linear algebra is, "a square matrix is either singular or nonsingular.")

A seemingly added complication in proof by contradiction is that we must be on the lookout for an "always false statement". The specific always false statement varies from proof to proof. Things like 0 = 1 or a number is both rational and irrational are obviously false. In linear algebra your logic bridge starting with hypothesis

$p \wedge (\sim q)$ may lead to some absurdity dealing with matrices or other linear algebra notion.

If we are using proof by contradiction, then we begin by assuming both p and $(\sim q)$ are true. This gives us an additional hypothesis to work with. We use forward building taking both p and $(\sim q)$ into account in order to use equivalences and implications to obtain simpler statements. Details are dependent upon the situation. We give two classic illustrations of proof by contradiction in Examples 7 and 8.

Example 7. Prove by contradiction:

If $x = \sqrt{2}$, then x is irrational.

Proof: Let p: $x = \sqrt{2}$, and let q: x is irrational.

Assume $p \wedge (\sim q)$ are true; that is, $x = \sqrt{2}$ and x is rational. The assumption x is rational implies that there is a fraction n/d such that

$$\sqrt{2} = n/d$$

We know that every fraction can be reduced to lowest terms so we can take the fraction n/d to be in lowest terms which is equivalent to saying that n and d have no common factors other than 1. Hence

$$p \wedge (\sim q) \Longrightarrow \underbrace{\sqrt{2} = n/d}_{b_1}$$

Using algebra we have

$$b_1 \Longleftrightarrow 2 = n^2/d^2 \Longleftrightarrow \underbrace{2d^2 = n^2}_{b_2}$$

Statement b_2 implies n^2 is even. The only way n^2 is even is for n to be even. So let n = 2k and we have

$$b_2 \Longrightarrow 2d^2 = n^2 = (2k)^2 \Longleftrightarrow 2d^2 = 4k^2 \Longleftrightarrow \underbrace{d^2 = 2k^2}_{b_3}$$

$b_3 \Longrightarrow d^2$ is even and as before this implies that d is even.

We now have that both n and d are even which implies that they have a common factor of 2. This contradicts the fact that n/d is in lowest terms. Thus our assumption $(\sim q)$ is invalid and it follows that $p \Longrightarrow q$.

■

Example 8. Prove by contradiction:

If S is the set of all prime numbers, then S has infinitely many members.

Proof: Let p: S is the set of all prime numbers
q: S has infinitely many members

Assume $p \wedge (\sim q)$ is true; that is, assume that S is the set of all primes and has only finitely many members. Since S has only a finite number of members, there is a largest member, call it n.

$p \wedge (\sim q) \Longrightarrow$ S contains all primes and there is the largest member n of S

This suggests that we try to construct a prime bigger than n or one that is not in S. We proceed as follows. Form the number

$$k = (n \times (n-1) \times \cdots \times 2 \times 1) + 1 = n! + 1$$

Number k is not evenly divisible by any of the numbers 2, 3, 4, ..., n since k divided by any of these will leave a remainder of 1. This implies k has no prime factors which are in S since the largest prime is n. The number k is either a prime greater than n or else k is not prime in which case it is divisible by a prime greater than n. In either case we contradict that S contains all the primes. Thus our assumption $(\sim q)$ is invalid and it follows that $p \Longrightarrow q$.

■

There is an important difference between direct proofs and indirect proofs. A direct proof (in the forward direction) starts with the hypothesis p and step-by-step builds conclusion q. We call this a constructive proof because q is actually built (by logical arguments). However, an indirect proof, say by contradiction, provides an argument that q follows from p without actually building q. We call this an existence proof. Note that in Example 7 we did not explicitly show $\sqrt{2}$ had a nonterminating nonrepeating decimal expansion, which is what a construction of an irrational number should look like. Similarly in Example 8 the infinite set of primes was not constructed. That is, we did not list its elements or give a way to determine all of its elements. (At present, such a construction is not known.) To see some of the power in an indirect proof, try to prove either Example 7 or 8 directly. A number of existence proofs appear in calculus.

Beware of the way that certain conditionals are stated. Often it is not explicitly stated that some statement is true for all objects of a certain type. For instance, the conditional statement in Example 1 could have been phrased,

If m and n are even, then m + n is even.

From the usual meaning of even, you are to infer that m and n are to be considered integers. In Example 2, the description "Let x be a real number." could have been omitted and you are then to infer from context (and experience) that we want real numbers not just integers. In Example 3, the result is to be valid for every angle θ. You cannot verify that the expression is true for a particular choice, say $\theta = 0$ or $\theta = \pi/2$, and then claim that it holds for all values of θ.

Most "proof situations" that you will encounter explicitly are of the form "prove $p \Longrightarrow q$" or "prove $p \Longleftrightarrow q$". The particular statement may be labeled a Theorem, Corollary, or Exercise. However, occasionally you may encounter a situation like

"Show that statement p is false."

In this case you need only find one example where statement p fails to be true. We call this determining a counterexample to p. There could be many counterexamples to p each as good as any other. Example 9 gives a classic use of a counterexample from calculus.

Example 9. Show that not every continuous function is differentiable.

Discussion: A simple counterexample used in calculus is to consider $f(x) = |x|$ on $[-1,1]$. Function f is continuous at every point of $[-1,1]$, but fails to be differentiable at $x = 0$. See your calculus book for details.

■

A related situation is,

Prove or disprove $p \Longrightarrow q$.

A fundamental decision that you must make is, do you think the statement is true or false. If you think it is false, then you try to construct a counterexample. If you think it is true, then you try to prove it using techniques discussed in this section. While there are no explicit guidelines for making this first decision there are several things to keep in mind.

- Think of equivalences for p.
- Is this the contrapositive of a previous result?
- Is this the converse of a previous result?
- Does the statement relate to any special cases discussed in the text?
- The appearance of words like every, all, never, and always are a signal to consider the corresponding statement very carefully.

Exercises 0.2

1. Use forward building to prove:

   If m and n are odd integers, then m + n is even.

2. Use forward building to prove:

   If m ,n, and k are integers and m divides n and m divides k, then m divides n + k.

   (Hint: x divides y $\Longleftrightarrow$ y = sx, for some integer s.)

3. Use backward building to prove:

   $|x - 3| < 2$ implies that $1 < x < 5$.

4. Prove: $|a - b| \geq |a| - |b|$ for any real numbers a and b. (Hint: Assume that you know $|a + b| \leq |a| + |b|$ and start with $|a| = |a + (b - b)|$.)

5. Prove: $\dfrac{\sin\theta}{\sec\theta} = \dfrac{1}{\tan\theta + \cot\theta}$

6. Prove: $\dfrac{1 - \cos\theta}{\sin\theta} = \dfrac{\sin\theta}{1 + \cos\theta}$

7. Prove: $\lim_{x \to 0} \dfrac{1}{x + 1} = 1$ using a δ-ϵ approach as in Example 4.

8. Use proof by contradiction to prove:

   If m^2 is odd, then m is odd.

9. The <u>inverse</u> of $p \Longrightarrow q$ is $\sim p \Longrightarrow \sim q$.

   (a) Show that $p \Longrightarrow q$ and its inverse are not equivalent.

   (b) State a relationship between the inverse, converse, and contrapositive of $p \Longrightarrow q$.

10. Prove: $(p \Longrightarrow q) \Longleftrightarrow ((p \wedge (\sim q)) \Longrightarrow c)$, where c is any statement that is always false.

11. Prove by contradiction that

    If $x > 0$, then $1/x > 0$.

12. Prove by contradiction that

    If x is irrational, then 3x is irrational.

Exercises 0.2

13. We have used the fundamental properties of even and odd integers several times. Prove by contradiction that

If 2 divides integer m, then m is even.

14. Prove or disprove: There are no consecutive integers m, n, and k so that $m^2 + n^2 = k^2$.

15. Let $\{a_n\}$ represent sequence $a_1, a_2, \ldots$. In addition suppose that a_i is positive for every i and that $\{a_n\}$ is convergent. Prove or disprove: $\lim_{n \to \infty} a_n = 0$.

16. Prove or disprove: If n is an even integer, then $n^2 + n + 4$ is even.

<u>Section 0.3</u> <u>Sets</u>

In Appendix A1 of the text there is a brief discussion of sets. We will assume that you have read Appendix A1 or are familiar with the concepts of <u>sets</u>, <u>subsets</u>, <u>equal sets</u>, and the <u>empty set</u>. In this manual we will use the following terminology and notation.

- Capital letters denote sets.

- Lower case letters denote elements or members of a set.

- The equivalent statements "x belongs to A", "x is an element of A", and "x is in A" are denoted

$$x \in A$$

- The equivalent statements "x does not belong to A", x is not an element of A", and "x is not in A" are denoted

$$x \notin A$$

- The statement "B is a subset of A" means each element of B is in A and is denoted

$$B \subseteq A$$

- The statement "B is not a subset of A" means there is at least one element of B that is not in A. We denote this by

$$B \not\subseteq A$$

- The empty set is denoted $\emptyset$.

- A set A equals a set B, denoted A = B, if each element of A is also in B and each element of B is also in A.

In our study of linear algebra we will use ideas about subsets and equal sets. Using the language of logic from Section 0.2 we have

$$\boxed{B \subseteq A} \text{ is equivalent to } \boxed{\text{if } x \in B, \text{ then } x \in A}$$

Hence to show that B is a subset of A we need to prove that every element of B is in A. The building blocks for such a proof will involve the definitions for membership to sets A and B. This type of proof will be used in Chapter 2 with vector spaces and subspaces. We illustrate the concept of subsets with Examples 1 - 5, which should be familiar from algebra and calculus.

<u>Example 1.</u> Let

N = all positive integers
Z = all integers
Q = all rational numbers, W = all irrational numbers
R = all real numbers

Section 0.3

The following set relations are well known.

(a) $N \subseteq Z$ (b) $Z \subseteq Q$ (c) $Q \subseteq R$ (d) $W \subseteq R$
(e) $W \not\subseteq Q$, see the proof of Example 7 in Section 0.2
(f) $N \subseteq Z \subseteq Q \subseteq R$, follows from (a), (b), and (c)

■

Example 2. Let k be a nonnegative integer and P_k denote the set of polynomials of degree k or less.

(a) $P_1 \subseteq P_2$

Proof: We show that

$$\text{if } \underbrace{f(x) \in P_1}_{p} \text{ then } \underbrace{f(x) \in P_2}_{q}.$$

$p \iff f = a_0 + a_1x$, for any real numbers a_0, a_1

$\iff f(x) = a_0 + a_1x + 0x^2$

$\implies f(x) \in P_2$, since it is of the form $a_0 + a_1x + a_2x^2$, where a_0, a_1, a_2 are real numbers

Thus $p \implies q$ or $P_1 \subseteq P_2$.

(b) $P_2 \subseteq P_3$ (Verify by following the pattern of proof in (a).)

(c) $P_1 \subseteq P_2 \subseteq \cdots \subseteq P_k \subseteq P_{k+1} \subseteq \cdots$

■

Example 3. Let

$C(a,b)$ = set of all continuous functions over interval (a,b)

$C^1(a,b)$ = set of all differentiable functions over interval (a,b)

Let P_k be defined as in Example 2.

(a) $P_k \subseteq C(a,b)$ for any k and any interval (a,b)

That is, every polynomial (of degree k or less) is continuous over any interval (a,b). This is shown in calculus.

(b) $P_k \subseteq C^1(a,b)$, for any k and any interval (a,b)

That is, every polynomial (of degree k or less) has a derivative at each point of (a,b). This is shown in calculus.

(c) $C^1(a,b) \subseteq C(a,b)$

That is, every function that is differentiable over (a,b) is continuous over (a,b). This is shown in calculus.

(d) $C(a,b) \not\subseteq C^1(a,b)$

That is, there exists a continuous function over (a,b) that is not differentiable at some point of (a,b). See Example 9 in Section 0.2 for a particular case. ■

Example 4. Let A = all lines not parallel to $x + y = 1$ and let B = all lines of slope $m = 1$. Prove $B \subseteq A$.

Proof: We show that if line $\underbrace{\ell \in B}_{p}$ then $\underbrace{\ell \in A}_{q}$.

$$p \Longrightarrow \underbrace{\text{slope of line } \ell \text{ is } 1}_{b_1} \Longrightarrow \underbrace{\ell \text{ does not have slope } -1}_{b_2}$$

$b_2 \Longrightarrow q$ since $x + y = 1 \Longleftrightarrow y = -x + 1$ which means member of A have slope $\neq -1$

Thus $p \Longrightarrow q$ or $B \subseteq A$. ■

Example 5. Let B = the intercepts of $y = x^2 - 1$ and A = the points on circle $x^2 + y^2 = 1$. Prove $B \subseteq A$.

Proof: We show that each element in B belongs to A. From algebra $B = \{(0,-1), (1,0), (-1,0)\}$. Using algebra it is easy to show that $(0,-1) \in A$, $(1,0) \in A$, and $(-1,0) \in A$. Thus $B \subseteq A$. ■

We continue our illustrations of subsets using concepts that appear in Chapter 1 in the text. For each of these examples we specify the section to which it is related.

Example 6. (Section 1.1) Let A be the set of all systems of 2 equations in 2 unknowns and let B be the set of pairs of perpendicular lines. We have $B \subseteq A$ since a pair of perpendicular lines can be represented by system

$$\begin{cases} y = mx + b \\ y = 1/m\ x + c \end{cases},\ m \neq 0 \quad \text{or} \quad \begin{cases} y = b \\ x = c \end{cases},\ m = 0$$

Such representations are systems of 2 equations in 2 unknowns. ■

Example 7. (Section 1.4) Let A be the set of all consistent systems of m equations in n unknowns, for any choice of positive integers m and n. Let B be the set of all pairs of distinct parallel lines. We have $B \not\subseteq A$. We see this as follows. A pair of distinct lines is represented by a system of 2 equations in 2 unknowns. Since the lines are parallel they do not intersect; hence there is no solution.

■

Example 8. (Section 1.4) Let A be the set of all consistent systems of m equations in n unknowns, for any choice of positive integers m and n. Let B be the set of all homogeneous systems of equations regardless of size. We have $B \subseteq A$ since every homogeneous system has the trivial solution. Hence we can say <u>every homogeneous system is consistent.</u>

■

Example 9. (Section 1.3) Let D be the set of $n \times n$ diagonal matrices and let S be the set of $n \times n$ symmetric matrices. We have $D \subseteq S$ since $D^T = D$. This result is true for each positive integer n, thus we can say <u>every diagonal matrix is symmetric.</u>

■

Example 10. (Section 1.3) Let S be the set of $n \times n$ symmetric matrices and let N be the set of $n \times n$ nonsingular (or invertible) matrices. Is $S \subseteq N$?

Discussion: We are asked whether $S \subseteq N$ is true or false. If we suspect it is true, then we must produce a proof. However, if we suspect it is false, then we must find a counterexample. The counterexample in this case would be an $n \times n$ symmetric matrix A for which there is no matrix B (of the same size) such that $AB = I_n$ Note that O_n is symmetric, but

$$O_n \times (\text{any } n \times n \text{ matrix}) = O_n \neq I_n.$$

Thus we have a counterexample. Hence $S \not\subseteq N$.

■

Example 10 illustrates the following:

$B \not\subseteq A$ $\iff$ there is at least one $x \in B$ such that $x \notin A$

A useful set of equivalences for equality of sets follows.

$$A = B \iff$$
$$(\text{every element in A is in B}) \wedge (\text{every element in B is in A}) \iff$$
$$(x \in A \Rightarrow x \in B) \wedge (x \in B \Rightarrow x \in A) \iff$$
$$(A \subseteq B) \wedge (B \subseteq A)$$

Thus we see that to prove equality of sets there are two things to be shown:

$$1.\ A \subseteq B \quad \text{and} \quad 2.\ B \subseteq A$$

Only if both are verified to be true is $A = B$.

The following discussion assumes that you have read Section 1.5.

A major goal in linear algebra is the study of linear systems of equations in which the coefficient matrix is nonsingular. It is important to be able to identify nonsingular matrices. What is obtained in parts of Chapters 2 - 5 are various alternative descriptions of the set of nonsingular matrices. The proof that we have a set of matrices equal to the set of nonsingular matrices will involve the two steps described above. The text emphasizes such "equivalent descriptions" in a box. Be on the look out for such equivalences. They should be studied carefully for they play a major role in theoretical as well as computational concepts.

Section 0.4 Mathematical Induction (Optional)

In Section 0.1 we introduced the ideas of elementary logic to provide a basis upon which to introduce techniques of proof. The discussion in Section 0.2 of techniques of proof can be succinctly summarized as follows:

From a stated hypothesis reason to a valid conclusion.

The term "reason" can be thought of as building a logic bridge. Our building blocks were other truths. The connection of one truth to the next formed our bridge from the hypothesis to the conclusion. This procedure is called deductive reasoning and is generally what is meant by a formal proof in mathematics. A new truth is deduced from other truths.

Another type of reasoning is known as inductive reasoning. Inductive reasoning involves collecting evidence from experiments or observations and using this information to formulate a general law or principle. Inductive reasoning attempts to go from the specific to the general, but even with large quantities of evidence the conclusion is not guaranteed. In general, mathematics rejects direct inductive reasoning. However, mathematics often uses an inductive process to formulate conjectures which are then subjected to rigorous deductive reasoning before they are accepted. Remember that a conjecture is a conclusion based on incomplete evidence — that is, a guess. Much can be gained from using experimental evidence to suggest conjectures and then applying deductive arguments to determine the truth or falsity of the conjecture. A number of important advances in science and engineering evolved in just this way.

The cycle of

experiment(s) — conjecture — check by deductive reasoning

is very important in mathematics. Here we introduce one method for performing the "deductive check" on certain conjectures that can be derived from experiments that involve using the natural numbers. (The natural numbers N are the set {1, 2, 3, ...}.) The process we describe below is called Mathematical Induction, but it is really a particular deductive checking procedure not reasoning by induction.

Experiments are often performed in mathematics to determine patterns of behavior. Patterns of behavior are often reformulated into mathematical properties and mathematical theorems. Experiments in mathematics often take the form of looking at special cases and trying to see some common pattern in order to derive a conjecture. The special cases in the patterns we explore are obtained by using the first "few" natural numbers.

Example 1. Determine a conjecture for a formula to compute the sum of consecutive natural numbers.

<u>Discussion:</u> The experiments for the cases of 1, 2, 3, 4, and 5 consecutive natural numbers are shown next.

$$
\begin{aligned}
&1 &&= 1\\
&1 + 2 &&= 3\\
&1 + 2 + 3 &&= 6\\
&1 + 2 + 3 + 4 &&= 10\\
&1 + 2 + 3 + 4 + 5 &&= 15
\end{aligned}
$$

Looking for a pattern in the sums is not easy. Try to connect the sum to the largest integer used in the set of addends. (We do not always have to start looking for a pattern at the "beginning".) Looking at

$$1 + 2 + \boxed{3} \qquad = \boxed{6}$$

$$1 + 2 + 3 + \boxed{4} \qquad = 10$$

we see that $\boxed{3}\,\boxed{4} = 12 = 2\,\boxed{6}$. Looking at

$$1 + 2 + 3 + \boxed{4} \qquad = \boxed{10}$$

$$1 + 2 + 3 + 4 + \boxed{5} \qquad = 15$$

we see that $\boxed{4}\,\boxed{5} = 20 = 2\,\boxed{10}$. Checking the first two sums we see that the same pattern holds. For

$$\boxed{1} \qquad = \boxed{1}$$

$$1 + \boxed{2} \qquad = 3$$

$\boxed{1}\,\boxed{2} = 2 = 2\,\boxed{1}$. Checking the second pair of sums we see that the same pattern holds. (Verify.) Next we adjoin another row to the experiment table above:

$$1 + 2 + 3 + 4 + 5 + 6 \quad = 21$$

Checking the last pair of rows:

$$\boxed{5}\,\boxed{6} = 2\,\boxed{15}$$

so the pattern holds here also. To formulate a conjecture we proceed as follows. Let the largest natural number used in a sum of consecutive natural numbers be k. Then we have

$$1 + 2 + \cdots + k-1 + \boxed{k} = S_k$$

$$1 + 2 + \cdots + k-1 + k + \boxed{k+1} = S_{k+1}$$

The pattern above suggests that

$$k(k + 1) = 2S_k$$

or equivalently

$$S_k = \frac{k(k + 1)}{2}$$

Using summation notation (see Section 1.2) we have

$$S_k = 1 + 2 + \cdots + k = \sum_{i=1}^{k} i$$

Hence our conjecture is

If we form the sum S_k of the first k consecutive natural numbers then $S_k = \sum_{i=1}^{k} i = \dfrac{k(k + 1)}{2}$.

We have only done the experiment and conjecture steps of our cycle. Before we can claim that the sum of the first k consecutive natural numbers is $k(k + 1)/2$ we must use deductive reasoning. (See Example 3.)

■

Example 2. Determine a conjecture for a relationship between the expressions $(1 + x)^k$ and $1 + kx$ where x is such that $1 + x > 0$ and k is any natural number.

Discussion: We experiment with the first few natural numbers k and various values of x.

Case k = 1: x any real number so that $x > -1$

$$(1 + x)^1 = 1 + 1x$$

Case k = 2:

$x = -.5$	$(1 + (-.5))^2 = .25$	$1 + 2(-.5) = 0$
$x = 0$	$(1 + 0)^2 = 1$	$1 + 2(0) = 1$
$x = 1$	$(1 + 1)^2 = 4$	$1 + 2(1) = 3$
$x = 15$	$(1 + 15)^2 = 256$	$1 + 2(15) = 31$

Summary: for the few case above, $(1 + x)^2 > 1 + 2x$

Case k = 3:

x = -.75	$(1 + (-.75))^3$ = .015625	1 + 3(-.75) = -1.25
x = -.1	$(1 + (-.1))^3$ = .970299	1 + 3(-.1) = .7
x = 2	$(1 + 2)^3$ = 27	1 + 3(2) = 7
x = 7	$(1 + 7)^3$ = 512	1 + 3(7) = 22
x = 20	$(1 + 20)^3$ = 9261	1 + 3(20) = 61

Summary: for the few cases above, $(1 + x)^3 > 1 + 3x$

This limited evidence suggests that we form the conjecture

If k is any natural number and $x > -1$
then $(1 + x)^k \geq 1 + kx$.

Before we can claim that this relationship is true we must use deductive reasoning to supply a proof. (See Example 4.) ∎

Note that the conjecture in both examples is stated as a conditional $p \Longrightarrow q$. The technique of Mathematical Induction which we define next is a method of proof for the special kind of conditional statements which appear in Examples 1 and 2. The principle of <u>Mathematical Induction</u> is stated as follows:

> If S is a set of natural numbers with
> (a) $1 \in S$
> and
> (b) $n \in S \Longrightarrow n+1 \in S$ for each natural number n
> then S = N, the set of all natural numbers.

The set S is the set of natural numbers for which the conjecture $p \Longrightarrow q$ is true. The method of mathematical induction says

<u>first:</u> show $p \Longrightarrow q$ for n = 1

<u>next:</u> assume $p \Longrightarrow q$ for arbitrary natural number n and **<u>prove</u>** that $p \Longrightarrow q$ is true for natural number n + 1.

If both steps are successful the principle of mathematical induction guarantees that $p \Longrightarrow q$ is true for <u>every</u> natural number. <u>Warning:</u> The proof for part (b) depends on the contents of the implication $p \Longrightarrow q$ and many times requires ingenuity. To illustrate the principle of mathematical induction we prove the conjectures developed in Examples 1 and 2. (Terminology: The name mathematical induction is often shortened to just <u>induction</u>.)

<u>Example 3.</u> Apply induction to prove the conjecture developed in Example 1. Let S be the set of all natural numbers k for which

$$\sum_{i=1}^{k} i = \frac{k(k + 1)}{2}$$

From the experiments in Example 1, we have 1, 2, 3, 4, 5 $\in$ S. The principle of induction says assume that n $\in$ S; that is,

$$\sum_{i=1}^{n} i = 1 + 2 + \cdots + n = \frac{n(n + 1)}{2} \tag{1}$$

Then we must verify that n + 1 $\in$ S; that is, we must show

$$\sum_{i=1}^{n+1} i = 1 + 2 + \cdots + n + n+1 = \frac{(n + 1)(n + 2)}{2} \tag{2}$$

(Note that n in the right-hand side of (1) was replaced by n + 1 to obtain the right-hand side of (2).) In order to verify that (2) is true we use (1) which is assumed to be true. (The expression in (1) is called the <u>induction hypothesis</u> for this conjecture.) Starting with the expression $\sum_{i=1}^{n+1} i$ we must show algebraically that we can produce the formula $\frac{(n + 1)(n + 2)}{2}$. We have

$$\begin{aligned}
\sum_{i=1}^{n+1} i &= \sum_{i=1}^{n} i + (n + 1) && \text{\{by properties of sums\}} \\
&= \frac{n(n + 1)}{2} + (n + 1) && \text{\{by (1)\}} \\
&= \frac{n(n + 1) + 2(n + 1)}{2} && \text{\{by algebra\}} \\
&= \frac{(n + 1)(n + 2)}{2} && \text{\{by factoring\}}
\end{aligned}$$

Hence we have shown that if n $\in$ S then n + 1 $\in$ S. Thus the principle of induction implies that S = N; that is, the formula

$$\sum_{i=1}^{k} i = \frac{k(k + 1)}{2}$$

is valid for all natural numbers k.

■

The deductive reasoning in the principle of mathematical induction is the proof we must supply to show part

$$(b): n \in S \Longrightarrow n + 1 \in S.$$

This step is itself a conditional statement that must be proven using its hypothesis, $n \in S$, and appropriate building blocks. The inductive hypothesis, $n \in S$, does not mean we are assuming what we want to prove. We are not assuming what we want to prove because we must supply a proof that $n + 1 \in S$. Thus to prove (b) we must adhere to the rules about proving conditionals which are stated in Table 8 in Section 0.1. From Table 8 we see that (b) is false only if n is in S but n + 1 is not. Hence if we can show that whenever $n \in S$ that it must follow that $n + 1 \in S$, then the conditional (b) is always true. In summary, assuming $n \in S$ does not assume what we must prove, namely that $n + 1 \in S$.

Example 4. Apply induction to prove the conjecture developed in Example 2. Let S be the set of all natural numbers k such that

$$(1 + x)^k \geq 1 + kx, \text{ whenever } 1 + x > 0$$

From the experiments in Example 2 we have that $1 \in S$ and we suspect that 2 and 3 belong to S. It is only a suspicion since we have not verified the conjecture for all $x > -1$ when $k = 2$ or 3. We next verify the conditional in (b). Assume that $n \in S$; that is,

$$(1 + x)^n \geq 1 + nx \text{ for all } x > -1$$

We must prove deductively that $n + 1 \in S$: that is, prove that

$$(1 + x)^{n+1} \geq 1 + (n+1)x \text{ for all } x > -1$$

We proceed as follows:

$$\begin{aligned}
(1 + x)^{n+1} &= (1 + x)^n(1 + x) && \{\text{by algebra}\}\\
&\geq (1 + nx)(1 + x) && \{\text{by the inductive hypothesis}\}\\
&= 1 + nx + x + nx^2 && \{\text{by algebra}\}\\
&= 1 + (n+1)x + nx^2 && \{\text{by algebra}\}\\
&\geq 1 + (n+1)x && \{\text{since } nx^2 > 0\}
\end{aligned}$$

Hence by the principle of induction

$$(1 + x)^k \geq 1 + kx$$

for all $x > -1$ and for all natural numbers k. ∎

Most cases in which you may need to use induction in linear algebra are already phrased as prove : $p \Longrightarrow q$. The experiments and conjecture formulation stages of the process we have described have been done. You must recognize that the special structure of inductive proofs is present. That is, the natural numbers play a role in p $\Longrightarrow$ q.The building blocks in proving part (b) of the principle of induction will most likely be facts about matrices or other linear algebra concepts rather than ordinary algebra facts as in Examples 3 and 4. Proof by induction appears in only a few places in this manual but it is an important mathematical technique. Places where induction can be used are listed next:

Section 1.4 Exercise 7; Supplementary Exercises for Chapter 1 Exercises 1(d) and 29; Section 4.4 Exercise 23; Supplementary Exercises for Chapter 5 Exercise 3.

For further reading on induction see the following sources.

[1] H. Burrows, et al, Mathematical Induction, FIAM Module, COMAP, 1989.

[2] S. Lay, <u>Analysis, An Introduction to Proof</u>, Prentice Hall, 1986.

[3] L. Swanson and R. Hansen, Mathematical Induction or "What Good is All This Stuff if We Are Going to Assume It's True Anyway?", Two Year College Mathematics Journal, v.12, 1981, pp8-12.

[4] B. Youse, <u>Mathematical Induction</u>, Prentice Hall, 1964.

Exercises 0.4

1. Determine a conjecture for a formula to compute the sum of consecutive odd integers. Try to prove your conjecture by induction

2. Determine a conjecture for a relationship between 2^k and $(k+1)!$ for natural numbers k. Try to prove your conjecture by induction.

3. Prove by induction that $\frac{d}{dx}(x^k) = k\,x^{k-1}$ for all natural numbers k.

4. Prove by induction that for every natural number k the expression $k^2 + k$ is divisible by 2.

5. Prove by induction that

$$\frac{1}{(1)(2)} + \frac{1}{(2)(3)} + \frac{1}{(3)(4)} + \cdots + \frac{1}{k(k+1)} = \frac{k}{k+1}$$

for all natural numbers k.

Solutions to Exercises

Section 0.1

1. (a) It is a statement which is false.
   (b) It is a true statement.
   (c) Not a statement since its truth depends on the value of x.
   (d) This is a question not a declarative sentence.
   (e) It is a true statement.
   (f) It is a true statement.

2. (a) The door is open and the tables are set.
   (b) The tables are set or the food is not ready.
   (c) The door is open and the tables are not set.
   (d) The door is open and the tables are set and the food is ready.

3. (a) $(\sim p) \wedge q$ (b) $r \vee q$ (c) $(\sim r) \wedge p$ (d) $\sim r$

4. (a)

p	q	$\sim q$	$p \wedge (\sim q)$
T	T	F	F
T	F	T	T
F	T	F	F
F	F	T	F

(b)

p	q	$\sim p$	$q \wedge (\sim p)$	$p \vee (q \wedge (\sim p))$
T	T	F	F	T
T	F	F	F	T
F	T	T	T	T
F	F	T	F	F

(c)

p	q	$\sim p$	$q \vee (\sim p)$	$p \wedge (q \vee (\sim p))$
T	T	F	T	T
T	F	F	F	F
F	T	T	T	F
F	F	T	T	F

(d)

p	q	$\sim q$	$p \vee (\sim q)$
T	T	F	T
T	F	T	T
F	T	F	F
F	F	T	T

5. (a) Divide both sides of the equation by 3.
   (b) Add -3 to both sides of the equation.
   (c) 2
   (d) $\dfrac{(x - 1)(x + 2)}{(x - 1)(x + 1)} = \dfrac{x + 2}{x + 1}, \quad x \neq 1$

6. (a) p: it rains   q: I will go to the movies
   (b) p: the diagonals of a rectangle are not perpendicular
       q: it is not a square
   (c) p: $x > 0$ and $y > 0$ q: $x + y > 0$
   (d) p: f(x) is a function on [a,b]
       q: f(x) is continuous on [a,b]
   (e) p: I will pay $5.00 for a cup of coffee   q: elephants fly
   (f) p: T is an equilateral triangle   q: all sides are equal
   (g) p: opposite sides are parallel
       q: a quadrilateral to be a parallelogram

7.

p	q	$\sim q$	$\sim p$	$(\sim q) \Longrightarrow (\sim p)$
T	T	F	F	T
T	F	T	F	F
F	T	F	T	T
F	F	T	T	T

The last column is identical to the last column of Table 8, hence the statements are equivalent.

8. (a) hypothesis p: two lines are perpendicular
       conclusion q: intersect at right angles
       contrapositive: If two lines do not intersect at right angles, then they are not perpendicular.
   (b) hypothesis p: f(x) is a polynomial
       conclusion q: f(x) is differentiable
       contrapositive: If f(x) is not differentiable then f(x) is not a polynomial.

9. (a) If two lines intersect at right angles, then they are perpendicular.
   (b) If f(x) is differentiable, then f(x) is a polynomial.

10. Show that the truth tables for $\sim(p \implies q)$ and $p \wedge (\sim q)$ are identical. Using Table 8 we have

$\sim(p \implies q)$
F
T
F
F

and

p	q	~q	$p \wedge (\sim q)$
T	T	F	F
T	F	T	T
F	T	F	F
F	F	F	F

Since the final columns are identical, the statements are equivalent.

11. Show that the truth tables for $\sim(p \wedge q)$ and $p \implies (\sim q)$ are identical.

p	q	$p \wedge q$	$\sim(p \wedge q)$
T	T	T	F
T	F	F	T
F	T	F	T
F	F	F	T

p	q	~q	$p \implies (\sim q)$
T	T	F	F
T	F	T	T
F	T	F	T
F	F	T	T

Since the final columns are identical, the statements are equivalent.

12. More than one equivalent statement may be possible. We present one such statement.

(a) evenly divisible by 2
(b) their sum is 90°.
(c) $|x| \leq 2$.
(d) n is odd.
(e) $a > 0$.

13. We construct the truth table for each part of the biconditional.

p	q	$p \Longleftrightarrow q$	$p \Longrightarrow q$	$q \Longrightarrow p$	$(p \Longrightarrow q) \wedge (q \Longrightarrow p)$
T	T	T	T	T	T
T	F	F	F	T	F
F	T	F	T	F	F
F	F	T	T	T	T

Since columns 3 and 6 are identical biconditional $p \Longleftrightarrow q$ is equivalent to conjunction $p \Longrightarrow q$ and $q \Longrightarrow p$.

Section 0.2

1. Prove $p \Longrightarrow q$ where p: m and n are odd, q: m + n is even.

Proof: $p \Longrightarrow \underbrace{m = 2k + 1,\ n = 2j + 1}_{b_1}$ for some integers k and j {since m and n are odd}

$$b_1 \Longrightarrow \underbrace{m + n = (2k + 1) + (2j + 1) = 2k + 2j + 2 = 2(k + j + 1)}_{b_2} \quad \text{\{by algebra\}}$$

$$b_2 \Longrightarrow \underbrace{m + n \text{ is even}}_{q} \quad \text{\{since it is a multiple of 2\}}$$

2. Prove $p \Longrightarrow q$ where p: (m divides n) $\wedge$ (m divides k)
q: m divdes n + k

Proof: $p \Longrightarrow \underbrace{(n = ma) \wedge (k = mb)}_{b_1}$, for some integers a and b {using the hint}

$$b_1 \Longrightarrow \underbrace{n + k = ma + mb = m(a + b)}_{b_2} \quad \text{\{by algebra\}}$$

$$b_2 \Longrightarrow \underbrace{m \text{ divides } n + k}_{q} \quad \text{\{by the hint\}}$$

3. Prove $p \Longrightarrow q$, where p: $|x - 3| < 2$ and q: $1 < x < 5$.

Proof: Use backward building.

What implies $1 < x < 5$, keeping in mind we want to involve the quantity $x - 3$?

Starting with $1 < x < 5$, to obtain an equivalent expression involving $x - 3$, subtract 3 from each piece of the inequality and simplify. Thus we have

$$q \Longleftarrow \underbrace{1 - 3 < x - 3 < 5 - 3 \iff -2 < x - 3 < 2}_{b_1}$$

Using properties of absolute values we have

$$b_1 \Longleftarrow \underbrace{|x - 3| < 2}_{p}$$

4. Use forward building starting with the hint:

$|a| = |a - 0| = | a - (b - b)|$

The objective is to show an inequality involving $|a - b|$, so we try to involve this quantity:

$|a - (b - b)| = |(a - b) + b|$

The hint also implies that we can use $|a + b| \le |a| + |b|$. We can interpret this as saying

$$|\text{quantity \#1} - \text{quantity \#2}| \le |\text{quantity \#1}| + |\text{quantity \#2}|$$

Do not let the symbols a and b prejudice you into thinking you can <u>only</u> use this hint with a and b. let $a - b$ = quantity #1 and let b = quantity #2. Then apply the hint. We obtain

$|(a - b) + b| \le |a - b| + |b|$

Putting things together we have

$|a| \le |a - b| + |b|$

but by algebra this is equivalent to

$|a| - |b| \le |a - b| \iff |a - b| \ge |a| - |b|$

This proof is more involved (tricky) than our examples or preceding exercises. Note that the first step is not obvious without the hint. Also note that we use another result, $|a + b| \le |a| + |b|$, called the <u>triangle inequality</u>. We had to interpret the meaning of it so we did not let the "generic use" of symbols a and b cloud the issue.

5. Using forward building we have

$$\frac{\sin\theta}{\sec\theta} \iff \frac{\sin\theta}{1/\cos\theta} \iff \sin\theta\cos\theta$$

Using backward building we have

$$\frac{1}{\tan\theta + \cot\theta} \iff \frac{1}{\sin\theta/\cos\theta + \cos\theta/\sin\theta} \iff \frac{1}{\dfrac{\sin^2\theta + \cos^2\theta}{\sin\theta\cos\theta}}$$

$$\iff \sin\theta\cos\theta$$

Since we used (algebraic) equivalence at each step the biconditional is proved.

6. To start with $\dfrac{1 - \cos\theta}{\sin\theta}$ and eventually obtain an expression with $1 + \cos\theta$ in the denominator, we must determine a way to introduce $1 + \cos\theta$ into the denominator. Here we try the equivalence of multiplying by 1 in the disguise $\dfrac{1 + \cos\theta}{1 + \cos\theta}$.

$$\frac{1 - \cos\theta}{\sin\theta} \iff \frac{(1 - \cos\theta)(1 + \cos\theta)}{\sin\theta(1 + \cos\theta)} \iff \frac{1 - \cos^2\theta}{\sin\theta(1 + \cos\theta)}$$

$$\iff \frac{\sin^2\theta}{\sin\theta(1 + \cos\theta)} \iff \frac{\sin\theta}{1 + \cos\theta}$$

Since we used (algebraic) equivalence at each step the biconditional is proved.

7. Let $\epsilon > 0$ be given. We must find $\delta > 0$ such that whenever x is chosen such that $0 < |x - 0| = |x| < \delta$, $\left|\dfrac{1}{1 + x} - 1\right| < \epsilon$. Here we let

p: $0 < |x| < \delta$ for some appropriate choice of δ (that in general depends on ϵ)

q: $\left|\dfrac{1}{1 + x} - 1\right| < \epsilon$

We start backwards to determine a δ that depends on ϵ.

$$q: \left|\frac{1}{1+x} - 1\right| < \epsilon \iff \underbrace{\frac{|x|}{|x+1|} < \epsilon}_{b_1}$$

Note that the term $|x|$, which is important to p, has appeared in this equivalent form of q. Near x = 0, $|x + 1|$ is near 1 and $|x|$ is near zero, of course. The next step is to move backwards from b_1 so that we leave term $|x|$ alone but replace the term $1/|x + 1|$ with a constant. Eventually as $x \to 0$ we can consider x in a small interval about 0. Suppose we take $-.01 < x < .01$ which is equivalent to $|x| < .01$. For $-.01 < x < .01$, $x + 1 = |x + 1|$. Hence we have $.99 < |x + 1| < 1.01$ which is equivalent to

$$\frac{1}{1.01} < \left|\frac{1}{x+1}\right| < \frac{1}{.99}$$

and implies $\left|\frac{1}{1+x}\right| < \frac{1}{.99}$ (for $|x| < .01$)

Thus we can see that b_1 is implied by b_2 where

$$b_1 \Longleftarrow \underbrace{\frac{1}{.99}|x| < \epsilon \text{ when } 0 < |x| < .01}_{b_2}$$

We see that

$$b_2 \iff \underbrace{|x| < .99\,\epsilon \text{ when } 0 < |x| < .01}_{b_3}$$

In b_3 we have two conditions that $|x|$ is to satisfy in order to imply b_2 which in turn implies b_1 which in turn implies q. We combine the two conditions into one as follows. Let $\delta = \min\{.01, .99\epsilon\}$, then

$$b_3 \Longleftarrow \underbrace{0 < |x| < \delta}_{p}$$

In summary, if we choose δ as defined above (or smaller), then $p \Longrightarrow q$.

8. Let p: m^2 is odd and q: m is odd. Prove, (~q) ⟹ (~p).

Proof: (~q) ⟺ m is even ⟹ $m = 2k$, for some integer k
⟹ $m^2 = (2k)^2 = 4k^2 = 2(2k^2)$
⟹ m^2 is even ⟺ (~p)

The reasons have been purposefully omitted. You supply reasons for each step.

9. (a)

p	q	p ⟹ q	(~p)	(~q)	(~p) ⟹ (~q)
T	T	T	F	F	T
T	F	F	F	T	T
F	T	T	T	F	F
F	F	T	T	T	T

Since columns 3 and 6 are not identical, p ⟹ q is not equivalent to its inverse.

(b) The converse of the inverse of p ⟹ q is the contrapositive of p ⟹ q.

10.

p	q	p ⟹ q	p	(~q)	p ∧ (~q)	c	(p ∧ (~q)) ⟹ c
T	T	T	T	F	F	F	T
T	F	F	T	T	T	F	F
F	T	T	F	F	F	F	T
F	F	T	F	T	F	F	T

Since columns 3 and 8 are identical we have (p ⟹ q) ⟺ ((p ∧ (~q)) ⟹ c).

11. Prove by contradiction that p ⟹ q where p: $x > 0$ and q: $1/x > 0$.

Proof: Assume p ∧ (~q). That is, $(x > 0) \wedge (1/x \leq 0)$.

p ∧ (~q) ⟹ since $x > 0$ we can multiply each side of $1/x \leq 0$ by x and preserve the inequality; $x\,(1/x) \leq x\,(0) = 0$

⟹ $1 \leq 0$, which is clearly a contradiction

It follows that assumption (~q) ⟺ $1/x \leq 0$ is false, so $1/x > 0$. That is, p ⟹ q.

12. Prove by contradiction that p ⟹ q, where p: x is irrational and q: 3x is irrational.

Proof: Assume $p \wedge (\sim q)$. That is, assume x is irrational and 3x is rational.

$p \wedge (\sim q)$ ⟹ x is irrational and $3x = m/n$, for some integers m and n

⟹ x is irrational and $x = m/3n$

⟹ x is irrational and rational, which is a contradiction

It follows that assumption $(\sim q)$ ⟺ 3x is rational is false, so 3x must be irrational. That is, p ⟹ q.

13. Prove by contradiction that p ⟹ q where p: 2 divides m and q: m is even.

Proof: Assume $p \wedge (\sim q)$. That is, assume 2 divides m and m is odd.

$p \wedge (\sim q)$ ⟹ ($m = 2k$, for some k)$\wedge$($m = 2j + 1$, for some j)

⟹ $2k = 2j + 1$ {note we cannot say even = odd here because we are trying to prove the characterization of even that says it is a multiple of 2.}

⟹ $2k - 2j = 1$ ⟹ $2(k - j) = 1$

⟹ the left side is negative (if $j > k$) or it is ≥ 2; in any case this is a contradiction that it is equal to 1

It follows that assumption, $(\sim q)$ ⟺ m is odd, is false. Hence m must be even. That is p ⟹ q.

14. The expression $m^2 + n^2 = k^2$ looks like something derived from the Pythagorean theorem for right triangles. A familiar right triangle is a 3,4,5 triangle. We see that $3^2 + 4^2 = 5^2$. Hence we have a counterexample.

15. A decreasing positive sequence $\{a_n\}$ could get closer and closer to a number L, but stay above it. Think about an easy number other than 0. Consider L = 1. We want a sequence to converge to 1, but stay above it. Hence $a_n = 1 +$ a little bit. Try something easy like $a_n = 1 + 1/n$. Check things carefully and you find that $\{1 + 1/n\}$ is decreasing, has positive terms, and converges to 1. Thus we have a counterexample.

16. To show something is even we show that it is a multiple of 2. We have

$$n^2 + n + 4$$
$$\qquad\uparrow\quad\uparrow$$
$$\quad\text{even even}$$

If n^2 is even we have sum of evens which is even because each term would have a factor of 2. Thus try to prove if n is even so is n^2. (See Exercise 8 for a similar problem.) Once we have done this, the result follows from the preceding argument.

Solutions 0.4

1. Experiments:

$$\begin{aligned} 1 &= 1 \\ 1 + 3 &= 4 \\ 1 + 3 + 5 &= 9 \\ 1 + 3 + 5 + 7 &= 16 \end{aligned}$$

The sums can be written as 1^2, 2^2, 3^2, 4^2 respectively.

Conjecture: If S_k is the sum of the first k odd integers then

$$S_k = 1 + 3 + \cdots + (2k - 1) = k^2$$

Proof by induction: Let S be the set of all natural numbers k for which the conjecture is true. From the experiments $1 \in S$. To show (b), we assume $n \in S$; that is,

$$S_n = 1 + 3 + \cdots + 2n-1 = n^2$$

We must show that $n+1 \in S$. We have

$$\begin{aligned} S_{n+1} &= 1 + 3 + \cdots + 2n-1 + 2(n+1)-1 \\ &= (1 + 3 + \cdots + 2n-1) + 2(n+1)-1 && \text{\{by algebra\}} \\ &= n^2 + 2(n+1)-1 && \text{\{by the induction hypothesis\}} \\ &= n^2 + 2n + 1 && \text{\{by algebra\}} \\ &= (n+1)^2 \end{aligned}$$

Thus by induction S = N.

2. Experiments:

$n = 1$	$2^1 = 2$	$(1 + 1)! = 2$
$n = 2$	$2^2 = 4$	$(2 + 1)! = 6$
$n = 3$	$2^3 = 8$	$(3 + 1)! = 24$
$n = 4$	$2^4 = 16$	$(4 + 1)! = 120$

Conjecture: If k is any natural number $2^k \le (k + 1)!$

Proof by induction: Let S be the set of all natural numbers k for which the conjecture is true. From the experiments $1 \in S$. To show (b), we assume $n \in S$; that is

$$2^n \le (n+1)!$$

We must show that $n+1 \in S$. We have

$$\begin{aligned} 2^{n+1} &= 2^n(2) && \{\text{by algebra}\} \\ &\le (n+1)!\ 2 && \{\text{by the induction hypothesis}\} \\ &\le (n+1)!(n+2) && \{\text{since } 2 \le n+1 \text{ when } n \ge 1\ \} \\ &= (n+2)! \end{aligned}$$

Thus by induction $S = N$.

3. Let S be the set of all natural numbers for which it is true that

$$\frac{d}{dx}(x^k) = k\ x^{k-1}$$

From calculus we have

$$\frac{d}{dx} x^1 = \frac{d}{dx} x = 1 = 1\ x^0$$

Hence $1 \in S$. Assume that $n \in S$; that is,

$$\frac{d}{dx} x^n = n\ x^{n-1}$$

We must show that $n+1 \in S$; that is, prove

$$\frac{d}{dx}(x^{n+1}) = (n+1)\ x^n$$

We have

$$\begin{aligned} \frac{d}{dx} x^{n+1} &= \frac{d}{dx}(x^n\ x) && \{\text{by algebra}\} \\ &= (x^n)\left[\frac{d}{dx} x\right] + \left[\frac{d}{dx} x^n\right]x && \{\text{product rule}\} \\ &= (x^n)(1) + (n\ x^{n-1})(x) && \{\text{since } 1 \in S \text{ and the induction hyp.}\} \\ &= x^n + nx^n && \{\text{by algebra}\} \\ &= (n+1)x^n \end{aligned}$$

Thus by induction $S = N$.

4. Let S be the set of all natural numbers k for which it is true that $k^2 + k$ is divisible by 2. If $k^2 + k$ is divisible by 2 then there exists some number m such that $k^2 + k = 2(m)$.

If k = 1, then $1^2 + 1 = 2$, which is certainly divisible by 2. Hence $1 \in S$. Assume $n \in S$; that is,

$$n^2 + n = 2(q)$$

for some number q. We must show that $n+1 \in S$; that is, prove that there exists a number r such that

$$(n+1)^2 + (n+1) = 2(r)$$

We have

$$\begin{aligned}(n+1)^2 + (n+1) &= n^2 + 2n + 1 + n + 1 && \{\text{by algebra}\}\\ &= (n^2 + n) + (2n + 2) && \{\text{by algebra}\}\\ &= 2q + 2(n + 1) && \{\text{by the induction hyp. and algebra}\}\\ &= 2(q + n + 1)\end{aligned}$$

Hence $(n + 1)^2 + (n + 1)$ is divisible by 2 and then by induction S = N.

5. Let S be the set of all natural numbers k for which it is true that

$$\frac{1}{(1)(2)} + \frac{1}{(2)(3)} + \frac{1}{(3)(4)} + \cdots + \frac{1}{k(k+1)} = \frac{k}{k+1}$$

If k = 1, then

$$\frac{1}{1(1 + 1)} = \frac{1}{2} = \frac{1}{1 + 1}$$

Hence $1 \in S$. Assume that $n \in S$; that is

$$\frac{1}{(1)(2)} + \frac{1}{(2)(3)} + \frac{1}{(3)(4)} + \cdots + \frac{1}{n(n+1)} = \frac{n}{n+1}$$

We must show that $n+1 \in S$; that is, prove

$$\frac{1}{(1)(2)} + \frac{1}{(2)(3)} + \frac{1}{(3)(4)} + \cdots + \frac{1}{n(n+1)} + \frac{1}{(n+1)(n+2)} = \frac{n+1}{n+1 + 1}$$

We have

$$\frac{1}{(1)(2)} + \frac{1}{(2)(3)} + \frac{1}{(3)(4)} + \cdots + \frac{1}{n(n+1)} + \frac{1}{(n+1)(n+2)}$$

$$= \left[\frac{1}{(1)(2)} + \frac{1}{(2)(3)} + \frac{1}{(3)(4)} + \cdots + \frac{1}{n(n+1)}\right] + \frac{1}{(n+1)(n+2)}$$

$$= \frac{n}{n+1} + \frac{1}{(n+1)(n+2)} = \frac{n(n+2) + 1}{(n+1)(n+2)} = \frac{n^2 + 2n + 1}{(n+1)(n+2)} = \frac{(n+1)^2}{(n+1)(n+2)}$$

$$= \frac{n+1}{n+2}$$

Thus by induction $S = N$.

EXERCISES 1.1

<<**Strategy:** In Exercises 1-13 we use the method of elimination in the following way. We first eliminate the variable x, then the variable y (if there are more than two variables). In some cases we eliminate the variables in a different order. We specifically point this out when we use such a variation. There are many ways to solve these problems, we present one following the guidelines stated above.>>

1.

$$\begin{aligned} x + 2y &= 8 \\ 3x - 4y &= 4 \end{aligned}$$

To eliminate x, we add -3 times the first equation to the second equation:

$$\begin{array}{c} -3(\,x + 2y = 8\,) \\ \underline{\quad 3x - 4y = 4 \quad} \end{array} \quad \text{----->} \quad \begin{array}{rcr} -3x - 6y &=& -24 \\ \underline{3x - 4y} &=& \underline{4} \\ -10y &=& -20 \\ y &=& 2. \end{array}$$

Substitute $y = 2$ into one of the original equations to solve for x. In equation $x + 2y = 8$ we obtain

$$\begin{aligned} x + (2)2 &= 8 \\ x + \quad 4 &= 8 \\ x &= 4. \end{aligned}$$

The solution of the system is $x = 4$ and $y = 2$.

3.

$$\begin{aligned} 3x + 2y + z &= 2 \\ 4x + 2y + 2z &= 8 \\ x - y + z &= 4 \end{aligned}$$

To eliminate x from the first and third equations, we add -3 times the third equation to the first equation:

$$\begin{array}{c} 3x + 2y + z = 2 \\ \underline{-3(\,x - y + z = 4\,)} \end{array} \quad \text{----->} \quad \begin{array}{rcr} 3x + 2y + z &=& 2 \\ \underline{-3x + 3y - 3z} &=& \underline{-12} \\ 5y - 2z &=& -10. \end{array}$$

To eliminate x from the second and third equations, we add -4 times the third equation to the second equation:

$$\begin{array}{c} 4x + 2y + 2z = 8 \\ \underline{-4(\,x - y + z = 4\,)} \end{array} \quad \text{----->} \quad \begin{array}{rcr} 4x + 2y + 2z &=& 8 \\ \underline{-4x + 4y - 4z} &=& \underline{-16} \\ 6y - 2z &=& -8. \end{array}$$

Thus we have the pair of equations

$$\begin{aligned} 5y - 2z &= -10 \\ 6y - 2z &= -8 \end{aligned}$$

to solve for y and z. In this case we observe that multiplying the second equation by (-1) and adding it to the first equation immediately eliminates z:

$$\begin{array}{r} 5y - 2z = -10 \\ \underline{-(6y - 2z = \ -8\ )} \end{array} \quad \longrightarrow \quad \begin{array}{r} 5y - 2z = -10 \\ \underline{-6y + 2z = \ \ 8} \\ -y \qquad = \ -2 \\ y = \ \ 2. \end{array}$$

Substitute y = 2 into the equation 5y - 2z = -10 and solve for z:

$$\begin{aligned} 5(2) - 2z &= -10 \\ 10 - 2z &= -10 \\ -2z &= -20 \\ z &= 10. \end{aligned}$$

Substitute y = 2 and z = 10 into one of the original equations to solve for the remaining variable x. Using the third equation we have

$$\begin{aligned} x - y + z &= 4 \\ x - 2 + 10 &= 4 \\ x &= -4. \end{aligned}$$

Thus the solution is x = -4, y = 2, and z = 10.

5.

$$\begin{aligned} 2x + 4y + 6z &= -12 \\ 2x - 3y - 4z &= 15 \\ 3x + 4y + 5z &= -8 \end{aligned}$$

To eliminate x from the first and second equations, add -1 times the first equation to the second equation:

$$\begin{array}{r} -(\ 2x + 4y + 6z = -12\ ) \\ \underline{2x - 3y - 4z = \ 15} \end{array} \quad \longrightarrow \quad \begin{array}{r} -2x - 4y - 6z = 12 \\ \underline{2x - 3y - 4z = 15} \\ -7y - 10z = 27. \end{array}$$

To eliminate x from the second and third equations, we multiply the second equation by 3 and then add -2 times the third equation to the new second equation:

$$\begin{array}{r} 3(\ 2x - 3y - 4z = 15\ ) \\ \underline{-2(\ 3x + 4y + 5z = -8\ )} \end{array} \quad \longrightarrow \quad \begin{array}{r} 6x - 9y - 12z = 45 \\ \underline{-6x - 8y - 10z = 16} \\ -17y - 22z = 61. \end{array}$$

Thus we have the pair of equations

$$\begin{aligned} -7y - 10z &= 27 \\ -17y - 22z &= 61 \end{aligned}$$

to solve for y and z. To eliminate y, we multiply the first equation by -17 and add it to 7 times the second equation:

$$\begin{array}{l} -17(\ -7y - 10z = 27\ ) \\ \underline{\ 7(-17y - 22z = 61\ )} \end{array} \quad \text{----->} \quad \begin{array}{r} 119y + 170z = -459 \\ \underline{-119y - 154z = 427} \\ 16z = -32 \\ z = -2. \end{array}$$

Substitute z = -2 into the equation -7y - 10 z = 27 and solve for y:

$$\begin{aligned} -7y - 10(-2) &= 27 \\ -7y + 20 &= 27 \\ -7y &= 7 \\ y &= -1. \end{aligned}$$

Substitute y = -1 and z = -2 into one of the original equations to solve for the remaining variable x. Using the first equation we have

$$\begin{aligned} 2x + 4y + 6z &= -12 \\ 2x + 4(-1) + 6(-2) &= -12 \\ 2x &= 4 \\ x &= 2. \end{aligned}$$

Thus the solution is x = 2, y = -1, and z = -2.

7.

$$\begin{aligned} x + 4y - z &= 12 \\ 3x + 8y - 2z &= 4 \end{aligned}$$

To eliminate x, we add -3 times the first equation to the second equation:

$$\begin{array}{l} -3(\ x + 4y - z = 12\ ) \\ \underline{\ 3x + 8y - 2z = 4\ } \end{array} \quad \text{----->} \quad \begin{array}{r} -3x - 12y + 3z = -36 \\ \underline{3x + 8y - 2z = 4} \\ -4y + z = -32. \end{array}$$

Since there are only two equations, but three variables, we solve the equation -4y + z = -32 for y:

$$y = (1/4)z + 8.$$

The variable z can be chosen to be any real number. To determine x we substitute this expression for y into one of the original equations:

$$\begin{aligned} x + 4y - z &= 12 \\ x + 4((1/4)z + 8) - z &= 12 \\ x + z + 32 - z &= 12 \\ x &= -20. \end{aligned}$$

Since z can be chosen arbitrarily, there are many solutions of this system. We express these in the form

$$x = -20,\ y = (1/4)z + 8,\ z = t,$$

where t is any real number.

Exercises 1.1

9.
$$\begin{aligned} x + y + 3z &= 12 \\ 2x + 2y + 6z &= 6 \end{aligned}$$

To eliminate x, we add -2 times the first equation to the second equation:

$$\begin{array}{r} -2(\ x + y + 3z = 12\ ) \\ \underline{2x + 2y + 6z = 6} \end{array} \quad \text{----->} \quad \begin{array}{r} -2x - 2y - 6z = -24 \\ \underline{2x + 2y + 6z = 6} \\ 0 = -18. \end{array}$$

The result 0 = 18 makes no sense. Thus this system of equations has no solution.

11.
$$\begin{aligned} 2x + 3y &= 13 \\ x - 2y &= 3 \\ 5x + 2y &= 27 \end{aligned}$$

To eliminate x from the first and second equations, we add -2 times the second equation to the first equation:

$$\begin{array}{r} 2x + 3y = 13 \\ \underline{-2(\ x - 2y = 3\ )} \end{array} \quad \text{----->} \quad \begin{array}{r} 2x + 3y = 13 \\ \underline{-2x + 4y = -6} \\ 7y = 7 \\ y = 1. \end{array}$$

To eliminate x from the second and third equations, we add -5 times the second equation to the third equation:

$$\begin{array}{r} -5(\ x - 2y = 3\ ) \\ \underline{5x + 2y = 27} \end{array} \quad \text{----->} \quad \begin{array}{r} -5x + 10y = -15 \\ \underline{5x + 2y = 27} \\ 12y = 12 \\ y = 1. \end{array}$$

Substituting y = 1 into equation x - 2y = 3, we obtain x = 5. Thus the solution of the linear system is x = 5, y = 1.

13.
$$\begin{aligned} x + 3y &= -4 \\ 2x + 5y &= -8 \\ x + 3y &= -5 \end{aligned}$$

Upon inspection of this system of equations we see that the first and third equations have identical left sides, but different right sides. Thus there are no choices for x and y that will satisfy these equations simultaneously. We verify this by adding -1 times the third equation to the first equation:

$$\begin{array}{r} x + 3y = -4 \\ \underline{-(x + 3y = -5\ )} \end{array} \quad \text{----->} \quad \begin{array}{r} x + 3y = -4 \\ \underline{-x - 3y = 5} \\ 0 = 1. \end{array}$$

Since this result makes no sense, this system of equations has no solution.

15.
$$\begin{aligned} 2x_1 - x_2 &= 5 \\ 4x_1 - 2x_2 &= t \end{aligned}$$

(a) To determine a value of t so that the linear system has a solution we begin by eliminating x_1. Multiply the first equation by 2 and then add (-1) times the second equation.

$$\begin{aligned} 2(\ 2x_1 - x_2 &= 5\ ) \\ \underline{-(\ 4x_1 - 2x_2} &\underline{= t\ )} \end{aligned} \quad \text{--->} \quad \begin{aligned} 4x_1 - 2x_2 &= 10 \\ \underline{-4x_1 + 2x_2} &\underline{= -t} \\ 0 &= 10 - t \end{aligned}$$

In order for the linear system to have a solution $0 = 10 - t$ must make sense, hence it follows that $t = 10$.

(b) In order for the linear system not to have a solution the relation $0 = 10 - t$ from part (a) must not make sense. If $t = 3$, then we would have $0 = 7$, which would imply the linear system has no solution.

(c) The choice $t = 3$ in part (b) was arbitrary. Any choice for t, other than $t = 10$, gives a relation which makes no sense. Hence there are infinitely many ways to choose a value for t in part (b).

17. The linear system

$$\begin{aligned} 2x + y - 2z &= -5 \\ 3y + z &= 7 \\ z &= 4 \end{aligned}$$

can be solved starting with the last equation which tells us that $z = 4$. Substituting $z = 4$ into the second equation we have

$$3y + 4 = 7 \quad \text{--->} \quad y = 1$$

Substituting $z = 4$ and $y = 1$ into the first equation gives

$$2x + 1 - 2(4) = -5 \quad \text{--->} \quad x = 1$$

Thus the solution is $x = 1$, $y = 1$, $z = 4$. (Such a linear system is called upper triangular.)

19. The values $x = 1$, $y = 2$, $z = r$ are a solution only if they satisfy each equation of the linear system. We substitute these values into each equation to determine if a single value for r exists. We have

$$\begin{aligned} &\text{First Equation:} & 2(1) + 3(2) - r &= 11 & \text{--->} \quad r &= -3 \\ &\text{Second Equation:} & 1 - 2 + 2r &= -7 & \text{--->} \quad r &= -3 \\ &\text{Third Equation:} & 4(1) + 2 - 2r &= 12 & \text{--->} \quad r &= -3 \end{aligned}$$

It follows that for $r = -3$ we have that $x = 1$, $y = 2$, $z = r$ is a solution.

21. Let the amount of low-sulfur fuel be denoted x_1 and the amount of high-sulfur fuel be denoted x_2. Convert all times to minutes. Then,

$$\begin{aligned}\text{blending plant requirements: } & 5x_1 + 4x_2 = 180 \\ \text{refining plant requirements: } & 4x_1 + 2x_2 = 120 \ .\end{aligned}$$

To eliminate x_1 we multiply the first equation by 4 and add -5 times the second equation to it:

$$\begin{array}{l} 4(5x_1 + 4x_2 = 180) \\ -5(4x_1 + 2x_2 = 120) \\ \hline \end{array} \quad \text{----->} \quad \begin{array}{rcr} 20x_1 + 16x_2 & = & 720 \\ -20x_1 - 10x_2 & = & -600 \\ \hline 6x_2 & = & 120 \\ x_2 & = & 20. \end{array}$$

To solve for x_1 we substitute $x_2 = 20$ into either of the original equations. Using the blending plant equation we have

$$\begin{aligned} 5x_1 + 4x_2 &= 180 \\ 5x_1 + 4(20) &= 180 \\ 5x_1 &= 100 \\ x_1 &= 20. \end{aligned}$$

Thus 20 tons of low-sulfur fuel and 20 tons of high-sulfur fuel should be manufactured so that the plants are fully utilized.

23. Let x_1 = the number of ounces of food A per meal,
x_2 = the number of ounces of food B per meal, and
x_3 = the number of ounces of food C per meal.

Then,

$$\begin{aligned}\text{protein requirements: } & 2x_1 + 3x_2 + 3x_3 = 25 \\ \text{fat requirements: } & 3x_1 + 2x_2 + 3x_3 = 24 \\ \text{carbohydrate requirements: } & 4x_1 + 1x_2 + 2x_3 = 21 \ .\end{aligned}$$

To eliminate x_1 from the first two equations, we multiply the first equation by 3 and add -2 times the second equation to it:

$$\begin{array}{l} 3(2x_1 + 3x_2 + 3x_3 = 25) \\ -2(3x_1 + 2x_2 + 3x_3 = 24) \\ \hline \end{array} \quad \text{----->} \quad \begin{array}{rcr} 6x_1 + 9x_2 + 9x_3 & = & 75 \\ -6x_1 - 4x_2 - 6x_3 & = & -48 \\ \hline 5x_2 + 3x_3 & = & 27. \end{array}$$

To eliminate x_1 from the second and third equations we multiply the second equation by 4 and add -3 times the third equation to it:

$$\begin{array}{rl} 4(3x_1 + 2x_2 + 3x_3 = 24\ ) & \\ -3(4x_1 + 1x_2 + 2x_3 = 21\ ) & \end{array} \;\;\text{----->}\;\; \begin{array}{rcrcrcr} 12x_1 & + & 8x_2 & + & 12x_3 & = & 96 \\ -12x_1 & - & 3x_2 & - & 6x_3 & = & -63 \\ \hline & & 5x_2 & + & 6x_3 & = & 33. \end{array}$$

Thus we have the pair of equations

$$\begin{aligned} 5x_2 + 3x_3 &= 27 \\ 5x_2 + 6x_3 &= 33 \end{aligned}$$

to solve for x_2 and x_3. To eliminate x_2 we add -1 times the second equation to the first equation:

$$\begin{array}{r} 5x_2 + 3x_3 = 27 \\ -(5x_2 + 6x_3 = 33\ ) \end{array} \;\;\text{----->}\;\; \begin{array}{rcrcr} 5x_2 & + & 3x_3 & = & 27 \\ -5x_2 & - & 6x_3 & = & -33 \\ \hline & & -3x_3 & = & -6 \\ & & x_3 & = & 2. \end{array}$$

Substitute $x_3 = 2$ into equation $5x_2 + 6x_3 = 27$ and solve for x_2. We obtain,

$$\begin{aligned} 5x_2 + 3x_3 &= 27 \\ 5x_2 + 3(2) &= 27 \\ 5x_2 &= 21 \\ x_2 &= 21/5 = 4.2\ . \end{aligned}$$

Substitute $x_2 = 4.2$ and $x_3 = 2$ into any of the original equations and solve for the remaining variable x_1. Using the equation for carbohydrate requirements we obtain,

$$\begin{aligned} 4x_1 + 1x_2 + 2x_3 &= 21 \\ 4x_1 + 4.2 + 2(2) &= 21 \\ 4x_1 &= 12.8 \\ x_1 &= 3.2\ . \end{aligned}$$

Thus each meal should contain 3.2 ounces of food A, 4.2 ounces of food B, and 2 ounces of food C.

T.1. Let a solution to the system of equations in (2) be given by

$$x_1 = s_1\ ,\ x_2 = s_2, \ldots,\ x_n = s_n.$$

Interchanging the position of two of the equations in (2) gives a system in which $x_j = s_j$, $j=1,2,\ldots,n$ still satisfies each equation.

T.3. If $s_1, s_2, \ldots, s_n$ is a solution to (2), then the pth and qth equations are satisfied:

$$a_{p1}s_1 + \cdots + a_{pn}s_n = b_p$$

$$a_{q1}s_1 + \cdots + a_{qn}s_n = b_q.$$

Thus, for any real number r,

$$(a_{p1} + ra_{q1})s_1 + \cdots + (a_{pn} + ra_{qn}) = b_p + rb_q$$

and so $s_1, \ldots, s_n$ is a solution to the new system.

Conversely, any solution to the new system is also a solution to the original system (2).

Exercises 1.2

1. (a) $a_{12} = -3$, $a_{22} = -5$, $a_{23} = 4$
   (b) $b_{11} = 4$, $b_{31} = 5$
   (c) $c_{13} = 2$, $c_{31} = 6$, $c_{33} = -1$

3. (a) $$\mathbf{A}(\mathbf{BD}) = \begin{bmatrix} 1 & 2 & 3 \\ 2 & 1 & 4 \end{bmatrix} \left(\begin{bmatrix} 1 & 0 \\ 2 & 1 \\ 3 & 2 \end{bmatrix} \begin{bmatrix} 3 & -2 \\ 2 & 4 \end{bmatrix} \right)$$

$$= \begin{bmatrix} 1 & 2 & 3 \\ 2 & 1 & 4 \end{bmatrix} \begin{bmatrix} (1)(3)+(0)(2) & (1)(-2)+(0)(4) \\ (2)(3)+(1)(2) & (2)(-2)+(1)(4) \\ (3)(3)+(2)(2) & (3)(-2)+(2)(4) \end{bmatrix} = \begin{bmatrix} 1 & 2 & 3 \\ 2 & 1 & 4 \end{bmatrix} \begin{bmatrix} 3 & -2 \\ 8 & 0 \\ 13 & 2 \end{bmatrix}$$

$$= \begin{bmatrix} (1)(3)+(2)(8)+(3)(13) & (1)(-2)+(2)(0)+(3)(2) \\ (2)(3)+(1)(8)+(4)(13) & (2)(-2)+(1)(0)+(4)(2) \end{bmatrix} = \begin{bmatrix} 58 & 4 \\ 66 & 4 \end{bmatrix}$$

$$(\mathbf{AB})\mathbf{D} = \left(\begin{bmatrix} 1 & 2 & 3 \\ 2 & 1 & 4 \end{bmatrix} \begin{bmatrix} 1 & 0 \\ 2 & 1 \\ 3 & 2 \end{bmatrix} \right) \begin{bmatrix} 3 & -2 \\ 2 & 4 \end{bmatrix}$$

$$= \begin{bmatrix} (1)(1)+(2)(2)+(3)(3) & (1)(0)+(2)(1)+(3)(2) \\ (2)(1)+(1)(2)+(4)(3) & (2)(0)+(1)(1)+(4)(2) \end{bmatrix} \begin{bmatrix} 3 & -2 \\ 2 & 4 \end{bmatrix}$$

$$= \begin{bmatrix} 14 & 8 \\ 16 & 9 \end{bmatrix} \begin{bmatrix} 3 & -2 \\ 2 & 4 \end{bmatrix} = \begin{bmatrix} (14)(3)+(8)(2) & (14)(-2)+(8)(4) \\ (16)(3)+(9)(2) & (16)(-2)+(9)(4) \end{bmatrix} = \begin{bmatrix} 58 & 4 \\ 66 & 4 \end{bmatrix}$$

(b) $$\mathbf{A}(\mathbf{C} + \mathbf{E}) = \begin{bmatrix} 1 & 2 & 3 \\ 2 & 1 & 4 \end{bmatrix} \left(\begin{bmatrix} 3 & -1 & 3 \\ 4 & 1 & 5 \\ 2 & 1 & 3 \end{bmatrix} + \begin{bmatrix} 2 & -4 & 5 \\ 0 & 1 & 4 \\ 3 & 2 & 1 \end{bmatrix} \right)$$

$$= \begin{bmatrix} 1 & 2 & 3 \\ 2 & 1 & 4 \end{bmatrix} \begin{bmatrix} 3+2 & -1+(-4) & 3+5 \\ 4+0 & 1+1 & 5+4 \\ 2+3 & 1+2 & 3+1 \end{bmatrix} = \begin{bmatrix} 1 & 2 & 3 \\ 2 & 1 & 4 \end{bmatrix} \begin{bmatrix} 5 & -5 & 8 \\ 4 & 2 & 9 \\ 5 & 3 & 4 \end{bmatrix}$$

$$= \begin{bmatrix} (1)(5)+(2)(4)+(3)(5) & (1)(-5)+(2)(2)+(3)(3) & (1)(8)+(2)(9)+(3)(4) \\ (2)(5)+(1)(4)+(4)(5) & (2)(-5)+(1)(2)+(4)(3) & (2)(8)+(1)(9)+(4)(4) \end{bmatrix}$$

$$= \begin{bmatrix} 28 & 8 & 38 \\ 34 & 4 & 41 \end{bmatrix}$$

$$\mathbf{AC} + \mathbf{AE} = \begin{bmatrix} 1 & 2 & 3 \\ 2 & 1 & 4 \end{bmatrix}\begin{bmatrix} 3 & -1 & 3 \\ 4 & 1 & 5 \\ 2 & 1 & 3 \end{bmatrix} + \begin{bmatrix} 1 & 2 & 3 \\ 2 & 1 & 4 \end{bmatrix}\begin{bmatrix} 2 & -4 & 5 \\ 0 & 1 & 4 \\ 3 & 2 & 1 \end{bmatrix}$$

$$= \begin{bmatrix} (1)(3)+(2)(4)+(3)(2) & (1)(-1)+(2)(1)+(3)(1) & (1)(3)+(2)(5)+(3)(3) \\ (2)(3)+(1)(4)+(4)(2) & (2)(-1)+(1)(1)+(4)(1) & (2)(3)+(1)(5)+(4)(3) \end{bmatrix}$$

$$+ \begin{bmatrix} (1)(2)+(2)(0)+(3)(3) & (1)(-4)+(2)(1)+(3)(2) & (1)(5)+(2)(4)+(3)(1) \\ (2)(2)+(1)(0)+(4)(3) & (2)(-4)+(1)(1)+(4)(2) & (2)(5)+(1)(4)+(4)(1) \end{bmatrix}$$

$$= \begin{bmatrix} 17 & 4 & 22 \\ 18 & 3 & 23 \end{bmatrix} + \begin{bmatrix} 11 & 4 & 16 \\ 16 & 1 & 18 \end{bmatrix} = \begin{bmatrix} 28 & 8 & 38 \\ 34 & 4 & 41 \end{bmatrix}$$

(c) $3\mathbf{A} + 2\mathbf{A} = 3\begin{bmatrix} 1 & 2 & 3 \\ 2 & 1 & 4 \end{bmatrix} + 2\begin{bmatrix} 1 & 2 & 3 \\ 2 & 1 & 4 \end{bmatrix} = \begin{bmatrix} 3 & 6 & 9 \\ 6 & 3 & 12 \end{bmatrix} + \begin{bmatrix} 2 & 4 & 6 \\ 4 & 2 & 8 \end{bmatrix}$

$$= \begin{bmatrix} 5 & 10 & 15 \\ 10 & 5 & 20 \end{bmatrix} = 5\mathbf{A}$$

(d) $\mathbf{DF} + \mathbf{AB} = \begin{bmatrix} 3 & -2 \\ 2 & 4 \end{bmatrix}\begin{bmatrix} -4 & 5 \\ 2 & 3 \end{bmatrix} + \begin{bmatrix} 1 & 2 & 3 \\ 2 & 1 & 4 \end{bmatrix}\begin{bmatrix} 1 & 0 \\ 2 & 1 \\ 3 & 2 \end{bmatrix}$

$$= \begin{bmatrix} (3)(-4)+(-2)(2) & (3)(5)+(-2)(3) \\ (2)(-4)+(4)(2) & (2)(5)+(4)(3) \end{bmatrix} +$$

$$\begin{bmatrix} (1)(1)+(2)(2)+(3)(3) & (1)(0)+(2)(1)+(3)(2) \\ (2)(0)+(1)(1)+(4)(2) & (2)(0)+(1)(1)+(4)(2) \end{bmatrix}$$

$$= \begin{bmatrix} -16 & 9 \\ 0 & 22 \end{bmatrix} + \begin{bmatrix} 14 & 8 \\ 16 & 9 \end{bmatrix} = \begin{bmatrix} -16+14 & 9+8 \\ 0+16 & 22+9 \end{bmatrix} = \begin{bmatrix} -2 & 17 \\ 16 & 31 \end{bmatrix}$$

(e) $\mathbf{EF} + 2\mathbf{A}$ is undefined since $\mathbf{E}$ is 3×3 and $\mathbf{F}$ is 2×2.

(f) $(-4)(3\mathbf{C}) = -4\left(3\begin{bmatrix} 3 & -1 & 3 \\ 4 & 1 & 5 \\ 2 & 1 & 3 \end{bmatrix}\right) = -4\begin{bmatrix} 9 & -3 & 9 \\ 12 & 3 & 15 \\ 6 & 3 & 9 \end{bmatrix} = \begin{bmatrix} -36 & 12 & -36 \\ -48 & -12 & -60 \\ -24 & -12 & -36 \end{bmatrix}$

$$(-12)\mathbf{C} = -12\begin{bmatrix} 3 & -1 & 3 \\ 4 & 1 & 5 \\ 2 & 1 & 3 \end{bmatrix} = \begin{bmatrix} -36 & 12 & -36 \\ -48 & -12 & -60 \\ -24 & -12 & -36 \end{bmatrix}$$

5. (a) $$\mathbf{A}^T = \begin{bmatrix} 1 & 2 & 3 \\ 2 & 1 & 4 \end{bmatrix}^T = \begin{bmatrix} 1 & 2 \\ 2 & 1 \\ 3 & 4 \end{bmatrix}$$

$$(\mathbf{A}^T)^T = \begin{bmatrix} 1 & 2 \\ 2 & 1 \\ 3 & 4 \end{bmatrix}^T = \begin{bmatrix} 1 & 2 & 3 \\ 2 & 1 & 4 \end{bmatrix}$$

(b) $$(\mathbf{C} + \mathbf{E})^T = \left(\begin{bmatrix} 3 & -1 & 3 \\ 4 & 1 & 5 \\ 2 & 1 & 3 \end{bmatrix} + \begin{bmatrix} 2 & -4 & 5 \\ 0 & 1 & 4 \\ 3 & 2 & 1 \end{bmatrix}\right)^T = \begin{bmatrix} 5 & -5 & 8 \\ 4 & 2 & 9 \\ 5 & 3 & 4 \end{bmatrix}^T$$

$$= \begin{bmatrix} 5 & 4 & 5 \\ -5 & 2 & 3 \\ 8 & 9 & 4 \end{bmatrix}$$

$$\mathbf{C}^T + \mathbf{E}^T = \begin{bmatrix} 3 & -1 & 3 \\ 4 & 1 & 5 \\ 2 & 1 & 3 \end{bmatrix}^T + \begin{bmatrix} 2 & -4 & 5 \\ 0 & 1 & 4 \\ 3 & 2 & 1 \end{bmatrix}^T$$

$$= \begin{bmatrix} 3 & 4 & 2 \\ -1 & 1 & 1 \\ 3 & 5 & 3 \end{bmatrix} + \begin{bmatrix} 2 & 0 & 3 \\ -4 & 1 & 2 \\ 5 & 4 & 1 \end{bmatrix} = \begin{bmatrix} 5 & 4 & 5 \\ -5 & 2 & 3 \\ 8 & 9 & 4 \end{bmatrix}$$

(c) $$(\mathbf{AB})^T = \left(\begin{bmatrix} 1 & 2 & 3 \\ 2 & 1 & 4 \end{bmatrix}\begin{bmatrix} 1 & 0 \\ 2 & 1 \\ 3 & 2 \end{bmatrix}\right)^T = \begin{bmatrix} 14 & 8 \\ 16 & 9 \end{bmatrix}^T = \begin{bmatrix} 14 & 16 \\ 8 & 9 \end{bmatrix}$$

$$\mathbf{B}^T\mathbf{A}^T = \begin{bmatrix} 1 & 0 \\ 2 & 1 \\ 3 & 2 \end{bmatrix}^T \begin{bmatrix} 1 & 2 & 3 \\ 2 & 1 & 4 \end{bmatrix}^T = \begin{bmatrix} 1 & 2 & 3 \\ 0 & 1 & 2 \end{bmatrix}\begin{bmatrix} 1 & 2 \\ 2 & 1 \\ 3 & 4 \end{bmatrix}$$

$$= \begin{bmatrix} (1)(1)+(2)(2)+(3)(3) & (1)(2)+(2)(1)+(3)(4) \\ (0)(1)+(1)(2)+(2)(3) & (0)(2)+(1)(1)+(2)(4) \end{bmatrix} = \begin{bmatrix} 14 & 16 \\ 8 & 9 \end{bmatrix}$$

Exercises 1.2

(d) $(2\mathbf{D}+3\mathbf{F})^T = \left(2\begin{bmatrix} 3 & -2 \\ 2 & 4 \end{bmatrix} + 3\begin{bmatrix} -4 & 5 \\ 2 & 3 \end{bmatrix}\right)^T = \left(\begin{bmatrix} 6 & -4 \\ 4 & 8 \end{bmatrix} + \begin{bmatrix} -12 & 15 \\ 6 & 9 \end{bmatrix}\right)^T$

$= \begin{bmatrix} -6 & 11 \\ 10 & 17 \end{bmatrix}^T = \begin{bmatrix} -6 & 10 \\ 11 & 17 \end{bmatrix}$

(e) $(2\mathbf{BC} + \mathbf{F})^T$ is undefined since the product **BC** is undefined. **B** is 3×2 and **C** is 3×3.

(f) $\mathbf{B}^T\mathbf{C} + \mathbf{A} = \begin{bmatrix} 1 & 0 \\ 2 & 1 \\ 3 & 2 \end{bmatrix}^T \begin{bmatrix} 3 & -1 & 3 \\ 4 & 1 & 5 \\ 2 & 1 & 3 \end{bmatrix} + \begin{bmatrix} 1 & 2 & 3 \\ 2 & 1 & 4 \end{bmatrix}$

$= \begin{bmatrix} 1 & 2 & 3 \\ 0 & 1 & 2 \end{bmatrix} \begin{bmatrix} 3 & -1 & 3 \\ 4 & 1 & 5 \\ 2 & 1 & 3 \end{bmatrix} + \begin{bmatrix} 1 & 2 & 3 \\ 2 & 1 & 4 \end{bmatrix}$

$= \begin{bmatrix} (1)(3)+(2)(4)+(3)(2) & (1)(-1)+(2)(1)+(3)(1) & (1)(3)+(2)(5)+(3)(3) \\ (0)(3)+(1)(4)+(2)(2) & (0)(-1)+(1)(1)+(2)(1) & (0)(3)+(1)(5)+(2)(3) \end{bmatrix}$

$+ \begin{bmatrix} 1 & 2 & 3 \\ 2 & 1 & 4 \end{bmatrix} = \begin{bmatrix} 17 & 4 & 22 \\ 8 & 3 & 11 \end{bmatrix} + \begin{bmatrix} 1 & 2 & 3 \\ 2 & 1 & 4 \end{bmatrix} = \begin{bmatrix} 18 & 6 & 25 \\ 10 & 4 & 15 \end{bmatrix}$

(g) $\mathbf{BB}^T = \begin{bmatrix} 1 & 2 & 3 \\ 2 & 5 & 8 \\ 3 & 8 & 13 \end{bmatrix}$ and $\mathbf{B}^T\mathbf{B} = \begin{bmatrix} 14 & 8 \\ 8 & 5 \end{bmatrix}$

7. $\mathbf{DI} = \begin{bmatrix} 3 & -2 \\ 2 & 4 \end{bmatrix}\begin{bmatrix} 1 & 0 \\ 0 & 1 \end{bmatrix} = \begin{bmatrix} (3)(1)+(-2)(0) & (3)(0)+(-2)(1) \\ (2)(1)+(4)(0) & (2)(0)+(4)(1) \end{bmatrix} = \begin{bmatrix} 3 & -2 \\ 2 & 4 \end{bmatrix}$

$\mathbf{ID} = \begin{bmatrix} 1 & 0 \\ 0 & 1 \end{bmatrix}\begin{bmatrix} 3 & -2 \\ 2 & 4 \end{bmatrix} = \begin{bmatrix} (1)(3)+(0)(2) & (1)(-2)+(0)(4) \\ (0)(3)+(1)(2) & (0)(-2)+(1)(4) \end{bmatrix} = \begin{bmatrix} 3 & -2 \\ 2 & 4 \end{bmatrix}$

We see that **DI** = **ID** = **D**.

9. Two matrices are equal if corresponding entries agree. Hence,

$$\begin{bmatrix} a+2b & 2a-b \\ 2c+d & c-2d \end{bmatrix} = \begin{bmatrix} 4 & -2 \\ 4 & -3 \end{bmatrix},$$

implies that we have the systems of of equations

$$\begin{aligned} a + 2b &= 4 \\ 2a - b &= -2 \end{aligned} \quad \text{and} \quad \begin{aligned} 2c + d &= 4 \\ c - 2d &= -3 \end{aligned} .$$

We use the method of elimination on each system to obtain

$$a = 0,\ b = 2,\ c = 1,\ d = 2.$$

11. $\mathbf{A0} = \begin{bmatrix} 2 & -3 & 5 \\ 6 & -5 & 4 \end{bmatrix} \begin{bmatrix} 0 & 0 \\ 0 & 0 \\ 0 & 0 \end{bmatrix}$

$$= \begin{bmatrix} (2)(0)+(-3)(0)+(5)(0) & (2)(0)+(-3)(0)+(5)(0) \\ (6)(0)+(-5)(0)+(4)(0) & (6)(0)+(-5)(0)+(4)(0) \end{bmatrix} = \begin{bmatrix} 0 & 0 \\ 0 & 0 \end{bmatrix} = \mathbf{0}$$

13. The corresponding linear system is

$$\begin{aligned} -2x - y + 4w &= 5 \\ -3x + 2y + 7z + 8w &= 3 \\ x + 2w &= 4 \\ 3x + z + 3w &= 6 . \end{aligned}$$

15. (a) The coefficient matrix is $\begin{bmatrix} 3 & -1 & 2 \\ 2 & 1 & 0 \\ 0 & 1 & 3 \\ 4 & 0 & -1 \end{bmatrix}$.

(b) The linear system in matrix form is

$$\begin{bmatrix} 3 & -1 & 2 \\ 2 & 1 & 0 \\ 0 & 1 & 3 \\ 4 & 0 & -1 \end{bmatrix} \begin{bmatrix} x \\ y \\ z \end{bmatrix} = \begin{bmatrix} 4 \\ 2 \\ 7 \\ 4 \end{bmatrix} .$$

(c) The augmented matrix is $\left[\begin{array}{ccc|c} 3 & -1 & 2 & 4 \\ 2 & 1 & 0 & 2 \\ 0 & 1 & 3 & 7 \\ 4 & 0 & -1 & 4 \end{array}\right]$.

Exercises 1.2

17. Multiply by the scalar, add the matrices on the left side, and then equate corresponding entries. We obtain the following linear systems.

(a) $\begin{aligned} x + 3y &= 3 \\ 2x \qquad &= 1 \\ x - \ y &= 4 \end{aligned}$ (b) $\begin{aligned} 2x + 3y \qquad\quad &= 0 \\ x - \ y + \ z &= 0 \\ 2y - \ z &= 0 \\ x + 2y + 3z &= 0 \end{aligned}$

19. (a) The second column of **AB** is obtained from **A** times the second column of **B**, which is denoted $\mathbf{B}_2$:

$$\mathbf{AB}_2 = \begin{bmatrix} 1 & -1 & 2 \\ 3 & 2 & 4 \\ 4 & -2 & 3 \\ 2 & 1 & 5 \end{bmatrix} \begin{bmatrix} 0 \\ 3 \\ 2 \end{bmatrix} = \begin{bmatrix} (1)(0)+(-1)(3)+(2)(2) \\ (3)(0)+(2)(3)+(4)(2) \\ (4)(0)+(-2)(3)+(3)(2) \\ (2)(0)+(1)(3)+(5)(2) \end{bmatrix} = \begin{bmatrix} 1 \\ 14 \\ 0 \\ 13 \end{bmatrix}.$$

(b) The fourth column of **AB** is obtained from **A** times the fourth column of **B**, which is denoted $\mathbf{B}_4$:

$$\mathbf{AB}_4 = \begin{bmatrix} 1 & -1 & 2 \\ 3 & 2 & 4 \\ 4 & -2 & 3 \\ 2 & 1 & 5 \end{bmatrix} \begin{bmatrix} 2 \\ 4 \\ 1 \end{bmatrix} = \begin{bmatrix} (1)(2)+(-1)(4)+(2)(1) \\ (3)(2)+(2)(4)+(4)(1) \\ (4)(2)+(-2)(4)+(3)(1) \\ (2)(2)+(1)(4)+(5)(1) \end{bmatrix} = \begin{bmatrix} 0 \\ 18 \\ 3 \\ 13 \end{bmatrix}.$$

21. $\mathbf{AB^T} = \begin{bmatrix} r & 1 & -2 \end{bmatrix} \begin{bmatrix} 1 \\ 3 \\ -1 \end{bmatrix} = r + 3 + 2 = 0$ only if $r = -5$.

23. $\mathbf{A^TA} = \begin{bmatrix} \mathbf{A}_1^{\mathbf{T}} \\ \mathbf{A}_2^{\mathbf{T}} \\ \mathbf{A}_3^{\mathbf{T}} \end{bmatrix} \begin{bmatrix} \mathbf{A}_1 & \mathbf{A}_2 & \mathbf{A}_3 \end{bmatrix} = \begin{bmatrix} \mathbf{A}_1^{\mathbf{T}}\mathbf{A}_1 & \mathbf{A}_1^{\mathbf{T}}\mathbf{A}_2 & \mathbf{A}_1^{\mathbf{T}}\mathbf{A}_3 \\ \mathbf{A}_2^{\mathbf{T}}\mathbf{A}_1 & \mathbf{A}_2^{\mathbf{T}}\mathbf{A}_2 & \mathbf{A}_2^{\mathbf{T}}\mathbf{A}_3 \\ \mathbf{A}_3^{\mathbf{T}}\mathbf{A}_1 & \mathbf{A}_3^{\mathbf{T}}\mathbf{A}_2 & \mathbf{A}_3^{\mathbf{T}}\mathbf{A}_3 \end{bmatrix} = \begin{bmatrix} 25 & 14 & -3 \\ 14 & 29 & 2 \\ -3 & 2 & 1 \end{bmatrix}$

25. The entries of the matrix product **AB** tell the manufacturer the cost of pollution control for each product P or Q. The matrix product is

$$\mathbf{AB} = \begin{array}{c} \text{Plant X} \quad \text{Plant Y} \\ \begin{bmatrix} 5350 & 6000 \\ 9350 & 8650 \end{bmatrix} \end{array} \begin{array}{l} \text{Product P} \\ \text{Product Q} \end{array}.$$

For example, it costs \$9350 to control the pollution from product Q in Plant X and \$6000 to control the pollution from product P in Plant Y.

27. (a) The total value of the cameras in New York is obtained by multiplying row 1 of **A** times column 1 of **B**:

$$\begin{bmatrix} 200 & 150 & 120 \end{bmatrix}\begin{bmatrix} 220 \\ 300 \\ 120 \end{bmatrix} = [(200)(220)+(150)(300)+(120)(120)] = 103400.$$

(b) The total value of the flash units in Los Angeles is obtained by multiplying row 2 of **A** times column 3 of **B**:

$$\begin{bmatrix} 50 & 40 & 25 \end{bmatrix}\begin{bmatrix} 100 \\ 120 \\ 250 \end{bmatrix} = [(50)(100)+(40)(120)+(25)(250)] = 16050.$$

T.1. Let $\mathbf{A} = [a_{ij}]$ be $m \times p$ and $\mathbf{B} = [b_{ij}]$ be $p \times n$.

(a) Let the ith row of **A** consist entirely of zeros, so $a_{ik} = 0$ for $k = 1,2,\ldots,p$. Then the (i,j) entry in **AB** is

$$\sum_{k=1}^{p} a_{ik}b_{kj} = 0 \text{ for } j=1,2,\ldots,n.$$

(b) Let the jth column of **B** consist entirely of zeros, so $b_{kj} = 0$ for $k = 1,2,\ldots,p$. Then again the (i,j) entry of **AB** is 0 for $i = 1,2,\ldots,m$.

T.3. If **A** and **B** are scalar matrices, then $\mathbf{C} = \mathbf{A} + \mathbf{B}$ is a scalar matrix with $c_{ii} = c = a + b = a_{ii} + b_{ii}$ and $\mathbf{C} = \mathbf{AB}$ is a scalar matrix with $c_{ii} = c = a \cdot b = a_{ii} \cdot b_{ii}$.

T.5. The jth column of **AB** is $\begin{bmatrix} \sum_k a_{1k}b_{kj} \\ \sum_k a_{2k}b_{k} \\ \vdots \\ \sum_k a_{mk}b_{kj} \end{bmatrix}$. The first entry is row 1 of **A** times $\mathbf{B}_j$; the second entry is row 2 of **A** times $\mathbf{B}_j$; and so on. Hence, column j of **AB** is **A** times $\mathbf{B}_j$.

Exercises 1.2

T.7. (i) $\sum_{i=1}^{n}(r_i+s_i)a_i = (r_1+s_1)a_1 + (r_2+s_2)a_2 + \cdots + (r_n+s_n)a_n$

$= r_1a_1 + s_1a_1 + r_2a_2 + s_2a_2 + \cdots + r_na_n + s_na_n$

$= (r_1a_1 + r_2a_2 + \cdots + r_na_n) + (s_1a_1 + s_2a_2 + \cdots + s_na_n)$

$= \sum_{i=1}^{n} r_ia_i + \sum_{i=1}^{n} s_ia_i .$

(ii) $\sum_{i=1}^{n} c(r_ia_i) = cr_1a_1 + cr_2a_2 + \cdots + cr_na_n$

$= c(r_1a_1 + r_2a_2 + \cdots + r_na_n) = c\sum_{i=1}^{n} r_ia_i .$

T.9. (a) True. $\sum_{i=1}^{n}(a_i + 1) = \sum_{i=1}^{n} a_i + \sum_{i=1}^{n} 1 = \sum_{i=1}^{n} a_i + n .$

(b) True. $\sum_{i=1}^{n}\sum_{j=1}^{m} 1 = \sum_{i=1}^{n}\left[\sum_{j=1}^{m} 1\right] = \sum_{i=1}^{n} m = nm .$

(c) True. $\left[\sum_{i=1}^{n} a_i\right]\left[\sum_{j=1}^{m} b_j\right] = a_1\sum_{j=1}^{m} b_j + a_2\sum_{j=1}^{m} b_j + \cdots + a_n\sum_{j=1}^{m} b_j$

$= (a_1 + a_2 + \cdots + a_n)\sum_{j=1}^{m} b_j$

$= \sum_{i=1}^{n} a_i\sum_{j=1}^{m} b_j = \sum_{i=1}^{n}\sum_{j=1}^{m} a_ib_j .$

ML.1. Once you have entered matrices **A** and **B** you can use the commands given below to see the items requested in parts (a) and (b).

(a) Commands: **A(2,3)**, **B(3,2)**, **B(1,2)**

(b) Use command **A(1,:)** for $row_1(A)$.
Use command **A(:,3)** for $col_3(A)$.
Use command **B(2,:)** for $row_2(B)$.
(In this context the colon means 'all'.)

(c) Matrix **B** in **format long** is

```
8.00000000000000       0.66666666666667
0.00497512437811      -3.20000000000000
0.00001000000000       4.33333333333333
```

ML.2. (a) Use command **size(H)** to see the size of matrix **H**.
(b) Just type **H** to see the contents of the matrix.
(c) Use command **rat(H,'s')** for rational display.
(d) Command **H(:,1:3)** displays the first 3 columns..
(e) Command **H(4:5,:)** displays the last 2 rows.

ML.3. (a) A*C

```
ans =

    4.5000    2.2500    3.7500
    1.5833    0.9167    1.5000
    0.9667    0.5833    0.9500
```

(b) A*B

```
■ Error using ==> *
Inner matrix dimensions must agree.
```

(c) A+C'

```
ans =

    5.0000    1.5000
    1.5833    2.2500
    2.4500    3.1667
```

(d) B*A - C'*A

```
■ Error using ==> *
Inner matrix dimensions must agree.
```

(e) (2*C - 6*A')*B'

```
■ Error using ==> *
Inner matrix dimensions must agree.
```

(f) A*C - C*A

```
■ Error using ==> -
Inner matrix dimensions must agree.
```

(g) A*A' + C'*C

```
ans =

   18.2500    7.4583   12.2833
    7.4583    5.7361    8.9208
   12.2833    8.9208   14.1303
```

Exercises 1.2

```
ML.4. aug =

          2       4       6    -12
          2      -3      -4     15
          3       4       5     -8

ML.5. aug =

          4      -3       2     -1     -5
          2       1      -3      0      7
         -1       4       1      2      8

ML.6. (a) R = A(2,:)

          R =

              3       2       4

          C = B(:,3)

          C =

             -1
             -3
              5

          V = R*C

          V =

             11

          V is the (2,3)-entry of the product A*B.

      (b) C = B(:,2)

          C =

              0
              3
              2

          V = A*C

          V =

              1
             14
              0
             13

          V is the column 2 of the product A*B.
```

(c) R = A(3,:)

```
R =

  4    -2     3

V = R*B

V =

  10     0    17     3
```

V is the row 3 of the product A*B.

ML.7. (a) diag([1 2 3 4])

```
ans =

     1     0     0     0
     0     2     0     0
     0     0     3     0
     0     0     0     4
```

(b) diag([0 1 1/2 1/3 1/4])

```
ans =

     0          0          0          0          0
     0     1.0000          0          0          0
     0          0     0.5000          0          0
     0          0          0     0.3333          0
     0          0          0          0     0.2500
```

(c) diag(5 5 5 5 5])

```
ans =

     5     0     0     0     0
     0     5     0     0     0
     0     0     5     0     0
     0     0     0     5     0
     0     0     0     0     5
```

Exercises 1.3

1. $\mathbf{A} + \mathbf{B} = \begin{bmatrix} 1 & 2 & -2 \\ 3 & 4 & 5 \end{bmatrix} + \begin{bmatrix} 2 & 0 & 1 \\ 3 & -2 & 5 \end{bmatrix} = \begin{bmatrix} 3 & 2 & -1 \\ 6 & 2 & 10 \end{bmatrix}$

$\mathbf{B} + \mathbf{A} = \begin{bmatrix} 2 & 0 & 1 \\ 3 & -2 & 5 \end{bmatrix} + \begin{bmatrix} 1 & 2 & -2 \\ 3 & 4 & 5 \end{bmatrix} = \begin{bmatrix} 3 & 2 & -1 \\ 6 & 2 & 10 \end{bmatrix}$

$$\mathbf{A} + (\mathbf{B} + \mathbf{C}) = \begin{bmatrix} 1 & 2 & -2 \\ 3 & 4 & 5 \end{bmatrix} + \left(\begin{bmatrix} 2 & 0 & 1 \\ 3 & -2 & 5 \end{bmatrix} + \begin{bmatrix} -4 & -6 & 1 \\ 2 & 3 & 0 \end{bmatrix}\right)$$

$$= \begin{bmatrix} 1 & 2 & -2 \\ 3 & 4 & 5 \end{bmatrix} + \begin{bmatrix} -2 & -6 & 2 \\ 5 & 1 & 5 \end{bmatrix} = \begin{bmatrix} -1 & -4 & 0 \\ 8 & 5 & 10 \end{bmatrix}$$

$$(\mathbf{A} + \mathbf{B}) + \mathbf{C} = \left(\begin{bmatrix} 1 & 2 & -2 \\ 3 & 4 & 5 \end{bmatrix} + \begin{bmatrix} 2 & 0 & 1 \\ 3 & -2 & 5 \end{bmatrix}\right) + \begin{bmatrix} -4 & -6 & 1 \\ 2 & 3 & 0 \end{bmatrix}$$

$$= \begin{bmatrix} 3 & 2 & -1 \\ 6 & 2 & 10 \end{bmatrix} + \begin{bmatrix} -4 & -6 & 1 \\ 2 & 3 & 0 \end{bmatrix} = \begin{bmatrix} -1 & -4 & 0 \\ 8 & 5 & 10 \end{bmatrix}$$

$$\mathbf{A} + \mathbf{O} = \begin{bmatrix} 1 & 2 & -2 \\ 3 & 4 & 5 \end{bmatrix} + \begin{bmatrix} 0 & 0 & 0 \\ 0 & 0 & 0 \end{bmatrix} = \begin{bmatrix} 1 & 2 & -2 \\ 3 & 4 & 5 \end{bmatrix}$$

$$\mathbf{A} + (-\mathbf{A}) = \begin{bmatrix} 1 & 2 & -2 \\ 3 & 4 & 5 \end{bmatrix} + \begin{bmatrix} -1 & -2 & 2 \\ -3 & -4 & -5 \end{bmatrix} = \begin{bmatrix} 0 & 0 & 0 \\ 0 & 0 & 0 \end{bmatrix}$$

3. $\mathbf{A}(\mathbf{B} + \mathbf{C}) = \begin{bmatrix} 1 & -3 \\ -3 & 4 \end{bmatrix}\left(\begin{bmatrix} 2 & -3 & 2 \\ 3 & -1 & -2 \end{bmatrix} + \begin{bmatrix} 0 & 1 & 2 \\ 1 & 3 & -2 \end{bmatrix}\right)$

$$= \begin{bmatrix} 1 & -3 \\ -3 & 4 \end{bmatrix}\begin{bmatrix} 2 & -2 & 4 \\ 4 & 2 & -4 \end{bmatrix} = \begin{bmatrix} -10 & -8 & 16 \\ 10 & 14 & -28 \end{bmatrix}$$

$$\mathbf{AB} + \mathbf{AC} = \begin{bmatrix} 1 & -3 \\ -3 & 4 \end{bmatrix}\begin{bmatrix} 2 & -3 & 2 \\ 3 & -1 & -2 \end{bmatrix} + \begin{bmatrix} 1 & -3 \\ -3 & 4 \end{bmatrix}\begin{bmatrix} 0 & 1 & 2 \\ 1 & 3 & -2 \end{bmatrix}$$

$$= \begin{bmatrix} -7 & 0 & 8 \\ 6 & 5 & -14 \end{bmatrix} + \begin{bmatrix} -3 & -8 & 8 \\ 4 & 9 & -14 \end{bmatrix} = \begin{bmatrix} -10 & -8 & 16 \\ 10 & 14 & -28 \end{bmatrix}$$

5. $\mathbf{A}(r\mathbf{B}) = \begin{bmatrix} 1 & 3 \\ 2 & -1 \end{bmatrix}\left(-3\begin{bmatrix} -1 & 3 & 2 \\ 1 & -3 & 4 \end{bmatrix}\right) = \begin{bmatrix} 1 & 3 \\ 2 & -1 \end{bmatrix}\begin{bmatrix} 3 & -9 & -6 \\ -3 & 9 & -12 \end{bmatrix}$

$= \begin{bmatrix} -6 & 18 & -42 \\ 9 & -27 & 0 \end{bmatrix}$

$r(\mathbf{AB}) = -3\left(\begin{bmatrix} 1 & 3 \\ 2 & -1 \end{bmatrix}\begin{bmatrix} -1 & 3 & 2 \\ 1 & -3 & 4 \end{bmatrix}\right) = -3\begin{bmatrix} 2 & -6 & 14 \\ -3 & 9 & 0 \end{bmatrix} = \begin{bmatrix} -6 & 18 & -42 \\ 9 & -27 & 0 \end{bmatrix}$

$(r\mathbf{A})\mathbf{B} = \left(-3\begin{bmatrix} 1 & 3 \\ 2 & -1 \end{bmatrix}\right)\begin{bmatrix} -1 & 3 & 2 \\ 1 & -3 & 4 \end{bmatrix} = \begin{bmatrix} -3 & -9 \\ -6 & 3 \end{bmatrix}\begin{bmatrix} -1 & 3 & 2 \\ 1 & -3 & 4 \end{bmatrix}$

$= \begin{bmatrix} -6 & 18 & -42 \\ 9 & -27 & 0 \end{bmatrix}$

7. $(\mathbf{AB})^T = \left(\begin{bmatrix} 1 & 3 & 2 \\ 2 & 1 & -3 \end{bmatrix}\begin{bmatrix} 3 & -1 \\ 2 & 4 \\ 1 & 2 \end{bmatrix}\right)^T = \begin{bmatrix} 11 & 15 \\ 5 & -4 \end{bmatrix}^T = \begin{bmatrix} 11 & 5 \\ 15 & -4 \end{bmatrix}$

$\mathbf{B}^T\mathbf{A}^T = \begin{bmatrix} 3 & -1 \\ 2 & 4 \\ 1 & 2 \end{bmatrix}^T\begin{bmatrix} 1 & 3 & 2 \\ 2 & 1 & -3 \end{bmatrix}^T = \begin{bmatrix} 3 & 2 & 1 \\ -1 & 4 & 2 \end{bmatrix}\begin{bmatrix} 1 & 2 \\ 3 & 1 \\ 2 & -3 \end{bmatrix} = \begin{bmatrix} 11 & 5 \\ 15 & -4 \end{bmatrix}$

9. $\mathbf{AB} = \begin{bmatrix} -2 & 3 \\ 2 & -3 \end{bmatrix}\begin{bmatrix} -1 & 3 \\ 2 & 0 \end{bmatrix} = \begin{bmatrix} 8 & -6 \\ -8 & 6 \end{bmatrix}$

$\mathbf{AC} = \begin{bmatrix} -2 & 3 \\ 2 & -3 \end{bmatrix}\begin{bmatrix} -4 & -3 \\ 0 & -4 \end{bmatrix} = \begin{bmatrix} 8 & -6 \\ -8 & 6 \end{bmatrix}$

11. (a) $\mathbf{A}^2 + 3\mathbf{A} = \mathbf{AA} + 3\mathbf{A} = \begin{bmatrix} 4 & 2 \\ 1 & 3 \end{bmatrix}\begin{bmatrix} 4 & 2 \\ 1 & 3 \end{bmatrix} + 3\begin{bmatrix} 4 & 2 \\ 1 & 3 \end{bmatrix}$

$= \begin{bmatrix} 18 & 14 \\ 7 & 11 \end{bmatrix} + \begin{bmatrix} 12 & 6 \\ 3 & 9 \end{bmatrix} = \begin{bmatrix} 30 & 20 \\ 10 & 20 \end{bmatrix}$

(b) $2\mathbf{A}^3 + 3\mathbf{A}^2 + 4\mathbf{A} + 5\mathbf{I}_2 = 2\mathbf{AAA} + 3\mathbf{AA} + 4\mathbf{A} + 5\mathbf{I}_2$

$$= 2\begin{bmatrix}4 & 2\\1 & 3\end{bmatrix}\begin{bmatrix}4 & 2\\1 & 3\end{bmatrix}\begin{bmatrix}4 & 2\\1 & 3\end{bmatrix} + 3\begin{bmatrix}4 & 2\\1 & 3\end{bmatrix}\begin{bmatrix}4 & 2\\1 & 3\end{bmatrix} + 4\begin{bmatrix}4 & 2\\1 & 3\end{bmatrix} + 5\begin{bmatrix}1 & 0\\0 & 1\end{bmatrix}$$

$$= 2\begin{bmatrix}18 & 14\\7 & 11\end{bmatrix}\begin{bmatrix}4 & 2\\1 & 3\end{bmatrix} + 3\begin{bmatrix}18 & 14\\7 & 11\end{bmatrix} + \begin{bmatrix}16 & 8\\4 & 12\end{bmatrix} + \begin{bmatrix}5 & 0\\0 & 5\end{bmatrix}$$

$$= 2\begin{bmatrix}86 & 78\\39 & 47\end{bmatrix} + \begin{bmatrix}54 & 42\\21 & 33\end{bmatrix} + \begin{bmatrix}21 & 8\\4 & 17\end{bmatrix}$$

$$= \begin{bmatrix}172 & 156\\78 & 94\end{bmatrix} + \begin{bmatrix}75 & 50\\25 & 50\end{bmatrix} = \begin{bmatrix}247 & 206\\103 & 144\end{bmatrix}$$

13. Compute $\mathbf{AX} = \begin{bmatrix}2 & 1\\1 & 2\end{bmatrix}\begin{bmatrix}1\\1\end{bmatrix} = \begin{bmatrix}3\\3\end{bmatrix}$. Comparing this with $\mathbf{X}$, we see that $\mathbf{AX} = 3\mathbf{X}$. Thus $r = 3$.

15. (a) The distribution of the market after 1 year is denoted $\mathbf{X}_1$ and is computed as follows:

$$\mathbf{X}_1 = \mathbf{AX}_0 = \begin{bmatrix}1/3 & 2/5\\2/3 & 3/5\end{bmatrix}\begin{bmatrix}2/3\\1/3\end{bmatrix} = \begin{bmatrix}2/9 + 2/15\\4/9 + 1/5\end{bmatrix} = \begin{bmatrix}16/45\\29/45\end{bmatrix}.$$

(b) The distribution of a market, $\mathbf{X}_0 = \begin{bmatrix}a\\b\end{bmatrix}$, is called **stable** if it has the same form from year to year. That is, if $\mathbf{AX}_0 = \mathbf{X}_0$. Since the market is controlled by two groups, we must have $a + b = 1$. We combine the requirements that $\mathbf{AX}_0 = \mathbf{X}_0$ and $a + b = 1$ as follows. The equation $\mathbf{AX}_0 = \mathbf{X}_0$ in matrix form is

$$\begin{bmatrix}1/3 & 2/5\\2/3 & 3/5\end{bmatrix}\begin{bmatrix}a\\b\end{bmatrix} = \begin{bmatrix}a\\b\end{bmatrix},$$

which is equivalent to the equations

$$\begin{aligned}(1/3)a + (2/5)b &= a\\(2/3)a + (3/5)b &= b\ .\end{aligned}$$

Combining like variables in each of the preceding equations we obtain

$$\begin{aligned}(-2/3)a + (2/5)b &= 0\\(\ 2/3)a - (2/5)b &= 0\ .\end{aligned}$$

We see that the first equation is -1 times the second. Hence, we use either of these equations together with the requirement that a + b = 1 to solve for a and b as follows:

$$\begin{aligned}(-2/3)a + (2/5)b &= 0\\ a + b &= 1\end{aligned} \quad .$$

To eliminate a we take (2/3) times the second equation and add the equations:

$$\begin{array}{l}(-2/3)a + (2/5)b = 0\\ \underline{(2/3)(\quad a + \quad b = 1\ )}\end{array} \quad \text{----->} \quad \begin{array}{r}(-2/3)a + (2/5)b = 0\\ \underline{(2/3)a + (2/3)b = 2/3}\\ \left[\frac{2}{5} + \frac{2}{3}\right]b = \frac{2}{3}\\ \frac{16}{15}b = \frac{2}{3}\\ b = \frac{5}{8}\ .\end{array}$$

To solve for a, substitute b = 5/8 into the equation a + b = 1. We obtain a = 3/8. Thus the stable distribution of the market is

$$\begin{bmatrix}3/8\\ 5/8\end{bmatrix} .$$

T.1. (b) The (i,j) entry of **A** + (**B** + **C**) is $a_{ij} + (b_{ij} + c_{ij})$, that of (**A** + **B**) + **C** is $(a_{ij} + b_{ij}) + c_{ij}$. The two entries are equal because of the associative law for addition of real numbers.

(d) For each (i,j) let $d_{ij} = -a_{ij}$, **D** = $[d_{ij}]$. Then **A** + **D** = **D** + **A** = **0**.

T.3. (b) The (i,j) entry of **A**(**B** + **C**) is $\sum_{k=1}^{p} a_{ik}(b_{kj}+c_{kj})$. Using properties of real numbers, we rearrange this expression as,

$$\sum_{k=1}^{p}(a_{ik}b_{kj}+a_{ik}c_{kj}) = \sum_{k=1}^{p}a_{ik}b_{kj} + \sum_{k=1}^{p}a_{ik}c_{kj}.$$

The last expression is just the (i,j) entry of **AB** plus the (i,j) entry of **BC**. Thus, it follows that the last expression is the (i,j) entry of **AB** + **BC**.

(c) The (i,j) entry of (**A** + **B**)**C** is $\sum_{k=1}^{p}(a_{ik}+b_{ik})c_{kj}$. Using properties of real numbers, we rearrange this expression as

$$\sum_{k=1}^{p}(a_{ik}c_{kj}+b_{ik}c_{kj}) = \sum_{k=1}^{p}a_{ik}c_{kj} + \sum_{k=1}^{p}b_{ik}c_{kj}.$$

The last expression is the (i,j) entry of **AC** plus the (i,j) entry of **BC**. Thus, it follows that the last expression is the (i,j) entry of **AB** + **BC**.

T.5. $$\mathbf{A}^p\mathbf{A}^q = \underbrace{\underbrace{(\mathbf{A}\cdot\mathbf{A}\cdots\mathbf{A})}_{p \text{ factors}} \cdot \underbrace{(\mathbf{A}\cdot\mathbf{A}\cdots\mathbf{A})}_{q \text{ factors}}}_{p+q \text{ factors}} = \mathbf{A}^{p+q},$$

$$(\mathbf{A}^p)^q = \underbrace{\mathbf{A}^p\cdot\mathbf{A}^p\cdots\mathbf{A}^p}_{q \text{ factors}} = \mathbf{A}^{\overbrace{p+p+\cdots+p}^{}}_{\ } = \mathbf{A}^{pq}.$$

(the exponent $p+p+\cdots+p$ has q summands)

T.7. For **A** and **B** $n \times n$ diagonal, to show **AB** = **BA** we show corresponding entries are equal.

$$(i,j) \text{ entry of } \mathbf{AB} = \text{row}_i(\mathbf{A}) * \text{col}_j(\mathbf{B})$$

$$= \begin{bmatrix} 0 & \cdot & \cdot & 0 & a_{ii} & 0 & \cdot & \cdot & 0 \end{bmatrix} \begin{bmatrix} 0 \\ \cdot \\ \cdot \\ 0 \\ b_{jj} \\ 0 \\ \cdot \\ \cdot \\ 0 \end{bmatrix} = \begin{cases} 0 & i \neq j \\ a_{ii}b_{ii} & i=j \end{cases}$$

Similarly, the (i,j) entry of **BA** = $\begin{cases} 0 & i \neq j \\ b_{ii}a_{ii} & i=j \end{cases}$. Since multiplication of real numbers is commutative, $a_{ii}b_{ii} = b_{ii}a_{ii}$. Thus corresponding entries of **AB** and **BA** are equal and **AB** = **BA**.

T.9. $$\mathbf{CA} = \begin{bmatrix} c_1 & c_2 & \dots & c_m \end{bmatrix} \begin{bmatrix} a_{11} & a_{12} & \cdots & a_{1n} \\ a_{21} & a_{22} & \cdots & a_{2n} \\ \cdot & \cdot & & \cdot \\ \cdot & \cdot & & \cdot \\ a_{m1} & a_{m2} & \cdots & a_{mn} \end{bmatrix}$$

$$= \begin{bmatrix} \sum_{k=1}^{m} c_k a_{k1} & \sum_{k=1}^{m} c_k a_{k2} & \cdots & \sum_{k=1}^{m} c_k a_{kn} \end{bmatrix}$$

$$= [c_1a_{11}+c_2a_{21}+\cdots+c_ma_{m1} \quad c_1a_{12}+c_2a_{22}+\cdots+c_ma_{m2} \quad \cdots$$

$$c_1a_{1n}+c_2a_{2n}+\cdots+c_ma_{mn}]$$

(Write this expression as a sum of rows with the first row containing the first summand from each entry, the second row the second summand, and so on.)

$$= [c_1a_{11} \quad c_1a_{12} \quad \cdots \quad c_1a_{1n}] + [c_2a_{21} \quad c_2a_{22} \quad \cdots \quad c_2a_{2n}]$$

$$+ \cdots + [c_ma_{m1} \quad c_ma_{m2} \quad \cdots \quad c_ma_{mn}]$$

(Factor the common term from each row.)

$$= c_1\mathbf{A}_1 + c_2\mathbf{A}_2 + \cdots + c_m\mathbf{A}_m = \sum_{j=1}^{m} c_j\mathbf{A}_j$$

T.11. For $p = 0$, $(c\mathbf{A})^0 = \mathbf{I}_n = 1\cdot\mathbf{I}_n = c^0\cdot\mathbf{A}^0$.

For $p = 1$, $c\mathbf{A} = c\mathbf{A}$.

Assume the result is true for $p = k$: $(c\mathbf{A})^k = c^k\mathbf{A}^k$. Then for $p = k+1$ we have,

$$(c\mathbf{A})^{k+1} = (c\mathbf{A})^k(c\mathbf{A}) = c^k\mathbf{A}^k\cdot c\mathbf{A} = c^k(\mathbf{A}^k c)\mathbf{A}$$

$$= c^k(c\mathbf{A}^k)\mathbf{A} = (c^k c)(\mathbf{A}^k\mathbf{A}) = c^{k+1}\mathbf{A}^{k+1} .$$

T.13. $(-1)a_{ij} = -a_{ij}$ (See Exercise T.1.)

T.15. $(\mathbf{A} - \mathbf{B})^T = (\mathbf{A} + (-1)\mathbf{B})^T$ (by Exercise T.13)

$= \mathbf{A}^T + ((-1)\mathbf{B})^T$ (by Thm 1.4(b))

$= \mathbf{A}^T + (-1)\mathbf{B}^T$ (by Thm 1.4(d))

$= \mathbf{A}^T - \mathbf{B}^T$ (by Exercise T.13)

T.17. If $\mathbf{A}$ is symmetric, then $\mathbf{A}^T = \mathbf{A}$. Thus $a_{ji} = a_{ij}$ for all i and j. Conversely, if $a_{ji} = a_{ij}$ for all i and j, then $\mathbf{A}^T = \mathbf{A}$ and $\mathbf{A}$ is symmetric.

Exercises 1.3

T.19. Given that $\mathbf{AA}^T = \mathbf{0}$, we have that each entry of $\mathbf{AA}^T$ is zero. In particular then, each diagonal entry of $\mathbf{AA}^T$ is zero. Hence

$$0 = \text{row}_i(\mathbf{A}) * \text{col}_i(\mathbf{A}^T) = \begin{bmatrix} a_{i1} & a_{i2} & \cdots & a_{in} \end{bmatrix} \begin{bmatrix} a_{i1} \\ a_{i2} \\ \cdot \\ \cdot \\ a_{in} \end{bmatrix} = \sum_{j=1}^{n} (a_{ij})^2$$

(Recall that $\text{col}_i(\mathbf{A}^T)$ is $\text{row}_i(\mathbf{A})$ written in column form.)

A sum of squares is zero only if each member of the sum is zero, hence $a_{i1} = a_{i2} = \cdots = a_{in} = 0$, which means that $\text{row}_i(\mathbf{A})$ consists of all zeros. The previous argument holds for each diagonal entry, hence each row of $\mathbf{A}$ contains all zeros. Thus it follows that $\mathbf{A} = \mathbf{0}$.

T.21. (a) $(\mathbf{A} + \mathbf{B})^T = \mathbf{A}^T + \mathbf{B}^T$ (by Thm 1.4(b))

$= \mathbf{A} + \mathbf{B}$ (since $\mathbf{A}$ and $\mathbf{B}$ are symmetric)

Thus $(\mathbf{A} + \mathbf{B})$ equals it transpose, hence is symmetric.

(b) Suppose that $\mathbf{AB}$ is symmetric. Then, $(\mathbf{AB})^T = \mathbf{AB}$. Thus

$\mathbf{AB} = (\mathbf{AB})^T = \mathbf{B}^T\mathbf{A}^T$ (by Thm. 1.4(c))

$= \mathbf{BA}$ (since $\mathbf{A}$ and $\mathbf{B}$ are symmetric)

which implies $\mathbf{A}$ and $\mathbf{B}$ commute.

Conversely, if $\mathbf{A}$ and $\mathbf{B}$ commute, then

$\mathbf{AB} = \mathbf{BA}$

$= \mathbf{B}^T\mathbf{A}^T$ (since $\mathbf{A}$ and $\mathbf{B}$ are symmetric)

$= (\mathbf{AB})^T$ (by Thm. 1.4(c))

Thus $\mathbf{AB}$ is symmetric.

T.25. (a) We show that $\mathbf{A} + \mathbf{A}^T$ equals its transpose.

$(\mathbf{A} + \mathbf{A}^T)^T = \mathbf{A}^T + (\mathbf{A}^T)^T$ (by Thm. 1.4(b))

$= \mathbf{A}^T + \mathbf{A}$ (by Thm. 1.4(a))

$= \mathbf{A} + \mathbf{A}^T$ (since matrix addition is commutative)

(b) We show $(\mathbf{A} - \mathbf{A}^T)$ equals $-(\mathbf{A} - \mathbf{A})^T$.

$$\begin{aligned}(\mathbf{A} - \mathbf{A}^T)^T &= \mathbf{A}^T - (\mathbf{A}^T)^T && \text{(by Thm. 1.4(b))}\\ &= \mathbf{A}^T - \mathbf{A} && \text{(by Thm. 1.4(a))}\\ &= -(\mathbf{A} - \mathbf{A}^T) && \text{(by Thm. 1.3(c))}\end{aligned}$$

T.27. If the diagonal entries of $\mathbf{A}$ are r, then since $r = r\cdot 1$, $\mathbf{A} = r\mathbf{I}_n$.

T.29. Suppose $r \neq 0$. Then the (i,j) entry of $r\mathbf{A}$ is ra_{ij}. Since $r \neq 0$, $a_{ij} = 0$ for all i and j. Thus $\mathbf{A} = \mathbf{0}$.

T.31. Suppose $\mathbf{A} = \begin{bmatrix} a & b \\ c & d \end{bmatrix}$ satisfies $\mathbf{AB} = \mathbf{BA}$ for any 2×2 matrix $\mathbf{B}$.

Choosing $\mathbf{B} = \begin{bmatrix} 1 & 0 \\ 0 & 0 \end{bmatrix}$ we get

$$\begin{bmatrix} a & b \\ c & d \end{bmatrix}\begin{bmatrix} 1 & 0 \\ 0 & 0 \end{bmatrix} = \begin{bmatrix} 1 & 0 \\ 0 & 0 \end{bmatrix}\begin{bmatrix} a & b \\ c & d \end{bmatrix}$$

$$\begin{bmatrix} a & 0 \\ c & 0 \end{bmatrix} = \begin{bmatrix} a & b \\ 0 & 0 \end{bmatrix}$$

which implies $b = c = 0$. Thus $\mathbf{A}$ is diagonal: $\mathbf{A} = \begin{bmatrix} a & 0 \\ 0 & d \end{bmatrix}$.

Next choosing $\mathbf{B} = \begin{bmatrix} 0 & 1 \\ 0 & 0 \end{bmatrix}$ we get $\mathbf{AB} = \begin{bmatrix} 0 & a \\ 0 & 0 \end{bmatrix} = \begin{bmatrix} 0 & d \\ 0 & 0 \end{bmatrix} = \mathbf{BA}$,

which implies $a = d$. Thus $\mathbf{A} = \begin{bmatrix} a & 0 \\ 0 & a \end{bmatrix}$ which is a scalar matrix.

ML.1. (a)

```
A^2

ans =

     0     1     0
     0     0     1
     1     0     0
```

```
A^3

ans =

     1     0     0
     0     1     0
     0     0     1
```

Thus k=3.

Exercises 1.3

(b) A^2

```
ans =

    -1     0     0     0
     0    -1     0     0
     0     0     1     0
     0     0     0     1
```

A^3

```
ans =

     0    -1     0     0
     1     0     0     0
     0     0     0     1
     0     0     1     0
```

A^4

```
ans =

     1     0     0     0
     0     1     0     0
     0     0     1     0
     0     0     0     1
```

A^5

```
ans =

     0     1     0     0
    -1     0     0     0
     0     0     0     1
     0     0     1     0
```

Thus k = 5.

ML.2. (a) A = tril(ones(5),-1)

```
A =

     0     0     0     0     0
     1     0     0     0     0
     1     1     0     0     0
     1     1     1     0     0
     1     1     1     1     0
```

A^2

```
ans =

     0     0     0     0     0
     0     0     0     0     0
     1     0     0     0     0
     2     1     0     0     0
     3     2     1     0     0
```

A^3

```
ans =

     0     0     0     0     0
     0     0     0     0     0
     0     0     0     0     0
     1     0     0     0     0
     3     1     0     0     0
```

```
A^4

ans =

     0     0     0     0     0
     0     0     0     0     0
     0     0     0     0     0
     0     0     0     0     0
     1     0     0     0     0

A^5

ans =

     0     0     0     0     0
     0     0     0     0     0
     0     0     0     0     0
     0     0     0     0     0
     0     0     0     0     0
```

Thus k = 5.

(b) This exercise uses the random number generator **rand**. The matrix **A** and the value of k may vary.

```
A = triu(fix(10*rand(7)),2)

A =

     0     0     0     0     0     2     8
     0     0     0     6     7     9     2
     0     0     0     0     3     7     4
     0     0     0     0     0     7     7
     0     0     0     0     0     0     4
     0     0     0     0     0     0     0
     0     0     0     0     0     0     0
```

Here A^3 is all zeros, so k = 3.

ML.3. (a) Define the vector of coefficients

```
v  =[1 -1 1 0 2];
```

then we have

```
polyvalm(v,A)

ans =

     0    -2     4
     4     0    -2
    -2     4     0
```

(b) Define the vector of coefficients

```
v =[1 -3 3 0];
```

then we have

```
polyvalm(v,A)

ans =

     0     0     0
     0     0     0
     0     0     0
```

ML.4. (a) `(A^2-7*A)*(A+3*eye(A))`

```
ans =

   -2.8770   -7.1070  -14.0160
   -4.9360   -5.0480  -14.0160
   -6.9090   -7.1070   -9.9840
```

(b) `(A - eye(A))^2+(A^3 + A)`

```
ans =

    1.3730    0.2430    0.3840
    0.2640    1.3520    0.3840
    0.1410    0.2430    1.6160
```

(c) Computing the powers of A as **A^2, A^3, ...** soon gives the impression that the sequence is converging to

```
    0.2273    0.2727    0.5000
    0.2273    0.2727    0.5000
    0.2273    0.2727    0.5000
```

Note that in terms of rational numbers we have that

```
rat(A^50,'s')

ans =

   5/22         3/11         1/2
   5/22         3/11         1/2
   5/22         3/11         1/2
```

ML.5. The sequence seems to be converging to

```
    1.0000    0.7500
         0         0
```

ML.6. The sequence is converging to the zero matrix.

ML.7. (a) A'*A

```
ans =

     2    -3    -1
    -3     9     2
    -1     2     6
```

A*A'

```
ans =

     6    -1    -3
    -1     6     4
    -3     4     5
```

$\mathbf{A^TA}$ and $\mathbf{AA^T}$ are not equal.

(b) B =A+A'

```
B =

     2    -3     1
    -3     2     4
     1     4     2
```

C = A-A'

```
C =

     0    -1     1
     1     0     0
    -1     0     0
```

Just observe that $\mathbf{B} = \mathbf{B^T}$ and that $\mathbf{C^T} = \mathbf{-C}$.

(c) B + C

```
ans =

     2    -4     2
    -2     2     4
     0     4     2
```

We see that $\mathbf{B+C} = \mathbf{2A}$.

Exercises 1.4

1. Matrices **A**, **E**, and **G** are in reduced row echelon form.

   Matrix **B** is not in reduced row echelon form since row 3 is nonzero and does not have a leading one.

   Matrix **C** is not in reduced row echelon form since column 2 contains a leading one, but the reminder of the entries in column 2 are not all zero.

   Matrix **D** is not in reduced row echelon form since row 4 is a zero row, but is not at the bottom of the matrix.

   Matrix **F** has its first row all zeros, but it does not appear at the bottom of the matrix.

   Matrix **H** is not in reduced row echelon form since the leading entry in row 3 is not a one.

3.
$$\mathbf{A} = \begin{bmatrix} 2 & 0 & 4 & 2 \\ 3 & -2 & 5 & 6 \\ -1 & 3 & 1 & 1 \end{bmatrix}$$

   (a) Interchanging the second and third rows of **A**, we obtain
$$\begin{bmatrix} 2 & 0 & 4 & 2 \\ -1 & 3 & 1 & 1 \\ 3 & -2 & 5 & 6 \end{bmatrix}.$$

   (b) Multiplying the second row of **A** by -4, we obtain
$$\begin{bmatrix} 2 & 0 & 4 & 2 \\ -12 & 8 & -20 & -24 \\ -1 & 3 & 1 & 1 \end{bmatrix}.$$

   (c) Adding 2 times the third row of **A** to the first row of **A**, we obtain
$$\begin{bmatrix} 0 & 6 & 6 & 4 \\ 3 & -2 & 5 & 6 \\ -1 & 3 & 1 & 1 \end{bmatrix}.$$

5. To find matrices that are row equivalent to $\begin{bmatrix} 4 & 3 & 7 & 5 \\ -1 & 2 & -1 & 3 \\ 2 & 0 & 1 & 4 \end{bmatrix}$ we apply elementary row operations. One such set of operations is:

multiply row 2 by 2 then interchange rows 2 and 3 to obtain

$$\begin{bmatrix} 4 & 3 & 7 & 5 \\ 2 & 0 & 1 & 4 \\ -2 & 4 & -2 & 6 \end{bmatrix}$$; add row 2 to row 1 then multiply row 2 by 4

to obtain $$\begin{bmatrix} 3 & 5 & 6 & 8 \\ -4 & 8 & -4 & 12 \\ 2 & 0 & 1 & 4 \end{bmatrix}$$; add 2 times row 2 to row 3 to obtain

$$\begin{bmatrix} 4 & 3 & 7 & 5 \\ -1 & 2 & -1 & 3 \\ 0 & 4 & -1 & 10 \end{bmatrix}.$$

7. $$\mathbf{A} = \begin{bmatrix} 1 & -2 & 0 & 2 \\ 2 & -3 & -1 & 5 \\ 1 & 3 & 2 & 5 \\ 1 & 1 & 0 & 2 \end{bmatrix}$$

The pivotal column is column 1. The pivot is the (1,1) entry. Add multiples of the first row of **A** to all the the other rows to make all entries in the pivotal column, except the pivot position, equal to zero. We use row operations,

-2 times row 1 added to row 2,
-1 times row 1 added to row 3, and
-1 times row 1 added to row 4 to obtain

$$\mathbf{A}_1 = \begin{bmatrix} 1 & -2 & 0 & 2 \\ 0 & 1 & -1 & 1 \\ 0 & 5 & 2 & 3 \\ 0 & 3 & 0 & 0 \end{bmatrix}.$$

Identify **B** as the 3×4 submatrix of $\mathbf{A}_1$ obtained by deleting row 1 of $\mathbf{A}_1$.

$$\begin{matrix} 1 & -2 & 0 & 2 \end{matrix}$$

$$\mathbf{B} = \begin{bmatrix} 0 & 1 & -1 & 1 \\ 0 & 5 & 2 & 3 \\ 0 & 3 & 0 & 0 \end{bmatrix}$$

We repeat the elimination procedure on matrix **B**. The pivotal column is column 2. The pivot is the (1,2) entry of **B**. Since it is 1, we add multiples of row 1 of **B** to make all entries in the pivotal column, except the pivot position, equal to zero. We use row operations,

-5 times row 1 added to row 2,
-3 times row 1 added to row 3, and
2 times row 1 added to row 1 of $\mathbf{A}_1$ to obtain

Exercises 1.4

$$
\begin{array}{cccc}
1 & 0 & -2 & 4
\end{array}
$$

$$
\mathbf{B}_1 = \begin{bmatrix} 0 & 1 & -1 & 1 \\ 0 & 0 & 7 & -2 \\ 0 & 0 & 3 & -3 \end{bmatrix} .
$$

Identify $\mathbf{C}$ as the 2×4 submatrix of $\mathbf{B}_1$ obtained by deleting row 1 of $\mathbf{B}_1$.

$$
\begin{array}{cccc}
1 & 0 & -2 & 4 \\
0 & 1 & -1 & 1
\end{array}
$$

$$
\mathbf{C} = \begin{bmatrix} 0 & 0 & 7 & -2 \\ 0 & 0 & 3 & -3 \end{bmatrix}
$$

We repeat the elimination process on matrix $\mathbf{C}$. The pivotal column is column 3. The pivot is the (1,3) entry. To obtain a 1 in the first row $\mathbf{C}$ in the pivot column we multiply row 1 by 1/7. We obtain

$$
\begin{array}{ccccl}
1 & 0 & -2 & 4 & \text{<-- first pivot row} \\
0 & 1 & -1 & 1 & \text{<-- second pivot row}
\end{array}
$$

$$
\mathbf{C}_1 = \begin{bmatrix} 0 & 0 & 1 & \frac{-2}{7} \\ 0 & 0 & 3 & -3 \end{bmatrix} .
$$

Add multiples of row 1 of $\mathbf{C}_1$ to the two rows outside of $\mathbf{C}_1$ (previous pivot rows) to introduce zeros above the pivot and similarly below the pivot row. We use row operations,

2 times row 1 of $\mathbf{C}_1$ added to the first pivot row,
1 times row 1 of **C**l added to the second pivot row, and
-3 times row 1 of $\mathbf{C}_1$ added to row 2 of $\mathbf{C}_1$ to obtain

$$
\begin{array}{cccc}
1 & 0 & 0 & \frac{24}{7} \\
0 & 1 & 0 & \frac{5}{7}
\end{array}
$$

$$
\mathbf{C}_2 = \begin{bmatrix} 0 & 0 & 1 & \frac{-2}{7} \\ 0 & 0 & 0 & \frac{-15}{7} \end{bmatrix} .
$$

Identify $\mathbf{D}$ as the 1×4 submatrix of $\mathbf{C}_2$ obtained by deleting row 1 of $\mathbf{C}_2$.

$$D = \begin{bmatrix} 1 & 0 & 0 & \frac{24}{7} \\ 0 & 1 & 0 & \frac{5}{7} \\ 0 & 0 & 1 & \frac{-2}{7} \\ 0 & 0 & 0 & \frac{-15}{7} \end{bmatrix}.$$

The pivotal column is column 4. The pivot is the (1,4) entry. To obtain a 1 in the pivot position, we multiply the first row of **D** by -7/15 to obtain

$$\begin{matrix} 1 & 0 & 0 & \frac{24}{7} & \text{<--- first pivot row} \\ 0 & 1 & 0 & \frac{5}{7} & \text{<--- second pivot row} \\ 0 & 0 & 1 & \frac{-2}{7} & \text{<--- third pivot row} \end{matrix}$$

$$\mathbf{D}_1 = [\, 0 \quad 0 \quad 0 \quad 1 \,].$$

Add multiples of the first row of $\mathbf{D}_1$ to the rows outside of $\mathbf{D}_1$ (previous pivot rows) to introduce zeros above the pivot element. We use row operations,

$\frac{-24}{7}$ times row 1 of $\mathbf{D}_1$ added to the first pivot row,

$\frac{-5}{7}$ times row 1 of $\mathbf{D}_1$ added to the second pivot row, and

$\frac{2}{7}$ times row 1 of $\mathbf{D}_1$ added to the third pivot row to obtain

$$\begin{matrix} 1 & 0 & 0 & 0 \\ 0 & 1 & 0 & 0 \\ 0 & 0 & 1 & 0 \end{matrix}$$

$$\mathbf{D}_2 = [\, 0 \quad 0 \quad 0 \quad 1 \,].$$

The matrix $\begin{bmatrix} 1 & 0 & 0 & 0 \\ 0 & 1 & 0 & 0 \\ 0 & 0 & 1 & 0 \\ 0 & 0 & 0 & 1 \end{bmatrix}$ is in reduced row echelon form and

is row equivalent to **A**.

9. (a)

$$\begin{aligned} x + y + 2z + 3w &= 13 \\ x - 2y + z + w &= 8 \\ 3x + y + z - w &= 1 \end{aligned}$$

We form the augmented matrix and use row operations to form equivalent systems.

Exercises 1.4

$$\left[\begin{array}{cccc|c} 1 & 1 & 2 & 3 & 13 \\ 1 & -2 & 1 & 1 & 8 \\ 3 & 1 & 1 & -1 & 1 \end{array}\right]$$

Applying row operations
-1 times row 1 added to row 2 and
-3 times row 1 added to row 3 we obtain,

$$\left[\begin{array}{cccc|c} 1 & 1 & 2 & 3 & 13 \\ 0 & -3 & -1 & -2 & -5 \\ 0 & -2 & -5 & -10 & -38 \end{array}\right].$$

Next multiply row 2 by -1/3 to obtain

$$\left[\begin{array}{cccc|c} 1 & 1 & 2 & 3 & 13 \\ 0 & 1 & \frac{1}{3} & \frac{2}{3} & \frac{5}{3} \\ 0 & -2 & -5 & -10 & -38 \end{array}\right].$$

Applying row operations
-1 times row 2 added to row 1 and
2 times row 2 added to row 3 we obtain,

$$\left[\begin{array}{cccc|c} 1 & 0 & \frac{5}{3} & \frac{7}{3} & \frac{34}{3} \\ 0 & 1 & \frac{1}{3} & \frac{2}{3} & \frac{5}{3} \\ 0 & 0 & \frac{-13}{3} & \frac{-26}{3} & \frac{-104}{3} \end{array}\right].$$

Multiply row 3 by -3/13 to put a 1 into the pivot position.
The result is the matrix

$$\left[\begin{array}{cccc|c} 1 & 0 & \frac{5}{3} & \frac{7}{3} & \frac{34}{3} \\ 0 & 1 & \frac{1}{3} & \frac{2}{3} & \frac{5}{3} \\ 0 & 0 & 1 & 2 & 8 \end{array}\right].$$

Applying row operations
$\frac{-5}{3}$ times row 3 added to row 1 and
$\frac{-1}{3}$ times row 3 added to row 2 we obtain,

$$\left[\begin{array}{cccc|c} 1 & 0 & 0 & -1 & -2 \\ 0 & 1 & 0 & 0 & -1 \\ 0 & 0 & 1 & 2 & 8 \end{array}\right].$$

This matrix is in reduced row echelon form and represents the equivalent linear system

$$\begin{aligned} x \quad\quad - w &= -2 \\ y \quad\quad &= -1 \\ z + 2w &= 8 . \end{aligned}$$

We solve each equation for the unknown that corresponds to a leading 1. Hence we have

$$\begin{aligned} x &= -2 + w \\ y &= -1 \\ z &= 8 - 2w. \end{aligned}$$

The unknown w can be chosen arbitrarily. Let $w = r$, where r is any real number. Then the solution of the linear system is given by

$$x = -2 + r, \quad y = -1, \quad z = 8 - 2r, \quad w = r.$$

Since r can be any real number, there are infinitely many solutions.

(b)
$$\begin{aligned} x + y + z &= 1 \\ x + y - 2z &= 3 \\ 2x + y + z &= 2 \end{aligned}$$

We form the augmented matrix and use row operations to form equivalent systems.

$$\left[\begin{array}{ccc|c} 1 & 1 & 1 & 1 \\ 1 & 1 & -2 & 3 \\ 2 & 1 & 1 & 2 \end{array}\right]$$

Applying row operations
-1 times row 1 added to row 2 and
-2 times row 1 added to row 3 we obtain,

$$\left[\begin{array}{ccc|c} 1 & 1 & 1 & 1 \\ 0 & 0 & -3 & 2 \\ 0 & -1 & -1 & 0 \end{array}\right].$$

The (2,2) entry is zero, hence we interchange rows 2 and 3 to obtain a nonzero pivot element. Thus the pivot element will be a -1. We next multiply row 2 by -1. The resulting matrix is

$$\left[\begin{array}{ccc|c} 1 & 1 & 1 & 1 \\ 0 & 1 & 1 & 0 \\ 0 & 0 & -3 & 2 \end{array}\right].$$

Applying row operation
-1 times row 2 added to row 1 we obtain,

$$\left[\begin{array}{ccc|c} 1 & 0 & 0 & 1 \\ 0 & 1 & 1 & 0 \\ 0 & 0 & -3 & 2 \end{array}\right].$$

Next we multiply row 3 by -1/3 to obtain

$$\left[\begin{array}{ccc|c} 1 & 0 & 0 & 1 \\ 0 & 1 & 1 & 0 \\ 0 & 0 & 1 & \frac{-2}{3} \end{array}\right].$$

We obtain the reduced row echelon form by applying row operation
-1 times row 3 added to row 2.
The final matrix is

$$\left[\begin{array}{ccc|c} 1 & 0 & 0 & 1 \\ 0 & 1 & 0 & \frac{2}{3} \\ 0 & 0 & 1 & \frac{-2}{3} \end{array}\right].$$

The unique solution of this system is

$x = 1$, $y = 2/3$, and $z = -2/3$.

(c)

$$\begin{aligned} 2x + y + z - 2w &= 1 \\ 3x - 2y + z - 6w &= -2 \\ x + y - z - w &= -1 \\ 6x + z - 9w &= -2 \\ 5x - y + 2z - 8w &= 3 \end{aligned}$$

We form the augmented matrix and use row operations to form equivalent systems.

$$\left[\begin{array}{rrrr|r} 2 & 1 & 1 & -2 & 1 \\ 3 & -2 & 1 & -6 & -2 \\ 1 & 1 & -1 & -1 & -1 \\ 6 & 0 & 1 & -9 & -2 \\ 5 & -1 & 2 & -8 & 3 \end{array}\right]$$

In an effort to keep the arithmetic simple, we first interchange rows 1 and 3. This operation gives us a 1 in the (1,1) entry to use as the first pivot. The resulting matrix is

$$\left[\begin{array}{rrrr|r} 1 & 1 & -1 & -1 & -1 \\ 3 & -2 & 1 & -6 & -2 \\ 2 & 1 & 1 & -2 & 1 \\ 6 & 0 & 1 & -9 & -2 \\ 5 & -1 & 2 & -8 & 3 \end{array}\right].$$

Applying row operations
-3 times row 1 added to row 2,
-2 times row 1 added to row 3,
-6 times row 1 added to row 4, and
-5 times row 1 added to row 5 we obtain,

$$\left[\begin{array}{rrrr|r} 1 & 1 & -1 & -1 & -1 \\ 0 & -5 & 4 & -3 & 1 \\ 0 & -1 & 3 & 0 & 3 \\ 0 & -6 & 7 & -3 & 4 \\ 0 & -6 & 7 & -3 & 8 \end{array}\right].$$

To simplify the arithmetic and avoid fractions, we interchange rows 2 and 3, and then multiply row 2 by -1. This puts a 1 into the (2,2) entry to use as the next pivot. The resulting matrix is,

$$\left[\begin{array}{rrrr|r} 1 & 1 & -1 & -1 & -1 \\ 0 & 1 & -3 & 0 & -3 \\ 0 & -5 & 4 & -3 & 1 \\ 0 & -6 & 7 & -3 & 4 \\ 0 & -6 & 7 & -3 & 8 \end{array}\right].$$

Applying row operations
-1 times row 2 added to row 1,
5 times row 2 added to row 3,
6 times row 2 added to row 4, and
6 times row 2 added to row 5 we obtain,

Exercises 1.4

$$\left[\begin{array}{cccc|c} 1 & 0 & 2 & -1 & 2 \\ 0 & 1 & -3 & 0 & -3 \\ 0 & 0 & -11 & -3 & -14 \\ 0 & 0 & -11 & -3 & -14 \\ 0 & 0 & -11 & -3 & -10 \end{array}\right].$$

We observe that the last three equations have identical coefficients, but the entries in the augmented column are not all the same. Thus we use row operations
-1 times row 3 added to row 4 and
-1 times row 3 added to row 5 to obtain

$$\left[\begin{array}{cccc|c} 1 & 0 & 2 & -1 & 2 \\ 0 & 1 & -3 & 0 & -3 \\ 0 & 0 & -11 & -3 & -14 \\ 0 & 0 & 0 & 0 & 0 \\ 0 & 0 & 0 & 0 & 4 \end{array}\right].$$

The fifth row is equivalent to the equation

$$0x + 0y + 0z + 0w = 4$$

which has no solution. Thus the linear system is inconsistent.

<<**Strategy:** In Exercises 11 and 13 we form the augmented matrix and then use row operations to reduce the system to row echelon form. We show how to choose values of a to obtain (a) no solution, (b) a unique solution, and (c) infinitely many solutions.>>

11.
$$\left[\begin{array}{ccc|c} 1 & 1 & -1 & 2 \\ 1 & 2 & 1 & 3 \\ 1 & 1 & a^2-5 & a \end{array}\right]$$

Applying row operations
-1 times row 1 added to row 2 and,
-1 times row 1 added to row 3 we obtain,

$$\left[\begin{array}{ccc|c} 1 & 1 & -1 & 2 \\ 0 & 1 & 2 & 1 \\ 0 & 0 & a^2-4 & a-2 \end{array}\right].$$

(a) There will be no solution if a is chosen so that row 3 has the form

$$[0 \quad 0 \quad 0 \mid \ *]$$

where $* \neq 0$. We see that $a^2-4 = 0$ if and only if $a = 2$ or $a = -2$. To have $a-2 \neq 0$, a must be chosen as -2.

(b) There will be a unique solution if a is chosen so that $a^2-4 \neq 0$. In this case the (3,3) entry can be used as pivot. Thus $a \neq 2$ and $a \neq -2$.

(c) There will be infinitely many solutions if the third row is a zero row. In that case variable z can be chosen arbitrarily. Thus a = 2.

13.
$$\left[\begin{array}{ccc|c} 1 & 1 & 1 & 2 \\ 1 & 2 & 1 & 3 \\ 1 & 1 & a^2-5 & a \end{array}\right]$$

Applying row operations
-1 times row 1 added to row 2 and
-1 times row 1 added to row 3 we obtain,

$$\left[\begin{array}{ccc|c} 1 & 1 & 1 & 2 \\ 0 & 1 & 0 & 1 \\ 0 & 0 & a^2-6 & a-2 \end{array}\right].$$

(a) There will be no solution if a is chosen so that row 3 has the form

$$[0 \quad 0 \quad 0 \mid \quad *]$$

where $* \neq 0$. We see that $a^2-6 = 0$ if and only if $a = \pm\sqrt{6}$. In either case $a-2 \neq 0$.

(b) There will be a unique solution if a is chosen so that $a^2-6 \neq 0$. In this case the (3,3) entry can be used as pivot. Thus $a \neq \pm\sqrt{6}$.

(c) There will be infinitely many solutions if the third row is a zero row. In that case variable z can be chosen arbitrarily. There is no way to choose a so that both a^2-6 and a-2 are zero.

<<**Strategy:** In Exercises 15 and 17 we put the augmented matrix in reduced row echelon form. Then determine its solution, if any exist.>>

15. (a) Applying row operation
-1 times row 1 added to row 2 we obtain,

$$\left[\begin{array}{ccc|c} 1 & 1 & 1 & 0 \\ 0 & 0 & -1 & 3 \\ 0 & 1 & 1 & 1 \end{array}\right].$$

Interchange rows 2 and 3 to get 1 into the (2,2) position to use as a pivot. The resulting matrix is

$$\left[\begin{array}{ccc|c} 1 & 1 & 1 & 0 \\ 0 & 1 & 1 & 1 \\ 0 & 0 & -1 & 3 \end{array}\right].$$

Applying row operations
-1 times row 2 added to row 1 and
-1 times row 3 we obtain,

$$\left[\begin{array}{ccc|c} 1 & 0 & 0 & -1 \\ 0 & 1 & 1 & 1 \\ 0 & 0 & 1 & -3 \end{array}\right].$$

Applying row operation
-1 times row 3 added to row 2 we obtain,

$$\left[\begin{array}{ccc|c} 1 & 0 & 0 & -1 \\ 0 & 1 & 0 & 4 \\ 0 & 0 & 1 & -3 \end{array}\right].$$

The linear system has a unique solution given by

$$x = -1, \quad y = 4, \quad z = -3.$$

(b) Applying row operations
-1 times row 1 added to row 2,
-1 times row 1 added to row 3,
-1 times row 1 added to row 4 and,
-1 times row 1 added to row 5 we obtain,

$$\left[\begin{array}{ccc|c} 1 & 2 & 3 & 0 \\ 0 & -1 & -2 & 0 \\ 0 & -1 & -1 & 0 \\ 0 & 1 & 0 & 0 \end{array}\right].$$

We multiply row 2 by -1, then use it as a pivot row with row operations
-2 times row 2 added to row 1,
1 times row 2 added to row 3, and
-1 times row 2 added to row 4 to obtain,

$$\left[\begin{array}{ccc|c} 1 & 0 & -1 & 0 \\ 0 & 1 & 2 & 0 \\ 0 & 0 & 1 & 0 \\ 0 & 0 & -2 & 0 \end{array}\right].$$

Applying row operations
1 times row 3 added to row 1,
-2 times row 3 added to row 2, and
2 times row 3 added to row 4 we obtain,

$$\left[\begin{array}{ccc|c} 1 & 0 & 0 & 0 \\ 0 & 1 & 0 & 0 \\ 0 & 0 & 1 & 0 \\ 0 & 0 & 0 & 0 \end{array}\right].$$

Thus this homogeneous linear system has only the trivial solution: x = 0, y = 0, z = 0.

17. (a) Applying row operations
-1 times row 1 added to row 2 and
-1 times row 1 added to row 3 we obtain,

$$\left[\begin{array}{cccc|c} 1 & 2 & 3 & 1 & 8 \\ 0 & 1 & -3 & 0 & -1 \\ 0 & -2 & -1 & 0 & -5 \end{array}\right].$$

Applying row operations
-2 times row 2 added to row 1 and
2 times row 2 added to row 3 we obtain,

$$\left[\begin{array}{cccc|c} 1 & 0 & 9 & 1 & 10 \\ 0 & 1 & -3 & 0 & -1 \\ 0 & 0 & -7 & 0 & -7 \end{array}\right].$$

Multiply row 3 by -1/7. The resulting matrix is

$$\left[\begin{array}{cccc|c} 1 & 0 & 9 & 1 & 10 \\ 0 & 1 & -3 & 0 & -1 \\ 0 & 0 & 1 & 0 & 1 \end{array}\right].$$

Next we use row operations
-9 times row 3 added to row 1 and
3 times row 3 added to row 2 to obtain,

$$\left[\begin{array}{cccc|c} 1 & 0 & 0 & 1 & 1 \\ 0 & 1 & 0 & 0 & 2 \\ 0 & 0 & 1 & 0 & 1 \end{array}\right].$$

This matrix represents the linear system

$$\begin{aligned} x \quad\quad + w &= 1 \\ y \quad\quad &= 2 \\ z \quad &= 1 . \end{aligned}$$

Thus we have x = 1 - w, y = 2, and z = 1. Let w = r, where r is any real number. Then the infinitely many solutions of this linear system have the form

x = 1 - r, y = 2, z = 1, w = r.

(b) Applying row operations
-2 times row 1 added to row 2,
-3 times row 1 added to row 3, and
-3 times row 1 added to row 4 we obtain,

$$\left[\begin{array}{ccc|c} 1 & -2 & 3 & 4 \\ 0 & 3 & -9 & -3 \\ 0 & 6 & -8 & -10 \\ 0 & 3 & -9 & -5 \end{array}\right].$$

Multiply row 2 by 1/3. The resulting matrix is

$$\left[\begin{array}{ccc|c} 1 & -2 & 3 & 4 \\ 0 & 1 & -3 & -1 \\ 0 & 6 & -8 & -10 \\ 0 & 3 & -9 & -5 \end{array}\right].$$

Applying row operations
2 times row 2 added to row 1,
-6 times row 2 added to row 3, and
-3 times row 2 added to row 4 we obtain,

$$\left[\begin{array}{ccc|c} 1 & 0 & -3 & 2 \\ 0 & 1 & -3 & -1 \\ 0 & 0 & 10 & -4 \\ 0 & 0 & 0 & -2 \end{array}\right].$$

We see that row 4 is in the form [0 0 0 | *] where * is not zero. This indicates that the corresponding equation has no solution. Thus this system has no solution.

19. The homogeneous system $(-4\mathbf{I}_3 - \mathbf{A})\mathbf{X} = \mathbf{0}$ has augmented matrix

$$\left[\begin{array}{ccc|c} -5 & 0 & -5 & 0 \\ -1 & -5 & -1 & 0 \\ 0 & -1 & 0 & 0 \end{array}\right].$$

Multiplying row 1 by -1/5 we obtain the matrix

$$\left[\begin{array}{ccc|c} 1 & 0 & 1 & 0 \\ -1 & -5 & -1 & 0 \\ 0 & -1 & 0 & 0 \end{array}\right].$$

Applying row operation
1 times row 1 added to row 2 we obtain,

$$\left[\begin{array}{ccc|c} 1 & 0 & 1 & 0 \\ 0 & -5 & 0 & 0 \\ 0 & -1 & 0 & 0 \end{array}\right].$$

Multiply row 2 by -1/5 and then add rows 2 and 3. The resulting matrix is

$$\left[\begin{array}{ccc|c} 1 & 0 & 1 & 0 \\ 0 & 1 & 0 & 0 \\ 0 & 0 & 0 & 0 \end{array}\right].$$

The corresponding linear system is

$$\begin{aligned} x \quad + z &= 0 \\ y \quad &= 0 . \end{aligned}$$

Thus $x = -z$ and $y = 0$. Let $z = r$, where r is any real number. Then the infinitely many nontrivial solutions of this linear system have the form

$$x = -r,\ y = 0,\ z = r \quad \text{where } r \neq 0.$$

21. Let: x_1 = the number of chairs made per week
x_2 = the number of coffee tables made per week
x_3 = the number of dining room tables made per week.
We convert all the times to minutes and develop an equation for use of the sanding bench, the staining bench, and the varnishing bench. The equations are respectively:

$$\begin{aligned} 10x_1 + 12x_2 + 15x_3 &= 960 \\ 6x_1 + 8x_2 + 12x_3 &= 660 \\ 12x_1 + 12x_2 + 18x_3 &= 1080 . \end{aligned}$$

We form the augmented matrix and use row operations to obtain the reduced row echelon form.

$$\left[\begin{array}{ccc|c} 10 & 12 & 15 & 960 \\ 6 & 8 & 12 & 660 \\ 12 & 12 & 18 & 1080 \end{array}\right]$$

Multiply the first row by 1/10 to put a 1 into the (1,1) position to use as a pivot. Writing the resulting matrix in decimal form we have

$$\left[\begin{array}{ccc|c} 1 & 1.2 & 1.5 & 96 \\ 6 & 8 & 12 & 660 \\ 12 & 12 & 18 & 1080 \end{array}\right].$$

Applying row operations
-6 times row 1 added to row 2 and
-12 times row 1 added to row 3 we obtain,

$$\left[\begin{array}{ccc|c} 1 & 1.2 & 1.5 & 96 \\ 0 & .8 & 3 & 84 \\ 0 & -2.4 & 0 & -72 \end{array}\right].$$

Multiply row 2 by 1/0.8 to put a 1 into the (2,2) position to use as a pivot. The resulting matrix is

Exercises 1.4

$$\left[\begin{array}{ccc|c} 1 & 1.2 & 1.5 & 96 \\ 0 & 1 & 3.75 & 105 \\ 0 & -2.4 & 0 & -72 \end{array}\right].$$

Applying row operations
-1.2 times row 2 added to row 1 and
2.4 times row 2 added to row 3 we obtain,

$$\left[\begin{array}{ccc|c} 1 & 0 & -3 & -30 \\ 0 & 1 & 3.75 & 105 \\ 0 & 0 & 9 & 180 \end{array}\right].$$

Multiply row 3 by 1/9 to put a 1 into the (3,3) position. The resulting matrix is

$$\left[\begin{array}{ccc|c} 1 & 0 & -3 & -30 \\ 0 & 1 & 3.75 & 105 \\ 0 & 0 & 1 & 20 \end{array}\right].$$

Applying row operations
3 times row 3 added to row 1 and
-3.75 times row 3 added to row 2 we obtain,

$$\left[\begin{array}{ccc|c} 1 & 0 & 0 & 30 \\ 0 & 1 & 0 & 30 \\ 0 & 0 & 1 & 20 \end{array}\right].$$

Thus we have $x_1 = 30$, $x_2 = 30$, and $x_3 = 20$.

T.1. Suppose that the leading entry of the ith row occurs in the jth column. Since leading entries of rows i+1, i+2,... are to the right of that of the ith row, and in any nonzero row, the leading entry is the first nonzero element, all entries in the jth column below the ith row must be zero.

T.3. The same set of elementary row operations that take **A** to **C** take the augmented matrix $[\mathbf{A}|\mathbf{0}]$ to $[\mathbf{C}|\mathbf{0}]$. Thus both systems have the same solution by Theorem 1.6.

T.5. If ad - bc = 0, the two rows of $\mathbf{A} = \begin{bmatrix} a & b \\ c & d \end{bmatrix}$ are multiples of one another: c[a b] = [ac bc] and a[c d] = [ac ad] and bc = ad. Any elementary row operation applied to **A** will produce a matrix with rows that are multiples of each other. In particular, elementary row operations cannot produce $\mathbf{I}_2$, and so $\mathbf{I}_2$ is not row equivalent to **A**. Thus we have shown that if ad - bc = 0, then **A** is not row equivalent to $\mathbf{I}_2$. Hence the following equivalent statement is true:

if A is row equivalent to I_2, then ad - bc $\neq$ 0.

We proceed to show that the converse of the preceding statement is also true. Assume that ad - bc $\neq$ 0. Then a and c are not both zero. Suppose a $\neq$ 0. The row operation

1/a times row 1 of **A** gives

$$\begin{bmatrix} 1 & \frac{b}{a} \\ c & d \end{bmatrix}.$$

Applying row operation

-c times row 1 added to row 2 we obtain,

$$\begin{bmatrix} 1 & \frac{b}{a} \\ 0 & d - \frac{bc}{a} \end{bmatrix}.$$

Multiplying row 2 by $\frac{a}{ad-bc}$ gives

$$\begin{bmatrix} 1 & \frac{b}{a} \\ 0 & 1 \end{bmatrix}.$$

Finally row operation $\frac{-b}{a}$ times row 2 added to row 1 gives $\mathbf{I}_2$. Hence, **A** is row equivalent to $\mathbf{I}_2$.

T.7. For any angle θ, cos θ and sin θ are not both zero. Assume that cos $\theta \neq 0$ and proceed as follows. The row operation

1/cos θ times row 1 gives

$$\begin{bmatrix} 1 & \frac{\sin\theta}{\cos\theta} \\ -\sin\theta & \cos\theta \end{bmatrix}.$$

Applying row operation

sin θ times row 1 added to row 2 we obtain

$$\begin{bmatrix} 1 & \frac{\sin\theta}{\cos\theta} \\ 0 & \cos\theta + \frac{\sin^2\theta}{\cos\theta} \end{bmatrix}.$$

Simplfying the (2,2)-entry we have

$$\cos\theta + \frac{\sin^2\theta}{\cos\theta} = \frac{\cos^2\theta + \sin^2\theta}{\cos\theta} = \frac{1}{\cos\theta}$$

hence our matrix is

$$\begin{bmatrix} 1 & \dfrac{\sin\theta}{\cos\theta} \\ 0 & \dfrac{1}{\cos\theta} \end{bmatrix}.$$

Applying row operations $\cos\theta$ times row 2 followed $-\sin\theta/\cos\theta$ times row 2 added to row 1 gives us $\mathbf{I}_2$. Hence the reduced row echelon form is the 2×2 identity matrix. (If $\cos\theta = 0$, then interchange rows and proceed in a similar manner.)

T.9. Let $\mathbf{A}$ be in reduced row echelon form and assume $\mathbf{A} \neq \mathbf{I}_n$. Thus there is at least one row of $\mathbf{A}$ without a leading 1. From the definition of reduced row echelon form, this row must be a zero row.

T.11. Since $\mathbf{X}_1$ and $\mathbf{X}_2$ are solutions of $\mathbf{AX} = \mathbf{0}$, we have that $\mathbf{AX}_1 = \mathbf{0}$ and $\mathbf{AX}_2 = \mathbf{0}$. We use the criteria that a solution must satisfy the system of equations in each of the following.

(a) $\mathbf{A}(\mathbf{X}_1 + \mathbf{X}_2) = \mathbf{AX}_1 + \mathbf{AX}_2 = \mathbf{0} + \mathbf{0} = \mathbf{0}$, thus $\mathbf{X}_1 + \mathbf{X}_2$ is a solution of $\mathbf{AX} = \mathbf{0}$.

(b) $\mathbf{A}(\mathbf{X}_1 - \mathbf{X}_2) = \mathbf{AX}_1 - \mathbf{AX}_2 = \mathbf{0} - \mathbf{0} = \mathbf{0}$, thus $\mathbf{X}_1 - \mathbf{X}_2$ is a solution of $\mathbf{AX} = \mathbf{0}$.

(c) $\mathbf{A}(r\mathbf{X}_1) = r(\mathbf{AX}_1) = r\mathbf{0} = \mathbf{0}$, thus $r\mathbf{X}_1$ is a solution of $\mathbf{AX} = \mathbf{0}$.

(d) $$\mathbf{A}(r\mathbf{X}_1 + s\mathbf{X}_1) = \mathbf{A}(r\mathbf{X}_1) + \mathbf{A}(s\mathbf{X}_1) = r(\mathbf{AX}_1) + s(\mathbf{AX}_1)$$
$$= r\mathbf{0} + s\mathbf{0} = \mathbf{0},$$
thus $r\mathbf{X}_1 + s\mathbf{X}_2$ is a solution of $\mathbf{AX} = \mathbf{0}$.

T.13. (a) $\mathbf{A}(\mathbf{X}_1 + \mathbf{Y}_1) = \mathbf{AX}_1 + \mathbf{AY}_1 = \mathbf{B} + \mathbf{0} = \mathbf{B}$.

(b) Let $\mathbf{X}_1$ be a particular solution to $\mathbf{AX} = \mathbf{B}$ and let $\mathbf{X}$ be any solution of this system. Let $\mathbf{Y}_1 = \mathbf{X} - \mathbf{X}_1$. Then $\mathbf{AY}_1 = \mathbf{A}(\mathbf{X} - \mathbf{X}_1) = \mathbf{AX} - \mathbf{AX}_1 = \mathbf{B} - \mathbf{B} = \mathbf{0}$. Thus $\mathbf{Y}_1$ is a solution of the homogeneous system $\mathbf{AX} = \mathbf{0}$. We have

$$\mathbf{X} = \mathbf{X} + \mathbf{X}_1 - \mathbf{X}_1 = \mathbf{X}_1 + (\mathbf{X} - \mathbf{X}_1) = \mathbf{X}_1 + \mathbf{Y}_1.$$

This implies that any solution of the system $\mathbf{AX} = \mathbf{B}$ can be written as a sum of a particular solution of this system and a solution of the associated homogeneous system $\mathbf{AX} = \mathbf{0}$.

ML.1. Enter A in to MATLAB and use the following MATLAB commands.

(a) A(1,:)=(1/4)*A(1,:)

```
A =

    1.0000    0.5000    0.5000
   -3.0000    1.0000    4.0000
    1.0000         0    3.0000
    5.0000   -1.0000    5.0000
```

(b) A(2,:)=3*A(1,:)+A(2,:)

```
A =

    1.0000    0.5000    0.5000
         0    2.5000    5.5000
    1.0000         0    3.0000
    5.0000   -1.0000    5.0000
```

(c) A(3,:)=-1*A(1,:)+A(3,:)

```
A =

    1.0000    0.5000    0.5000
         0    2.5000    5.5000
         0   -0.5000    2.5000
    5.0000   -1.0000    5.0000
```

(d) A(4,:)=-5*A(1,:)+A(4,:)

```
A =

    1.0000    0.5000    0.5000
         0    2.5000    5.5000
         0   -0.5000    2.5000
         0   -3.5000    2.5000
```

(e) temp=A(2,:)

```
temp =

         0    2.5000    5.5000
```

A(2,:)=A(4,:)

```
A =

    1.0000    0.5000    0.5000
         0   -3.5000    2.5000
         0   -0.5000    2.5000
         0   -3.5000    2.5000
```

```
A(4,:)=temp

A =

    1.0000    0.5000    0.5000
         0   -3.5000    2.5000
         0   -0.5000    2.5000
         0    2.5000    5.5000
```

ML.2. Enter the matrix A into MATLAB and use the following MATLAB commands. We use the rat command to display the matrix A in rational form at each stage.

```
A=[1/2 1/3 1/4 1/5;1/3 1/4 1/5 1/6;1 1/2 1/3 1/4]

A =

    0.5000    0.3333    0.2500    0.2000
    0.3333    0.2500    0.2000    0.1667
    1.0000    0.5000    0.3333    0.2500

rat(A,'s')

ans =

     1/2          1/3          1/4          1/5
     1/3          1/4          1/5          1/6
      1           1/2          1/3          1/4
```

(a) `A(1,:)=2*A(1,:)`

```
A =

    1.0000    0.6667    0.5000    0.4000
    0.3333    0.2500    0.2000    0.1667
    1.0000    0.5000    0.3333    0.2500

rat(A,'s')

ans =

      1           2/3          1/2          2/5
     1/3          1/4          1/5          1/6
      1           1/2          1/3          1/4
```

(b) `A(2,:)=(-1/3)*A(1,:)+A(2,:)`

```
A =

    1.0000    0.6667    0.5000    0.4000
         0    0.0278    0.0333    0.0333
    1.0000    0.5000    0.3333    0.2500
```

```
rat(A,'s')

ans =

     1              2/3             1/2             2/5
     0              1/36            1/30            1/30
     1              1/2             1/3             1/4
```

(c)
```
A(3,:)=-1*A(1,:)+A(3,:)

A =

    1.0000    0.6667    0.5000    0.4000
         0    0.0278    0.0333    0.0333
         0   -0.1667   -0.1667   -0.1500

rat(A,'s')

ans =

     1              2/3             1/2             2/5
     0              1/36            1/30            1/30
     0             -1/6            -1/6            -3/20
```

(d)
```
temp=A(2,:)

temp =

         0    0.0278    0.0333    0.0333

A(2,:)=A(3,:)

A =

    1.0000    0.6667    0.5000    0.4000
         0   -0.1667   -0.1667   -0.1500
         0   -0.1667   -0.1667   -0.1500

A(3,:)=temp

A =

    1.0000    0.6667    0.5000    0.4000
         0   -0.1667   -0.1667   -0.1500
         0    0.0278    0.0333    0.0333

rat(A,'s')

ans =

     1              2/3             1/2             2/5
     0             -1/6            -1/6            -3/20
     0              1/36            1/30            1/30
```

Exercises 1.4

ML.3. Enter **A** into MATLAB, then type **reduce(A)**. Use the menu to select row operations. There are many different sequences of row operations that can be used to obtain the reduced row echelon form. However, the reduced row echelon form is unique and is

```
ans =

     1     0     0
     0     1     0
     0     0     1
     0     0     0
```

ML.4. Enter **A** into MATLAB, then type **reduce(A)**. Use the menu to select row operations. There are many different sequences of row operations that can be used to obtain the reduced row echelon form. However, the reduced row echelon form is unique and is

```
ans =

    1.0000         0         0    0.0500
         0    1.0000         0   -0.6000
         0         0    1.0000    1.5000

rat(ans,'s')

ans =

     1             0             0             1/20
     0             1             0            -3/5
     0             0             1             3/2
```

ML.5. Enter the augmented matrix **AUG** into MATLAB. Then use command **reduce(AUG)** to construct row operations to obtain the reduced row echelon form. We obtain

```
ans =

     1     0     0    -1    -2
     0     1     0     0    -1
     0     0     1     2     8
```

We write the equations equivalent to rows of the reduced row echelon form and use back substitution to determine the solution. The last row corresponds to equation

$$z + 2w = 8$$

Hence we can choose w arbitrarily, w = r, r any real number. Then

$$z = 8 - 2r$$

The second row corresponds to the equation

$$y = -1$$

The first row corresponds to the equation

$$x - w = -2$$

hence

$$x = -2 + w = -2 + r.$$

Thus the solution is given by

$$\begin{aligned} x &= -2 + r \\ y &= -1 \\ z &= 8 - 2r \\ w &= r \end{aligned}$$

ML.6. Enter the augmented matrix **AUG** into MATLAB. Then use command **reduce(AUG)** to construct row operations to obtain the reduced row echelon form

```
ans =

     1     0     1     2     0
     0     1     2     0     0
     0     0     0     0     1
```

The last row is equivalent to the equation

$$0x + 0y + 0z + 0w = 1$$

which is clearly impossible. Thus the system is inconsistent.

ML.7. Enter the augmented matrix **AUG** into MATLAB. Then use command **reduce(AUG)** to construct row operations to obtain the reduced row echelon form. We obtain

```
ans =

     1     0     0     0
     0     1     0     0
     0     0     1     0
     0     0     0     0
```

It follows that this system has only the trivial solution.

ML.8. Enter the augmented matrix **AUG** into MATLAB. Then use command **reduce(AUG)** to construct row operations to obtain the reduced row echelon form. We obtain

```
ans =

     1     0    -1     0
     0     1     2     0
     0     0     0     0
```

The second row corresponds to the equation

$$y + 2z = 0$$

Hence we can choose z arbitrarily. Set z = r, any real number, then

$$y = -2r$$

The first row corresponds to the equation

$$x - z = 0$$

which is the same as x = z = r. Hence the solution of this system is

$$\begin{aligned} x &= r \\ y &= -2r \\ z &= r \end{aligned}$$

ML.9. After entering **A** into MATLAB, use command **reduce(5*eye(A)-A)**. Selecting row operations, we can show that the reduced row echelon form of 5**I** - **A** is

$$\begin{matrix} 1 & -1/2 \\ 0 & 0 \end{matrix}$$

Thus the solution of the homogeneous system is

$$\mathbf{x} = \begin{bmatrix} .5r \\ r \end{bmatrix}$$

Hence for any real number r, not zero, we obtain a nontrivial solution.

ML.10. After entering **A** into MATLAB, use command **reduce(-4*eye(A)-A)**. Selecting row operations, we can show that the reduced row echelon form of -4**I** - **A** is

$$\begin{matrix} 1 & 1 \\ 0 & 0 \end{matrix}$$

Thus the solution of the homogeneous system is

$$\mathbf{x} = \begin{bmatrix} -r \\ r \end{bmatrix}$$

Hence for any real number r, not zero, we obtain a nontrivial solution.

ML.11. For linear system enter the augmented matrix **AUG** and use command **rref**. Then write out the solution.

For 15(a)

rref(AUG)

ans =

```
1    0    0   -1
0    1    0    4
0    0    1   -3
```

It follows that there is a unique solution x = -1, y = 4, z = -3.

For 15(b)

rref(AUG)

ans =

```
1    0    0    0
0    1    0    0
0    0    1    0
0    0    0    0
```

It follows that the only solution is the trivial solution.

For 16(a)

rref(AUG)

ans =

```
1    0   -1    0
0    1    2    0
0    0    0    0
```

It follows that x = r, y = -2r, z = r, where r is any real number.

Exercises 1.4

For 16(b)

```
rref(AUG)

ans =

     1     0     0     1
     0     1     0     2
     0     0     1     2
     0     0     0     0
     0     0     0     0
```

It follows that there is unique solution x = 1, y = 2, and z = 2.

ML.12. (a)

```
A=[1 1 1;1 1 0;0 1 1];

B=[0 3 1]';

X=A\B

X =

    -1
     4
    -3
```

(b)

```
A=[1 1 1;1 1 -2;2 1 1];

B=[1 3 2]';

X=A\B

X =

    1.0000
    0.6667
   -0.6667
```

ML.13.

```
A=[1 2 3;4 5 6;7 8 9];

B=[1 0 0]';

rref([A B])

ans =

     1     0    -1     0
     0     1     2     0
     0     0     0     1
```

This augmented matrix implies that the system is inconsistent. We can also infer that the coefficient matrix is singular.

```
X=A\B

Warning: Matrix is close to singular or badly scaled.
         Results may be inaccurate. RCOND = 2.937385e-018


X =

   1.0e+015 *

    3.1522
   -6.3044
    3.1522
```

Each element of the solution displayed using \ is huge. This, together with the warning, suggests that errors due to using computer arithmetic were magnified in the solution process. MATLAB uses an LU-factorization procedure when \ is used to solve linear systems (see Section 8.3), while **rref** actually rounds values before displaying them.

Exercises 1.5

<<**Strategy:** Exercises 1 and 3 use the method of Examples 2 and 3. Represent $\mathbf{A}^{-1}$ as a matrix of unknowns which are to be determined. From the equation $\mathbf{AA}^{-1} = \mathbf{I}$ construct a system of equations to determine the entries of $\mathbf{A}^{-1}$ by equating corresponding entries.>>

1. $\mathbf{A} = \begin{bmatrix} 2 & 1 \\ -2 & 3 \end{bmatrix}$. Let $\mathbf{A}^{-1} = \begin{bmatrix} a & b \\ c & d \end{bmatrix}$. Forming the product $\mathbf{AA}^{-1}$ and setting it equal to $\mathbf{I}_2$ we obtain

$$\begin{bmatrix} 2a+c & 2b+d \\ -2a+3c & -2b+3d \end{bmatrix} = \begin{bmatrix} 1 & 0 \\ 0 & 1 \end{bmatrix}.$$

Equating corresponding entries, we obtain the two linear systems

$$\begin{aligned} 2a + c &= 1 \\ -2a + 3c &= 0 \end{aligned} \qquad \text{and} \qquad \begin{aligned} 2b + d &= 0 \\ -2b + 3d &= 1 \end{aligned}.$$

Adding the equations in the first system gives $4c = 1$, which implies $c = 1/4$. Substituting $c = 1/4$ into either equation in the first system gives $a = 3/8$. Adding the equations in the second system gives $4d = 1$, which implies $d = 1/4$. Then, substituting $d = 1/4$ into either equation in the second system gives $b = -1/8$. Thus we have

$$\begin{bmatrix} a & b \\ c & d \end{bmatrix} = \begin{bmatrix} \frac{3}{8} & \frac{-1}{8} \\ \frac{1}{4} & \frac{1}{4} \end{bmatrix}$$

In addition,

$$\begin{bmatrix} a & b \\ c & d \end{bmatrix} \begin{bmatrix} 2 & 1 \\ -2 & 3 \end{bmatrix} = \mathbf{I}_2,$$

hence we conclude that $\mathbf{A}$ is nonsingular and that

$$\mathbf{A}^{-1} = \begin{bmatrix} \frac{3}{8} & \frac{-1}{8} \\ \frac{1}{4} & \frac{1}{4} \end{bmatrix}.$$

3. $\mathbf{A} = \begin{bmatrix} 1 & 1 \\ 3 & 4 \end{bmatrix}$. Let $\mathbf{A}^{-1} = \begin{bmatrix} a & b \\ c & d \end{bmatrix}$. Forming the product $\mathbf{AA}^{-1}$ and setting it equal to $\mathbf{I}_2$ we obtain

$$\begin{bmatrix} a+c & b+d \\ 3a+4c & 3b+4d \end{bmatrix} = \begin{bmatrix} 1 & 0 \\ 0 & 1 \end{bmatrix} .$$

Equating corresponding entries, we obtain the two linear systems

$$\begin{aligned} a + c &= 1 \\ 3a + 4c &= 0 \end{aligned} \qquad \text{and} \qquad \begin{aligned} b + d &= 0 \\ 3b + 4d &= 1 . \end{aligned}$$

To eliminate a in the first system, we take -3 times the first equation and add it to the second. The result is c = -3. Substitute c = -3 into either equation in the first system to obtain a = 4. To eliminate b in the second system, we take -3 times the first equation and add it to the second. The result is d = 1. Substituting d = 1 into either equation in the second system gives b = -1. Thus we have

$$\begin{bmatrix} a & b \\ c & d \end{bmatrix} = \begin{bmatrix} 4 & -1 \\ -3 & 1 \end{bmatrix} .$$

In addition,

$$\begin{bmatrix} 4 & -1 \\ -3 & 1 \end{bmatrix}\begin{bmatrix} 1 & 1 \\ 3 & 4 \end{bmatrix} = \mathbf{I}_2$$

hence we conclude that **A** is nonsingular and that

$$\mathbf{A}^{-1} = \begin{bmatrix} 4 & -1 \\ -3 & 1 \end{bmatrix} .$$

<<**Strategy**: Exercises 5, 7, and 9 use the method of Examples 5 and 6.>>

5. (a) $\mathbf{A} = \begin{bmatrix} 1 & 3 \\ -2 & 6 \end{bmatrix}$. Form the matrix $[\mathbf{A}|\mathbf{I}_2] = \left[\begin{array}{cc|cc} 1 & 3 & 1 & 0 \\ -2 & 6 & 0 & 1 \end{array}\right]$.

Applying row operation 2 times row 1 added to row 2, we obtain

$$\left[\begin{array}{cc|cc} 1 & 3 & 1 & 0 \\ 0 & 12 & 2 & 1 \end{array}\right].$$

Multiply row 2 by 1/12, then take -3 times row 2 added to row 1. The result is the row equivalent matrices

$$\left[\begin{array}{cc|cc} 1 & 3 & 1 & 0 \\ 0 & 1 & \frac{1}{6} & \frac{1}{12} \end{array}\right] \quad \text{and} \quad \left[\begin{array}{cc|cc} 1 & 0 & \frac{1}{2} & \frac{-1}{4} \\ 0 & 1 & \frac{1}{6} & \frac{1}{12} \end{array}\right].$$

Exercises 1.5

Thus $\mathbf{A}^{-1} = \begin{bmatrix} \frac{1}{2} & \frac{-1}{4} \\ \frac{1}{6} & \frac{1}{12} \end{bmatrix}$.

(b) $\mathbf{A} = \begin{bmatrix} 1 & 2 & 3 \\ 1 & 1 & 2 \\ 0 & 1 & 2 \end{bmatrix}$. Form the matrix $[\mathbf{A}|\mathbf{I}_3] = \left[\begin{array}{ccc|ccc} 1 & 2 & 3 & 1 & 0 & 0 \\ 1 & 1 & 2 & 0 & 1 & 0 \\ 0 & 1 & 2 & 0 & 0 & 1 \end{array}\right]$.

Applying row operation -1 times row 1 added to row 2, we obtain

$$\left[\begin{array}{ccc|ccc} 1 & 2 & 3 & 1 & 0 & 0 \\ 0 & -1 & -1 & -1 & 1 & 0 \\ 0 & 1 & 2 & 0 & 0 & 1 \end{array}\right].$$

Multiply row 2 by -1, then use row operations -2 times row 2 added to row 1 and -1 times row 2 added to row 3. The result is the row equivalent matrices

$$\left[\begin{array}{ccc|ccc} 1 & 2 & 3 & 1 & 0 & 0 \\ 0 & 1 & 1 & 1 & -1 & 0 \\ 0 & 1 & 2 & 0 & 0 & 1 \end{array}\right] \text{ and } \left[\begin{array}{ccc|ccc} 1 & 0 & 1 & -1 & 2 & 0 \\ 0 & 1 & 1 & 1 & -1 & 0 \\ 0 & 0 & 1 & -1 & 1 & 1 \end{array}\right].$$

Applying row operations -1 times row 3 added to row 1 and -1 times row 3 added to row 2, we obtain

$$\left[\begin{array}{ccc|ccc} 1 & 0 & 0 & 0 & 1 & -1 \\ 0 & 1 & 0 & 2 & -2 & -1 \\ 0 & 0 & 1 & -1 & 1 & 1 \end{array}\right]. \text{ Thus } \mathbf{A}^{-1} = \begin{bmatrix} 0 & 1 & -1 \\ 2 & -2 & -1 \\ -1 & 1 & 1 \end{bmatrix}.$$

(c) $\mathbf{A} = \begin{bmatrix} 1 & 1 & 1 & 1 \\ 1 & 2 & -1 & 2 \\ 1 & -1 & 2 & 1 \\ 1 & 3 & 3 & 2 \end{bmatrix}$. Form the matrix

$[\mathbf{A}|\mathbf{I}_4] = \left[\begin{array}{cccc|cccc} 1 & 1 & 1 & 1 & 1 & 0 & 0 & 0 \\ 1 & 2 & -1 & 2 & 0 & 1 & 0 & 0 \\ 1 & -1 & 2 & 1 & 0 & 0 & 1 & 0 \\ 1 & 3 & 3 & 2 & 0 & 0 & 0 & 1 \end{array}\right]$. Applying row

operations -1 times row 1 added to row j, for j=2,3,4, we obtain

$$\left[\begin{array}{cccc|cccc} 1 & 1 & 1 & 1 & 1 & 0 & 0 & 0 \\ 0 & 1 & -2 & 1 & -1 & 1 & 0 & 0 \\ 0 & -2 & 1 & 0 & -1 & 0 & 1 & 0 \\ 0 & 2 & 2 & 1 & -1 & 0 & 0 & 1 \end{array}\right].$$

Applying row operations
-1 times row 2 added to row 1,
2 times row 2 added to row 3, and
-2 times row 2 added to row 4, we obtain

$$\left[\begin{array}{cccc|cccc} 1 & 0 & 3 & 0 & 2 & -1 & 0 & 0 \\ 0 & 1 & -2 & 1 & -1 & 1 & 0 & 0 \\ 0 & 0 & -3 & 2 & -3 & 2 & 1 & 0 \\ 0 & 0 & 6 & -1 & 1 & -2 & 0 & 1 \end{array}\right].$$

Multiply row 3 by -1/3, then apply row operations
-3 times row 3 added to row 1,
2 times row 3 added to row 2, and
-6 times row 3 added to row 4, we obtain

$$\left[\begin{array}{cccc|cccc} 1 & 0 & 3 & 0 & 2 & -1 & 0 & 0 \\ 0 & 1 & -2 & 1 & -1 & 1 & 0 & 0 \\ 0 & 0 & 1 & \frac{-2}{3} & 1 & \frac{-2}{3} & \frac{-1}{3} & 0 \\ 0 & 0 & 6 & -1 & 1 & -2 & 0 & 1 \end{array}\right] \text{ and}$$

$$\left[\begin{array}{cccc|cccc} 1 & 0 & 0 & 2 & -1 & 1 & 1 & 0 \\ 0 & 1 & 0 & \frac{-1}{3} & 1 & \frac{-1}{3} & \frac{-2}{3} & 0 \\ 0 & 0 & 1 & \frac{-2}{3} & 1 & \frac{-2}{3} & \frac{-1}{3} & 0 \\ 0 & 0 & 0 & 3 & -5 & 2 & 2 & 1 \end{array}\right].$$

Multiply row 4 by 1/3, and then apply row operations
-2 times row 4 added to row 1,
$\frac{1}{3}$ times row 4 added to row 2, and
$\frac{2}{3}$ times row 4 added to row 3.

The result is

$$\left[\begin{array}{cccc|cccc} 1 & 0 & 0 & 0 & \frac{7}{3} & \frac{-1}{3} & \frac{-1}{3} & \frac{-2}{3} \\ 0 & 1 & 0 & 0 & \frac{4}{9} & \frac{-1}{9} & \frac{-4}{9} & \frac{1}{9} \\ 0 & 0 & 1 & 0 & \frac{-1}{9} & \frac{-2}{9} & \frac{1}{9} & \frac{2}{9} \\ 0 & 0 & 0 & 1 & \frac{-5}{3} & \frac{2}{3} & \frac{2}{3} & \frac{1}{3} \end{array}\right].$$

Thus $\mathbf{A}^{-1} = \begin{bmatrix} \frac{7}{3} & \frac{-1}{3} & \frac{-1}{3} & \frac{-2}{3} \\ \frac{4}{9} & \frac{-1}{9} & \frac{-4}{9} & \frac{1}{9} \\ \frac{-1}{9} & \frac{-2}{9} & \frac{1}{9} & \frac{2}{9} \\ \frac{-5}{3} & \frac{2}{3} & \frac{2}{3} & \frac{1}{3} \end{bmatrix}$.

7. (a) $\mathbf{A} = \begin{bmatrix} 1 & 3 \\ 2 & 4 \end{bmatrix}$. Form the matrix $[\mathbf{A}|\mathbf{I}_2] = \left[\begin{array}{cc|cc} 1 & 3 & 1 & 0 \\ 2 & 4 & 0 & 1 \end{array}\right]$.

Applying row operation -2 times row 1 added to row 2, we obtain

$$\left[\begin{array}{cc|cc} 1 & 3 & 1 & 0 \\ 0 & -2 & -2 & 1 \end{array}\right].$$

Multiply row 2 by -1/2, then apply row operation -3 times row 2 added to row 1. The result is the row equivalent matrices

$$\left[\begin{array}{cc|cc} 1 & 3 & 1 & 0 \\ 0 & 1 & 1 & \frac{-1}{2} \end{array}\right] \text{ and }$$

$$\left[\begin{array}{cc|cc} 1 & 0 & -2 & \frac{3}{2} \\ 0 & 1 & 1 & \frac{-1}{2} \end{array}\right].$$

Thus $\mathbf{A}^{-1} = \begin{bmatrix} -2 & \frac{3}{2} \\ 1 & \frac{-1}{2} \end{bmatrix}$.

(b) $\mathbf{A} = \begin{bmatrix} 1 & 1 & 1 & 1 \\ 1 & 3 & 1 & 2 \\ 1 & 2 & -1 & 1 \\ 5 & 9 & 1 & 6 \end{bmatrix}$. Form the matrix

$$[\mathbf{A}|\mathbf{I}_4] = \left[\begin{array}{cccc|cccc} 1 & 1 & 1 & 1 & 1 & 0 & 0 & 0 \\ 1 & 3 & 1 & 2 & 0 & 1 & 0 & 0 \\ 1 & 2 & -1 & 1 & 0 & 0 & 1 & 0 \\ 5 & 9 & 1 & 6 & 0 & 0 & 0 & 1 \end{array}\right].$$

Applying the row operations
-1 times row 1 added to row 2,
-1 times row 1 added to row 3, and
-5 times row 1 added to row 4, we obtain

$$\left[\begin{array}{cccc|cccc} 1 & 1 & 1 & 1 & 1 & 0 & 0 & 0 \\ 0 & 2 & 0 & 1 & -1 & 1 & 0 & 0 \\ 0 & 1 & -2 & 0 & -1 & 0 & 1 & 0 \\ 0 & 4 & -4 & 1 & -5 & 0 & 0 & 1 \end{array}\right].$$

Interchange rows 2 and 3 to get a 1 into the (2,2) position to use as a pivot (and avoid fractions). Then apply row operations
-1 times row 2 added to row 1,
-2 times row 2 added to row 3, and
-4 times row 2 added to row 4.
The matrix that results is

$$\left[\begin{array}{cccc|cccc} 1 & 0 & 3 & 1 & 2 & 0 & -1 & 0 \\ 0 & 1 & -2 & 0 & -1 & 0 & 1 & 0 \\ 0 & 0 & 4 & 1 & 1 & 1 & -2 & 0 \\ 0 & 0 & 4 & 1 & -1 & 0 & -4 & 1 \end{array}\right].$$

Applying the row operation
-1 times row 3 added to row 4
gives us a matrix with a zero row that is row equivalent to **A**. Thus **A** has no inverse. Matrix **A** is singular.

(c) $\mathbf{A} = \begin{bmatrix} 1 & 2 & 1 \\ 1 & 3 & 2 \\ 1 & 0 & 1 \end{bmatrix}$. Form the matrix $[\mathbf{A}|\mathbf{I}_3] = \left[\begin{array}{ccc|ccc} 1 & 2 & 1 & 1 & 0 & 0 \\ 1 & 3 & 2 & 0 & 1 & 0 \\ 1 & 0 & 1 & 0 & 0 & 1 \end{array}\right]$.

Applying row operations

-1 times row 1 added to row 2, and
-1 times row 1 added to row 3, we obtain

$$\left[\begin{array}{rrr|rrr} 1 & 2 & 1 & 1 & 0 & 0 \\ 0 & 1 & 1 & -1 & 1 & 0 \\ 0 & -2 & 0 & -1 & 0 & 1 \end{array}\right].$$

Applying row operations
-2 times row 2 added to row 1 and
2 times row 2 added to row 3, we obtain

$$\left[\begin{array}{rrr|rrr} 1 & 0 & -1 & 3 & -2 & 0 \\ 0 & 1 & 1 & -1 & 1 & 0 \\ 0 & 0 & 2 & -3 & 2 & 1 \end{array}\right].$$

Multiply row 3 by 1/2, then apply row operations
1 times row 3 added to row 1 and
-1 times row 3 added to row 2, we obtain

$$\left[\begin{array}{rrr|rrr} 1 & 0 & 0 & \frac{3}{2} & -1 & \frac{1}{2} \\ 0 & 1 & 0 & \frac{1}{2} & 0 & \frac{-1}{2} \\ 0 & 0 & 1 & \frac{-3}{2} & 1 & \frac{1}{2} \end{array}\right].$$

Thus $\mathbf{A}^{-1} = \begin{bmatrix} \frac{3}{2} & -1 & \frac{1}{2} \\ \frac{1}{2} & 0 & \frac{-1}{2} \\ \frac{-3}{2} & 1 & \frac{1}{2} \end{bmatrix}$.

9. (a) $\mathbf{A} = \begin{bmatrix} 1 & 2 & -3 & 1 \\ -1 & 3 & -3 & -2 \\ 2 & 0 & 1 & 5 \\ 3 & 1 & -2 & 5 \end{bmatrix}$.

Form the matrix $[\mathbf{A}|\mathbf{I}_4] = \left[\begin{array}{rrrr|rrrr} 1 & 2 & -3 & 1 & 1 & 0 & 0 & 0 \\ -1 & 3 & -3 & -2 & 0 & 1 & 0 & 0 \\ 2 & 0 & 1 & 5 & 0 & 0 & 1 & 0 \\ 3 & 1 & -2 & 5 & 0 & 0 & 0 & 1 \end{array}\right]$.

Applying row operations
1 times row 1 added to row 2,
-2 times row 1 added to row 3, and
-3 times row 1 added to row 4, we obtain

$$\left[\begin{array}{rrrr|rrrr} 1 & 2 & -3 & 1 & 1 & 0 & 0 & 0 \\ 0 & 5 & -6 & -1 & 1 & 1 & 0 & 0 \\ 0 & -4 & 7 & 3 & -2 & 0 & 1 & 0 \\ 0 & -5 & 7 & 2 & -3 & 0 & 0 & 1 \end{array}\right].$$

In order to continue the reduction, we want to obtain a 1 to use as a pivot in the (2,2) entry. Obviously, we can multiply row 2 by 1/5 to achieve this goal. However, to avoid fractions, we observe that if we add row 3 to row 2 the (2,2) entry will then contain a 1. The result is

$$\left[\begin{array}{rrrr|rrrr} 1 & 2 & -3 & 1 & 1 & 0 & 0 & 0 \\ 0 & 1 & 1 & 2 & -1 & 1 & 1 & 0 \\ 0 & -4 & 7 & 3 & -2 & 0 & 1 & 0 \\ 0 & -5 & 7 & 2 & -3 & 0 & 0 & 1 \end{array}\right].$$

Performing row operations

-2 times row 2 added to row 1,
4 times row 2 added to row 3, and
5 times row 2 added to row 4, we obtain

$$\left[\begin{array}{rrrr|rrrr} 1 & 0 & -5 & -3 & 3 & -2 & -2 & 0 \\ 0 & 1 & 1 & 2 & -1 & 1 & 1 & 0 \\ 0 & 0 & 11 & 11 & -6 & 4 & 5 & 0 \\ 0 & 0 & 12 & 12 & -8 & 5 & 5 & 1 \end{array}\right].$$

Next we see that $\frac{-12}{11}$ times row 3 added to row 4 will produce zeros in the first four entries of row 4. Thus **A** will be row equivalent to a matrix with a zero row. Hence, **A** is singular.

(b) $\mathbf{A} = \begin{bmatrix} 3 & 1 & 2 \\ 2 & 1 & 2 \\ 1 & 2 & 2 \end{bmatrix}$. Form the matrix $[\mathbf{A}|\mathbf{I}_3] = \left[\begin{array}{rrr|rrr} 3 & 1 & 2 & 1 & 0 & 0 \\ 2 & 1 & 2 & 0 & 1 & 0 \\ 1 & 2 & 2 & 0 & 0 & 1 \end{array}\right]$.

To avoid fractions and simplify the arithmetic, interchange rows 1 and 3. This puts a 1 into the (1,1) entry to use as a pivot. Then apply row operations

-2 times row 1 added to row 2 and
-3 times row 1 added to row 3.

The result is the row equivalent matrices

$$\left[\begin{array}{rrr|rrr} 1 & 2 & 2 & 0 & 0 & 1 \\ 2 & 1 & 2 & 0 & 1 & 0 \\ 3 & 1 & 2 & 1 & 0 & 0 \end{array}\right] \text{ and } \left[\begin{array}{rrr|rrr} 1 & 2 & 2 & 0 & 0 & 1 \\ 0 & -3 & -2 & 0 & 1 & -2 \\ 0 & -5 & -4 & 1 & 0 & -3 \end{array}\right].$$

Multiply row 2 by -1/3 to get a 1 into the (2,2) position.

$$\left[\begin{array}{ccc|ccc} 1 & 2 & 2 & 0 & 0 & 1 \\ 0 & 1 & \frac{2}{3} & 0 & \frac{-1}{3} & \frac{2}{3} \\ 0 & -5 & -4 & 1 & 0 & -3 \end{array}\right].$$

Applying row operations
-2 times row 2 added to row 1 and
5 times row 2 added to row 3, we obtain

$$\left[\begin{array}{ccc|ccc} 1 & 0 & \frac{2}{3} & 0 & \frac{2}{3} & \frac{-1}{3} \\ 0 & 1 & \frac{2}{3} & 0 & \frac{-1}{3} & \frac{2}{3} \\ 0 & 0 & \frac{-2}{3} & 1 & \frac{-5}{3} & \frac{1}{3} \end{array}\right].$$

Add row 3 to both rows 1 and 2 to "zero out" above the (3,3) entry. Then multiply row 3 by $\frac{-3}{2}$. The result is

$$\left[\begin{array}{ccc|ccc} 1 & 0 & 0 & 1 & -1 & 0 \\ 0 & 1 & 0 & 1 & -2 & 1 \\ 0 & 0 & 1 & \frac{-3}{2} & \frac{5}{2} & \frac{-1}{2} \end{array}\right].$$

Thus $\mathbf{A}^{-1} = \begin{bmatrix} 1 & -1 & 0 \\ 1 & -2 & 1 \\ \frac{-3}{2} & \frac{5}{2} & \frac{-1}{2} \end{bmatrix}$.

(c) $\mathbf{A} = \begin{bmatrix} 1 & 2 & 3 \\ 1 & 1 & 2 \\ 1 & 1 & 0 \end{bmatrix}$. Form the matrix $[\mathbf{A}|\mathbf{I}_3] = \left[\begin{array}{ccc|ccc} 1 & 2 & 3 & 1 & 0 & 0 \\ 1 & 1 & 2 & 0 & 1 & 0 \\ 1 & 1 & 0 & 0 & 0 & 1 \end{array}\right]$.

Applying row operations
-1 times row 1 added to row 2 and
-1 times row 1 added to row 3, we obtain

$$\left[\begin{array}{ccc|ccc} 1 & 2 & 3 & 1 & 0 & 0 \\ 0 & -1 & -1 & -1 & 1 & 0 \\ 0 & -1 & -3 & -1 & 0 & 1 \end{array}\right].$$

Multiply row 2 by -1, then apply row operations
-2 times row 2 added to row 1 and
1 times row 2 added to row 3.
The result is the row equivalent matrices

$$\left[\begin{array}{ccc|ccc} 1 & 2 & 3 & 1 & 0 & 0 \\ 0 & 1 & 1 & 1 & -1 & 0 \\ 0 & -1 & -3 & -1 & 0 & 1 \end{array}\right] \text{ and } \left[\begin{array}{ccc|ccc} 1 & 0 & 1 & -1 & 2 & 0 \\ 0 & 1 & 1 & 1 & -1 & 0 \\ 0 & 0 & -2 & 0 & -1 & 1 \end{array}\right].$$

Multiply row 3 by -1/2, then apply row operations
-1 times row 3 added to row 1 and
-1 times row 3 added to row 2.
The end result is the matrix

$$\left[\begin{array}{ccc|ccc} 1 & 0 & 0 & -1 & \frac{3}{2} & \frac{1}{2} \\ 0 & 1 & 0 & 1 & \frac{-3}{2} & \frac{1}{2} \\ 0 & 0 & 1 & 0 & \frac{1}{2} & \frac{-1}{2} \end{array}\right].$$

Thus $\mathbf{A}^{-1} = \begin{bmatrix} -1 & \frac{3}{2} & \frac{1}{2} \\ 1 & \frac{-3}{2} & \frac{1}{2} \\ 0 & \frac{1}{2} & \frac{-1}{2} \end{bmatrix}$.

<<**Strategy:** For Exercise 11 we use Theorem 1.12. Row reduce the coefficient matrix and determine whether it is row equivalent to $\mathbf{I}_n$ or not. If it is not row equivalent to $\mathbf{I}_n$, then the system has a nontrivial solution.>>

11. (a) The linear system in matrix form is $\mathbf{AX} = \begin{bmatrix} 1 & 2 & 3 \\ 0 & 2 & 2 \\ 1 & 2 & 3 \end{bmatrix}\begin{bmatrix} x \\ y \\ z \end{bmatrix} = \mathbf{0}$.

Applying operation -1 times row 1 added to row 3 to **A**, we obtain the row equivalent matrix

$$\begin{bmatrix} 1 & 2 & 3 \\ 0 & 2 & 2 \\ 0 & 0 & 0 \end{bmatrix}.$$

Thus **A** is row equivalent to a matrix with a zero row. Hence **A** is singular and the linear system **AX** = **0** has a nontrivial solution.

(b) The linear system in matrix form is $\mathbf{AX} = \begin{bmatrix} 2 & 1 & -1 \\ 1 & -2 & -3 \\ -3 & -1 & 2 \end{bmatrix}\begin{bmatrix} x \\ y \\ z \end{bmatrix} = \mathbf{0}$.

Interchange rows 1 and 2 of **A**. Then apply row operations
-2 times row 1 added to row 2 and
3 times row 1 added to row 3.
The result is the row equivalent matrices

$$\begin{bmatrix} 1 & -2 & -3 \\ 2 & 1 & -1 \\ -3 & -1 & 2 \end{bmatrix} \text{ and } \begin{bmatrix} 1 & -2 & -3 \\ 0 & 5 & 5 \\ 0 & -7 & -7 \end{bmatrix}.$$

Multiply row 2 by 1/5, then apply row operation 7 times row 2 added to row 3. The result is the matrix

$$\begin{bmatrix} 1 & -2 & -3 \\ 0 & 1 & 1 \\ 0 & 0 & 0 \end{bmatrix}.$$

Thus **A** is row equivalent to a matrix with a zero row. Hence **A** is singular. It follows that the linear system **AX** = **0** has a nontrivial solution.

13. To find **A**, given $\mathbf{A}^{-1}$, we find the inverse of $\mathbf{A}^{-1}$. This follows from Theorem 1.9(a); $(\mathbf{A}^{-1})^{-1} = \mathbf{A}$. Using the technique from Example 5, we proceed as follows. Form the matrix

$[\mathbf{A}^{-1}|\mathbf{I}_2] = \left[\begin{array}{cc|cc} 2 & 3 & 1 & 0 \\ 1 & 4 & 0 & 1 \end{array}\right]$. Use row operations to obtain the reduced row echelon form. Interchange rows 1 and 2, then apply the row operation -2 times row 1 added to row 2. The result is the matrix

$$\left[\begin{array}{cc|cc} 1 & 4 & 0 & 1 \\ 0 & -5 & 1 & -2 \end{array}\right].$$

Multiply row 2 by -1/5, then apply row operation -4 times row 2 added to row 1. The result is

$$\left[\begin{array}{cc|cc} 1 & 0 & \frac{4}{5} & \frac{-3}{5} \\ 0 & 1 & \frac{-1}{5} & \frac{2}{5} \end{array}\right].$$

Thus $\mathbf{A} = \begin{bmatrix} \frac{4}{5} & \frac{-3}{5} \\ \frac{-1}{5} & \frac{2}{5} \end{bmatrix}$.

15. Let **A** be a matrix with a row of zeros. Then for any matrix **B**, such that **BA** is defined, the product **BA** also has a row of zeros. (See Exercise T.5 in Section 1.2.) Hence there does not exist a matrix **B** such that **BA** = $\mathbf{I}_n$, since $\mathbf{I}_n$ has no row of zeros. It follows that **A** has no inverse matrix, hence **A** is singular. A similar argument holds for a column of zeros using the product **AB**.

17. Following Example 7, the input matrix $\mathbf{X} = \mathbf{A}^{-1}\mathbf{B}$. We compute $\mathbf{A}^{-1}$ and form the product for each output matrix **B**. Form the matrix $[\mathbf{A}|\mathbf{I}_3] = \left[\begin{array}{ccc|ccc} 2 & 1 & 3 & 1 & 0 & 0 \\ 3 & 2 & -1 & 0 & 1 & 0 \\ 2 & 1 & 1 & 0 & 0 & 1 \end{array}\right]$. Multiply row 1 by 1/2, then apply row operations
    -3 times row 1 added to row 2 and
    -2 times row 1 added to row 3.
The result is the matrix

$$\left[\begin{array}{ccc|ccc} 1 & \frac{1}{2} & \frac{3}{2} & \frac{1}{2} & 0 & 0 \\ 0 & \frac{1}{2} & \frac{-11}{2} & \frac{-3}{2} & 1 & 0 \\ 0 & 0 & -2 & -1 & 0 & 1 \end{array}\right].$$

Multiply row 2 by 2, then apply row operation $\frac{-1}{2}$ times row 2 added to row 1. The result is the matrix

$$\left[\begin{array}{ccc|ccc} 1 & 0 & 7 & 2 & -1 & 0 \\ 0 & 1 & -11 & -3 & 2 & 0 \\ 0 & 0 & -2 & -1 & 0 & 1 \end{array}\right].$$

Multiply row 3 by -1/2, then apply row operations
    -7 times row 3 added to row 1 and
    11 times row 3 added to row 2.

The result is matrix

$$\left[\begin{array}{ccc|ccc} 1 & 0 & 0 & \frac{-3}{2} & -1 & \frac{7}{2} \\ 0 & 1 & 0 & \frac{5}{2} & 2 & \frac{-11}{2} \\ 0 & 0 & 1 & \frac{1}{2} & 0 & \frac{-1}{2} \end{array}\right].$$

Thus $\mathbf{A}^{-1} = \begin{bmatrix} \frac{-3}{2} & -1 & \frac{7}{2} \\ \frac{5}{2} & 2 & \frac{-11}{2} \\ \frac{1}{2} & 0 & \frac{-1}{2} \end{bmatrix}.$

(a) For output matrix $\mathbf{B} = \begin{bmatrix} 30 \\ 20 \\ 10 \end{bmatrix}$, the input matrix is

$$\mathbf{X} = \mathbf{A}^{-1}\mathbf{B} = \begin{bmatrix} -30 \\ 60 \\ 10 \end{bmatrix}.$$

(b) For output matrix $\mathbf{B} = \begin{bmatrix} 12 \\ 8 \\ 14 \end{bmatrix}$, the input matrix is

$$\mathbf{X} = \mathbf{A}^{-1}\mathbf{B} = \begin{bmatrix} 23 \\ -31 \\ -1 \end{bmatrix}.$$

19. Suppose that $\mathbf{A}$ is symmetric and nonsingular. Then $\mathbf{A}^T = \mathbf{A}$ and $\mathbf{A}^{-1}$ exists. We show that $\mathbf{A}^{-1}$ is symmetric by verifing that $(\mathbf{A}^{-1})^T = \mathbf{A}^{-1}$. By Theorem 1.9(c) $(\mathbf{A}^{-1})^T = (\mathbf{A}^T)^{-1}$. However, $\mathbf{A}^T = \mathbf{A}$, thus it follows that $(\mathbf{A}^{-1})^T = (\mathbf{A})^{-1}$.

21. The homogeneous system will have a nontrivial solution provided the coefficient matrix $\mathbf{A} = \begin{bmatrix} \lambda-1 & 2 \\ 2 & \lambda-1 \end{bmatrix}$ is singular.

We row reduce $\mathbf{A}$ and determine values of λ so that $\mathbf{A}$ has a zero row. Interchange rows 1 and 2 and multiply the new row 1 by 1/2. The resulting matrix is

$$\begin{bmatrix} 1 & \frac{\lambda-1}{2} \\ \lambda-1 & 2 \end{bmatrix}.$$

Applying row operation $-(\lambda-1)$ times row 1 added to row 2, we obtain

$$\begin{bmatrix} 1 & \frac{\lambda-1}{2} \\ 0 & \frac{-(\lambda-1)(\lambda-1)}{2} + 2 \end{bmatrix} = \begin{bmatrix} 1 & \frac{\lambda-1}{2} \\ 0 & \frac{-\lambda^2 + 2\lambda + 3}{2} \end{bmatrix}.$$

The second row will be a zero row provided $-(\lambda^2 - 2\lambda - 3) = -(\lambda - 3)(\lambda + 1) = 0$. That is, if $\lambda = 3$ or $\lambda = -1$.

23. Form the matrix $[\mathbf{D}|\mathbf{I}_3]$. Applying row operations $\frac{1}{4}$ times row 1, $\frac{-1}{2}$ times row 2, and $\frac{1}{3}$ times row 3 we obtain,

$$\left[\begin{array}{ccc|ccc} 1 & 0 & 0 & \frac{1}{4} & 0 & 0 \\ 0 & 1 & 0 & 0 & \frac{-1}{2} & 0 \\ 0 & 0 & 1 & 0 & 0 & \frac{1}{3} \end{array}\right]. \text{ Thus } \mathbf{D}^{-1} = \begin{bmatrix} \frac{1}{4} & 0 & 0 \\ 0 & \frac{-1}{2} & 0 \\ 0 & 0 & \frac{1}{3} \end{bmatrix}.$$

25. $\mathbf{X} = \mathbf{A}^{-1}\mathbf{B} = \begin{bmatrix} 19 \\ 23 \end{bmatrix}.$

T.1. $\mathbf{B}$ is nonsingular, so $\mathbf{B}^{-1}$ exists, and

$$\mathbf{A} = \mathbf{A}\mathbf{I}_n = \mathbf{A}(\mathbf{B}\mathbf{B}^{-1}) = (\mathbf{A}\mathbf{B})\mathbf{B}^{-1} = \mathbf{0}\mathbf{B}^{-1} = \mathbf{0}.$$

T.3. $\mathbf{A}$ is row equivalent to a matrix $\mathbf{B}$ in reduced row echelon form which, by Theorem 1.11, is not $\mathbf{I}_n$. Thus $\mathbf{B}$ has fewer than n nonzero rows, and fewer than n unknowns corresponding to the pivotal columns of $\mathbf{B}$. Choose one of the free unknowns -- unknowns not corresponding to pivotal columns of $\mathbf{B}$. Assign any nonzero value to that unknown. This leads to a nontrivial solution of the homogeneous system $\mathbf{AX} = \mathbf{0}$.

Exercises 1.5

T.5. For any angle θ, $\cos\theta$ and $\sin\theta$ are never simultaneously zero. Thus at least one element in column 1 is not zero. Assume $\cos\theta \neq 0$. (If $\cos\theta = 0$, then interchange rows 1 and 2 and proceed in a manner similar to that described below.) To show that the matrix is nonsingular and determine its inverse, we put

$$\left[\begin{array}{cc|cc} \cos\theta & \sin\theta & 1 & 0 \\ -\sin\theta & \cos\theta & 0 & 1 \end{array}\right]$$

into reduced row echelon form. (From Exercise T.& in Section 1.4 we have that **A** is nonsingular, but we need to determine its inverse.Apply row operations $1/\cos\theta$ times row 1 and $\sin\theta$ times row 1 added to row 2 to obtain

$$\left[\begin{array}{cc|cc} 1 & \sin\theta/\cos\theta & 1/\cos\theta & 0 \\ 0 & \sin^2\theta/\cos\theta + \cos\theta & \sin\theta/\cos\theta & 1 \end{array}\right]$$

Since $\sin^2\theta/\cos\theta + \cos\theta = \dfrac{\sin^2\theta + \cos^2\theta}{\cos\theta} = 1/\cos\theta$, the (2,2)-element is not zero. Applying row operations $\cos\theta$ times row 2 and $-\sin\theta/\cos\theta$ times row 2 added to row 1 we obtain

$$\left[\begin{array}{cc|cc} 1 & 0 & \cos\theta & -\sin\theta \\ 0 & 1 & \sin\theta & \cos\theta \end{array}\right]$$

It follows that the matrix is nonsingular and its inverse is

$$\begin{bmatrix} \cos\theta & -\sin\theta \\ \sin\theta & \cos\theta \end{bmatrix}.$$

T.7. Let $\mathbf{X}_1$ be one solution of $\mathbf{AX} = \mathbf{B}$. Since **A** is singular, the homogeneous system $\mathbf{AX} = \mathbf{0}$ has a nontrivial solution $\mathbf{Y}_0$. (See Exercise T.3.) Then for any real number r, $\mathbf{Y}_1 = r\mathbf{Y}_0$ is also a solution to $\mathbf{AX} = \mathbf{0}$. Finally, by Exercise T.9(a) of Section 1.4, for each of the infinitely many matrices $\mathbf{Y}_1$, the matrix $\mathbf{X} = \mathbf{X}_1 + \mathbf{Y}_1$ is a solution to the nonhomogeneous system $\mathbf{AX} = \mathbf{B}$.

ML.1. We use that **A** is nonsingular if **rref(A)** is the identity matrix.

(a) A=[1 2;-2 1];

```
rref(A)

ans =

     1     0
     0     1
```

Thus **A** is nonsingular.

(b)
```
A=[1 2 3;4 5 6;7 8 9];

rref(A)

ans =

     1     0    -1
     0     1     2
     0     0     0
```

Thus **A** is singular.

(c)
```
A=[1 2 3;4 5 6;7 8 0];

rref(A)

ans =

     1     0     0
     0     1     0
     0     0     1
```

Thus **A** is nonsingular.

ML.2. We use that **A** is nonsingular if **rref(A)** is the identity matrix.

(a)
```
A=[1 2;2 4];

rref(A)

ans =

     1     2
     0     0
```

Thus **A** is singular.

(b)
```
A=[1 0 0;0 1 0;1 1 1];

rref(A)

ans =

     1     0     0
     0     1     0
     0     0     1
```

Thus **A** is nonsingular.

(c)
```
A=[1 2 1;0 1 2;1 0 0];

rref(A)
```

```
ans =

     1     0     0
     0     1     0
     0     0     1
```

Thus **A** is nonsingular.

ML.3. (a) `A=[1 3;1 2];`

```
rref([A eye(A)])

ans =

     1     0    -2     3
     0     1     1    -1
```

Thus $\mathbf{A}^{-1} = \begin{bmatrix} -2 & 3 \\ 1 & -1 \end{bmatrix}$.

(b) `A=[1 1 2;2 1 1;1 2 1];`

```
rref([A eye(A)])

ans =

    1.0000         0         0   -0.2500    0.7500   -0.2500
         0    1.0000         0   -0.2500   -0.2500    0.7500
         0         0    1.0000    0.7500   -0.2500   -0.2500

rat(ans,'s')

ans =

     1        0        0     -1/4      3/4     -1/4
     0        1        0     -1/4     -1/4      3/4
     0        0        1      3/4     -1/4     -1/4
```

Thus $\mathbf{A}^{-1} = \begin{bmatrix} -1/4 & 3/4 & -1/4 \\ -1/4 & -1/4 & 3/4 \\ 3/4 & -1/4 & -1/4 \end{bmatrix}$.

ML.4. (a) `A=[2 1;2 3];`

```
rref([A eye(A)])

ans =

    1.0000         0    0.7500   -0.2500
         0    1.0000   -0.5000    0.5000
```

```
rat(ans,'s')

ans =

     1         0        3/4      -1/4
     0         1       -1/2       1/2
```

Thus $\mathbf{A^{-1}} = \begin{bmatrix} 3/4 & -1/4 \\ -1/2 & 1/2 \end{bmatrix}$.

(b) `A=[1 -1 2;0 2 1;1 0 0];`

```
rref([A eye(A)])

ans =

    1.0000         0         0         0         0    1.0000
         0    1.0000         0   -0.2000    0.4000    0.2000
         0         0    1.0000    0.4000    0.2000   -0.4000

rat(ans,'s')

ans =

     1         0         0         0         0         1
     0         1         0      -1/5       2/5       1/5
     0         0         1       2/5       1/5      -2/5
```

Thus $\mathbf{A^{-1}} = \begin{bmatrix} 0 & 0 & 1 \\ -1/5 & 2/5 & 1/5 \\ 2/5 & 1/5 & -2/5 \end{bmatrix}$

ML.5. We experiment choosing successive values of t then computing the **rref** of **(t*eye(A) - A)**.

(a) `A=[1 3;3 1];`

```
t=1;rref(t*eye(A)-A)      {Use the up arrow key to
                           recall and then revise it
                           for use below.}
ans =

     1     0
     0     1

t=2;rref(t*eye(A)-A)

ans =

     1     0
     0     1
```

```
t=3;rref(t*eye(A)-A)

ans =

     1     0
     0     1

t=4;rref(t*eye(A)-A)

ans =

     1    -1
     0     0
```

Thus t = 4.

(b)
```
A=[4 1 2;1 4 1;0 0 -4];
t=1;rref(t*eye(A)-A)

ans =

     1     0     0
     0     1     0
     0     0     1

t=2;rref(t*eye(A)-A)

ans =

     1     0     0
     0     1     0
     0     0     1

t=3;rref(t*eye(A)-A)

ans =

     1     1     0
     0     0     1
     0     0     0
```

Thus t = 3.

Chapter 1 Supplementary Exercises

1. $2\mathbf{A} + \mathbf{BC} = 2\begin{bmatrix} 1 & 2 \\ 3 & -2 \end{bmatrix} + \begin{bmatrix} 3 & -5 \\ 2 & 4 \end{bmatrix}\begin{bmatrix} 4 & 1 \\ 3 & 2 \end{bmatrix} = \begin{bmatrix} 2 & 4 \\ 6 & -4 \end{bmatrix} + \begin{bmatrix} -3 & -7 \\ 20 & 10 \end{bmatrix}$

$= \begin{bmatrix} -1 & -3 \\ 26 & 6 \end{bmatrix}$

3. $\mathbf{A}^T + \mathbf{B}^T\mathbf{C} = \begin{bmatrix} 1 & 3 \\ 2 & -2 \end{bmatrix} + \begin{bmatrix} 3 & 2 \\ -5 & 4 \end{bmatrix}\begin{bmatrix} 4 & 1 \\ 3 & 2 \end{bmatrix} = \begin{bmatrix} 1 & 3 \\ 2 & -2 \end{bmatrix} + \begin{bmatrix} 18 & 7 \\ -8 & 3 \end{bmatrix}$

$= \begin{bmatrix} 19 & 10 \\ -6 & 1 \end{bmatrix}$

5. (a) The augmented matrix is $\left[\begin{array}{cccc|c} 1 & 2 & -1 & 1 & 7 \\ 2 & -1 & 0 & 2 & -8 \end{array}\right]$.

   (b) The corresponding linear system is

$$\begin{aligned} 3x + 2y &= -4 \\ 5x + y &= 2 \\ 3x + 2y &= 6 . \end{aligned}$$

7. We form the augmented matrix for the linear system.

$$\left[\begin{array}{ccc|c} 1 & 1 & -1 & 5 \\ 2 & 1 & 1 & 2 \\ 1 & -1 & -2 & 3 \end{array}\right]$$

Applying row operations
-2 times row 1 added to row 2 and
-1 times row 1 added to row 3, we obtain

$$\left[\begin{array}{ccc|c} 1 & 1 & -1 & 5 \\ 0 & -1 & 3 & -8 \\ 0 & -2 & -1 & -2 \end{array}\right].$$

Applying row operations
-1 times row 2,
-1 times row 2 added to row 1, and
2 times row 2 added to row 3, we obtain

$$\left[\begin{array}{ccc|c} 1 & 0 & 2 & -3 \\ 0 & 1 & -3 & 8 \\ 0 & 0 & -7 & 14 \end{array}\right].$$

Finally we apply row operations
multiply row 3 by -1/7, then
-2 times row 3 added to row 1 and
3 times row 3 added to row 2, to obtain

$$\left[\begin{array}{ccc|c} 1 & 0 & 0 & 1 \\ 0 & 1 & 0 & 2 \\ 0 & 0 & 1 & -2 \end{array}\right].$$

Thus we see that the solution of the linear system is

$$x = 1,\ y = 2,\ \text{and}\ z = -2.$$

9. We form the augmented matrix and apply row operations to obtain a simpler equivalent system. We have

$$\left[\begin{array}{ccc|c} 1 & 1 & -1 & 3 \\ 1 & -1 & 3 & 4 \\ 1 & 1 & a^2-10 & a \end{array}\right].$$

Applying row operations
-1 times row 1 added to row 2 and
-1 times row 1 added to row 3, we obtain

$$\left[\begin{array}{ccc|c} 1 & 1 & -1 & 3 \\ 0 & -2 & 4 & 1 \\ 0 & 0 & a^2-9 & a-3 \end{array}\right].$$

(a) There will be no solutions if a is chosen so that row 3 has the form
$[\ 0 \quad 0 \quad 0 \quad | \quad *]$
where $* \neq 0$. We see that $a^2 - 9 = 0$ if and only if $a = \pm 3$. If $a = -3$, then $a - 3 \neq 0$. Thus there will be no solutions if $a = -3$.

(b) There will be a unique solution in the case that a is chosen so that $a^2 - 9$ is not zero. In that case the (3,3) entry can be used as a pivot. Thus there will be a unique solution if $a \neq 3$ or -3.

(c) There will be infinitely many solutions if the third row is a zero row. This is the case if $a = 3$.

11. To determine all the solutions of $(\lambda \mathbf{I}_3 - \mathbf{A})\mathbf{X} = \mathbf{0}$ we row reduce the augmented matrix

$$[\lambda \mathbf{I}_3 - \mathbf{A} \mid \mathbf{0}] = [-4\mathbf{I}_3 - \mathbf{A} \mid \mathbf{0}] = \left[\begin{array}{ccc|c} -1 & -3 & -1 & 0 \\ 0 & 0 & -2 & 0 \\ 0 & 0 & 0 & 0 \end{array}\right].$$

Applying row operations
-1 times row 1 and

$\frac{-1}{2}$ times row 2, we obtain

$$\left[\begin{array}{ccc|c} 1 & 3 & 1 & 0 \\ 0 & 0 & 1 & 0 \\ 0 & 0 & 0 & 0 \end{array}\right].$$

Finally applying
-1 times row 2 added to row 1 gives

$$\left[\begin{array}{ccc|c} 1 & 3 & 0 & 0 \\ 0 & 0 & 1 & 0 \\ 0 & 0 & 0 & 0 \end{array}\right].$$

The corresponding linear system is

$$\begin{aligned} x + 3y \quad &= 0 \\ z &= 0 . \end{aligned}$$

Let y = r, where r is any real number, then the solution is given by

$$x = -3r, \ y = r, \text{ and } z = 0.$$

13. We form the matrix $\left[\begin{array}{ccc|ccc} 1 & 2 & 3 & 1 & 0 & 0 \\ 2 & 5 & 3 & 0 & 1 & 0 \\ 1 & 0 & 8 & 0 & 0 & 1 \end{array}\right]$. We apply row operations to obtain the reduced row echelon form. Applying row operations
-2 times row 1 added to row 2 and
-1 times row 1 added to row 3, we obtain

$$\left[\begin{array}{ccc|ccc} 1 & 2 & 3 & 1 & 0 & 0 \\ 0 & 1 & -3 & -2 & 1 & 0 \\ 0 & -2 & 5 & -1 & 0 & 1 \end{array}\right].$$

Applying row operations
-2 times row 2 added to row 1 and
2 times row 2 added to row 3 gives

$$\left[\begin{array}{ccc|ccc} 1 & 0 & 9 & 5 & -2 & 0 \\ 0 & 1 & -3 & -2 & 1 & 0 \\ 0 & 0 & -1 & -5 & 2 & 1 \end{array}\right].$$

Applying row operations
-1 times row 3,
3 times row 3 added to row 2, and
-9 times row 3 added to row 1, we obtain

$$\left[\begin{array}{ccc|ccc} 1 & 0 & 0 & -40 & 16 & 9 \\ 0 & 1 & 0 & 13 & -5 & -3 \\ 0 & 0 & 1 & 5 & -2 & -1 \end{array}\right].$$

Thus the matrix has an inverse given by $\begin{bmatrix} -40 & 16 & 9 \\ 13 & -5 & -3 \\ 5 & -2 & -1 \end{bmatrix}$.

15. We row reduce the corresponding augmented matrix

$$\left[\begin{array}{ccc|c} 1 & -1 & 3 & 0 \\ 1 & 2 & -3 & 0 \\ 2 & 1 & 0 & 0 \end{array}\right].$$

Applying row operations
-1 times row 1 added to row 2 and
-2 times row 1 added to row 3, we obtain

$$\left[\begin{array}{ccc|c} 1 & -1 & 3 & 0 \\ 0 & 3 & -6 & 0 \\ 0 & 3 & -6 & 0 \end{array}\right].$$

Then adding (-1) times row 2 to row 3 gives $\left[\begin{array}{ccc|c} 1 & -1 & 3 & 0 \\ 0 & 3 & -6 & 0 \\ 0 & 0 & 0 & 0 \end{array}\right]$.

We see that the linear system equivalent to this last matrix will have 2 equations in 3 unknowns. Hence at least one of the variables can be chosen arbitrarily. Thus the system has a nontrivial solution.

17. $\mathbf{X} = \mathbf{A}^{-1}\mathbf{B} = \begin{bmatrix} 1 & 2 & 0 \\ 0 & 1 & 0 \\ 3 & 1 & -1 \end{bmatrix}\begin{bmatrix} 2 \\ 1 \\ 3 \end{bmatrix} = \begin{bmatrix} 4 \\ 1 \\ 4 \end{bmatrix}$.

19. Assuming that $\mathbf{A}^4 = \mathbf{0}$, we verify that

$$(\mathbf{I}_4 - \mathbf{A})^{-1} = \mathbf{I}_4 + \mathbf{A} + \mathbf{A}^2 + \mathbf{A}^3$$

by showing that

$$(\mathbf{I}_4 - \mathbf{A})(\mathbf{I}_4 + \mathbf{A} + \mathbf{A}^2 + \mathbf{A}^3) = (\mathbf{I}_4 + \mathbf{A} + \mathbf{A}^2 + \mathbf{A}^3)(\mathbf{I}_4 - \mathbf{A}) = \mathbf{I}_4.$$

We have, $(\mathbf{I}_4 - \mathbf{A})(\mathbf{I}_4 + \mathbf{A} + \mathbf{A}^2 + \mathbf{A}^3) =$

$$\mathbf{I}_4 + \mathbf{A} + \mathbf{A}^2 + \mathbf{A}^3 - \mathbf{A} - \mathbf{A}^2 - \mathbf{A}^3 - \mathbf{A}^4 = \mathbf{I}_4$$

Similarly, $(\mathbf{I}_4 + \mathbf{A} + \mathbf{A}^2 + \mathbf{A}^3)(\mathbf{I}_4 - \mathbf{A}) =$

$$\mathbf{I}_4 - \mathbf{A} + \mathbf{A} - \mathbf{A}^2 + \mathbf{A}^2 - \mathbf{A}^3 + \mathbf{A}^3 - \mathbf{A}^4 = \mathbf{I}_4.$$

21. Form the augmented matrix $\left[\begin{array}{cc|c} 1 & 1 & 3 \\ 5 & 5 & a \end{array}\right]$. Applying row operation -5 times row 1 added to row 2 gives the equivalent linear system

$$\left[\begin{array}{cc|c} 1 & 1 & 3 \\ 0 & 0 & a-15 \end{array}\right].$$

(a) For $a \neq 15$, the linear system is inconsistent, hence has no solution.

(b) There are no values of a such that the system has exactly one solution.

(c) There are infinitely many solutions for $a = 15$.

23. The coefficent matrix is $\begin{bmatrix} 1-a & 0 & 1 \\ 0 & -a & 1 \\ 0 & 1 & -a \end{bmatrix}$. Applying row operation a times row 3 added to row 2 gives the matrix

$$\begin{bmatrix} 1-a & 0 & 0 \\ 0 & 0 & 1-a^2 \\ 0 & 1 & -a \end{bmatrix}.$$

The corresponding homogeneous system will have a nontrivial solution provided $1-a^2 = 0$; that is, for $a = 1$ or $a = -1$.

25. (a) For $k = 1$, $\mathbf{B} = \begin{bmatrix} b_1 \\ 0 \end{bmatrix}$. ($b_1$ arbitrary.)

For $k = 2$, $\mathbf{B} = \begin{bmatrix} b_{11} & b_{12} \\ 0 & 0 \end{bmatrix}$. ($b_{11}$ and b_{12} arbitrary.)

For $k = 3$, $\mathbf{B} = \begin{bmatrix} b_{11} & b_{12} & b_{13} \\ 0 & 0 & 0 \end{bmatrix}$. ($b_{11}$, b_{12}, and b_{13} arbitrary.)

For $k = 4$, $\mathbf{B} = \begin{bmatrix} b_{11} & b_{12} & b_{13} & b_{14} \\ 0 & 0 & 0 & 0 \end{bmatrix}$. ($b_{11}$, b_{12}, b_{13}, and b_{14} arbitrary.)

(b) The answers are not unique. The only requirement is that row 2 of **B** have all zero entries.

27. (a) Let $\mathbf{A} = \begin{bmatrix} a & b \\ c & d \end{bmatrix}$. Then

$$\mathbf{A}^2 = \begin{bmatrix} a^2+bc & ab+bd \\ ac+dc & bc+d^2 \end{bmatrix} = \begin{bmatrix} 1 & 1 \\ 0 & 1 \end{bmatrix} = \mathbf{B}.$$

Equating corresponding entries we have

$$\begin{array}{ll} a^2 + bc = 1 & b(a + d) = 1 \\ c(a + d) = 0 & bc + d^2 = 1. \end{array}$$

It follows that $a + d \neq 0$ and $c = 0$. Hence we have

$$a^2 = 1, \qquad b = \frac{1}{a + d}, \qquad d^2 = 1.$$

One solution is

$$\mathbf{A} = \begin{bmatrix} 1 & \frac{1}{2} \\ 0 & 1 \end{bmatrix} \text{ and another is } \mathbf{A} = \begin{bmatrix} -1 & \frac{-1}{2} \\ 0 & -1 \end{bmatrix}.$$

(b) One solution is $\mathbf{A} = \mathbf{B}$.

(c) One solution is $\mathbf{A} = \mathbf{I}_4$.

(d) Let $\mathbf{A} = \begin{bmatrix} a & b \\ c & d \end{bmatrix}$. Then

$$\mathbf{A}^2 = \begin{bmatrix} a^2+bc & ab+bd \\ ac+dc & bc+d^2 \end{bmatrix} = \begin{bmatrix} 0 & 1 \\ 0 & 0 \end{bmatrix} = \mathbf{B}.$$

Equating corresponding entries we have

$$\begin{array}{ll} a^2 + bc = 0 & b(a + d) = 1 \\ c(a + d) = 0 & bc + d^2 = 0. \end{array}$$

It follows that $a + d \neq 0$ and $c = 0$. Thus

$$\mathbf{A}^2 = \begin{bmatrix} a^2 & b(a+d) \\ 0 & d^2 \end{bmatrix} = \begin{bmatrix} 0 & 1 \\ 0 & 0 \end{bmatrix}.$$

Again equating corresponding entries, we have

$$a = 0, \qquad b = \frac{1}{a+d}, \qquad d = 0.$$

Hence, $a = d = 0$, which is a contradiction. $\mathbf{B}$ has no square root.

29. Inspect the sequence $\mathbf{A}^2$, $\mathbf{A}^3$, $\mathbf{A}^4$, $\mathbf{A}^5$, and higher powers if need be, to discover a pattern. We have

$$\mathbf{A}^2 = \begin{bmatrix} 1 & 3/4 \\ 0 & 1/4 \end{bmatrix}, \quad \mathbf{A}^3 = \begin{bmatrix} 1 & 7/8 \\ 0 & 1/8 \end{bmatrix},$$

$$\mathbf{A}^4 = \begin{bmatrix} 1 & 15/16 \\ 0 & 1/16 \end{bmatrix}, \quad \mathbf{A}^5 = \begin{bmatrix} 1 & 31/32 \\ 0 & 1/32 \end{bmatrix}$$

It appears that $\mathbf{A}^n = \begin{bmatrix} 1 & (2^n-1)/2^n \\ 0 & 1/2^n \end{bmatrix}$.

T.1. (a) $\operatorname{Tr}(c\mathbf{A}) = \sum_{i=1}^{n} ca_{ii} = c\sum_{i=1}^{n} a_{ii} = c\operatorname{Tr}(\mathbf{A})$

(b) $\operatorname{Tr}(\mathbf{A} + \mathbf{B}) = \sum_{i=1}^{n}(a_{ii} + b_{ii}) = \sum_{i=1}^{n} a_{ii} + \sum_{i=1}^{n} b_{ii}$

$= \operatorname{Tr}(\mathbf{A}) + \operatorname{Tr}(\mathbf{B})$

(c) Let $\mathbf{AB} = \mathbf{C} = (c_{ij})$. Then

$$\operatorname{Tr}(\mathbf{AB}) = \operatorname{Tr}(\mathbf{C}) = \sum_{i=1}^{n} c_{ii} = \sum_{i=1}^{n}\sum_{k=1}^{n} a_{ik}b_{ki}$$

$$= \sum_{k=1}^{n}\sum_{i=1}^{n} b_{ki}a_{ik} = \operatorname{Tr}(\mathbf{BA})$$

(d) $\operatorname{Tr}(\mathbf{A}^T) = \sum_{i=1}^{n} a_{ii} = \operatorname{Tr}(\mathbf{A})$

(e) $\operatorname{Tr}(\mathbf{A}^T\mathbf{A})$ = sum of the diagonal entries of $\mathbf{A}^T\mathbf{A}$

$$= \sum_{i=1}^{n} \operatorname{row}_i(\mathbf{A}^T)\operatorname{col}_i(\mathbf{A}) = \sum_{i=1}^{n} [\operatorname{col}_i(\mathbf{A})]^T \operatorname{col}_i(\mathbf{A})$$

$$= \sum_{i=1}^{n} (a_{1i}{}^2 + a_{2i}{}^2 + \cdots + a_{ni}{}^2) \ge 0$$

T.3. If $\mathbf{AX} = \mathbf{0}$ for all $n \times 1$ matrices $\mathbf{X}$, then $\mathbf{AE}_j = \mathbf{0}$, $j=1,2,\ldots,n$, where $\mathbf{E}_j$ = column j of $\mathbf{I}_n$. But then

$$\mathbf{AE}_j = \begin{bmatrix} a_{1j} \\ a_{2j} \\ \cdot \\ \cdot \\ a_{nj} \end{bmatrix} = \mathbf{0}.$$

Hence column j of $\mathbf{A} = \mathbf{0}$ for each j and it follows that $\mathbf{A} = \mathbf{O}$.

T.5. If **AX** = **BX** for all n × 1 matrices **X**, then $\mathbf{AE_j} = \mathbf{BE_j}$, j = 1,2,...,n, where $\mathbf{E_j}$ = column j of $\mathbf{I_n}$. But then

$$\mathbf{AE_j} = \begin{bmatrix} a_{ij} \\ a_{2j} \\ \cdot \\ \cdot \\ a_{nj} \end{bmatrix} = \mathbf{BE_j} = \begin{bmatrix} b_{ij} \\ b_{2j} \\ \cdot \\ \cdot \\ b_{nj} \end{bmatrix}.$$

Hence column j of **A** = column j of **B** for each j and it follows that **A** = **B**.

T.7. $(\mathbf{A}^k)^T = (\underbrace{\mathbf{A}\cdot\mathbf{A}\cdots\mathbf{A}}_{k\text{-times}})^T = \underbrace{\mathbf{A}^T\cdot\mathbf{A}^T\cdots\mathbf{A}^T}_{k\text{-times}} = (\mathbf{A}^T)^k$

T.9. Assume that **A** is upper triangular. If **A** is nonsingular then **A** is row equivalent to $\mathbf{I_n}$. Since **A** is upper triangular this can occur only if $a_{ii} \neq 0$ because in the reduction process we must perform the row operations

$$\frac{1}{a_{ii}} * \text{ ith row of } \mathbf{A}.$$

The steps are reversible.

T.11. Let **A** and **B** be row equivalent n × n matrices. Then there exists a finite number of row operations which when applied to **A** yield **B** and vice versa. If **B** is nonsingular, then **B** is row equivalent to $\mathbf{I_n}$. Thus **A** is also row equivalent to $\mathbf{I_n}$, hence **A** is nonsingular. We repeat the argument with **A** and **B** interchanged to prove the converse.

T.13. If **A** is skew symmetric, $\mathbf{A}^T = -\mathbf{A}$. Thus $a_{ii} = -a_{ii}$, so $a_{ii} = 0$.

T.15. (a) $\mathbf{I_n}^2 = \mathbf{I_n}$ and $\mathbf{O_n}^2 = \mathbf{O_n}$.

(b) One such matrix is $\begin{bmatrix} 0 & 0 \\ 0 & 1 \end{bmatrix}$.

(c) If $\mathbf{A}^2 = \mathbf{A}$ and $\mathbf{A}^{-1}$ exists, then $\mathbf{A}^{-1}(\mathbf{A}^2) = \mathbf{A}^{-1}\mathbf{A}$. Simplifying gives $\mathbf{A} = \mathbf{I_n}$.

(d) To find the values of k so that k**A** is idempotent when **A** is idempotent we proceed as follows.

$$\begin{aligned} k\mathbf{A} &= (k\mathbf{A})^2 && \{\text{since } k\mathbf{A} \text{ is to be idempotent}\} \\ &= k^2\mathbf{A}^2 && \{\text{using properties of matrix algebra}\} \\ &= k^2\mathbf{A} && \{\text{since } \mathbf{A} \text{ is idempotent}\} \end{aligned}$$

This equality is equivalent to

$$k^2\mathbf{A} - k\mathbf{A} = (k^2 - k)\mathbf{A} = \mathbf{0}$$

For $\mathbf{A} \neq \mathbf{0}$, $k^2 - k = 0$ provided $k = 0$, or $k = 1$. If $\mathbf{A} = \mathbf{0}$, then k can be any value.

T.17. We have that $\mathbf{A}^2 = \mathbf{A}$ and $\mathbf{B}^2 = \mathbf{B}$.

(a) $(\mathbf{AB})^2 = \mathbf{ABAB} = \mathbf{A}(\mathbf{BA})\mathbf{B} = \mathbf{A}(\mathbf{AB})\mathbf{B}$ {since $\mathbf{AB} = \mathbf{BA}$}
$= \mathbf{A}^2\mathbf{B}^2 = \mathbf{AB}$ {since $\mathbf{A}$ and $\mathbf{B}$ are idempotent}

(b) $(\mathbf{A}^T)^2 = \mathbf{A}^T\mathbf{A}^T = (\mathbf{AA})^T$ {by properties of the transpose}
$= (\mathbf{A}^2)^T = \mathbf{A}^T$ {since $\mathbf{A}$ is idempotent}

(c) If $\mathbf{A}$ and $\mathbf{B}$ are $n \times n$ and idempotent, then $\mathbf{A} + \mathbf{B}$ need not be idempotent. For example, let

$$\mathbf{A} = \begin{bmatrix} 1 & 1 \\ 0 & 0 \end{bmatrix} \text{ and } \mathbf{B} = \begin{bmatrix} 0 & 0 \\ 1 & 1 \end{bmatrix}.$$

Both $\mathbf{A}$ and $\mathbf{B}$ are idempotent and $\mathbf{C} = \mathbf{A} + \mathbf{B} = \begin{bmatrix} 1 & 1 \\ 1 & 1 \end{bmatrix}$.

However, $\mathbf{C}^2 = \begin{bmatrix} 2 & 2 \\ 2 & 2 \end{bmatrix} \neq \mathbf{C}$.

T.19. Assume $\mathbf{A}$ is nonsingular, then $\mathbf{A}^{-1}$ exists. Hence we can multiply $\mathbf{AX} = \mathbf{B}$ by $\mathbf{A}^{-1}$ on the left on both sides, obtaining

$$\mathbf{A}^{-1}(\mathbf{AX}) = \mathbf{A}^{-1}\mathbf{B}$$

or

$$\mathbf{I}_n\mathbf{X} = \mathbf{X} = \mathbf{A}^{-1}\mathbf{B}.$$

Thus $\mathbf{AX} = \mathbf{B}$ has a unique solution.

Assume $\mathbf{AX} = \mathbf{B}$ has a unique solution for every $\mathbf{B}$. Since $\mathbf{AX} = \mathbf{0}$ has a solution $\mathbf{X} = \mathbf{0}$, Theorem 1.12 implies that $\mathbf{A}$ is nonsingular.

T.21. It is not true that $\mathbf{XY}^T$ must be equal to $\mathbf{YX}^T$. For example, let $\mathbf{X} = \begin{bmatrix} 1 \\ 2 \end{bmatrix}$ and $\mathbf{Y} = \begin{bmatrix} 4 \\ 5 \end{bmatrix}$. Then $\mathbf{XY}^T = \begin{bmatrix} 4 & 5 \\ 8 & 10 \end{bmatrix}$ and

$$\mathbf{YX}^T = \begin{bmatrix} 4 & 8 \\ 5 & 10 \end{bmatrix}.$$

T.23. Prove: $\mathbf{XY}^T$ is row equivalent to $\mathbf{O}_n$ or a matrix with n-1 zero rows.

<u>Proof:</u> The outer product of $\mathbf{X}$ and $\mathbf{Y}$ can be written in the form

$$\mathbf{XY}^T = \begin{bmatrix} x_1[y_1 \ y_2 \ \cdots \ y_n] \\ x_2[y_1 \ y_2 \ \cdots \ y_n] \\ \cdot \\ \cdot \\ x_n[y_1 \ y_2 \ \cdots \ y_n] \end{bmatrix}$$

If either $\mathbf{X} = \mathbf{O}$ or $\mathbf{Y} = \mathbf{O}$, then $\mathbf{XY}^T = \mathbf{O}_n$. Thus assume that there is at least one nonzero component in $\mathbf{X}$, say x_i, and at least one nonzero component in $\mathbf{Y}$, say y_j. Then $(1/x_i)\text{Row}_i(\mathbf{XY}^T)$ makes the ith row exactly $\mathbf{Y}^T$. Since all the the other rows are multiples of $\mathbf{Y}^T$, row operations of the form $-x_kR_i + R_p$, for $p \neq i$, can be performed to zero out everything but the ith row. It follows that either $\mathbf{XY}^T$ is row equivalent to $\mathbf{O}_n$ or a matrix with n-1 zero rows.

T.25. Let $\mathbf{B} = \begin{bmatrix} a & b \\ c & d \end{bmatrix}$. Then $\mathbf{AB} = \mathbf{BA}$ implies that

$$\begin{bmatrix} 2 & 0 \\ -1 & 1 \end{bmatrix} \begin{bmatrix} a & b \\ c & d \end{bmatrix} = \begin{bmatrix} a & b \\ c & d \end{bmatrix} \begin{bmatrix} 2 & 0 \\ -1 & 1 \end{bmatrix}.$$

Performing the multiplication we have

$$\begin{bmatrix} 2a & 2b \\ -a+c & -b+d \end{bmatrix} = \begin{bmatrix} 2a-b & b \\ 2c-d & d \end{bmatrix}$$

and equating corresponding entries gives us the equations $2a = 2a - b$, $2b = b$, $-a + c = 2c - d$, $-b + d = d$. Solving these equations we have that

$$a = r,\ b = 0,\ d = s,\ c = d - a = s - r,$$

where r and s are any real numbers. Thus $\mathbf{B} = \begin{bmatrix} r & 0 \\ s-r & s \end{bmatrix}$.

Exercises 2.1

1. S = {1, 2, 3, 4, 5}
   (a) Permutation 52134 has 5 inversions: 52, 51, 53, 54, 21.

   (b) Permutation 45213 has 7 inversions: 42, 41, 43, 52, 51, 53, 21.

   (c) Permutation 42135 has 4 inversions: 42, 41, 43, 21.

   (d) Permutation 13542 has 4 inversions: 32, 54, 52, 42.

   (e) Permutation 35241 has 7 inversions: 32, 31, 52, 54, 51, 21, 41.

   (f) Permutation 12345 has no inversions.

3. S = {1, 2, 3, 4, 5}
   (a) Since 25431 is an odd permutation (seven inversions: 21, 54, 53, 51, 43, 41, 31) a - sign is associated with it.

   (b) Since 31245 is an even permutation (two inversions: 31, 32) a + sign is associated with it.

   (c) Since 21345 is an odd permutation (one inversion: 21) a - sign is associated with it.

   (d) Since 52341 is an odd permutation (seven inversions: 52, 53, 54, 51, 21, 31, 41) a - sign is associated with it.

   (e) Since 34125 is an even permutation (four inversions: 31, 32, 41, 42) a + sign is associated with it.

   (f) Since 41253 is an even permutation (four inversions; 41, 42, 43, 53) a + sign is associated with it.

5. (a) $\begin{vmatrix} 2 & -1 \\ 3 & 2 \end{vmatrix} = (2)(2)-(-1)(3) = 7$ as in Example 5.

   (b) $\begin{vmatrix} 0 & 3 & 0 \\ 2 & 0 & 0 \\ 0 & 0 & -5 \end{vmatrix} = (0)(0)(-5) + (3)(0)(0) + (0)(2)(0)$

   $- (0)(0)(0) - (3)(2)(-5) - (0)(0)(0) = 30$
   as in Example 6. (See Equation (3).)

   (c) $\begin{vmatrix} 4 & 2 & 0 \\ 0 & -2 & 5 \\ 0 & 0 & 3 \end{vmatrix} = (4)(-2)(3) + (2)(5)(0) + (0)(0)(0)$

   $- (4)(5)(0) -(2)(0)(3) - (0)(-2)(0) = -24$
   as in Example 6. (See Equation (3).)

(d) $$\begin{vmatrix} 4 & 2 & 2 & 0 \\ 2 & 0 & 0 & 0 \\ 3 & 0 & 0 & 1 \\ 0 & 0 & 1 & 0 \end{vmatrix} = (4)(0)(0)(0) + (4)(0)(0)(1) + (4)(0)(1)(0)$$

$$\begin{aligned}
&+ (2)(2)(1)(1) + (2)(0)(3)(0) + (2)(0)(0)(0) + (2)(2)(0)(0) \\
&+ (2)(0)(1)(0) + (2)(0)(3)(0) + (0)(2)(0)(0) + (0)(0)(0)(1) \\
&+ (0)(0)(0)(0) - (4)(0)(1)(1) - (4)(0)(0)(0) - (4)(0)(0)(0) \\
&- (2)(2)(0)(0) - (2)(0)(1)(0) - (2)(0)(3)(1) - (2)(2)(1)(0) \\
&- (2)(0)(3)(0) - (2)(0)(0)(0) - (0)(0)(0)(1) - (0)(0)(0)(0) \\
&- (0)(0)(3)(0) = 4
\end{aligned}$$

7. Let $A = [a_{ij}]$ be a 4×4 matrix. Using Equation (2) we have

$$\det(A) = \begin{vmatrix} a_{11} & a_{12} & a_{13} & a_{14} \\ a_{21} & a_{22} & a_{23} & a_{24} \\ a_{31} & a_{32} & a_{33} & a_{34} \\ a_{41} & a_{42} & a_{43} & a_{44} \end{vmatrix}$$

$$\begin{aligned}
&= a_{11}a_{22}a_{33}a_{44} + a_{11}a_{24}a_{32}a_{43} + a_{11}a_{23}a_{34}a_{42} + a_{12}a_{21}a_{34}a_{43} \\
&+ a_{12}a_{23}a_{31}a_{44} + a_{12}a_{24}a_{33}a_{41} + a_{13}a_{21}a_{32}a_{44} + a_{13}a_{22}a_{34}a_{41} \\
&+ a_{13}a_{24}a_{31}a_{42} + a_{14}a_{21}a_{33}a_{42} + a_{14}a_{22}a_{31}a_{43} + a_{14}a_{23}a_{32}a_{41} \\
&- a_{11}a_{22}a_{34}a_{43} - a_{11}a_{23}a_{32}a_{44} - a_{11}a_{24}a_{33}a_{42} - a_{12}a_{21}a_{33}a_{44} \\
&- a_{12}a_{23}a_{34}a_{41} - a_{12}a_{24}a_{31}a_{43} - a_{13}a_{21}a_{34}a_{42} - a_{13}a_{22}a_{31}a_{44} \\
&- a_{13}a_{24}a_{32}a_{41} - a_{14}a_{21}a_{32}a_{43} - a_{14}a_{22}a_{33}a_{41} - a_{14}a_{23}a_{31}a_{42}
\end{aligned}$$

To construct the preceding terms, take the set $S = \{1, 2, 3, 4\}$, write out the 24 permutations of S, and determine which are even and odd. Then use these permutations to fill in the blanks in 24 products of the form

$$a_{1\_}a_{2\_}a_{3\_}a_{4\_} \quad .$$

The sign of each term is determined by whether the permutation is even or odd.

9. Given that $|\mathbf{A}| = \begin{vmatrix} a_1 & a_2 & a_3 \\ b_1 & b_2 & b_3 \\ c_1 & c_2 & c_3 \end{vmatrix} = 3$. Matrix **B** is obtained from **A** by performing the row operations $2\mathbf{r}_2+\mathbf{r}_1\to\mathbf{r}_1$ and $-3\mathbf{r}_3+\mathbf{r}_1\to\mathbf{r}_1$. Thus by Theorem 2.6, $|\mathbf{B}| = |\mathbf{A}| = 3$. Matrix **C** is obtained from **A** by performing the operation $3\mathbf{c}_2\to\mathbf{c}_2$. Thus by Theorem 2.5, $|\mathbf{C}| = 3|\mathbf{A}| = 9$. Matrix **D** is obtained from **A** by performing the operation $\mathbf{r}_2\leftrightarrow\mathbf{r}_3$. Thus by Theorem 2.2, $|\mathbf{D}| = -|\mathbf{A}| = -3$.

11. (a) $\begin{vmatrix} \lambda-1 & 2 \\ 3 & \lambda-2 \end{vmatrix} = (\lambda-1)(\lambda-2) - (2)(3) = \lambda^2 - 3\lambda - 4$

(b) $|\lambda \mathbf{I}_2 - \mathbf{A}| = \begin{vmatrix} \lambda-4 & -2 \\ 1 & \lambda-1 \end{vmatrix} = (\lambda-4)(\lambda-1) - (-2)(1) = \lambda^2 - 5\lambda + 6$

13. (a) $\begin{vmatrix} \lambda-1 & 2 \\ 3 & \lambda-2 \end{vmatrix} = \lambda^2 - 3\lambda - 4 = 0$. Factoring, we have $\lambda^2 - 3\lambda - 4 = (\lambda-4)(\lambda+1) = 0$. Thus, $\lambda = 4$ or $\lambda = -1$.

(b) $|\lambda \mathbf{I}_2 - \mathbf{A}| = \begin{vmatrix} \lambda-4 & -2 \\ 1 & \lambda-1 \end{vmatrix} = \lambda^2 - 5\lambda + 6 = 0$. Factoring, we have $\lambda^2 - 5\lambda + 6 = (\lambda-3)(\lambda-2) = 0$. Thus, $\lambda = 3$ or $\lambda = 2$.

<<**Strategy:** In Exercises 15 and 17 we compute the determinant of 3×3 matrices using Equation (3) unless a theorem can be used to obtain the result more efficiently. For the 4×4 matrices, we use a sequence of row and column operations to obtain an upper triangular matrix as in Example 16. Many different sets of row and column operations can be used, we present one such set.>>

15. (a) $\begin{vmatrix} 4 & -3 & 5 \\ 5 & 2 & 0 \\ 2 & 0 & 4 \end{vmatrix} = (4)(2)(4) + (-3)(0)(2) + (5)(5)(0)$

$- (4)(0)(0) - (-3)(5)(4) - (5)(2)(2) = 72$.

(b) $\begin{vmatrix} 2 & 0 & 1 & 4 \\ 3 & 2 & -4 & -2 \\ 2 & 3 & -1 & 0 \\ 11 & 8 & -4 & 6 \end{vmatrix}_{\mathbf{c}_1 \leftrightarrow \mathbf{c}_4} = (-1)\begin{vmatrix} 4 & 0 & 1 & 2 \\ -2 & 2 & -4 & 3 \\ 0 & 3 & -1 & 2 \\ 6 & 8 & -4 & 11 \end{vmatrix}_{3\mathbf{r}_2+\mathbf{r}_4 \to \mathbf{r}_4}$

$= (-1)\begin{vmatrix} 4 & 0 & 1 & 2 \\ -2 & 2 & -4 & 3 \\ 0 & 3 & -1 & 2 \\ 0 & 14 & -16 & 20 \end{vmatrix}_{\mathbf{r}_1 \leftrightarrow \mathbf{r}_2} = \begin{vmatrix} -2 & 2 & -4 & 3 \\ 4 & 0 & 1 & 2 \\ 0 & 3 & -1 & 2 \\ 0 & 14 & -16 & 20 \end{vmatrix}_{2\mathbf{r}_1+\mathbf{r}_2 \to \mathbf{r}_2}$

$$= \begin{vmatrix} -2 & 2 & -4 & 3 \\ 0 & 4 & -7 & 8 \\ 0 & 3 & -1 & 2 \\ 0 & 14 & -16 & 20 \end{vmatrix}_{\frac{1}{4}r_2 \to r_2} = (4)\begin{vmatrix} -2 & 2 & -4 & 3 \\ 0 & 1 & \frac{-7}{4} & 2 \\ 0 & 3 & -1 & 2 \\ 0 & 14 & -16 & 20 \end{vmatrix}_{-3r_2+r_3 \to r_3}$$

$$=(4)\begin{vmatrix} -2 & 2 & -4 & 3 \\ 0 & 1 & \frac{-7}{4} & 2 \\ 0 & 0 & \frac{17}{4} & -4 \\ 0 & 14 & -16 & 20 \end{vmatrix}_{-14r_2+r_4 \to r_4} =(4)\begin{vmatrix} -2 & 2 & -4 & 3 \\ 0 & 1 & \frac{-7}{4} & 2 \\ 0 & 0 & \frac{17}{4} & -4 \\ 0 & 0 & \frac{17}{2} & -8 \end{vmatrix}_{\frac{1}{2}r_4 \to r_4}$$

$$=(2)(4)\begin{vmatrix} -2 & 2 & -4 & 3 \\ 0 & 1 & \frac{-7}{4} & 2 \\ 0 & 0 & \frac{17}{4} & -4 \\ 0 & 0 & \frac{17}{4} & -4 \end{vmatrix} =(4)(2)(0)=0 \text{ since two rows are equal.}$$

(c) $\begin{vmatrix} 4 & 1 & 2 \\ 0 & 2 & 3 \\ 0 & 0 & -3 \end{vmatrix} = (4)(2)(-3) = -24$ by Theorem 2.7.

17\. (a)

$$\begin{vmatrix} 4 & 2 & 3 & -4 \\ 3 & -2 & 1 & 5 \\ -2 & 0 & 1 & -3 \\ 8 & -2 & 6 & 4 \end{vmatrix} \frac{1}{4}\mathbf{r}_1 \to \mathbf{r}_1 = (4)\begin{vmatrix} 1 & \frac{1}{2} & \frac{3}{4} & -1 \\ 3 & -2 & 1 & 5 \\ -2 & 0 & 1 & -3 \\ 8 & -2 & 6 & 4 \end{vmatrix} \begin{matrix} -3\mathbf{r}_1+\mathbf{r}_2\to\mathbf{r}_2 \\ 2\mathbf{r}_1+\mathbf{r}_3\to\mathbf{r}_3 \\ -8\mathbf{r}_1+\mathbf{r}_4\to\mathbf{r}_4 \end{matrix}$$

$$= (4)\begin{vmatrix} 1 & \frac{1}{2} & \frac{3}{4} & -1 \\ 0 & \frac{-7}{2} & \frac{-5}{4} & 8 \\ 0 & 1 & \frac{5}{2} & -5 \\ 0 & -6 & 0 & 12 \end{vmatrix} \frac{-2}{7}\mathbf{r}_2\to\mathbf{r}_2 = (4)\left(\frac{-7}{2}\right)\begin{vmatrix} 1 & \frac{1}{2} & \frac{3}{4} & -1 \\ 0 & 1 & \frac{5}{14} & \frac{-16}{7} \\ 0 & 1 & \frac{5}{2} & -5 \\ 0 & -6 & 0 & 12 \end{vmatrix} \begin{matrix} -1\mathbf{r}_2+\mathbf{r}_3\to\mathbf{r}_3 \\ 6\mathbf{r}_2+\mathbf{r}_4\to\mathbf{r}_4 \end{matrix}$$

$$= (4)\left(\frac{-7}{2}\right)\begin{vmatrix} 1 & \frac{1}{2} & \frac{3}{4} & -1 \\ 0 & 1 & \frac{5}{14} & \frac{-16}{7} \\ 0 & 0 & \frac{15}{7} & \frac{-19}{7} \\ 0 & 0 & \frac{15}{7} & \frac{-12}{7} \end{vmatrix} -1\mathbf{r}_3+\mathbf{r}_4\to\mathbf{r}_4 = (4)\left(\frac{-7}{2}\right)\begin{vmatrix} 1 & \frac{1}{2} & \frac{3}{4} & -1 \\ 0 & 1 & \frac{5}{14} & \frac{-16}{7} \\ 0 & 0 & \frac{15}{7} & \frac{19}{7} \\ 0 & 0 & 0 & 1 \end{vmatrix}$$

$$= (4)\left(\frac{-7}{2}\right)(1)(1)\left(\frac{15}{7}\right)(1) = -30$$

(b)

$$\begin{vmatrix} 1 & 3 & -4 \\ -2 & 1 & 2 \\ -9 & 15 & 0 \end{vmatrix} \begin{matrix} 2\mathbf{r}_1+\mathbf{r}_2\to\mathbf{r}_2 \\ 9\mathbf{r}_1+\mathbf{r}_3\to\mathbf{r}_3 \end{matrix} = \begin{vmatrix} 1 & 3 & -4 \\ 0 & 7 & -6 \\ 0 & 42 & -36 \end{vmatrix} -6\mathbf{r}_2+\mathbf{r}_3\to\mathbf{r}_3$$

$$= \begin{vmatrix} 1 & 3 & -4 \\ 0 & 7 & -6 \\ 0 & 0 & 0 \end{vmatrix} = (1)(7)(0) = 0$$

(c)

$$\begin{vmatrix} 1 & 1 & 2 \\ 0 & 2 & -2 \\ 0 & 0 & 3 \end{vmatrix} = (1)(2)(3) = 6 \text{ by Theorem 2.7.}$$

Exercises 2.1

19. (a) $\det(\mathbf{A}) = \begin{vmatrix} 1 & -2 & 3 \\ -2 & 3 & 1 \\ 0 & 1 & 0 \end{vmatrix}_{2\mathbf{r}_1+\mathbf{r}_2\to\mathbf{r}_2} = \begin{vmatrix} 1 & -2 & 3 \\ 0 & -1 & 7 \\ 0 & 1 & 0 \end{vmatrix}_{\mathbf{r}_2+\mathbf{r}_3\to\mathbf{r}_3}$

$$= \begin{vmatrix} 1 & -2 & 3 \\ 0 & -1 & 7 \\ 0 & 0 & 7 \end{vmatrix} = (1)(-1)(7) = -7$$

$$\det(\mathbf{B}) = \begin{vmatrix} 1 & 0 & 2 \\ 3 & -2 & 5 \\ 2 & 1 & 3 \end{vmatrix}_{\substack{-3\mathbf{r}_1+\mathbf{r}_2\to\mathbf{r}_2 \\ -2\mathbf{r}_1+\mathbf{r}_3\to\mathbf{r}_3}} = \begin{vmatrix} 1 & 0 & 2 \\ 0 & -2 & -1 \\ 0 & 1 & -1 \end{vmatrix}_{\frac{1}{2}\mathbf{r}_2+\mathbf{r}_3\to\mathbf{r}_3}$$

$$= \begin{vmatrix} 1 & 0 & 2 \\ 0 & -2 & -1 \\ 0 & 0 & \frac{-3}{2} \end{vmatrix} = (1)(-2)(\frac{-3}{2}) = 3$$

$$\det(\mathbf{AB}) = \begin{vmatrix} 1 & 7 & 1 \\ 9 & -5 & 14 \\ 3 & -2 & 5 \end{vmatrix}_{\substack{-9\mathbf{r}_1+\mathbf{r}_2\to\mathbf{r}_2 \\ -3\mathbf{r}_1+\mathbf{r}_3\to\mathbf{r}_3}} = \begin{vmatrix} 1 & 7 & 1 \\ 0 & -68 & 5 \\ 0 & -23 & 2 \end{vmatrix}_{\frac{-23}{68}\mathbf{r}_2+\mathbf{r}_3\to\mathbf{r}_3}$$

$$= \begin{vmatrix} 1 & 7 & 1 \\ 0 & -68 & 5 \\ 0 & 0 & \frac{21}{68} \end{vmatrix} = (1)(-68)(\frac{21}{68}) = -21$$

Thus we see that $\det(\mathbf{A})\det(\mathbf{B}) = (-7)(3) = -21 = \det(\mathbf{AB})$.

(b) $\det(\mathbf{A}) = \begin{vmatrix} 2 & 3 & 6 \\ 0 & 3 & 2 \\ 0 & 0 & -4 \end{vmatrix} = (2)(3)(-4) = -24$

$$\det(\mathbf{B}) = \begin{vmatrix} 3 & 0 & 0 \\ 4 & 5 & 0 \\ 2 & 1 & -2 \end{vmatrix} = (3)(5)(-2) = -30$$

$$\det(\mathbf{AB}) = \begin{vmatrix} 30 & 21 & -12 \\ 16 & 17 & -4 \\ -8 & -4 & 8 \end{vmatrix}_{\frac{-1}{8}\mathbf{r}_3 \to \mathbf{r}_3} = (-8)\begin{vmatrix} 30 & 21 & -12 \\ 16 & 17 & -4 \\ 1 & \frac{1}{2} & -1 \end{vmatrix}_{\mathbf{r}_1 \leftrightarrow \mathbf{r}_3}$$

$$=(-8)(-1)\begin{vmatrix} 1 & \frac{1}{2} & -1 \\ 16 & 17 & -4 \\ 30 & 21 & -12 \end{vmatrix}_{\substack{-16\mathbf{r}_1+\mathbf{r}_2 \to \mathbf{r}_2 \\ -30\mathbf{r}_1+\mathbf{r}_3 \to \mathbf{r}_3}} =(8)\begin{vmatrix} 1 & \frac{1}{2} & -1 \\ 0 & 9 & 12 \\ 0 & 6 & 18 \end{vmatrix}_{\frac{-6}{9}\mathbf{r}_2+\mathbf{r}_3 \to \mathbf{r}_3}$$

$$=(8)\begin{vmatrix} 1 & \frac{1}{2} & -1 \\ 0 & 9 & 12 \\ 0 & 0 & 10 \end{vmatrix} = (8)(1)(9)(10) = 720$$

Thus we have $\det(\mathbf{A})\det(\mathbf{B}) = (-24)(-30) = 720 = \det(\mathbf{AB})$

21. Given, $|\mathbf{A}| = 2$ and $|\mathbf{B}| = -3$. Then

$$|\mathbf{A}^{-1}\mathbf{B}^T| = |\mathbf{A}^{-1}||\mathbf{B}^T| = \frac{1}{|\mathbf{A}|}\,|\mathbf{B}| = \frac{-3}{2}$$

T.1. If adjacent numbers j_i and j_{i+1} are interchanged, all inversions between numbers distinct from j_i and j_{i+1} remain unchanged, and all inversions between one of j_i, j_{i+1} and some other number also remain unchanged. If originally $j_i < j_{i+1}$, then after interchange there is one additional inversion due to $j_{i+1}j_i$. If originally $j_i > j_{i+1}$, then after interchange there is one fewer inversion.

Suppose j_p and j_q are separated by k intervening numbers. Then k interchanges of adjacent numbers will move j_p next to j_q. One interchange switches j_p and j_q. Finally, k interchanges of adjacent numbers takes j_q back to j_p's original position. The total number of interchanges is the odd number 2k+1.

T.3. By definition $c\mathbf{A} = [ca_{ij}]$. Thus each of the n rows of A have been multiplied by c. Applying Theorem 2.5 n times we have

$\det(c\mathbf{A}) = c^n\det(\mathbf{A})$.

T.5. By Theorem 2.8, $\det(\mathbf{AB}) = \det(\mathbf{A})\det(\mathbf{B})$. Thus if $\det(\mathbf{AB}) = 0$, then $\det(\mathbf{A})\det(\mathbf{B}) = 0$ and either $\det(\mathbf{A}) = 0$ or $\det(\mathbf{B}) = 0$.

T.7. From the definition of the determinant as,

$$\det(\mathbf{A}) = \Sigma(\pm)\ a_{1j_1}a_{2j_2}\cdots a_{nj_n}$$

there will be exactly one nonzero term. Thus, $\det(\mathbf{A}) \neq 0$.

T.9. (a) If $\mathbf{A} = \mathbf{A}^{-1}$, then $\mathbf{I}_n = \mathbf{AA}^{-1} = \mathbf{AA}$. Taking determinants we have

$$1 = \det(\mathbf{A})\det(\mathbf{A}) = \det(\mathbf{A})^2.$$

Thus $\det(\mathbf{A}) = \pm 1$.

(b) If $\mathbf{A}^T = \mathbf{A}^{-1}$, then $\mathbf{I}_n = \mathbf{AA}^{-1} = \mathbf{AA}^T$. Taking determinants and using $\det(\mathbf{A}^T) = \det(\mathbf{A})$ we have

$$1 = \det(\mathbf{A})\det(\mathbf{A}^T) = \det(\mathbf{A})\det(\mathbf{A}) = \det(\mathbf{A})^2.$$

Thus $\det(\mathbf{A}) = \pm 1$.

T.11. $\det(\mathbf{A}^T\mathbf{B}^T) = \det(\mathbf{A}^T)\det(\mathbf{B}^T) = \det(\mathbf{A})\det(\mathbf{B}^T) = \det(\mathbf{A}^T)\det(\mathbf{B})$. (We have used that the determinant of a matrix and its transpose are equal; Theorem 2.1.)

T.13. If $\mathbf{A}$ is nonsingular, by Corollary 2.1, $\det(\mathbf{A}) \neq 0$. By Theorem 2.7, $\det(\mathbf{A}) = a_{11}a_{22}\cdots a_{nn}$. Thus $a_{jj} \neq 0$ for $j = 1,2,\ldots,n$.

Conversely, if $a_{jj} \neq 0$ for $j = 1,2,\ldots,n$, then row operations $(1/a_{jj})\mathbf{r}_j \rightarrow \mathbf{r}_j$ give an equivalent matrix with 1's on the diagonal. It follows that there are row operations that will make $\mathbf{A}$ row equivalent to $\mathbf{I}_n$. Thus by Theorem 1.11 $\mathbf{A}$ is nonsingular.

T.15. If $\mathbf{A}^n = \mathbf{0}$ for some positive integer n, then $\det(\mathbf{A}^n) = \det(\mathbf{0}) = 0$. Repeated application of Theorem 2.8 gives $\det(\mathbf{A}^n) = \det(\mathbf{A})^n$. Thus $\det(\mathbf{A})^n = 0$. The real number whose nth power is zero is zero. Hence $\det(\mathbf{A}) = 0$.

ML.1. There a many sequences of row operations that can be used. Here we record the value of the determinant so you may check your result. (a) $\det(\mathbf{A}) = -18$ (b) $\det(\mathbf{A}) = 5$

ML.2. There a many sequences of row operations that can be used.

Here we record the value of the determinant so you may check your result. (a) det(**A**) = -9 (b) det(**A**) = 5

ML.3. (a)
```
A=[1 -1 1;1 1 -1;-1 1 1];
det(A)

ans =

     4
```

(b)
```
A=[1 2 3 4;2 3 4 5;3 4 5 6;4 5 6 7];
det(A)

ans =

     0
```

ML.4. (a)
```
A=[2 3 0;4 1 0;0 0 5];
det(5*eye(A)-A)

ans =

     0
```

(b)
```
A=[1 1;5 2];
det((3*eye(A)-A)^2)

ans =

     9
```

(c)
```
A=[1 1 0;0 1 0;1 0 1];
det(inverse(A)*A)

ans =

     1
```

ML.5.
```
A=[5 2;-1 2];
t=1;
det(t*eye(A)-A)

ans =

     6
t=2;
det(t*eye(A)-A)

ans =

     2
```

Exercises 2.1

```
t=3;
det(t*eye(A)-A)

ans =

     0
```

Exercises 2.2

1. $\mathbf{A} = \begin{bmatrix} 1 & 0 & -2 \\ 3 & 1 & 4 \\ 5 & 2 & -3 \end{bmatrix}$. The cofactors of **A** are:

$$A_{11} = (-1)^2\begin{vmatrix} 1 & 4 \\ 2 & -3 \end{vmatrix} = -11, \quad A_{12} = (-1)^3\begin{vmatrix} 3 & 4 \\ 5 & -3 \end{vmatrix} = 29,$$

$$A_{13} = (-1)^4\begin{vmatrix} 3 & 1 \\ 5 & 2 \end{vmatrix} = 1, \quad A_{21} = (-1)^3\begin{vmatrix} 0 & -2 \\ 2 & -3 \end{vmatrix} = -4,$$

$$A_{22} = (-1)^4\begin{vmatrix} 1 & -2 \\ 5 & -3 \end{vmatrix} = 7, \quad A_{23} = (-1)^5\begin{vmatrix} 1 & 0 \\ 5 & 2 \end{vmatrix} = -2,$$

$$A_{31} = (-1)^4\begin{vmatrix} 0 & -2 \\ 1 & 4 \end{vmatrix} = 2, \quad A_{32} = (-1)^5\begin{vmatrix} 1 & -2 \\ 3 & 4 \end{vmatrix} = -10,$$

$$A_{33} = (-1)^6\begin{vmatrix} 1 & 0 \\ 3 & 1 \end{vmatrix} = 1 .$$

<<**Strategy:** To evaluate the determinants in Exercises 3 and 5 we use expansion by cofactors as in Theorem 2.9. To reduce the work involved, we choose a row or column to expand about that has one or more zeros, if possible. We present one approach for the computations. Many others are possible.>>

3. (a) $\mathbf{A} = \begin{bmatrix} 1 & 2 & 3 \\ -1 & 5 & 2 \\ 3 & 2 & 0 \end{bmatrix}$. Since row 3 contains a zero, we use expansion by cofactors about row 3. Thus we need only cofactors A_{31} and A_{32}.

$$A_{31} = (-1)^4\begin{vmatrix} 2 & 3 \\ 5 & 2 \end{vmatrix} = -11, \quad A_{32} = (-1)^5\begin{vmatrix} 1 & 3 \\ -1 & 2 \end{vmatrix} = -5$$

Then $\det(\mathbf{A}) = a_{31}A_{31} + a_{32}A_{32} + a_{33}A_{33}$

$= 3(-11) + 2(-5) + 0(A_{33}) = -43$.

(b) $\mathbf{A} = \begin{bmatrix} 4 & -4 & 2 & 1 \\ 1 & 2 & 0 & 3 \\ 2 & 0 & 3 & 4 \\ 0 & -3 & 2 & 1 \end{bmatrix}$. Since column 1 contains a zero, we use expansion by cofactors about column 1. Thus we need only cofactors A_{11}, A_{21}, and A_{31}.

$$A_{11} = (-1)^2\begin{vmatrix} 2 & 0 & 3 \\ 0 & 3 & 4 \\ -3 & 2 & 1 \end{vmatrix}.$$ We evaluate this determinant using expansion by cofactors about the first row.

$$A_{11} = (-1)^2\left[(-1)^2(2)\begin{vmatrix} 3 & 4 \\ 2 & 1 \end{vmatrix} + (-1)^4(3)\begin{vmatrix} 0 & 3 \\ -3 & 2 \end{vmatrix}\right]$$

$$= (2)(-5)+(3)(9) = 17 .$$

$$A_{21} = (-1)^3\begin{vmatrix} -4 & 2 & 1 \\ 0 & 3 & 4 \\ -3 & 2 & 1 \end{vmatrix}.$$ We evaluate this determinant using expansion by cofactors about the first column.

$$A_{21} = (-1)^3\left[(-1)^2(-4)\begin{vmatrix} 3 & 4 \\ 2 & 1 \end{vmatrix} + (-1)^4(-3)\begin{vmatrix} 2 & 1 \\ 3 & 4 \end{vmatrix}\right]$$

$$= (-1)[(-4)(-5) + (-3)(5)] = -5 .$$

$$A_{31} = (-1)^4\begin{vmatrix} -4 & 2 & 1 \\ 2 & 0 & 3 \\ -3 & 2 & 1 \end{vmatrix}.$$ We evaluate this determinant using expansion by cofactors about the second row.

$$A_{31} = (-1)^4\left[(-1)^3(2)\begin{vmatrix} 2 & 1 \\ 2 & 1 \end{vmatrix} + (-1)^5(3)\begin{vmatrix} -4 & 2 \\ -3 & 2 \end{vmatrix}\right]$$

$$= (-1)^4[(-1)^3(2)(0) + (-1)^5(3)(-2)] = 6 .$$

Then $\det(\mathbf{A}) = a_{11}A_{11} + a_{21}A_{21} + a_{31}A_{31} + a_{41}A_{41}$

$$= (4)(17) + (1)(-5) + (2)(6) + (0)A_{41} = 75 .$$

(c) $\mathbf{A} = \begin{bmatrix} 4 & -2 & 0 \\ 0 & 2 & 4 \\ -1 & -1 & -3 \end{bmatrix}$. Since row 1 contains a zero we use expansion by cofactors about row 1. Thus we need only cofactors A_{11} and A_{12}.

$$A_{11} = (-1)^2\begin{vmatrix} 2 & 4 \\ -1 & -3 \end{vmatrix} = -2, \quad A_{12} = (-1)^3\begin{vmatrix} 0 & 4 \\ -1 & -3 \end{vmatrix} = -4 .$$

Then $\det(\mathbf{A}) = a_{11}A_{11} + a_{12}A_{12} + a_{13}A_{13}$

$$= (4)(-2) + (-2)(-4) + (0)A_{13} = 0 .$$

5. (a) $\mathbf{A} = \begin{bmatrix} 3 & 1 & 2 & -1 \\ 2 & 0 & 3 & -7 \\ 1 & 3 & 4 & -5 \\ 0 & -1 & 1 & -5 \end{bmatrix}$. Expand by cofactors about column 1.

We need only cofactors A_{11}, A_{21}, and A_{31}.

$A_{11} = (-1)^2 \begin{vmatrix} 0 & 3 & -7 \\ 3 & 4 & -5 \\ -1 & 1 & -5 \end{vmatrix}$. Expand by cofactors about row 1.

$$A_{11} = (-1)^2 \left[(-1)^3(3) \begin{vmatrix} 3 & -5 \\ -1 & -5 \end{vmatrix} + (-1)^4(-7) \begin{vmatrix} 3 & 4 \\ -1 & 1 \end{vmatrix} \right]$$

$$= (-1)(3)(-20) + (-7)(7) = 11 .$$

$A_{21} = (-1)^3 \begin{vmatrix} 1 & 2 & -1 \\ 3 & 4 & -5 \\ -1 & 1 & -5 \end{vmatrix}$. Expand by cofactors about row 1.

$$A_{21} = (-1)^3 \left[(-1)^2(1) \begin{vmatrix} 4 & -5 \\ 1 & -5 \end{vmatrix} + (-1)^3(2) \begin{vmatrix} 3 & -5 \\ -1 & -5 \end{vmatrix} + (-1)^4(-1) \begin{vmatrix} 3 & 4 \\ -1 & 1 \end{vmatrix} \right]$$

$$= (-1)[(-15) + (-2)(-20) + (-7)] = -18 .$$

$A_{31} = (-1)^4 \begin{vmatrix} 1 & 2 & -1 \\ 0 & 3 & -7 \\ -1 & 1 & -5 \end{vmatrix}$. Expand by cofactors about column 1.

$$A_{31} = (-1)^4 \left[(-1)^2(1) \begin{vmatrix} 3 & -7 \\ 1 & -5 \end{vmatrix} + (-1)^4(-1) \begin{vmatrix} 2 & -1 \\ 3 & -7 \end{vmatrix} \right] = -8 + 11 = 3.$$

Then $\det(\mathbf{A}) = a_{11}A_{11} + a_{21}A_{21} + a_{31}A_{31} + a_{41}A_{41}$

$= (3)(11) + (2)(-18) + (1)(3) + (0)A_{41} = 0$.

(b) $\mathbf{A} = \begin{bmatrix} 3 & 1 & 0 \\ 3 & 2 & 1 \\ 0 & 1 & -1 \end{bmatrix}$. Expand by cofactors about row 1.

Thus we need only cofactors A_{11} and A_{12}.

$A_{11} = (-1)^2 \begin{vmatrix} 2 & 1 \\ 1 & -1 \end{vmatrix} = -3$, $A_{12} = (-1)^3 \begin{vmatrix} 3 & 1 \\ 0 & -1 \end{vmatrix} = 3$

Then $\det(\mathbf{A}) = a_{11}A_{11} + a_{12}A_{12} + a_{13}A_{13}$
$= (3)(-3) + (1)(3) + (0)A_{13} = -6$.

(c) $\mathbf{A} = \begin{bmatrix} 3 & -3 & 0 \\ 2 & 0 & 2 \\ 2 & 1 & -3 \end{bmatrix}$. Expand by cofactors about row 1.

Thus we need only cofactors A_{11} and A_{12}.

$A_{11} = (-1)^2 \begin{vmatrix} 0 & 2 \\ 1 & -3 \end{vmatrix} = -2$, $A_{12} = (-1)^3 \begin{vmatrix} 2 & 2 \\ 2 & -3 \end{vmatrix} = 10$

Then $\det(\mathbf{A}) = a_{11}A_{11} + a_{12}A_{12} + a_{13}A_{13}$

$= (3)(-2) + (-3)(10) + (0)A_{13} = -36$.

7. $\mathbf{A} = \begin{bmatrix} -2 & 3 & 0 \\ 4 & 1 & -3 \\ 2 & 0 & 1 \end{bmatrix}$, then $A_{12} = (-1)^3 \begin{vmatrix} 4 & -3 \\ 2 & 1 \end{vmatrix} = -10$,

$A_{22} = (-1)^4 \begin{vmatrix} -2 & 0 \\ 2 & 1 \end{vmatrix} = -2$, and $A_{32} = (-1)^5 \begin{vmatrix} -2 & 0 \\ 4 & -3 \end{vmatrix} = -6$. Thus we have $a_{11}A_{12} + a_{21}A_{22} + a_{31}A_{32} = (-2)(-10)+(4)(-2)+(2)(-6) = 0$.

9. $\mathbf{A} = \begin{bmatrix} 6 & 2 & 8 \\ -3 & 4 & 1 \\ 4 & -4 & 5 \end{bmatrix}$

(a) adj $\mathbf{A} = \begin{bmatrix} A_{11} & A_{21} & A_{31} \\ A_{12} & A_{22} & A_{32} \\ A_{13} & A_{23} & A_{33} \end{bmatrix}$ where $A_{11} = (-1)^2 \begin{vmatrix} 4 & 1 \\ -4 & 5 \end{vmatrix} = 24$,

$A_{12} = (-1)^3 \begin{vmatrix} -3 & 1 \\ 4 & 5 \end{vmatrix} = 19$, $A_{13} = (-1)^4 \begin{vmatrix} -3 & 4 \\ 4 & -4 \end{vmatrix} = -4$,

$A_{21} = (-1)^3 \begin{vmatrix} 2 & 8 \\ -4 & 5 \end{vmatrix} = -42$, $A_{22} = (-1)^4 \begin{vmatrix} 6 & 8 \\ 4 & 5 \end{vmatrix} = -2$,

$A_{23} = (-1)^5 \begin{vmatrix} 6 & 2 \\ 4 & -4 \end{vmatrix} = 32$, $A_{31} = (-1)^4 \begin{vmatrix} 2 & 8 \\ 4 & 1 \end{vmatrix} = -30$,

$A_{32} = (-1)^5 \begin{vmatrix} 6 & 8 \\ -3 & 1 \end{vmatrix} = -30$, and $A_{33} = (-1)^6 \begin{vmatrix} 6 & 2 \\ -3 & 4 \end{vmatrix} = 30$.

Thus $\text{adj } \mathbf{A} = \begin{bmatrix} 24 & -42 & -30 \\ 19 & -2 & -30 \\ -4 & 32 & 30 \end{bmatrix}$.

(b) $\det(\mathbf{A}) = \begin{vmatrix} 6 & 2 & 8 \\ -3 & 4 & 1 \\ 4 & -4 & 5 \end{vmatrix}$. Expand by cofactors about row 1.

$$\det(\mathbf{A}) = (-1)^2(6)\begin{vmatrix} 4 & 1 \\ -4 & 5 \end{vmatrix} + (-1)^3(2)\begin{vmatrix} -3 & 1 \\ 4 & 5 \end{vmatrix} + (-1)^4(8)\begin{vmatrix} -3 & 4 \\ 4 & -4 \end{vmatrix} =$$

$$= (6)(24) + (-2)(-19) + (8)(-4) = 150 .$$

(c) $$(\text{adj } \mathbf{A})\mathbf{A} = \begin{bmatrix} 24 & -42 & -30 \\ 19 & -2 & -30 \\ -4 & 32 & 30 \end{bmatrix}\begin{bmatrix} 6 & 2 & 8 \\ -3 & 4 & 1 \\ 4 & -4 & 5 \end{bmatrix} = \begin{bmatrix} 150 & 0 & 0 \\ 0 & 150 & 0 \\ 0 & 0 & 150 \end{bmatrix}$$

$$= 150\ \mathbf{I}_3 = \det(\mathbf{A})\ \mathbf{I}_3$$

<<**Strategy:** In Exercises 11 and 13 we first compute $|\mathbf{A}|$. If $|\mathbf{A}| \neq 0$, then we compute $\mathbf{A}^{-1} = \dfrac{1}{\det(\mathbf{A})}\text{adj}(\mathbf{A})$.>>

11. (a) $\mathbf{A} = \begin{bmatrix} 1 & 2 & -3 \\ -4 & -5 & 2 \\ -1 & 1 & -7 \end{bmatrix}$. To find $\det(\mathbf{A})$ expand by cofactors about the first row. $\det(\mathbf{A}) = \begin{vmatrix} 1 & 2 & -3 \\ -4 & -5 & 2 \\ -1 & 1 & -7 \end{vmatrix} = (-1)^2(1)\begin{vmatrix} -5 & 2 \\ 1 & -7 \end{vmatrix}$

$+ (-1)^3(2)\begin{vmatrix} -4 & 2 \\ -1 & -7 \end{vmatrix} + (-1)^4(-3)\begin{vmatrix} -4 & -5 \\ -1 & 1 \end{vmatrix} = 33 - 60 + 27 = 0$

Thus A is singular and has no inverse.

(b) $\mathbf{A} = \begin{bmatrix} 2 & 3 \\ -1 & 2 \end{bmatrix}$. Then $\det(\mathbf{A}) = 7$ and the cofactors are, $A_{11} = (-1)^2(2) = 2$, $A_{12} = (-1)^3(-1) = 1$, $A_{21} = (-1)^3(3) = -3$, and $A_{22} = (-1)^4(2) = 2$. Hence

$$\text{adj } \mathbf{A} = \begin{bmatrix} 2 & -3 \\ 1 & 2 \end{bmatrix} \text{ and } \mathbf{A}^{-1} = \frac{1}{\det(\mathbf{A})}(\text{adj } \mathbf{A}) = \begin{bmatrix} \frac{2}{7} & \frac{-3}{7} \\ \frac{1}{7} & \frac{2}{7} \end{bmatrix}.$$

(c) $\mathbf{A} = \begin{bmatrix} 4 & 0 & 2 \\ 0 & 3 & 4 \\ 0 & 1 & -2 \end{bmatrix}$. To find $\det(\mathbf{A})$, expand by cofactors about column 1. $\det(\mathbf{A}) = (-1)^2(4)\begin{vmatrix} 3 & 4 \\ 1 & -2 \end{vmatrix} = (4)(-10) = -40.$

The cofactors are $A_{11} = (-1)^2\begin{vmatrix} 3 & 4 \\ 1 & -2 \end{vmatrix} = -10,$

$A_{12} = (-1)^3\begin{vmatrix} 0 & 4 \\ 0 & -2 \end{vmatrix} = 0,\ A_{13} = (-1)^4\begin{vmatrix} 0 & 3 \\ 0 & 1 \end{vmatrix} = 0,$

$A_{21} = (-1)^3\begin{vmatrix} 0 & 2 \\ 1 & -2 \end{vmatrix} = 2,\ A_{22} = (-1)^4\begin{vmatrix} 4 & 2 \\ 0 & -2 \end{vmatrix} = -8,$

$A_{23} = (-1)^5\begin{vmatrix} 4 & 0 \\ 0 & 1 \end{vmatrix} = -4,\ A_{31} = (-1)^4\begin{vmatrix} 0 & 2 \\ 3 & 4 \end{vmatrix} = -6,$

$A_{32} = (-1)^5\begin{vmatrix} 4 & 2 \\ 0 & 4 \end{vmatrix} = -16$, and $A_{33} = (-1)^6\begin{vmatrix} 4 & 0 \\ 0 & 3 \end{vmatrix} = 12.$

Thus it follows that

$$\text{adj } \mathbf{A} = \begin{bmatrix} -10 & 2 & -6 \\ 0 & -8 & -16 \\ 0 & -4 & 12 \end{bmatrix} \text{ and } \mathbf{A}^{-1} = \frac{1}{\det(\mathbf{A})}(\text{adj } \mathbf{A}) = \begin{bmatrix} \frac{1}{4} & \frac{-1}{20} & \frac{3}{20} \\ 0 & \frac{1}{5} & \frac{2}{5} \\ 0 & \frac{1}{10} & \frac{-3}{10} \end{bmatrix}.$$

13. (a) $\mathbf{A} = \begin{bmatrix} -3 & 1 \\ 2 & 0 \end{bmatrix}$. Then $\det(\mathbf{A}) = -2$ and the cofactors are, $A_{11} = (-1)^2(0) = 0$, $A_{12} = (-1)^3(2) = -2$, $A_{21} = (-1)^3(1) =$

-1, and $A_{22} = (-1)^4(-3) = -3$. Thus

$$\text{adj } \mathbf{A} = \begin{bmatrix} 0 & -1 \\ -2 & -3 \end{bmatrix} \text{ and } \mathbf{A}^{-1} = \frac{1}{\det(\mathbf{A})} (\text{adj } \mathbf{A}) = \begin{bmatrix} 0 & \frac{1}{2} \\ 1 & \frac{3}{2} \end{bmatrix}.$$

(b) $\mathbf{A} = \begin{bmatrix} 4 & 0 & 0 \\ 0 & -3 & 0 \\ 0 & 0 & 2 \end{bmatrix}$. Then $\det(\mathbf{A}) = (4)(-3)(2) = -24$ and the cofactors are $A_{11} = (-1)^2\begin{vmatrix} -3 & 0 \\ 0 & 2 \end{vmatrix} = -6$, $A_{12} = (-1)^3\begin{vmatrix} 0 & 0 \\ 0 & 2 \end{vmatrix}$ $= 0$, $A_{13} = (-1)^4\begin{vmatrix} 0 & -3 \\ 0 & 0 \end{vmatrix} = 0$, $A_{21} = (-1)^3\begin{vmatrix} 0 & 0 \\ 0 & 2 \end{vmatrix} = 0$,

$A_{22} = (-1)^4\begin{vmatrix} 4 & 0 \\ 0 & 2 \end{vmatrix} = 8$, $A_{23} = (-1)^5\begin{vmatrix} 4 & 0 \\ 0 & 0 \end{vmatrix} = 0$,

$A_{31} = (-1)^4\begin{vmatrix} 0 & 0 \\ -3 & 0 \end{vmatrix} = 0$, $A_{32} = (-1)^5\begin{vmatrix} 4 & 0 \\ 0 & 0 \end{vmatrix} = 0$,

and $A_{33} = (-1)^6\begin{vmatrix} 4 & 0 \\ 0 & -3 \end{vmatrix} = -12$. Thus $\text{adj } \mathbf{A} = \begin{bmatrix} -6 & 0 & 0 \\ 0 & 8 & 0 \\ 0 & 0 & -12 \end{bmatrix}$

and $\mathbf{A}^{-1} = \dfrac{1}{\det(\mathbf{A})} (\text{adj } \mathbf{A}) = \begin{bmatrix} \frac{1}{4} & 0 & 0 \\ 0 & \frac{-1}{3} & 0 \\ 0 & 0 & \frac{1}{2} \end{bmatrix}$.

(c) $\mathbf{A} = \begin{bmatrix} 0 & 2 & 1 & 3 \\ 2 & -1 & 3 & 4 \\ -2 & 1 & 5 & 2 \\ 0 & 1 & 0 & 2 \end{bmatrix}$. To find $\det(\mathbf{A})$ expand by cofactors about column 1.

$$\det(\mathbf{A}) = (-1)^3(2)\begin{vmatrix} 2 & 1 & 3 \\ 1 & 5 & 2 \\ 1 & 0 & 2 \end{vmatrix} + (-1)^4(-2)\begin{vmatrix} 2 & 1 & 3 \\ -1 & 3 & 4 \\ 1 & 0 & 2 \end{vmatrix}$$

$= (-2)(5) + (-2)(9) = -28$. The cofactors are:

Exercises 2.2

$$A_{11} = (-1)^2\begin{vmatrix} -1 & 3 & 4 \\ 1 & 5 & 2 \\ 1 & 0 & 2 \end{vmatrix} = -30, \quad A_{12} = (-1)^3\begin{vmatrix} 2 & 3 & 4 \\ -2 & 5 & 2 \\ 0 & 0 & 2 \end{vmatrix} = -32,$$

$$A_{13} = (-1)^4\begin{vmatrix} 2 & -1 & 4 \\ -2 & 1 & 2 \\ 0 & 1 & 2 \end{vmatrix} = -12, \quad A_{14} = (-1)^5\begin{vmatrix} 2 & -1 & 3 \\ -2 & 1 & 5 \\ 0 & 1 & 0 \end{vmatrix} = 16,$$

$$A_{21} = (-1)^3\begin{vmatrix} 2 & 1 & 3 \\ 1 & 5 & 2 \\ 1 & 0 & 2 \end{vmatrix} = -5, \quad A_{22} = (-1)^4\begin{vmatrix} 0 & 1 & 3 \\ -2 & 5 & 2 \\ 0 & 0 & 2 \end{vmatrix} = 4,$$

$$A_{23} = (-1)^5\begin{vmatrix} 0 & 2 & 3 \\ -2 & 1 & 2 \\ 0 & 1 & 2 \end{vmatrix} = -2, \quad A_{24} = (-1)^6\begin{vmatrix} 0 & 2 & 1 \\ -2 & 1 & 5 \\ 0 & 1 & 0 \end{vmatrix} = -2,$$

$$A_{31} = (-1)^4\begin{vmatrix} 2 & 1 & 3 \\ -1 & 3 & 4 \\ 1 & 0 & 2 \end{vmatrix} = 9, \quad A_{32} = (-1)^5\begin{vmatrix} 0 & 1 & 3 \\ 2 & 3 & 4 \\ 0 & 0 & 2 \end{vmatrix} = 4,$$

$$A_{33} = (-1)^6\begin{vmatrix} 0 & 2 & 3 \\ 2 & -1 & 4 \\ 0 & 1 & 2 \end{vmatrix} = -2, \quad A_{34} = (-1)^7\begin{vmatrix} 0 & 2 & 1 \\ 2 & -1 & 3 \\ 0 & 1 & 0 \end{vmatrix} = -2,$$

$$A_{41} = (-1)^5\begin{vmatrix} 2 & 1 & 3 \\ -1 & 3 & 4 \\ 1 & 5 & 2 \end{vmatrix} = 46, \quad A_{42} = (-1)^6\begin{vmatrix} 0 & 1 & 3 \\ 2 & 3 & 4 \\ -2 & 5 & 2 \end{vmatrix} = 36,$$

$$A_{43} = (-1)^7\begin{vmatrix} 0 & 2 & 3 \\ 2 & -1 & 4 \\ -2 & 1 & 2 \end{vmatrix} = 24, \quad A_{44} = (-1)^8\begin{vmatrix} 0 & 2 & 1 \\ 2 & -1 & 3 \\ -2 & 1 & 5 \end{vmatrix} = -32.$$

Thus $\text{adj}\,\mathbf{A} = \begin{bmatrix} -30 & -5 & 9 & 46 \\ -32 & 4 & 4 & 36 \\ -12 & -2 & -2 & 24 \\ 16 & -2 & -2 & -32 \end{bmatrix}$ and

$$\mathbf{A}^{-1} = \frac{1}{\det(\mathbf{A})}\ (\text{adj }\mathbf{A}) = \begin{bmatrix} \frac{15}{14} & \frac{5}{28} & \frac{-9}{28} & \frac{-23}{14} \\ \frac{8}{7} & \frac{-1}{7} & \frac{-1}{7} & \frac{-9}{7} \\ \frac{3}{7} & \frac{1}{14} & \frac{1}{14} & \frac{-6}{7} \\ \frac{-4}{7} & \frac{1}{14} & \frac{1}{14} & \frac{8}{7} \end{bmatrix}.$$

<<**Strategy:** In Exercise 15 the determinants involved can be computed using either row and column operations as in Section 2.1 or expansion by cofactors. When possible we use the special forms for the determinants of 2×2 and 3×3 matrices as discussed in Examples 5 and 6 of Section 2.1 respectively.>>

15. (a) $\mathbf{A} = \begin{bmatrix} 4 & 3 & -5 \\ -2 & -1 & 3 \\ 4 & 6 & -2 \end{bmatrix}$. Using Example 6 of Section 2.1,

det($\mathbf{A}$) = 0. Thus $\mathbf{A}$ is singular.

(b) $\mathbf{A} = \begin{bmatrix} 1 & 3 & -1 & 2 \\ 2 & -6 & 4 & 1 \\ 3 & 5 & -1 & 3 \\ 4 & -6 & 5 & 2 \end{bmatrix}$. Using row operations we have,

$$\det(\mathbf{A}) = \begin{vmatrix} 1 & 3 & -1 & 2 \\ 2 & -6 & 4 & 1 \\ 3 & 5 & -1 & 3 \\ 4 & -6 & 5 & 2 \end{vmatrix} \begin{matrix} \\ \\ -2\mathbf{r}_1+\mathbf{r}_2 \rightarrow \mathbf{r}_2 \\ -3\mathbf{r}_1+\mathbf{r}_3 \rightarrow \mathbf{r}_3 \\ -4\mathbf{r}_1+\mathbf{r}_4 \rightarrow \mathbf{r}_4 \end{matrix} = \begin{vmatrix} 1 & 3 & -1 & 2 \\ 0 & -12 & 6 & -3 \\ 0 & -4 & 2 & -3 \\ 0 & -18 & 9 & -6 \end{vmatrix}.$$

Then expanding by cofactors about the first column,

det($\mathbf{A}$) = $(-1)^2(1)\begin{vmatrix} -12 & 6 & -3 \\ -4 & 2 & -3 \\ -18 & 9 & -6 \end{vmatrix}$. Using Example 6 from

Section 2.1 to compute the previous determinant we obtain det($\mathbf{A}$) = 0 and hence $\mathbf{A}$ is singular.

(c) $\mathbf{A} = \begin{bmatrix} 2 & 2 & -4 \\ 1 & 5 & 2 \\ 3 & 7 & -2 \end{bmatrix}$. Using Example 6 from Section 2.1 we

obtain det($\mathbf{A}$) = 0. Thus $\mathbf{A}$ is singular.

(d) $\mathbf{A} = \begin{bmatrix} 0 & 1 & 2 \\ 1 & 2 & 0 \\ 1 & 3 & 4 \end{bmatrix}$. Using Example 6 Section 2.1, $|\mathbf{A}| = -2$.

Thus **A** is nonsingular.

17. (a) To find all the values of λ for which $\begin{vmatrix} \lambda-1 & -4 \\ 0 & \lambda-4 \end{vmatrix} = 0$,

compute the determinant and solve the resulting polynomial equation.

$$\begin{vmatrix} \lambda-1 & -4 \\ 0 & \lambda-4 \end{vmatrix} = (\lambda-1)(\lambda-4) = 0$$

Thus $\lambda = 1$ or $\lambda = 4$.

(b) To find all the values of λ for which $|\lambda\mathbf{I}_3 - \mathbf{A}| = 0$,

where $\mathbf{A} = \begin{bmatrix} -3 & -1 & -3 \\ 0 & 3 & 0 \\ -2 & -1 & -2 \end{bmatrix}$, compute the determinant and solve

the resulting polynomial equation.

$$\det(\lambda\mathbf{I}_3 - \mathbf{A}) = \begin{vmatrix} \lambda+3 & 1 & 3 \\ 0 & \lambda-3 & 0 \\ 2 & 1 & \lambda+2 \end{vmatrix} = (\lambda+3)(\lambda-3)(\lambda+2) - 6(\lambda-3)$$

$$= \lambda^3 + 2\lambda^2 - 15\lambda = \lambda(\lambda^2 + 2\lambda - 15) = \lambda(\lambda+5)(\lambda-3) = 0$$

Thus $\lambda = 0$, $\lambda = -5$, or $\lambda = 3$.

<<**Strategy:** In Exercise 19, write the linear system in matrix form $\mathbf{AX} = \mathbf{0}$ then compute $\det(\mathbf{A})$. If $\det(\mathbf{A}) = 0$, then the system has a nontrivial solution.>>

19. (a) $\mathbf{AX} = \begin{bmatrix} 1 & 1 & -1 \\ 2 & 1 & 2 \\ 3 & -1 & 1 \end{bmatrix}\begin{bmatrix} x \\ y \\ z \end{bmatrix} = \mathbf{0}$. Then $\det(\mathbf{A}) = 12$, using

Example 6 in Section 2.1. Thus the linear system has only the trivial solution.

(b) $\mathbf{AX} = \begin{bmatrix} 1 & 1 & 2 & 1 \\ 2 & -1 & 1 & -1 \\ 3 & 1 & 2 & 3 \\ 2 & -1 & -1 & 1 \end{bmatrix}\begin{bmatrix} x \\ y \\ z \\ w \end{bmatrix} = \mathbf{0}$. Then

$$\det(\mathbf{A}) = \begin{vmatrix} 1 & 1 & 2 & 1 \\ 2 & -1 & 1 & -1 \\ 3 & 1 & 2 & 3 \\ 2 & -1 & -1 & 1 \end{vmatrix} \begin{matrix} \\ \\ -2\mathbf{r}_1+\mathbf{r}_2\to\mathbf{r}_2 \\ -3\mathbf{r}_1+\mathbf{r}_3\to\mathbf{r}_3 \\ -2\mathbf{r}_1+\mathbf{r}_4\to\mathbf{r}_4 \end{matrix} = \begin{vmatrix} 1 & 1 & 2 & 1 \\ 0 & -3 & -3 & -3 \\ 0 & -2 & -4 & 0 \\ 0 & -3 & -5 & -1 \end{vmatrix}$$

. Using expansion by cofactors about the first column we have

$$\det(\mathbf{A}) = (-1)^2(1)\begin{vmatrix} -3 & -3 & -3 \\ -2 & -4 & 0 \\ -3 & -5 & -1 \end{vmatrix} = 0.$$

Thus the linear system has nontrivial solutions.

<<**Strategy:** In Exercises 21 and 23 first write the linear system in matrix form **AX** = **B**. Then use Cramer's Rule to solve the linear system, provided det(**A**) ≠ 0. The determinants of 3 × 3 matrices are done using Example 6 in Section 2.1>>

21. $$\mathbf{AX} = \begin{bmatrix} 1 & 1 & 1 & -2 \\ 0 & 2 & 1 & 3 \\ 2 & 1 & -1 & 2 \\ 1 & -1 & 0 & 1 \end{bmatrix}\begin{bmatrix} x \\ y \\ z \\ w \end{bmatrix} = \begin{bmatrix} -4 \\ 4 \\ 5 \\ 4 \end{bmatrix}$$

. Using row operations we have

$$\det(\mathbf{A}) = \begin{vmatrix} 1 & 1 & 1 & -2 \\ 0 & 2 & 1 & 3 \\ 2 & 1 & -1 & 2 \\ 1 & -1 & 0 & 1 \end{vmatrix} \begin{matrix} \\ \\ -2\mathbf{r}_1+\mathbf{r}_3\to\mathbf{r}_3 \\ -1\mathbf{r}_1+\mathbf{r}_4\to\mathbf{r}_4 \end{matrix} = \begin{vmatrix} 1 & 1 & 1 & -2 \\ 0 & 2 & 1 & 3 \\ 0 & -1 & -3 & 6 \\ 0 & -2 & -1 & 3 \end{vmatrix}$$

. Expanding by cofactors about the first column we have

$$\det(\mathbf{A}) = (-1)^2(1)\begin{vmatrix} 2 & 1 & 3 \\ -1 & -3 & 6 \\ -2 & -1 & 3 \end{vmatrix} = -30.$$

Then

$$\det(\mathbf{A}_1) = \begin{vmatrix} -4 & 1 & 1 & -2 \\ 4 & 2 & 1 & 3 \\ 5 & 1 & -1 & 2 \\ 4 & -1 & 0 & 1 \end{vmatrix} \mathbf{c}_1\leftrightarrow\mathbf{c}_3 = (-1)\begin{vmatrix} 1 & 1 & -4 & -2 \\ 1 & 2 & 4 & 3 \\ -1 & 1 & 5 & 2 \\ 0 & -1 & 4 & 1 \end{vmatrix} \begin{matrix} \\ \\ -1\mathbf{r}_1+\mathbf{r}_2\to\mathbf{r}_2 \\ 1\mathbf{r}_1+\mathbf{r}_3\to\mathbf{r}_3 \end{matrix}$$

$$= (-1)\begin{vmatrix} 1 & 1 & -4 & -2 \\ 0 & 1 & 8 & 5 \\ 0 & 2 & 1 & 0 \\ 0 & -1 & 4 & 1 \end{vmatrix} = (-1)\begin{vmatrix} 1 & 8 & 5 \\ 2 & 1 & 0 \\ -1 & 4 & 1 \end{vmatrix} = -30,$$

Exercises 2.2

$$\det(\mathbf{A}_2) = \begin{vmatrix} 1 & -4 & 1 & -2 \\ 0 & 4 & 1 & 3 \\ 2 & 5 & -1 & 2 \\ 1 & 4 & 0 & 1 \end{vmatrix} \begin{matrix} \\ \\ -2\mathbf{r}_1+\mathbf{r}_3 \to \mathbf{r}_3 \\ -1\mathbf{r}_1+\mathbf{r}_4 \to \mathbf{r}_4 \end{matrix} = \begin{vmatrix} 1 & -4 & 1 & -2 \\ 0 & 4 & 1 & 3 \\ 0 & 13 & -3 & 6 \\ 0 & 8 & -1 & 3 \end{vmatrix} = \begin{vmatrix} 4 & 1 & 3 \\ 13 & -3 & 6 \\ 8 & -1 & 3 \end{vmatrix} = 30,$$

$$\det(\mathbf{A}_3) = \begin{vmatrix} 1 & 1 & -4 & -2 \\ 0 & 2 & 4 & 3 \\ 2 & 1 & 5 & 2 \\ 1 & -1 & 4 & 1 \end{vmatrix} \begin{matrix} \\ \\ -2\mathbf{r}_1+\mathbf{r}_3 \to \mathbf{r}_3 \\ -1\mathbf{r}_1+\mathbf{r}_4 \to \mathbf{r}_4 \end{matrix} = \begin{vmatrix} 1 & 1 & -4 & -2 \\ 0 & 2 & 4 & 3 \\ 0 & -1 & 13 & 6 \\ 0 & -2 & 8 & 3 \end{vmatrix} = \begin{vmatrix} 2 & 4 & 3 \\ -1 & 13 & 6 \\ -2 & 8 & 3 \end{vmatrix} = 0,$$

$$\det(\mathbf{A}_4) = \begin{vmatrix} 1 & 1 & 1 & -4 \\ 0 & 2 & 1 & 4 \\ 2 & 1 & -1 & 5 \\ 1 & -1 & 0 & 4 \end{vmatrix} \begin{matrix} \\ \\ -2\mathbf{r}_1+\mathbf{r}_3 \to \mathbf{r}_3 \\ -1\mathbf{r}_1+\mathbf{r}_4 \to \mathbf{r}_4 \end{matrix} = \begin{vmatrix} 1 & 1 & 1 & -4 \\ 0 & 2 & 1 & 4 \\ 0 & -1 & -3 & 13 \\ 0 & -2 & -1 & 8 \end{vmatrix} = \begin{vmatrix} 2 & 1 & 4 \\ -1 & -3 & 13 \\ -2 & -1 & 8 \end{vmatrix} = -60.$$

Thus, $x = \dfrac{\det(\mathbf{A}_1)}{\det(\mathbf{A})} = \dfrac{-30}{-30} = 1,\ y = \dfrac{\det(\mathbf{A}_2)}{\det(\mathbf{A})} = \dfrac{30}{-30} = -1,$

$z = \dfrac{\det(\mathbf{A}_3)}{\det(\mathbf{A})} = \dfrac{0}{-30} = 0$, and $w = \dfrac{\det(\mathbf{A}_4)}{\det(\mathbf{A})} = \dfrac{-60}{-30} = 2.$

23. $\mathbf{AX} = \begin{bmatrix} 2 & 3 & 7 \\ -2 & 0 & -4 \\ 1 & 2 & 4 \end{bmatrix} \begin{bmatrix} x \\ y \\ z \end{bmatrix} = \begin{bmatrix} 2 \\ 0 \\ 0 \end{bmatrix}$. Then $\det(\mathbf{A}) = \begin{vmatrix} 2 & 3 & 7 \\ -2 & 0 & -4 \\ 1 & 2 & 4 \end{vmatrix} = 0.$

Thus the linear system has no solution.

T.1. Let **A** be upper triangular. Then

$$\det(\mathbf{A}) = \begin{vmatrix} a_{11} & a_{12} & \cdots & a_{1n} \\ 0 & a_{22} & \cdots & a_{2n} \\ \cdot & \cdot & & \cdot \\ \cdot & \cdot & & \cdot \\ \cdot & \cdot & & \cdot \\ 0 & 0 & \cdots & a_{nn} \end{vmatrix} \text{(Cofactor expansion about column 1.)}$$

$$= a_{11} \begin{vmatrix} a_{22} & \cdots & a_{2n} \\ 0 & \cdots & \\ 0 & \cdots & a_{nn} \end{vmatrix} \text{(Cofactor expansion about column 1.)}$$

$$= a_{11}a_{22} \begin{vmatrix} a_{33} & \cdots & a_{3n} \\ 0 & \cdots & \\ 0 & \cdots & a_{nn} \end{vmatrix} = \cdots = a_{11}a_{22} \cdots a_{nn} .$$

If **A** is lower triangular, then expansion by cofactors about successive rows will give the result in a similar fashion.

T.3. Let $\mathbf{M}_{ij}$ be the $(n-1) \times (n-1)$ submatrix of $\mathbf{A}$ obtained by deleting the ith row and jth column. Since $\mathbf{A}$ is symmetric, the ith row of $\mathbf{A}$ and the ith column of $\mathbf{A}$ contain the same entries. Thus

$$\mathbf{M}_{ji}{}^T = \mathbf{M}_{ij}$$

and hence

$$\det(\mathbf{M}_{ij}) = \det(\mathbf{M}_{ji}{}^T) = \det(\mathbf{M}_{ji}).$$

The (i,j) entry of adj $\mathbf{A}$ is

$$\mathbf{A}_{ji} = (-1)^{j+i}\det(\mathbf{M}_{ji}) = (-1)^{j+i}\det(\mathbf{M}_{ij}) = \mathbf{A}_{ij}$$

which is the (j,i) entry of adj $\mathbf{A}$. It follows that adj $\mathbf{A}$ is symmetric.

T.5. By Theorem 2.12, $\mathbf{A}$ nonsingular if and only if $\det(\mathbf{A}) \neq 0$. Since $\det(\mathbf{A}) = ad - bc$, A is nonsingular if and only if $ad - bc \neq 0$. By Corollary 2.2, $\mathbf{A}^{-1} = \frac{1}{\det(\mathbf{A})}(\text{adj } \mathbf{A})$

$$= \frac{1}{ad-bc}\begin{bmatrix} d & -b \\ -c & a \end{bmatrix}.$$

T.7. We show that if $\mathbf{A}$ is singular, then $\mathbf{A}(\text{adj } \mathbf{A}) = \mathbf{0}$. If $\mathbf{A} = \mathbf{0}$ then adj $\mathbf{A} = \mathbf{0}$ and is singular. Suppose $\mathbf{A} \neq \mathbf{0}$, but is singular. By Theorem 2.11, $\mathbf{A}(\text{adj } \mathbf{A}) = \det(\mathbf{A})\mathbf{I}_n = 0$. Were adj $\mathbf{A}$ nonsingular, it would have an inverse, and

$$\mathbf{A} = \mathbf{A}(\text{adj } \mathbf{A})(\text{adj } \mathbf{A})^{-1} = \mathbf{0}(\text{adj } \mathbf{A}) = \mathbf{0}.$$

But this a contradiction.

T.9. By Corollary 2.3, there is a nontrivial solution if and only if the determinant of the coefficient matrix is zero. Thus

$$\begin{vmatrix} a-\lambda & b \\ c & d-\lambda \end{vmatrix} = (a-\lambda)(d-\lambda) - bc = 0.$$

T.11. Since the entries of $\mathbf{A}$ are integers, the cofactors of $\mathbf{A}$ are integers. Thus adj $\mathbf{A}$ is a matrix of integers. By Corollary 2.2, $\mathbf{A}^{-1} = \frac{1}{\det(\mathbf{A})}(\text{adj } \mathbf{A}) = \pm 1\,(\text{adj } \mathbf{A})$, hence $\mathbf{A}^{-1}$ is also a matrix of integers.

Exercises 2.2

ML.1. We present a sample of the cofactors.

```
A=[1 0 -2;3 1 4;5 2 -3];

cofactor(1,1,A)        cofactor(2,3,A)        cofactor(3,1,A)

ans =                  ans =                  ans =

   -11                    -2                     2
```

ML.2.

```
A=[1 5 0;2 -1 3;3 2 1];
cofactor(2,1,A)

ans =

   -5

cofactor(2,2,A)

ans =

     1

cofactor(2,3,A)

ans =

    13
```

ML.3.

```
A=  [4 0 -1;-2 2 -1;0 4 -3];
detA = 4*cofactor(1,1,A)+(-1)*cofactor(1,3,A)

detA =

     0
```

We can check this using the **det** command.

ML.4.

```
A=[-1 2 0 0;2 -1 2 0;0 2 - 1 2;0 0 2 -1];
```

(Use expansion about the first column.)

```
detA=-1*cofactor(1,1,A)+2*cofactor(2,1,A)

detA =

     5
```

ML.5. Before using the expression for the inverse in Corollary 2.2, check the value of the determinant to avoid division by zero.

(a)
```
A=[1 2 -3;-4 -5 2;-1 1 -7];
det(A)

ans =

     0
```

The matrix is singular.

(b)
```
A=[2 3;-1 2];
det(A)

ans =

    7

invA=(1/det(A))*adjoint(A)

invA =

    0.2857   -0.4286
    0.1429    0.2857
```

To see the inverse with rational entries proceed as follows.

```
rat(invA,'s')

ans =

      2/7          -3/7
      1/7           2/7
```

(c)
```
A  =[4 0 2;0 3 4;0 1 -2];
det(A)

ans =

    -40

invA=(1/det(A))*adjoint(A)

invA =

      0.2500   -0.0500    0.1500
           0    0.2000    0.4000
           0    0.1000   -0.3000

rat(invA,'s')

ans =

      1/4          -1/20          3/20
       0            1/5           2/5
       0            1/10         -3/10
```

Chapter 2 Supplementary Exercises

1. Use Equation (2).

(a) $\begin{vmatrix} 0 & 2 & 0 \\ 0 & 0 & -3 \\ 4 & 0 & 0 \end{vmatrix} = (0)(0)(0) - (0)(-3)(0) - (2)(0)(0)$

$+ (2)(-3)(4) + (0)(0)(0) - (0)(0)(4) = -24$

(b) $\begin{vmatrix} 3 & 0 & 0 & 0 \\ 0 & -2 & 0 & 0 \\ 0 & 4 & 1 & 0 \\ 3 & 2 & -1 & -4 \end{vmatrix} = (3)(-2)(1)(-4) + (3)(0)(4)(-1)$

$+ (3)(0)(0)(2) + (0)(0)(0)(-1) + (0)(0)(0)(-4) + (0)(0)(1)(3)$
$+ (0)(0)(4)(-4) + (0)(-2)(0)(3) + (0)(0)(0)(2) + (0)(0)(1)(2)$
$+ (0)(-2)(0)(-1) + (0)(0)(4)(3) - (3)(-2)(0)(-1) - (3)(0)(4)(-4)$
$- (3)(0)(1)(2) - (0)(0)(1)(-4) - (0)(0)(0)(3) - (0)(0)(0)(-1)$
$- (0)(0)(0)(2) - (0)(-2)(0)(-4) - (0)(0)(4)(3) - (0)(0)(4)(-1)$
$- (0)(-2)(1)(3) - (0)(0)(0)(2) = 24$

(See also Exercise 7 in Section 2.1.)

3. $\mathbf{A}$ is 4×4 and $\det(\mathbf{A}) = 5$.

(a) Since $\det(\mathbf{A}) \neq 0$, Theorem 2.12 implies $\mathbf{A}$ is nonsingular. Then Corollary 2.1 implies that $\det(\mathbf{A}^{-1}) = \dfrac{1}{\det(\mathbf{A})} = \dfrac{1}{5}$.

(b) By Theorem 2.5, $\det(2\mathbf{A}) = 2^4\det(\mathbf{A}) = (16)(5) = 80$.

(c) By Theorem 2.5 and part (a) we have that $\det(2\mathbf{A}^{-1}) = 2^4\det(\mathbf{A}^{-1}) = (16)\,\dfrac{1}{5} = \dfrac{16}{5}$.

(d) By Corollary 2.1 and part (b), $\det((2\mathbf{A})^{-1}) = \dfrac{1}{\det(2\mathbf{A})} = \dfrac{1}{80}$.

5. Expanding by cofactors about row 3, we have $\begin{vmatrix} \lambda+2 & -1 & 3 \\ 2 & \lambda-1 & 2 \\ 0 & 0 & \lambda+4 \end{vmatrix} =$

$(\lambda+4)(-1)^6\begin{vmatrix} \lambda+2 & -1 \\ 2 & \lambda-1 \end{vmatrix} = (\lambda+4)((\lambda+2)(\lambda-1) + 2) = (\lambda+4)(\lambda^2+\lambda-2+2)$
$= (\lambda+4)(\lambda^2+\lambda) = (\lambda+4)\lambda(\lambda+1) = 0$. Thus $\lambda = -4$, $\lambda = 0$, or $\lambda = -1$.

7. $\begin{vmatrix} 3 & 2 & -1 & 1 \\ 4 & 1 & 1 & 0 \\ -1 & 2 & 3 & 4 \\ -2 & 3 & 5 & 1 \end{vmatrix}_{c_1 \leftrightarrow c_4} = (-1)\begin{vmatrix} 1 & 2 & -1 & 3 \\ 0 & 1 & 1 & 4 \\ 4 & 2 & 3 & -1 \\ 1 & 3 & 5 & -2 \end{vmatrix}_{\substack{-4r_1+r_3 \to r_3 \\ -1r_1+r_4 \to r_4}}$

$= (-1)\begin{vmatrix} 1 & 2 & -1 & 3 \\ 0 & 1 & 1 & 4 \\ 0 & -6 & 7 & -13 \\ 0 & 1 & 6 & -5 \end{vmatrix}_{\substack{6r_2+r_3 \to r_3 \\ -1r_2+r_4 \to r_4}}$

$= (-1)\begin{vmatrix} 1 & 2 & -1 & 3 \\ 0 & 1 & 1 & 4 \\ 0 & 0 & 13 & 11 \\ 0 & 0 & 5 & -9 \end{vmatrix}_{\frac{-5}{13}r_3+r_4 \to r_4} = (-1)\begin{vmatrix} 1 & 2 & -1 & 3 \\ 0 & 1 & 1 & 4 \\ 0 & 0 & 13 & 11 \\ 0 & 0 & 0 & \frac{-172}{13} \end{vmatrix}$

$= 172$

9. Since row 1 contains a zero, expand by cofactors about row 1. First compute the cofactors using Equation (3) of Section 2.1.

$\mathbf{A}_{11} = (-1)^2\begin{vmatrix} 0 & 3 & 2 \\ 1 & 5 & -2 \\ 3 & 2 & -3 \end{vmatrix} = -35$, $\mathbf{A}_{12} = (-1)^3\begin{vmatrix} -1 & 3 & 2 \\ 4 & 5 & -2 \\ 1 & 2 & -3 \end{vmatrix} = -47$,

$\mathbf{A}_{13} = (-1)^4\begin{vmatrix} -1 & 0 & 2 \\ 4 & 1 & -2 \\ 1 & 3 & -3 \end{vmatrix} = 19$

Then $\det(\mathbf{A}) = a_{11}\mathbf{A}_{11} + a_{12}\mathbf{A}_{12} + a_{13}\mathbf{A}_{13} = (3)(-35) + (2)(-47) + (-1)(19) = -218$.

11. We first use Equation (3) of Section 2.1 to compute $\det(\mathbf{A}) = 5$. Thus, $\mathbf{A}$ is nonsingular and we proceed by finding adj $\mathbf{A}$.

The cofactors are: $\mathbf{A}_{11} = (-1)^2\begin{vmatrix} 1 & 2 \\ 1 & 2 \end{vmatrix} = 0$,

$\mathbf{A}_{12} = (-1)^3\begin{vmatrix} 0 & 2 \\ -1 & 2 \end{vmatrix} = -2$, $\mathbf{A}_{13} = (-1)^4\begin{vmatrix} 0 & 1 \\ -1 & 1 \end{vmatrix} = 1$,

$$\mathbf{A}_{21} = (-1)^3\begin{vmatrix} -1 & 3 \\ 1 & 2 \end{vmatrix} = 5,\ \mathbf{A}_{22} = (-1)^4\begin{vmatrix} 2 & 3 \\ -1 & 2 \end{vmatrix} = 7,$$

$$\mathbf{A}_{23} = (-1)^5\begin{vmatrix} 2 & -1 \\ -1 & 1 \end{vmatrix} = -1,\ \mathbf{A}_{31} = (-1)^4\begin{vmatrix} -1 & 3 \\ 1 & 2 \end{vmatrix} = -5,$$

$$\mathbf{A}_{32} = (-1)^5\begin{vmatrix} 2 & 3 \\ 0 & 2 \end{vmatrix} = -4,\ \mathbf{A}_{33} = (-1)^6\begin{vmatrix} 2 & -1 \\ 0 & 1 \end{vmatrix} = 2. \text{ Thus,}$$

$$\mathbf{A}^{-1} = \frac{1}{\det(\mathbf{A})}(\text{adj } \mathbf{A}) = \frac{1}{5}\begin{bmatrix} 0 & 5 & -5 \\ -2 & 7 & -4 \\ 1 & -1 & 2 \end{bmatrix}.$$

13. By Corollary 2.3, the homogeneous system $\mathbf{AX} = \mathbf{0}$ has only the trivial solution when $\det(\mathbf{A}) \neq 0$. We compute $\det(\mathbf{A})$, set it equal to zero, and determine the values of λ for which there are nontrivial solutions. Then for all other values of λ there is only the trivial solution.

$$\begin{vmatrix} \lambda & 0 & 1 \\ 1 & \lambda-1 & 0 \\ 0 & 0 & \lambda+1 \end{vmatrix} = \lambda(\lambda-1)(\lambda+1) = 0 \text{ when } \lambda = 0,\ \lambda = 1, \text{ or } \lambda = -1.$$

Thus there is only the trivial solution to $\mathbf{AX} = \mathbf{0}$ for $\lambda \neq 0, 1, -1$.

15. (a) $$\begin{vmatrix} a & 1 & b \\ b & 1 & c \\ c & 1 & a \end{vmatrix}_{-1\mathbf{c}_3+\mathbf{c}_1\to\mathbf{c}_1} = \begin{vmatrix} a-b & 1 & b \\ b-c & 1 & c \\ c-a & 1 & a \end{vmatrix}_{\mathbf{c}_1+\mathbf{c}_3\to\mathbf{c}_3}$$

$$= \begin{vmatrix} a-b & 1 & a \\ b-c & 1 & b \\ c-a & 1 & c \end{vmatrix}$$

(b) $$\begin{vmatrix} 1 & a & a^2 \\ 1 & b & b^2 \\ 1 & c & c^2 \end{vmatrix}_{\substack{-\mathbf{r}_1+\mathbf{r}_2\to\mathbf{r}_2 \\ -\mathbf{r}_1+\mathbf{r}_3\to\mathbf{r}_3}}$$

$$= \begin{vmatrix} 1 & a & a^2 \\ 0 & b-a & (b+a)(b-a) \\ 0 & c-a & (c+a)(c-a) \end{vmatrix}_{\substack{-a\mathbf{c}_1+\mathbf{c}_2\to\mathbf{c}_2 \\ -a^2\mathbf{c}_1+\mathbf{c}_3\to\mathbf{c}_3}}$$

$$= \begin{vmatrix} 1 & 0 & 0 \\ 0 & b-a & (b+a)(b-a) \\ 0 & c-a & (c+a)(c-a) \end{vmatrix}_{-(a+b+c)\mathbf{c}_2+\mathbf{c}_3 \to \mathbf{c}_3}$$

$$= \begin{vmatrix} 1 & 0 & 0 \\ 0 & b-a & -c(b-a) \\ 0 & c-a & -b(c-a) \end{vmatrix}_{\substack{a\mathbf{c}_1+\mathbf{c}_2 \to \mathbf{c}_2 \\ bc\mathbf{c}_1+\mathbf{c}_3 \to \mathbf{c}_3}}$$

$$= \begin{vmatrix} 1 & a & bc \\ 0 & b-a & -c(b-a) \\ 0 & c-a & -b(c-a) \end{vmatrix}_{\substack{\mathbf{r}_1+\mathbf{r}_2 \to \mathbf{r}_2 \\ \mathbf{r}_1+\mathbf{r}_3 \to \mathbf{r}_3}} = \begin{vmatrix} 1 & a & bc \\ 1 & b & ca \\ 1 & c & ab \end{vmatrix}$$

17. Let $\mathbf{A} = \begin{bmatrix} a-2 & 2 \\ a-2 & a+2 \end{bmatrix}$. **A** is nonsingular if and only if

$$\det(\mathbf{A}) = \begin{vmatrix} a-2 & 2 \\ a-2 & a+2 \end{vmatrix} = (a-2)a \neq 0.$$

Thus **A** is nonsingular for $a \neq 0$ and $a \neq 2$.

T.1. If rows i and j are proportional with $ta_{ik} = a_{jk}$, for $k = 1,2,\ldots,n$ then

$$|\mathbf{A}| = |\mathbf{A}|_{-t\mathbf{r}_i+\mathbf{r}_j \to \mathbf{r}_j} = 0$$

since this row operation makes row j all zeros.

T.3. $$|\mathbf{Q} - n\mathbf{I}_n| = \begin{vmatrix} 1-n & 1 & \cdot & \cdot & \cdot & 1 \\ 1 & 1-n & \cdot & \cdot & \cdot & 1 \\ \cdot & \cdot & & & & \cdot \\ \cdot & \cdot & & & & \cdot \\ 1 & 1 & \cdot & \cdot & \cdot & 1-n \end{vmatrix}_{\substack{\mathbf{r}_i+\mathbf{r}_1 \to \mathbf{r}_1 \\ i=2,3,\ldots,n}}$$

$$= \begin{vmatrix} 0 & 0 & \cdot & \cdot & \cdot & 0 \\ 1 & 1-n & \cdot & \cdot & \cdot & 1 \\ \cdot & \cdot & & & & \cdot \\ \cdot & \cdot & & & & \cdot \\ 1 & 1 & \cdot & \cdot & \cdot & 1-n \end{vmatrix} = 0$$

T.5. From Theorem 2.11, **A** (adj **A**) = $\det(\mathbf{A})\mathbf{I}_n$. Since **A** is singular, $\det(\mathbf{A}) = 0$, and then **A** (adj **A**) = **0**.

T.7. Compute $\begin{vmatrix} \mathbf{A} & \mathbf{0} \\ \mathbf{0} & \mathbf{B} \end{vmatrix}$ by expanding about the first column and expand the resulting (n-1) × (n-1) determinants about the first column, etc.

T.9. Since det(**A**) ≠ 0, **A** is nonsingular. Hence the solution to $\mathbf{AX} = \mathbf{B}$ is $\mathbf{X} = \mathbf{A}^{-1}\mathbf{B}$. By Exercise T.11 in Section 2.2 matrix $\mathbf{A}^{-1}$ has only integer entries. It follows that the product $\mathbf{A}^{-1}*\mathbf{B}$ has only integer entries.

Exercises 3.1

1.

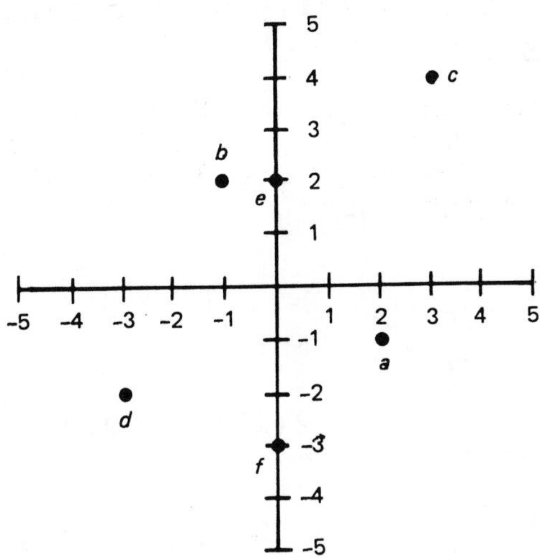

3. The vector $\begin{bmatrix} -2 \\ 5 \end{bmatrix}$ is to have its tail at $(x_1, x_2) = (3,2)$. Let (y_1, y_2) represent the head of the vector. Then we must have $\begin{bmatrix} y_1 - x_1 \\ y_2 - x_2 \end{bmatrix} = \begin{bmatrix} -2 \\ 5 \end{bmatrix}$. Setting $x_1 = 3$ and $x_2 = 2$ and solving for y_1 and y_2 we have $y_1 = 1$ and $y_2 = 7$. The head of the vector is at (1,7).

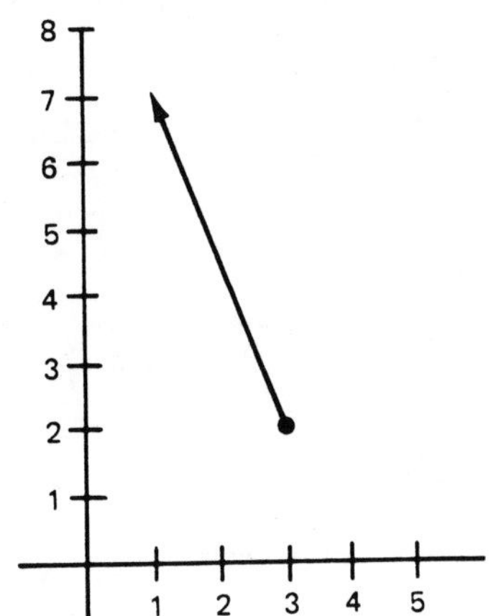

Exercises 3.1

<<In Exercises 5 and 7 we use the following vector operations: for $\mathbf{X} = (x_1,x_2)$, $\mathbf{Y} = (y_1,y_2)$, and scalar c, $\mathbf{X} + \mathbf{Y} = (x_1+y_1,x_2+y_2)$, $\mathbf{X} - \mathbf{Y} = (x_1-y_1,x_2-y_2)$, and $c\mathbf{X} = (cx_1,cx_2)$.>>

5. (a) $\mathbf{X} = (2,3)$, $\mathbf{Y} = (-2,5)$.
$\mathbf{X} + \mathbf{Y} = (2+(-2),3+5) = (0,8)$.
$\mathbf{X} - \mathbf{Y} = (2-(-2),3-5) = (4,-2)$.
$2\mathbf{X} = (2(2),2(3)) = (4,6)$.
$3\mathbf{X} - 2\mathbf{Y} = (3(2),3(3)) - (2(-2),2(5)) = (6,9) - (-4,10)$
$= (6-(-4),9-10) = (10,-1)$.

(b) $\mathbf{X} = (0,3)$, $\mathbf{Y} = (3,2)$.
$\mathbf{X} + \mathbf{Y} = (0+3,3+2) = (3,5)$.
$\mathbf{X} - \mathbf{Y} = (0-3,3-2) = (-3,1)$.
$2\mathbf{X} = (2(0),2(3)) = (0,6)$.
$3\mathbf{X} - 2\mathbf{Y} = (3(0),3(3)) - (2(3),2(2)) = (0,9) - (6,4)$
$= (0-6,9-4) = (-6,5)$.

(c) $\mathbf{X} = (2,6)$, $Y = (3,2)$.
$\mathbf{X} + \mathbf{Y} = (2+3,6+2) = (5,8)$.
$\mathbf{X} - \mathbf{Y} = (2-3,6-2) = (-1,4)$.
$2\mathbf{X} = (2(2),2(6)) = (4,12)$.
$3\mathbf{X} - 2\mathbf{Y} = (3(2),3(6)) - (2(3),2(2)) = (6,18) - (6,4)$
$= (6-6,18-4) = (0,14)$.

7. Let $\mathbf{X} = (1,2)$, $\mathbf{Y} = (-3,4)$, $\mathbf{Z} = (x,4)$, and $\mathbf{U} = (-2,y)$. Determine x or y in each of the following cases.
(a) Require $\mathbf{Z} = 2\mathbf{X}$. Then $(x,4) = (2(1),2(2)) = (2,4)$. Hence, $x = 2$.

(b) Require $\frac{3}{2}\mathbf{U} = \mathbf{Y}$. Then $(\frac{3}{2}(-2),\frac{3}{2}y) = (-3,4)$. Hence, $\frac{3}{2}y = 4$ and $y = \frac{8}{3}$.

(c) Require $\mathbf{Z} + \mathbf{U} = \mathbf{X}$. Then $(x+(-2),4+y) = (1,2)$. Hence $x - 2 = 1$, so $x = 3$ and $y + 4 = 2$, so $y = -2$.

9. Use Equation (1) to find the length of a vector.

(a) $\|(1,2)\| = \sqrt{(1)^2 + (2)^2} = \sqrt{1 + 4} = \sqrt{5}$.

(b) $\|(-3,-4)\| = \sqrt{(-3)^2 + (-4)^2} = \sqrt{9 + 16} = \sqrt{25} = 5$.

(c) $\|(0,2)\| = \sqrt{(0)^2 + (2)^2} = \sqrt{4} = 2$.

(d) $\|(-4,3)\| = \sqrt{(-4)^2 + (3)^2} = \sqrt{16 + 9} = \sqrt{25} = 5$.

11. The distance between points is determined using Equation (2).

(a) $P(2,3)$, $Q(3,4)$.

Distance from P to Q = $\sqrt{(3-2)^2 + (4-3)^2} = \sqrt{2}$.

(b) $P(0,0)$, $Q(3,4)$.

Distance from P to Q = $\sqrt{(3-0)^2 \ (4-0)^2} = \sqrt{25} = 5$.

(c) $P(-3,2)$, $Q(0,1)$.

Distance from P to Q = $\sqrt{(0-(-3))^2 + (1-2)^2} = \sqrt{10}$.

(d) $P(0,3)$, $Q(2,0)$.

Distance from P to Q = $\sqrt{(2-0)^2 + (0-3)^2} = \sqrt{13}$.

13. (a) $\mathbf{X} = (3,4)$. $\|\mathbf{X}\| = \sqrt{(3)^2 + (4)^2} = \sqrt{25} = 5$. Then a unit vector in the direction of $\mathbf{X}$ is given by

$$\mathbf{U} = \frac{1}{\|\mathbf{X}\|}\mathbf{X} = \frac{1}{5}(3,4) = (\frac{3}{5}, \frac{4}{5}).$$

(b) $\mathbf{X} = (-2,-3)$. $\|\mathbf{X}\| = \sqrt{(-2)^2 + (-3)^2} = \sqrt{13}$. Then a unit vector in the direction of $\mathbf{X}$ is given by

$$\mathbf{U} = \frac{1}{\|\mathbf{X}\|}\mathbf{X} = \frac{1}{\sqrt{13}}(-2,-3) = (\frac{-2}{\sqrt{13}}, \frac{-3}{\sqrt{13}}).$$

(c) $\mathbf{X} = (5,0)$. $\|\mathbf{X}\| = \sqrt{(5)^2 + (0)^2} = \sqrt{25} = 5$. Then a unit vector in the direction of $\mathbf{X}$ is given by

$$\mathbf{U} = \frac{1}{\|\mathbf{X}\|}\mathbf{X} = \frac{1}{5}(5,0) = (1,0).$$

15. (a) $\mathbf{X} = (1,2)$, $\mathbf{Y} = (2,-3)$. $\mathbf{X}\cdot\mathbf{Y} = (1)(2) + (2)(-3) = -4$.

(b) $\mathbf{X} = (1,0)$, $\mathbf{Y} = (0,1)$. $\mathbf{X}\cdot\mathbf{Y} = (1)(0) + (0)(1) = 0$.

(c) $\mathbf{X} = (-3,-4)$, $\mathbf{Y} = (4,-3)$. $\mathbf{X}\cdot\mathbf{Y} = (-3)(4) + (-4)(-3) = 0$.

(d) $\mathbf{X} = (2,1)$, $\mathbf{Y} = (-2,-1)$. $\mathbf{X}\cdot\mathbf{Y} = (2)(-2) + (1)(-1) = -5$.

17. Use the results from Exercise 15 and Equation (7).

(a) $\cos\theta = \dfrac{-4}{\|\mathbf{X}\|\ \|\mathbf{Y}\|} = \dfrac{-4}{\sqrt{5}\sqrt{13}} = \dfrac{-4}{\sqrt{65}}$.

(b) $\cos\theta = \dfrac{0}{\|\mathbf{X}\|\ \|\mathbf{Y}\|} = 0$.

(c) $\cos\theta = \dfrac{0}{\|\mathbf{X}\|\ \|\mathbf{Y}\|} = 0.$

(d) $\cos\theta = \dfrac{-5}{\|\mathbf{X}\|\ \|\mathbf{Y}\|} = \dfrac{-5}{\sqrt{5}\ \sqrt{5}} = -1.$

19. $\mathbf{i} = (1,0)$ and $\mathbf{j} = (0,1)$.
(a) $\mathbf{i}\cdot\mathbf{i} = (1,0)\cdot(1,0) = 1$ and $\mathbf{j}\cdot\mathbf{j} = (0,1)\cdot(0,1) = 1.$

(b) $\mathbf{i}\cdot\mathbf{j} = (1,0)\cdot(0,1) = 0.$

21. Two vectors are parallel if one is a scalar multiple of the other. We determine a so that

$$k(a,4) = (2,5)$$

We have $k(a,4) = (ka,4k) = (2,5)$ and equating corresponding elements we have

$$ka = 2 \quad \text{and} \quad 4k = 5$$

It follows that $k = 5/4$ and hence $a = 2/k = 8/5$.

23. (a) $(1,3) = (1,0) + (0,3) = (1,0) + 3(0,1) = 1\mathbf{i} + 3\mathbf{j}.$

(b) $(-2,-3) = (-2,0) + (0,-3) = -2(1,0) + (-3)(0,1)$
$= -2\mathbf{i} + -3\mathbf{j}.$

(c) $(-2,0) = -2(1,0) = -2\mathbf{i}.$

(d) $(0,3) = 3(0,1) = 3\mathbf{j}.$

25. Represent the force of 300 pounds along the negative y-axis by the vector $-300\mathbf{j}$ and the force of 400 pounds along the negative x-axis by the vector $-400\mathbf{i}$. The resultant force is the direction of the vector $-400\mathbf{i} - 300\mathbf{j}$. The magnitude of the resultant force is

$$\|-400\mathbf{i} - 300\mathbf{j}\| = \sqrt{(-400)^2 + (-300)^2} = \sqrt{250000} = 500 \text{ pounds.}$$

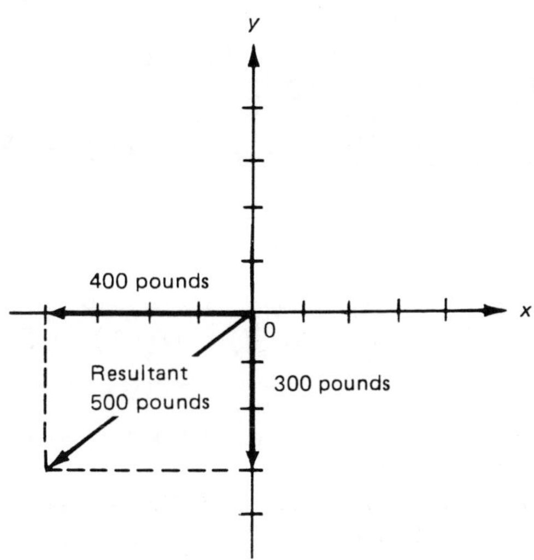

T.1. Locate the point A on the x-axis which is x units from the origin. Construct a perpendicular to the x-axis through A. Locate B on the y-axis y units from the origin. Construct a perpendicular to the y-axis through B. The intersection of those two perpendiculars is the desired point in the plane.

T.3. Let $\mathbf{X} = (x_1,x_2)$. Then

$$\mathbf{X} + (-1)\mathbf{X} = (x_1,x_2) + (-x_1,-x_2) = (x_1-x_1,x_2-x_2) = (0,0) = \mathbf{0}.$$

T.5. We show that **U** has length 1 and is a positive scalar multiple of **X**.

$\|\mathbf{U}\| = \left\|\frac{1}{\|\mathbf{X}\|}\mathbf{X}\right\| = \frac{1}{\|\mathbf{X}\|}\cdot\|\mathbf{X}\| = 1$, and since $\frac{1}{\|\mathbf{X}\|} > 0$, the result follows.

T.7. Let $\mathbf{X} = (x_1,x_2)$, $\mathbf{Y} = (y_1,y_2)$, $\mathbf{Z} = (z_1,z_2)$, and c be a scalar.

(a) $\mathbf{X}\cdot\mathbf{X} = (x_1,x_2)\cdot(x_1,x_2) = x_1^2 + x_2^2 = \|\mathbf{X}\|^2 > 0$ provided that $\mathbf{X} \neq \mathbf{0}$. $\mathbf{X}\cdot\mathbf{X} = 0$ if and only if $x_1^2 + x_2^2 = 0$. Which is true if and only if $x_1 = x_2 = 0$; that is, $\mathbf{X} = \mathbf{0}$.

(b) $\mathbf{X}\cdot\mathbf{Y} = (x_1,x_2)\cdot(y_1,y_2) = x_1y_1 + x_2y_2 = y_1x_1 + y_2x_2$
$= (y_1,y_2)\cdot(x_1,x_2) = \mathbf{Y}\cdot\mathbf{X}.$

(c) $(\mathbf{X} + \mathbf{Y})\cdot\mathbf{Z} = ((x_1,x_2) + (y_1,y_2))\cdot(z_1,z_2)$
$= (x_1+y_1,x_2+y_2)\cdot(z_1,z_2) = (x_1+y_1)z_1 + (x_2+y_2)z_2$
$= x_1z_1 + y_1z_1 + x_2z_2 + y_2z_2$
$= (x_1z_1 + x_2z_2) + (y_1z_1 + y_2z_2)$
$= (x_1,x_2)\cdot(z_1,z_2) + (y_1,y_2)\cdot(z_1,z_2) = \mathbf{X}\cdot\mathbf{Z} + \mathbf{Y}\cdot\mathbf{Z}$.

(d) $(c\mathbf{X})\cdot\mathbf{Y} = (cx_1,cx_2)\cdot(y_1,y_2) = cx_1y_1 + cx_2y_2$
$= x_1cy_1 + x_2cy_2 = (x_1,x_2)\cdot(cy_1,cy_2) = \mathbf{X}\cdot(c\mathbf{Y})$.

$\mathbf{X}\cdot(c\mathbf{Y}) = (x_1,x_2)\cdot(cy_1,cy_2) = cx_1y_1 + cx_2y_2$
$= c(x_1y_1 + x_2y_2) = c(\mathbf{X}\cdot\mathbf{Y})$.

T.9. If **X** and **Y** are parallel, then there exists a nonzero scalar k so that **Y** = k**X**. Thus

$$\cos\theta = \frac{\mathbf{X}\cdot\mathbf{Y}}{\|\mathbf{X}\|\ \|\mathbf{Y}\|} = \frac{\mathbf{X}\cdot k\mathbf{X}}{\|\mathbf{X}\|\ \|k\mathbf{Y}\|} = \frac{k(\mathbf{X}\cdot\mathbf{X})}{\|\mathbf{X}\|\ \sqrt{(k\mathbf{X}\cdot k\mathbf{X})}}$$

$$= \frac{k\|\mathbf{X}\|^2}{\|\mathbf{X}\|\ \sqrt{k^2}\ \|\mathbf{X}\|} = \frac{k}{\sqrt{k^2}} = \frac{k}{\pm k} = \pm 1$$

Exercises 3.2

1. (a) $\mathbf{X} = (1,2,-3)$, $\mathbf{Y} = (0,1,-2)$.
$\mathbf{X} + \mathbf{Y} = (1+0,2+1,(-3)+(-2)) = (1,3,-5)$.
$\mathbf{X} - \mathbf{Y} = (1-0,2-1,-3-(-2)) = (1,1,-1)$.
$2\mathbf{X} = (2(1),2(2),2(-3)) = (2,4,-6)$.
$3\mathbf{X} - 2\mathbf{Y} = (3(1),3(2),3(-3)) - (2(0),2(1),2(-2))$
$= (3,6,-9) - (0,2,-4) = (3,4,-5)$.

(b) $\mathbf{X} = (4,-2,1,3)$, $\mathbf{Y} = (-1,2,5,-4)$.
$\mathbf{X} + \mathbf{Y} = (4+(-1),(-2)+2,1+5,3+(-4)) = (3,0,6,-1)$.
$\mathbf{X} - \mathbf{Y} = (4-(-1),-2-2,1-5,3-(-4)) = (5,-4,-4,7)$.
$2\mathbf{X} = (2(4),2(-2),2(1),2(3)) = (8,-4,2,6)$.
$3\mathbf{X} - 2\mathbf{Y} = (3(4),3(-2),3(1),3(3)) - (2(-1),2(2),2(5),2(-4))$
$= (12,-6,3,9) - (-2,4,10,-8) = (14,-10,-7,17)$.

3. $\mathbf{X} = \begin{bmatrix} 1 \\ -2 \\ 3 \end{bmatrix}$, $\mathbf{Y} = \begin{bmatrix} -3 \\ -1 \\ 3 \end{bmatrix}$, $\mathbf{Z} = \begin{bmatrix} x \\ -1 \\ y \end{bmatrix}$, $\mathbf{U} = \begin{bmatrix} 3 \\ u \\ 2 \end{bmatrix}$. Determine x,y,u in each of the following cases.

(a) Require $\mathbf{Z} = \frac{1}{2}\mathbf{X}$. Then $\begin{bmatrix} x \\ -1 \\ y \end{bmatrix} = \frac{1}{2}\begin{bmatrix} 1 \\ -2 \\ 3 \end{bmatrix} = \begin{bmatrix} 1/2 \\ -1 \\ 3/2 \end{bmatrix}$. Hence, $x = \frac{1}{2}$ and $y = \frac{3}{2}$.

(b) Require $\mathbf{Z} + \mathbf{Y} = \mathbf{X}$. Then $\begin{bmatrix} x-3 \\ -2 \\ y+3 \end{bmatrix} = \begin{bmatrix} 1 \\ -2 \\ 3 \end{bmatrix}$. Hence, $x-3 = 1$, $y+3 = 3$, and it follows that $x = 4$ and $y = 0$.

(c) Require $\mathbf{Z} + \mathbf{U} = \mathbf{Y}$. Then $\begin{bmatrix} x+3 \\ -1+u \\ y+2 \end{bmatrix} = \begin{bmatrix} -3 \\ -1 \\ 3 \end{bmatrix}$. Hence, $x+3 = -3$, $-1+u = -1$, $y+2 = 3$, and it follows that $x = -6$, $u = 0$, and $y = 1$.

5. $\mathbf{X} = (4,5,-2,3)$, $\mathbf{Y} = (3,-2,0,1)$, $\mathbf{Z} = (-3,2,-5,3)$, $c = 2$, and $d = 3$.
(a) **Verify** $\mathbf{X} + \mathbf{Y} = \mathbf{Y} + \mathbf{X}$.
$\mathbf{X} + \mathbf{Y} = (4+3,5+(-2),-2+0,3+1) = (7,3,-2,4)$.
$\mathbf{Y} + \mathbf{X} = (3+4,-2+5,0+(-2),1+3) = (7,3,-2,4)$.

Exercises 3.2

(b) Verify **X** + (**Y** + **Z**) = (**X** + **Y**) + **Z**.
X + (**Y** + **Z**) = **X** + (3+(-3),-2+2,0+(-5),1+3)
= (4,5,-2,3) + (0,0,-5,4) = (4,5,-7,7).
(**X** + **Y**) + **Z** = (4+3,5+(-2),-2+0,3+1) + **Z**
= (7,3,-2,4) + (-3,2,-5,3) = (4,5,-7,7).

(c) Verify **X** + **0** = **0** + **X** = **X**.
X + **0** = (4+0,5+0,-2+0,3+0) = (4,5,-2,3) = **X**.
0 + **X** = (0+4,0+5,0+(-2),0+3) = (4,5,-2,3) = **X**.

(d) Verify **X** + (-**X**) = **0**.
X + (-**X**) = (4-4,5-5,-2-(-2),3-3) = (0,0,0,0) = **0**.

(e) Verify c(**X** + **Y**) = c**X** + c**Y**.
c(**X** + **Y**) = 2(4+3,5+(-2),-2+0,3+1) = 2(7,3,-2,4)
= (14,6,-4,8).
c**X** + c**Y** = 2(4,5,-2,3) + 2(3,-2,0,1)
= (8,10,-4,6) + (6,-4,0,2) = (14,6,-4,8).

(f) Verify (c+d)**X** = c**X** + d**X**.
(c+d)**X** = 5**X** = 5(4,5,-2,3) = (20,25,-10,15).
c**X** + d**X** = 2**X** + 3**X** = (8,10,-4,6) + (12,15,-6,9)
= (20,25,-10,15).

(g) Verify c(d**X**) = (cd)**X**.
c(d**X**) = 2(3**X**) = 2(12,15,-6,9) = (24,30,-12,18).
(cd)**X** = 6**X** = (24,30,-12,18).

(h) Verify 1**X** = **X**.
1**X** = (4,5,-2,3) = **X**.

7.

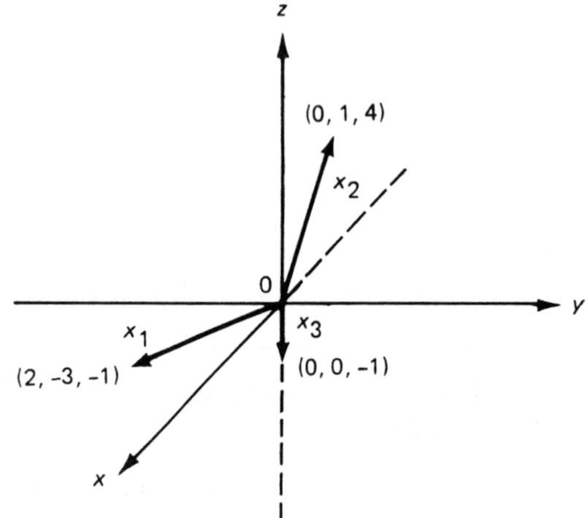

9. Let the head of the vector be (x,y,z). Then we have (3,4,-1) = (x-1,y-(-2),z-3). Hence, x-1 = 3, y-(-2) = 4, z-3 = -1, and it follows that x = 4, y = 2, and z = 2.

11. (a) $\mathbf{X} = (2,3,4)$.

$\|\mathbf{X}\| = \sqrt{2^2 + 3^2 + 4^2} = \sqrt{29}$.

(b) $\mathbf{X} = (0,-1,2,3)$.

$\|\mathbf{X}\| = \sqrt{0^2 + (-1)^2 + 2^2 + 3^2} = \sqrt{14}$.

(c) $\mathbf{X} = (-1,-2,0)$.

$\|\mathbf{X}\| = \sqrt{(-1)^2 + (-2)^2 + 0^2} = \sqrt{5}$.

(d) $\mathbf{X} = (1,2,-3,-4)$.

$\|\mathbf{X}\| = \sqrt{1^2 + 2^2 + (-3)^2 + (-4)^2} = \sqrt{30}$.

13. (a) Let $\mathbf{X} = (1,1,0)$ and $\mathbf{Y} = (2,-3,1)$.

$\|\mathbf{X} - \mathbf{Y}\| = \sqrt{(1-2)^2 + (1-(-3))^2 + (0-1)^2} = \sqrt{18}$.

(b) Let $\mathbf{X} = (4,2,-1,6)$ and $\mathbf{Y} = (4,3,1,5)$.

$\|\mathbf{X} - \mathbf{Y}\| = \sqrt{(4-4)^2 + (2-3)^2 + (-1-1)^2 + (6-5)^2} = \sqrt{6}$.

(c) Let $\mathbf{X} = (0,2,3)$ and $\mathbf{Y} = (1,2,-4)$.

$\|\mathbf{X} - \mathbf{Y}\| = \sqrt{(0-1)^2 + (2-2)^2 + (3-(-4))^2} = \sqrt{50}$.

(d) Let $\mathbf{X} = (3,4,0,1)$ and $\mathbf{Y} = (2,2,1,-1)$.

$\|\mathbf{X} - \mathbf{Y}\| = \sqrt{(3-2)^2 + (4-2)^2 + (0-1)^2 \ (1+1)^2} = \sqrt{10}$.

15. (a) $\mathbf{X} = (2,3,1)$ and $\mathbf{Y} = (3,-2,0)$.
$\mathbf{X}\cdot\mathbf{Y} = (2)(3) + (3)(-2) + (1)(0) = 0$.

(b) $\mathbf{X} = (1,2,-1,3)$ and $\mathbf{Y} = (0,0,-1,-2)$.
$\mathbf{X}\cdot\mathbf{Y} = (1)(0) + (2)(0) + (-1)(-1) + (3)(-2) = -5$.

(c) $\mathbf{X} = (2,0,1)$ and $\mathbf{Y} = (2,2,-1)$.
$\mathbf{X}\cdot\mathbf{Y} = (2)(2) + (0)(2) + (1)(-1) = 3$.

(d) $\mathbf{X} = (0,4,2,3)$ and $\mathbf{Y} = (0,-1,2,0)$.
$\mathbf{X}\cdot\mathbf{Y} = (0)(0) + (4)(-1) + (2)(2) + (3)(0) = 0$.

17. $\mathbf{X}\cdot\mathbf{Y} = 0$ provided that $(a,2,1,a)\cdot(a,-1,-2,-3) = 0$. Computing the inner product gives equation

$$a^2 - 2 - 2 - 3a = 0 \leftrightarrow a^2 - 3a - 4 = 0 \leftrightarrow (a - 4)(a + 1) = 0$$

Hence $a = 4$ or $a = -1$.

Exercises 3.2

19. $\mathbf{X} = (1,2,3)$, $\mathbf{Y} = (1,2,-4)$.
Verify $|\mathbf{X}\cdot\mathbf{Y}| \le \|\mathbf{X}\|\ \|\mathbf{Y}\|$.
$|\mathbf{X}\cdot\mathbf{Y}| = |(1)(1) + (2)(2) + (3)(-4)| = |-7| = 7$.
$\|\mathbf{X}\| = \sqrt{1^2 + 2^2 + 3^2} = \sqrt{14}$.
$\|\mathbf{Y}\| = \sqrt{1^2 + 2^2 + (-4)^2} = \sqrt{21}$.
Hence $\|\mathbf{X}\|\ \|\mathbf{Y}\| = \sqrt{14}\sqrt{21} = \sqrt{294} = 7\sqrt{6}$. It follows that
$|\mathbf{X}\cdot\mathbf{Y}| = 7 \le \|\mathbf{X}\|\ \|\mathbf{Y}\| = 7\sqrt{6}$.

21. (a) $\mathbf{X} = (2,3,1)$ and $\mathbf{Y} = (3,-2,0)$.
$\mathbf{X}\cdot\mathbf{Y} = 0$, $\|\mathbf{X}\| = \sqrt{14}$, $\|\mathbf{Y}\| = \sqrt{13}$.

$$\cos\theta = \frac{0}{\sqrt{14}\sqrt{13}} = 0.$$

(b) $\mathbf{X} = (1,2,-1,3)$ and $\mathbf{Y} = (0,0,-1,-2)$.
$\mathbf{X}\cdot\mathbf{Y} = -5$, $\|\mathbf{X}\| = \sqrt{15}$, $\|\mathbf{Y}\| = \sqrt{5}$.

$$\cos\theta = \frac{-5}{\sqrt{15}\sqrt{5}} = \frac{-5}{\sqrt{75}} = \frac{-5}{5\sqrt{3}} = \frac{-1}{\sqrt{3}}.$$

(c) $\mathbf{X} = (2,0,1)$ and $\mathbf{Y} = (2,2,-1)$.
$\mathbf{X}\cdot\mathbf{Y} = 3$, $\|\mathbf{X}\| = \sqrt{5}$, $\|\mathbf{Y}\| = \sqrt{9} = 3$.

$$\cos\theta = \frac{3}{\sqrt{5}\,3} = \frac{1}{\sqrt{5}}.$$

(d) $\mathbf{X} = (0,4,2,3)$ and $\mathbf{Y} = (0,-1,2,0)$.
$\mathbf{X}\cdot\mathbf{Y} = 0$, $\|\mathbf{X}\| = \sqrt{29}$, $\|\mathbf{Y}\| = \sqrt{5}$.

$$\cos\theta = \frac{0}{\sqrt{29}\sqrt{5}} = 0.$$

23. $\mathbf{X}_1 = (4,2,6,-8)$, $\mathbf{X}_2 = (-2,3,-1,-1)$, $\mathbf{X}_3 = (-2,-1,-3,4)$,
$\mathbf{X}_4 = (1,0,0,2)$, $\mathbf{X}_5 = (1,2,3,-4)$, $\mathbf{X}_6 = (0,-3,1,0)$.

(a) To determine which vectors are orthogonal, we compute the dot products $\mathbf{X}_s\cdot\mathbf{X}_t$ for s and t = 1,2,...,6. If the dot product is zero the vectors are orthogonal. We display the results in the following table.

·	X_1	X_2	X_3	X_4	X_5	X_6
X_1	120	0	-60	-12	58	0
X_2	0	15	0	-4	5	-10
X_3	-60	0	30	6	-29	0
X_4	-12	-4	6	5	-7	0
X_5	58	5	-29	-7	30	-3
X_6	0	-10	0	0	-3	10

Thus we see that the following pairs of vectors are orthogonal: $\mathbf{X}_1$ and $\mathbf{X}_2$, $\mathbf{X}_1$ and $\mathbf{X}_6$, $\mathbf{X}_2$ and $\mathbf{X}_3$, $\mathbf{X}_3$ and $\mathbf{X}_6$, and $\mathbf{X}_4$ and $\mathbf{X}_6$.

(b) Vector $\mathbf{X}_s$ and $\mathbf{X}_t$ are parallel, provided

$$|\mathbf{X}_s \cdot \mathbf{X}_t| = \|\mathbf{X}_s\| \ \|\mathbf{X}_t\|.$$

The products $\|\mathbf{X}_s\| \ \|\mathbf{X}_t\|$ are displayed in the following table.

	$\|\mathbf{X}_1\|$	$\|\mathbf{X}_2\|$	$\|\mathbf{X}_3\|$	$\|\mathbf{X}_4\|$	$\|\mathbf{X}_5\|$	$\|\mathbf{X}_6\|$
$\|\mathbf{X}_1\|$	120	$30\sqrt{2}$	60	$10\sqrt{6}$	60	$20\sqrt{3}$
$\|\mathbf{X}_2\|$	$30\sqrt{2}$	15	$15\sqrt{2}$	$5\sqrt{3}$	$15\sqrt{2}$	$5\sqrt{6}$
$\|\mathbf{X}_3\|$	60	$15\sqrt{2}$	30	$5\sqrt{6}$	30	$10\sqrt{3}$
$\|\mathbf{X}_4\|$	$10\sqrt{6}$	$5\sqrt{3}$	$5\sqrt{6}$	5	$5\sqrt{6}$	$5\sqrt{2}$
$\|\mathbf{X}_5\|$	60	$15\sqrt{2}$	30	$5\sqrt{6}$	30	$10\sqrt{3}$
$\|\mathbf{X}_6\|$	$20\sqrt{3}$	$5\sqrt{6}$	$10\sqrt{3}$	$5\sqrt{2}$	$10\sqrt{3}$	10

Compare the entries of this table with the absolute value of the corresponding entries in the table in part (a). Naturally every vector is parallel to itself (see the diagonal entries). The only nondiagonal entries that are equal are in the (1,3) and (3,1) positions. Thus, $\mathbf{X}_1$ and $\mathbf{X}_3$ are parallel.

(c) Vectors $\mathbf{X}_s$ and $\mathbf{X}_t$ are in the same direction provided

$$\mathbf{X}_s \cdot \mathbf{X}_t = \|\mathbf{X}_s\| \ \|\mathbf{X}_t\|.$$

Compare corresponding entries from the tables in parts (a) and (b). Naturally a vector is in the same direction as itself (see the diagonal entries), but no pair of distinct vectors are in the same direction.

25. (a) $\mathbf{X} = (2,-1,3)$.

$\|\mathbf{X}\| = \sqrt{14}$, $\mathbf{U} = \frac{1}{\|\mathbf{X}\|}\mathbf{X} = (\frac{2}{\sqrt{14}}, \frac{-1}{\sqrt{14}}, \frac{3}{\sqrt{14}})$.

(b) $\mathbf{X} = (1,2,3,4)$.

$\|\mathbf{X}\| = \sqrt{30}$, $\mathbf{U} = \frac{1}{\|\mathbf{X}\|}\mathbf{X} = (\frac{1}{\sqrt{30}}, \frac{2}{\sqrt{30}}, \frac{3}{\sqrt{30}}, \frac{4}{\sqrt{30}})$.

(c) $\mathbf{X} = (0,1,-1)$.

$\|\mathbf{X}\| = \sqrt{2}$, $\mathbf{U} = \frac{1}{\|\mathbf{X}\|}\mathbf{X} = (0, \frac{1}{\sqrt{2}}, \frac{-1}{\sqrt{2}})$.

(d) $\mathbf{X} = (0,-1,2,-1)$.

$\|\mathbf{X}\| = \sqrt{6}$, $\mathbf{U} = \frac{1}{\|\mathbf{X}\|}\mathbf{X} = (0, \frac{-1}{\sqrt{6}}, \frac{2}{\sqrt{6}}, \frac{-1}{\sqrt{6}})$.

27. (a) $(1,2,-3) = (1,0,0) + (0,2,0) + (0,0,-3) = (1,0,0) + 2(0,1,0) - 3(0,0,1) = \mathbf{i} + 2\mathbf{j} - 3\mathbf{k}$.

(b) $(2,3,-1) = (2,0,0) + (0,3,0) + (0,0,-1) = 2(1,0,0) + 3(0,1,0) - 1(0,0,1) = 2\mathbf{i} + 3\mathbf{j} - \mathbf{k}$.

(c) $(0,1,2) = (0,0,0) + (0,1,0) + (0,0,2) = 0(1,0,0) + (0,1,0) + 2(0,0,1) = 0\mathbf{i} + \mathbf{j} + 2\mathbf{k}$.

(d) $(0,0,-2) = -2(0,0,1) = -2\mathbf{k}$.

29. A triangle with vertices $P_1(2,3,-4)$, $P_2(3,1,2)$, $P_3(-3,0,4)$ is isosceles provided two of the sides are of equal length.

$\|\overrightarrow{P_1P_2}\| = ((3-2)^2 + (1-3)^2 + (2-(-4))^2)^{1/2} = (41)^{1/2}$

$\|\overrightarrow{P_1P_3}\| = ((-3-2)^2 + (0-3)^2 + (4-(-4))^2)^{1/2} = (98)^{1/2}$

$\|\overrightarrow{P_2P_3}\| = ((-3-3)^2 + (0-1)^2 + (4-2)^2)^{1/2} = (41)^{1/2}$

Thus, the side connecting P_1 to P_2 and the side connecting P_2 to P_3 have the same length.

31. The "new salaries" = "old salaries" + "the increase". Thus,

"new salaries" = $\mathbf{S} + .08\mathbf{S} = 1.08\mathbf{S}$.

33. The average daily price of stock is $.5(\mathbf{L} + \mathbf{H})$.

T.1. (a) $\mathbf{X} + \mathbf{Y} = \begin{bmatrix} x_1 + y_1 \\ \cdot \\ \cdot \\ \cdot \\ x_n + y_n \end{bmatrix} = \begin{bmatrix} y_1 + x_1 \\ \cdot \\ \cdot \\ \cdot \\ y_n + x_n \end{bmatrix} = \mathbf{Y} + \mathbf{X}$.

(b) $\mathbf{X} + (\mathbf{Y}+\mathbf{Z}) = \begin{bmatrix} x_1 + (y_1+z_1) \\ \cdot \\ \cdot \\ \cdot \\ x_n + (y_n+z_n) \end{bmatrix} = \begin{bmatrix} (x_1+y_1) + z_1 \\ \cdot \\ \cdot \\ \cdot \\ (x_n+y_n) + z_n \end{bmatrix} = (\mathbf{X}+\mathbf{Y}) + \mathbf{Z}.$

(c) $\mathbf{X} + \mathbf{0} = \begin{bmatrix} x_1 + 0 \\ \cdot \\ \cdot \\ \cdot \\ x_n + 0 \end{bmatrix} = \begin{bmatrix} x_1 \\ \cdot \\ \cdot \\ \cdot \\ x_n \end{bmatrix} = \mathbf{X}.$

(d) $\mathbf{X} + (-\mathbf{X}) = \begin{bmatrix} x_1 + (-x_1) \\ \cdot \\ \cdot \\ \cdot \\ x_n + (-x_n) \end{bmatrix} = \begin{bmatrix} 0 \\ \cdot \\ \cdot \\ \cdot \\ 0 \end{bmatrix} = \mathbf{0}.$

(e) $c(\mathbf{X} + \mathbf{Y}) = \begin{bmatrix} c(x_1 + y_1) \\ \cdot \\ \cdot \\ \cdot \\ c(x_n + y_n) \end{bmatrix} = \begin{bmatrix} cx_1 + cy_1 \\ \cdot \\ \cdot \\ \cdot \\ cx_n + cy_n \end{bmatrix} = \begin{bmatrix} cx_1 \\ \cdot \\ \cdot \\ \cdot \\ cx_n \end{bmatrix} + \begin{bmatrix} cy_1 \\ \cdot \\ \cdot \\ \cdot \\ cy_n \end{bmatrix}$

$= c\mathbf{X} + c\mathbf{Y}.$

(g) $c(d\mathbf{X}) = \begin{bmatrix} c(dx_1) \\ \cdot \\ \cdot \\ \cdot \\ c(dx_n) \end{bmatrix} = \begin{bmatrix} (cd)x_1 \\ \cdot \\ \cdot \\ \cdot \\ (cd)x_n \end{bmatrix} = (cd)\mathbf{X}.$

(h) $1\mathbf{X} = \begin{bmatrix} 1x_1 \\ \cdot \\ \cdot \\ \cdot \\ 1x_n \end{bmatrix} = \begin{bmatrix} x_1 \\ \cdot \\ \cdot \\ \cdot \\ x_n \end{bmatrix} = \mathbf{X}.$

T.3. The origin **O** and the head of the vector **X**, call it P, are opposite vertices of a parallelepiped with faces parallel to the coordinate planes (see Figure).

Exercises 3.2

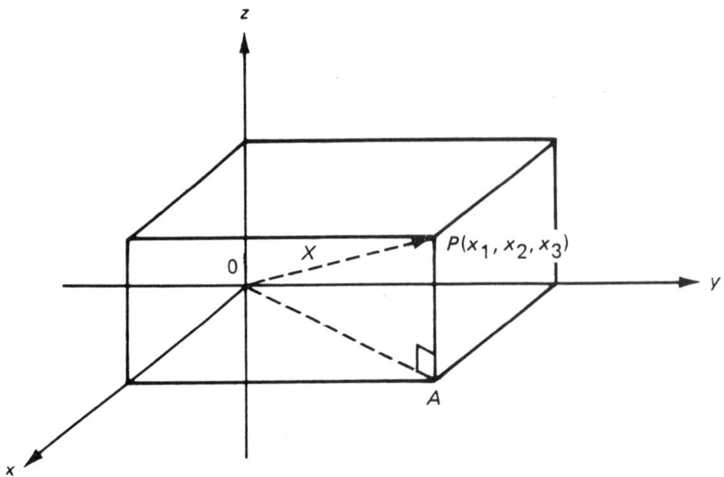

The face diagonal OA has length $\sqrt{x_1^2 + x_2^2}$ by one application of the Pythagorean Theorem. By a second application, the body diagonal has length

$$\|\mathbf{X}\| = OP = \sqrt{\left[\sqrt{x_1^2 + x_2^2}\right]^2 + x_3^2} = \sqrt{x_1^2 + x_2^2 + x_3^2}.$$

T.5. Given that $\mathbf{Z}\cdot\mathbf{X} = \mathbf{Z}\cdot\mathbf{Y} = 0$. Let r and s be scalars. Then $\mathbf{Z}\cdot(r\mathbf{X} + s\mathbf{Y}) = r(\mathbf{Z}\cdot\mathbf{X}) + s(\mathbf{Z}\cdot\mathbf{Y}) = r(0) + s(0) = 0$. Thus, $\mathbf{Z}$ is orthogonal to $r\mathbf{X} + s\mathbf{Y}$. (See also Exercise T.8 in Section 3.1.)

T.7. $\mathbf{X}\cdot(\mathbf{Y} + \mathbf{Z}) = (\mathbf{Y} + \mathbf{Z})\cdot\mathbf{X}$ (by part (b) of Theorem 3.3)

$= \mathbf{Y}\cdot\mathbf{X} + \mathbf{Z}\cdot\mathbf{X}$ (by part (c) of Theorem 3.3)

$= \mathbf{X}\cdot\mathbf{Y} + \mathbf{X}\cdot\mathbf{Z}$ (by part (b) of Theorem 3.3)

T.9. $\|c\mathbf{X}\| = \sqrt{(cx_1)^2 + (cx_2)^2 + \cdots + (cx_n)^n}$

$= \sqrt{c^2\,(x_1^2 + x_2^2 + \cdots + x_n^2)}$

$= \sqrt{c^2}\,\sqrt{x_1^2 + x_2^2 + \cdots + x_n^2} = |c|\ \|\mathbf{X}\|.$

T.11. Consider $\mathbf{X}$ and $\mathbf{Y}$, $n \times 1$ matrices with entries x_j and y_j respectively. Then

$$\mathbf{X}\cdot\mathbf{Y} = x_1y_1 + x_2y_2 + \cdots + x_ny_n = [x_1\ x_2\ \ldots\ x_n]\begin{bmatrix} y_1 \\ y_2 \\ \cdot \\ \cdot \\ y_n \end{bmatrix} = \mathbf{X}^T\mathbf{Y}.$$

T.13. From the solution to Exercise T.10 we have

$$\|\mathbf{X} + \mathbf{Y}\|^2 = \|\mathbf{X}\|^2 + \|\mathbf{Y}\|^2 + 2(\mathbf{X}\cdot\mathbf{Y}).$$

Replace **Y** by **-Y** and we have

$$\|\mathbf{X} - \mathbf{Y}\|^2 = \|\mathbf{X}\|^2 + \|\mathbf{Y}\|^2 - 2(\mathbf{X}\cdot\mathbf{Y}).$$

Add these two equations to obtain

$$\|\mathbf{X} + \mathbf{Y}\|^2 + \|\mathbf{X} - \mathbf{Y}\|^2 = 2\|\mathbf{X}\|^2 + 2\|\mathbf{Y}\|^2.$$

T.15. From the solution to Exercise T.13 we have

$$\|\mathbf{X} + \mathbf{Y}\|^2 = \|\mathbf{X}\|^2 + \|\mathbf{Y}\|^2 + 2(\mathbf{X}\cdot\mathbf{Y})$$

and

$$\|\mathbf{X} - \mathbf{Y}\|^2 = \|\mathbf{X}\|^2 + \|\mathbf{Y}\|^2 - 2(\mathbf{X}\cdot\mathbf{Y}).$$

Forming the difference $\|\mathbf{X} + \mathbf{Y}\|^2 - \|\mathbf{X} - \mathbf{Y}\|^2$ gives

$$\|\mathbf{X} + \mathbf{Y}\|^2 - \|\mathbf{X} - \mathbf{Y}\|^2 = 4(\mathbf{X}\cdot\mathbf{Y})$$

and it follows that

$$\mathbf{X}\cdot\mathbf{Y} = (1/4)\|\mathbf{X} + \mathbf{Y}\|^2 - (1/4)\|\mathbf{X} - \mathbf{Y}\|^2 .$$

ML.2. (a)
```
X=[2 2 -1]';norm(X)
ans =
     3
```

(b)
```
Y=[0 4 -3 0]';norm(Y)
ans =
     5
```

Exercises 3.2

(c) Z = [1 0 1 0 3]';norm(Z)

```
ans =

     3.3166
```

ML.3. (a) X=[2 0 3]';Y=[2 -1 1]';norm(X-Y)

```
ans =

     2.2361
```

(b) X=[2 0 0 1];Y=[2 5 -1 3];norm(X-Y)

```
ans =

     5.4772
```

(c) X=[1 0 4 3];Y=[-1 1 2 2];norm(X-Y)

```
ans =

     3.1623
```

ML.4. Enter A, B, and C as points and construct vectors vAB,vBC, and vCA. Then determine the lengths of the vectors.

```
A==[1 3 -2];B=[4 -1 0];C=[1 1 2];
vAB=B-A

vAB =

     3    -4     2

norm(vAB)

ans =

     5.3852

vBC=C-B

vBC =

    -3     2     2

norm(vBC)
```

```
ans =

     4.1231

vCA=A-C

vCA =

     0     2    -4

norm(vCA)

ans =

     4.4721
```

ML.5. (a)
```
X=[5 4 -4];Y=[3 2 1];
dot(X,Y)

ans =

    19
```

(b)
```
X=[3 -1 0 2];Y=[-1 2 -5 -3];
dot(X,Y)

ans =

   -11
```

(c)
```
X=[1 2 3 4 5];
dot(X,-X)

ans =

   -55
```

ML.8. (a)
```
X=[3 2 4 0];Y=[0 2 -1 0];
ang=dot(X,Y)/(norm(X)*norm(Y))

ang =

     0
```

(b)
```
X=[2 2 -1];Y=[2 0 1];
ang=dot(X,Y)/(norm(X)*norm(Y))

ang =

     0.4472

degrees=ang*(180/pi)
```

```
degrees =

        25.6235
```

(c)
```
X=[1 0 0 2];Y=[0 3 -4 0];
ang=dot(X,Y)/(norm(X)*norm(Y))

ang =

     0
```

ML.9. (a)
```
X=[2 2 -1]';
unit=X/norm(X)

unit =

       0.6667
       0.6667
      -0.3333

rat(unit,'s')

ans =

     2/3
     2/3
    -1/3
```

(b)
```
Y=[0 4 -3 0]';
unit=Y/norm(Y)

unit =

            0
       0.8000
      -0.6000
            0

rat(unit,'s')

ans =

      0
     4/5
    -3/5
      0
```

(c)
```
Z=[1 0 1 0 3]';
unit=Z/norm(Z)

unit =

    0.3015
         0
    0.3015
         0
    0.9045
```

Exercises 3.3

1. (a) $\mathbf{X} = 2\mathbf{i} + 3\mathbf{j} + 4\mathbf{k}$, $\mathbf{Y} = -\mathbf{i} + 3\mathbf{j} - \mathbf{k}$.

$$\mathbf{X} \times \mathbf{Y} = \begin{vmatrix} \mathbf{i} & \mathbf{j} & \mathbf{k} \\ 2 & 3 & 4 \\ -1 & 3 & -1 \end{vmatrix} = \begin{vmatrix} 3 & 4 \\ 3 & -1 \end{vmatrix}\mathbf{i} - \begin{vmatrix} 2 & 4 \\ -1 & -1 \end{vmatrix}\mathbf{j} + \begin{vmatrix} 2 & 3 \\ -1 & 3 \end{vmatrix}\mathbf{k}$$

$$= -15\mathbf{i} - 2\mathbf{j} + 9\mathbf{k}.$$

(b) $\mathbf{X} = (1,0,1)$, $\mathbf{Y} = (2,3,-1)$.

$$\mathbf{X} \times \mathbf{Y} = \begin{vmatrix} \mathbf{i} & \mathbf{j} & \mathbf{k} \\ 1 & 0 & 1 \\ 2 & 3 & -1 \end{vmatrix} = \begin{vmatrix} 0 & 1 \\ 3 & -1 \end{vmatrix}\mathbf{i} - \begin{vmatrix} 1 & 1 \\ 2 & -1 \end{vmatrix}\mathbf{j} + \begin{vmatrix} 1 & 0 \\ 2 & 3 \end{vmatrix}\mathbf{k}$$

$$= -3\mathbf{i} + 3\mathbf{j} + 3\mathbf{k}.$$

(c) $\mathbf{X} = \mathbf{i} - \mathbf{j} + 2\mathbf{k}$, $\mathbf{Y} = 3\mathbf{i} - 4\mathbf{j} + \mathbf{k}$.

$$\mathbf{X} \times \mathbf{Y} = \begin{vmatrix} \mathbf{i} & \mathbf{j} & \mathbf{k} \\ 1 & -1 & 2 \\ 3 & -4 & 1 \end{vmatrix} = \begin{vmatrix} -1 & 2 \\ -4 & 1 \end{vmatrix}\mathbf{i} - \begin{vmatrix} 1 & 2 \\ 3 & 1 \end{vmatrix}\mathbf{j} + \begin{vmatrix} 1 & -1 \\ 3 & -4 \end{vmatrix}\mathbf{k}$$

$$= 7\mathbf{i} + 5\mathbf{j} - \mathbf{k}.$$

(d) $\mathbf{X} = (2,-1,1)$, $\mathbf{Y} = -2\mathbf{X}$.

$\mathbf{X} \times \mathbf{Y} = \mathbf{X} \times (-2\mathbf{X}) = -2(\mathbf{X} \times \mathbf{X})$ (by Theorem 3.6(d))

$= -2\,\mathbf{0} = \mathbf{0}$ (by Theorem 3.6(e))

3. Let $\mathbf{X} = \mathbf{i} + 2\mathbf{j} - 3\mathbf{k}$, $\mathbf{Y} = 2\mathbf{i} + 3\mathbf{j} + \mathbf{k}$, $\mathbf{Z} = 2\mathbf{i} - \mathbf{j} + 2\mathbf{k}$, $c = -3$.

(a) Verify $\mathbf{X} \times \mathbf{Y} = -(\mathbf{Y} \times \mathbf{X})$.

$$\mathbf{X} \times \mathbf{Y} = \begin{vmatrix} \mathbf{i} & \mathbf{j} & \mathbf{k} \\ 1 & 2 & -3 \\ 2 & 3 & 1 \end{vmatrix} = \begin{vmatrix} 2 & -3 \\ 3 & 1 \end{vmatrix}\mathbf{i} - \begin{vmatrix} 1 & -3 \\ 2 & 1 \end{vmatrix}\mathbf{j} + \begin{vmatrix} 1 & 2 \\ 2 & 3 \end{vmatrix}\mathbf{k}$$

$$= 11\mathbf{i} - 7\mathbf{j} - \mathbf{k}.$$

$$-(\mathbf{Y} \times \mathbf{X}) = -\begin{vmatrix} \mathbf{i} & \mathbf{j} & \mathbf{k} \\ 2 & 3 & 1 \\ 1 & 2 & -3 \end{vmatrix} = -\begin{vmatrix} 3 & 1 \\ 2 & -3 \end{vmatrix}\mathbf{i} + \begin{vmatrix} 2 & 1 \\ 1 & -3 \end{vmatrix}\mathbf{j} - \begin{vmatrix} 2 & 3 \\ 1 & 2 \end{vmatrix}\mathbf{k}$$

$$= 11\mathbf{i} - 7\mathbf{j} - \mathbf{k}.$$

(b) Verify $\mathbf{X} \times (\mathbf{Y} + \mathbf{Z}) = \mathbf{X} \times \mathbf{Y} + \mathbf{X} \times \mathbf{Z}$.

$\mathbf{X} \times (\mathbf{Y} + \mathbf{Z}) = (\mathbf{i} + 2\mathbf{j} - 3\mathbf{k}) \times (4\mathbf{i} + 2\mathbf{j} + 3\mathbf{k})$

$$= \begin{vmatrix} \mathbf{i} & \mathbf{j} & \mathbf{k} \\ 1 & 2 & -3 \\ 4 & 2 & 3 \end{vmatrix} = \begin{vmatrix} 2 & -3 \\ 2 & 3 \end{vmatrix}\mathbf{i} - \begin{vmatrix} 1 & -3 \\ 4 & 3 \end{vmatrix}\mathbf{j} + \begin{vmatrix} 1 & 2 \\ 4 & 2 \end{vmatrix}\mathbf{k}$$

$$= 12\mathbf{i} - 15\mathbf{j} - 6\mathbf{k}.$$

$$\mathbf{X} \times \mathbf{Y} + \mathbf{X} \times \mathbf{Z} = \begin{vmatrix} \mathbf{i} & \mathbf{j} & \mathbf{k} \\ 1 & 2 & -3 \\ 2 & 3 & 1 \end{vmatrix} + \begin{vmatrix} \mathbf{i} & \mathbf{j} & \mathbf{k} \\ 1 & 2 & -3 \\ 2 & -1 & 2 \end{vmatrix}$$

$$= (11\mathbf{i} - 7\mathbf{j} - \mathbf{k}) + \begin{vmatrix} 2 & -3 \\ -1 & 2 \end{vmatrix}\mathbf{i} - \begin{vmatrix} 1 & -3 \\ 2 & 2 \end{vmatrix}\mathbf{j} + \begin{vmatrix} 1 & 2 \\ 2 & -1 \end{vmatrix}\mathbf{k}$$

$$= (11\mathbf{i} - 7\mathbf{j} - \mathbf{k}) + (\mathbf{i} - 8\mathbf{j} - 5\mathbf{k}) = 12\mathbf{i} - 15\mathbf{j} - 6\mathbf{k}.$$

(c) Verify $(\mathbf{X} + \mathbf{Y}) \times \mathbf{Z} = \mathbf{X} \times \mathbf{Z} + \mathbf{Y} \times \mathbf{Z}$.
$(\mathbf{X} + \mathbf{Y}) \times \mathbf{Z} = (3\mathbf{i} + 5\mathbf{j} - 2\mathbf{k}) \times (2\mathbf{i} - \mathbf{j} + 2\mathbf{k})$

$$= \begin{vmatrix} \mathbf{i} & \mathbf{j} & \mathbf{k} \\ 3 & 5 & -2 \\ 2 & -1 & 2 \end{vmatrix} = \begin{vmatrix} 5 & -2 \\ -1 & 2 \end{vmatrix}\mathbf{i} - \begin{vmatrix} 3 & -2 \\ 2 & 2 \end{vmatrix}\mathbf{j} + \begin{vmatrix} 3 & 5 \\ 2 & -1 \end{vmatrix}\mathbf{k}$$

$$= 8\mathbf{i} - 10\mathbf{j} - 13\mathbf{k}.$$

$$\mathbf{X} \times \mathbf{Z} + \mathbf{Y} \times \mathbf{Z} = \begin{vmatrix} \mathbf{i} & \mathbf{j} & \mathbf{k} \\ 1 & 2 & -3 \\ 2 & -1 & 2 \end{vmatrix} + \begin{vmatrix} \mathbf{i} & \mathbf{j} & \mathbf{k} \\ 2 & 3 & 1 \\ 2 & -1 & 2 \end{vmatrix}$$

$$= (\mathbf{i} - 8\mathbf{j} - 5\mathbf{k}) + \begin{vmatrix} 3 & 1 \\ -1 & 2 \end{vmatrix}\mathbf{i} - \begin{vmatrix} 2 & 1 \\ 2 & 2 \end{vmatrix}\mathbf{j} + \begin{vmatrix} 2 & 3 \\ 2 & -1 \end{vmatrix}\mathbf{k}$$

$$= (\mathbf{i} - 8\mathbf{j} - 5\mathbf{k}) + (7\mathbf{i} - 2\mathbf{j} - 8\mathbf{k}) = 8\mathbf{i} - 10\mathbf{j} - 13\mathbf{k}.$$

(d) Verify $c(\mathbf{X} \times \mathbf{Y}) = (c\mathbf{X}) \times \mathbf{Y} = \mathbf{X} \times (c\mathbf{Y})$.
$-3(\mathbf{X} \times \mathbf{Y}) = -3(11\mathbf{i} - 7\mathbf{j} - \mathbf{k}) = -33\mathbf{i} + 21\mathbf{j} + 3\mathbf{k}.$
$(-3\mathbf{X}) \times \mathbf{Y} = (-3\mathbf{i} - 6\mathbf{j} + 9\mathbf{k}) \times (2\mathbf{i} + 3\mathbf{j} + \mathbf{k})$

$$= \begin{vmatrix} \mathbf{i} & \mathbf{j} & \mathbf{k} \\ -3 & -6 & 9 \\ 2 & 3 & 1 \end{vmatrix} = \begin{vmatrix} -6 & 9 \\ 3 & 1 \end{vmatrix}\mathbf{i} - \begin{vmatrix} -3 & 9 \\ 2 & 1 \end{vmatrix}\mathbf{j} + \begin{vmatrix} -3 & -6 \\ 2 & 3 \end{vmatrix}\mathbf{k}$$

$$= -33\mathbf{i} + 21\mathbf{j} + 3\mathbf{k}.$$

$\mathbf{X} \times (c\mathbf{Y}) = (\mathbf{i} + 2\mathbf{j} - 3\mathbf{k}) \times (-6\mathbf{i} - 9\mathbf{j} - 3\mathbf{k})$

$$= \begin{vmatrix} \mathbf{i} & \mathbf{j} & \mathbf{k} \\ 1 & 2 & -3 \\ -6 & -9 & -3 \end{vmatrix} = \begin{vmatrix} 2 & -3 \\ -9 & -3 \end{vmatrix}\mathbf{i} - \begin{vmatrix} 1 & -3 \\ -6 & -3 \end{vmatrix}\mathbf{j} + \begin{vmatrix} 1 & 2 \\ -6 & -9 \end{vmatrix}\mathbf{k}$$

$$= -33\mathbf{i} + 21\mathbf{j} + 3\mathbf{k}.$$

5. $\mathbf{X} = \mathbf{i} - \mathbf{j} + 2\mathbf{k}$, $\mathbf{Y} = 2\mathbf{i} + 2\mathbf{j} - \mathbf{k}$, $\mathbf{Z} = \mathbf{i} + \mathbf{j} - \mathbf{k}$.
(a) Verify $(\mathbf{X} \times \mathbf{Y})\cdot\mathbf{Z} = \mathbf{X}\cdot(\mathbf{Y} \times \mathbf{Z})$.

Exercises 3.3

$$(\mathbf{X} \times \mathbf{Y})\cdot\mathbf{Z} = \begin{vmatrix} \mathbf{i} & \mathbf{j} & \mathbf{k} \\ 1 & -1 & 2 \\ 2 & 2 & -1 \end{vmatrix}\cdot(\mathbf{i} + \mathbf{j} - \mathbf{k})$$

$$= \left[\begin{vmatrix} -1 & 2 \\ 2 & -1 \end{vmatrix}\mathbf{i} - \begin{vmatrix} 1 & 2 \\ 2 & -1 \end{vmatrix}\mathbf{j} + \begin{vmatrix} 1 & -1 \\ 2 & 2 \end{vmatrix}\mathbf{k}\right]\cdot(\mathbf{i} + \mathbf{j} - \mathbf{k})$$

$$= (-3\mathbf{i} + 5\mathbf{j} + 4\mathbf{k})\cdot(\mathbf{i} + \mathbf{j} - \mathbf{k}) = -3 + 5 - 4 = -2.$$

$$\mathbf{X}\cdot(\mathbf{Y} \times \mathbf{Z}) = (\mathbf{i} - \mathbf{j} + 2\mathbf{k})\cdot\begin{vmatrix} \mathbf{i} & \mathbf{j} & \mathbf{k} \\ 2 & 2 & -1 \\ 1 & 1 & -1 \end{vmatrix}$$

$$= (\mathbf{i} - \mathbf{j} + 2\mathbf{k})\cdot\left[\begin{vmatrix} 2 & -1 \\ 1 & -1 \end{vmatrix}\mathbf{i} - \begin{vmatrix} 2 & -1 \\ 1 & -1 \end{vmatrix}\mathbf{j} + \begin{vmatrix} 2 & 2 \\ 1 & 1 \end{vmatrix}\mathbf{k}\right]$$

$$= (\mathbf{i} - \mathbf{j} + 2\mathbf{k})\cdot(-\mathbf{i} + \mathbf{j}) = -1 - 1 = -2.$$

(b) Verify $\mathbf{X} \times (\mathbf{Y} \times \mathbf{Z}) = (\mathbf{X}\cdot\mathbf{Z})\mathbf{Y} - (\mathbf{X}\cdot\mathbf{Y})\mathbf{Z}$.

$$\mathbf{X} \times (\mathbf{Y} \times \mathbf{Z}) = (\mathbf{i} - \mathbf{j} + 2\mathbf{k})\times(-\mathbf{i} + \mathbf{j}) = \begin{vmatrix} \mathbf{i} & \mathbf{j} & \mathbf{k} \\ 1 & -1 & 2 \\ -1 & 1 & 0 \end{vmatrix}$$

$$= \begin{vmatrix} -1 & 2 \\ 1 & 0 \end{vmatrix}\mathbf{i} - \begin{vmatrix} 1 & 2 \\ -1 & 0 \end{vmatrix}\mathbf{j} + \begin{vmatrix} 1 & -1 \\ -1 & 1 \end{vmatrix}\mathbf{k} = -2\mathbf{i} - 2\mathbf{j}.$$

$$(\mathbf{X}\cdot\mathbf{Z})\mathbf{Y} - (\mathbf{X}\cdot\mathbf{Y})\mathbf{Z} = -2\mathbf{Y} + 2\mathbf{Z} = -2\mathbf{i} - 2\mathbf{j}.$$

7. Use the results from Exercise 2 for $\mathbf{X} \times \mathbf{Y}$.

(a) $\mathbf{X}\cdot(\mathbf{X} \times \mathbf{Y}) = (\mathbf{i} - \mathbf{j} + 2\mathbf{k})\cdot(-4\mathbf{i} + 4\mathbf{j} + 4\mathbf{k}) = -4 - 4 + 8 = 0.$
$\mathbf{Y}\cdot(\mathbf{X} \times \mathbf{Y}) = (3\mathbf{i} + \mathbf{j} + 2\mathbf{k})\cdot(-4\mathbf{i} + 4\mathbf{j} + 4\mathbf{k})$
$= -12 + 4 + 8 = 0.$

(b) $\mathbf{X}\cdot(\mathbf{X} \times \mathbf{Y}) = (2\mathbf{i} + \mathbf{j} - 2\mathbf{k})\cdot(3\mathbf{i} - 8\mathbf{j} - \mathbf{k}) = 6 - 8 + 2 = 0.$
$\mathbf{Y}\cdot(\mathbf{X} \times \mathbf{Y}) = (\mathbf{i} + 3\mathbf{k})\cdot(3\mathbf{i} - 8\mathbf{j} - \mathbf{k}) = 3 - 0 - 3 = 0.$

(c) $\mathbf{X}\cdot(\mathbf{X} \times \mathbf{Y}) = (2\mathbf{j} + \mathbf{k})\cdot(0\mathbf{i} + 0\mathbf{j} + 0\mathbf{k}) = 0.$
$\mathbf{Y}\cdot(\mathbf{X} \times \mathbf{Y}) = (6\mathbf{j} + 3\mathbf{k})\cdot(0\mathbf{i} + 0\mathbf{j} + 0\mathbf{k}) = 0.$

(d) $\mathbf{X}\cdot(\mathbf{X} \times \mathbf{Y}) = (4\mathbf{i} - 2\mathbf{k})\cdot(4\mathbf{i} + 4\mathbf{j} + 8\mathbf{k}) = 16 + 0 - 16 = 0.$
$\mathbf{Y}\cdot(\mathbf{X} \times \mathbf{Y}) = (2\mathbf{j} - \mathbf{k})\cdot(4\mathbf{i} + 4\mathbf{j} + 8\mathbf{k}) = 0 + 8 - 8 = 0.$

9. Let a triangle have vertices $\mathbf{X}_1 = (1,-2,3)$, $\mathbf{X}_2 = (-3,1,4)$, $\mathbf{X}_3 = (0,4,3)$. Its area is given by

$$A_T = \frac{1}{2}\,\|(\mathbf{X}_2 - \mathbf{X}_1)\times(\mathbf{X}_3 - \mathbf{X}_1)\| = \frac{1}{2}\,\|(-4,3,1)\times(-1,6,0)\|$$

$$= \frac{1}{2}\left\|\begin{vmatrix} \mathbf{i} & \mathbf{j} & \mathbf{k} \\ -4 & 3 & 1 \\ -1 & 6 & 0 \end{vmatrix}\right\| = \frac{1}{2}\,\|-6\mathbf{i} + \mathbf{j} - 21\mathbf{k}\|$$

$$= \frac{1}{2}\sqrt{(-6)^2 + 1^2 + (-21)^2} = \frac{1}{2}\sqrt{478}.$$

11. $A_P = \|(\mathbf{X}_2 - \mathbf{X}_1)\times(\mathbf{X}_3 - \mathbf{X}_1)\| = \|-5\mathbf{i} - 5\mathbf{j} - 10\mathbf{k}\|$
$= \sqrt{(-5)^2 + (-5)^2 + (-10)^2} = \sqrt{150} = 5\sqrt{6}.$

13. Volume $= |\mathbf{X}\cdot(\mathbf{Y}\times\mathbf{Z})| = |(\mathbf{i} - 2\mathbf{j} + 4\mathbf{k})\cdot(3\mathbf{i} - 4\mathbf{j} + 7\mathbf{k})|$
$= |3 + 8 + 28| = 39.$

T.1. Let $\mathbf{X} = (x_1,x_2,x_3)$, $\mathbf{Y} = (y_1,y_2,y_3)$, $\mathbf{Z} = (z_1,z_2,z_3)$, and c be a scalar.

(a) Prove $\mathbf{X} \times \mathbf{Y} = -(\mathbf{Y} \times \mathbf{X})$.

$$\mathbf{X} \times \mathbf{Y} = \begin{vmatrix} \mathbf{i} & \mathbf{j} & \mathbf{k} \\ x_1 & x_2 & x_3 \\ y_1 & y_2 & y_3 \end{vmatrix} \underset{\mathbf{r}_2 \leftrightarrow \mathbf{r}_3}{=} -\begin{vmatrix} \mathbf{i} & \mathbf{j} & \mathbf{k} \\ y_1 & y_2 & y_3 \\ x_1 & x_2 & x_3 \end{vmatrix} = -(\mathbf{Y} \times \mathbf{X}).$$

(b) Prove $\mathbf{X} \times (\mathbf{Y} + \mathbf{Z}) = \mathbf{X} \times \mathbf{Y} + \mathbf{X} \times \mathbf{Y}$.

$$\mathbf{X} \times (\mathbf{Y} + \mathbf{Z}) = \begin{vmatrix} \mathbf{i} & \mathbf{j} & \mathbf{k} \\ x_1 & x_2 & x_3 \\ y_1+z_1 & y_2+z_2 & y_3+z_3 \end{vmatrix}$$

$$= \begin{vmatrix} \mathbf{i} & \mathbf{j} & \mathbf{k} \\ x_1 & x_2 & x_3 \\ y_1 & y_2 & y_3 \end{vmatrix} + \begin{vmatrix} \mathbf{i} & \mathbf{j} & \mathbf{k} \\ x_1 & x_2 & x_3 \\ y_1 & y_2 & y_3 \end{vmatrix} = \mathbf{X} \times \mathbf{Y} + \mathbf{X} \times \mathbf{Z}.$$

(c) Prove $(\mathbf{X} + \mathbf{Y}) \times \mathbf{Z} = \mathbf{X} \times \mathbf{Z} + \mathbf{Y} \times \mathbf{Z}$.

$$\begin{aligned} (\mathbf{X} + \mathbf{Y}) \times \mathbf{Z} &= -[\mathbf{Z} \times (\mathbf{X} + \mathbf{Y})] && \text{(by part (a))} \\ &= -[\mathbf{Z} \times \mathbf{X} + \mathbf{Z} \times \mathbf{Y}] && \text{(by part (b))} \\ &= -(\mathbf{Z} \times \mathbf{X}) - (\mathbf{Z} \times \mathbf{Y}) \\ &= \mathbf{X} \times \mathbf{Z} + \mathbf{Y} \times \mathbf{Z}. && \text{(by part (a))} \end{aligned}$$

(d) Prove $c(\mathbf{X} \times \mathbf{Y}) = (c\mathbf{X}) \times \mathbf{Y} = \mathbf{X} \times (c\mathbf{Y})$.

$$c(\mathbf{X} \times \mathbf{Y}) = c\begin{vmatrix} \mathbf{i} & \mathbf{j} & \mathbf{k} \\ x_1 & x_2 & x_3 \\ y_1 & y_2 & y_3 \end{vmatrix} = \begin{vmatrix} \mathbf{i} & \mathbf{j} & \mathbf{k} \\ cx_1 & cx_2 & cx_3 \\ y_1 & y_2 & y_3 \end{vmatrix} \quad \text{(by Theorem 2.5)}$$

$$= (c\mathbf{X}) \times \mathbf{Y}.$$

$$c(\mathbf{X} \times \mathbf{Y}) = c\begin{vmatrix} \mathbf{i} & \mathbf{j} & \mathbf{k} \\ x_1 & x_2 & x_3 \\ y_1 & y_2 & y_3 \end{vmatrix} = \begin{vmatrix} \mathbf{i} & \mathbf{j} & \mathbf{k} \\ x_1 & x_2 & x_3 \\ cy_1 & cy_2 & cy_3 \end{vmatrix} \quad \text{(by Theorem 2.5)}$$

$$= \mathbf{X} \times (c\mathbf{Y}).$$

(e) Prove $\mathbf{X} \times \mathbf{X} = \mathbf{0}$.

$$\mathbf{X} \times \mathbf{X} = \begin{vmatrix} \mathbf{i} & \mathbf{j} & \mathbf{k} \\ x_1 & x_2 & x_3 \\ x_1 & x_2 & x_3 \end{vmatrix} = \mathbf{0} \quad \text{(by Theorem 2.2).}$$

(f) Prove $\mathbf{0} \times \mathbf{X} = \mathbf{X} \times \mathbf{0} = \mathbf{0}$.

$$\mathbf{0} \times \mathbf{X} = \begin{vmatrix} \mathbf{i} & \mathbf{j} & \mathbf{k} \\ 0 & 0 & 0 \\ x_1 & x_2 & x_3 \end{vmatrix} = \mathbf{0} \quad \text{(by Theorem 2.4).}$$

$$\mathbf{X} \times \mathbf{0} = -(\mathbf{0} \times \mathbf{X}) = -\mathbf{0} = \mathbf{0} \quad \text{(by part (a)).}$$

(g) Prove $(\mathbf{X} \times \mathbf{Y}) \times \mathbf{Z} = (\mathbf{Z}\cdot\mathbf{X})\mathbf{Y} - (\mathbf{Z}\cdot\mathbf{Y})\mathbf{X}$.
First let $\mathbf{Z} = \mathbf{i}$. Then,

$$(\mathbf{X} \times \mathbf{Y}) \times \mathbf{i}$$
$$= [(x_2y_3 - x_3y_2)\mathbf{i} + (x_3y_1 - x_1y_3)\mathbf{j} + (x_1y_2 - x_2y_1)\mathbf{k}] \times \mathbf{i} \quad \text{(using Equation (1))}$$
$$= (x_2y_3 - x_3y_2)(\mathbf{i} \times \mathbf{i}) + (x_3y_1 - x_1y_3)(\mathbf{j} \times \mathbf{i}) + (x_1y_2 - x_2y_1)(\mathbf{k} \times \mathbf{i}) \quad \text{(by part (c))}$$
$$= (x_1y_2 - x_2y_1)\mathbf{j} - (x_3y_1 - x_1y_3)\mathbf{k} \quad \text{(using Example 2)}$$

$$(\mathbf{i}\cdot\mathbf{X})\mathbf{Y} - (\mathbf{i}\cdot\mathbf{Y})\mathbf{X} = x_1\mathbf{Y} - y_1\mathbf{X}$$
$$= x_1y_1\mathbf{i} + x_1y_2\mathbf{j} + x_1y_3\mathbf{k} - y_1x_1\mathbf{i} - y_1x_2\mathbf{j} - y_1x_3\mathbf{k}$$
$$= (x_1y_2 - x_2y_1)\mathbf{j} - (x_3y_1 - x_1y_3)\mathbf{k}.$$

Thus equality holds when $\mathbf{Z} = \mathbf{i}$. Similarly it holds when $\mathbf{Z} = \mathbf{j}$ and when $\mathbf{Z} = \mathbf{k}$. Let $\mathbf{Z} = z_1\mathbf{i} + z_2\mathbf{j} + z_3\mathbf{k}$ and let $\mathbf{Q} = \mathbf{X} \times \mathbf{Y}$. Then,

$$\mathbf{Q} \times \mathbf{Z} = \mathbf{Q} \times (z_1\mathbf{i} + z_2\mathbf{j} + z_3\mathbf{k})$$
$$= \mathbf{Q} \times (z_1\mathbf{i}) + \mathbf{Q} \times (z_2\mathbf{j}) + \mathbf{Q} \times (z_3\mathbf{k}) \quad \text{(by part (b))}$$
$$= z_1(\mathbf{Q} \times \mathbf{i}) + z_2(\mathbf{Q} \times \mathbf{j}) + z_3(\mathbf{Q} \times \mathbf{k}) \quad \text{(by part (d))}$$
$$= z_1[(\mathbf{i}\cdot\mathbf{X})\mathbf{Y} - (\mathbf{i}\cdot\mathbf{Y})\mathbf{X}] + z_2[(\mathbf{j}\cdot\mathbf{X})\mathbf{Y} - (\mathbf{j}\cdot\mathbf{Y})\mathbf{X}] + z_3[(\mathbf{k}\cdot\mathbf{X})\mathbf{Y} - (\mathbf{k}\cdot\mathbf{Y})\mathbf{X}] \quad \text{(using the first part of this proof)}$$
$$= [z_1(\mathbf{i}\cdot\mathbf{X}) + z_2(\mathbf{j}\cdot\mathbf{X}) + z_3(\mathbf{k}\cdot\mathbf{X})]\mathbf{Y} - [z_1(\mathbf{i}\cdot\mathbf{Y}) + z_2(\mathbf{j}\cdot\mathbf{Y}) + z_3(\mathbf{k}\cdot\mathbf{Y})]\mathbf{X}$$
$$= (\mathbf{Z}\cdot\mathbf{X})\mathbf{Y} - (\mathbf{Z}\cdot\mathbf{Y})\mathbf{X}.$$

T.3. Prove $\mathbf{X} \times (\mathbf{Y} \times \mathbf{Z}) = (\mathbf{X}\cdot\mathbf{Z})\mathbf{Y} - (\mathbf{X}\cdot\mathbf{Y})\mathbf{Z}$.

$$\begin{aligned}\mathbf{X} \times (\mathbf{Y} \times \mathbf{Z}) &= -(\mathbf{Y} \times \mathbf{Z}) \times \mathbf{X} && \text{(by Theorem 3.6(a))}\\ &= -[(\mathbf{X}\cdot\mathbf{Y})\mathbf{Z} - (\mathbf{X}\cdot\mathbf{Z})\mathbf{Y}] && \text{(by Theorem 3.6(g))}\\ &= (\mathbf{X}\cdot\mathbf{Z})\mathbf{Y} - (\mathbf{X}\cdot\mathbf{Y})\mathbf{Z}.\end{aligned}$$

T.5. Vectors **X** and **Y** are parallel provided one is a scalar multiple of the other. If $\mathbf{Y} = c\mathbf{X}$, then $\mathbf{X} \times \mathbf{Y} = \mathbf{X} \times (c\mathbf{X}) = c(\mathbf{X} \times \mathbf{X}) = \mathbf{0}$ by Theorem 3.6(e). Conversely, assume $\mathbf{X} \times \mathbf{Y} = \mathbf{0}$. Then $0 = \|\mathbf{X} \times \mathbf{Y}\| = \|\mathbf{X}\|\ \|\mathbf{Y}\| \sin\theta$ by Equation (7). Assuming that neither **X** nor **Y** is the zero vector, it follows that $\sin\theta = 0$. Hence the angle between **X** and **Y** is either 0 or π radians. Thus **X** and **Y** are parallel.

T.7. Prove $(\mathbf{X} \times \mathbf{Y}) \times \mathbf{Z} + (\mathbf{Y} \times \mathbf{Z}) \times \mathbf{X} + (\mathbf{Z} \times \mathbf{X}) \times \mathbf{Y} = \mathbf{0}$.

$$\begin{aligned}(\mathbf{X} \times \mathbf{Y}) \times \mathbf{Z} + (\mathbf{Y} \times \mathbf{Z}) \times \mathbf{X} + (\mathbf{Z} \times \mathbf{X}) \times \mathbf{Y} &= [(\mathbf{Z}\cdot\mathbf{X})\mathbf{Y} - (\mathbf{Z}\cdot\mathbf{Y})\mathbf{X}]\\ &+ [(\mathbf{X}\cdot\mathbf{Y})\mathbf{Z} - (\mathbf{X}\cdot\mathbf{Z})\mathbf{Y}] + [(\mathbf{Y}\cdot\mathbf{Z})\mathbf{X} - (\mathbf{Y}\cdot\mathbf{X})\mathbf{Z}] = \mathbf{0},\end{aligned}$$

using Theorem 3.6(g).

ML.1. (a)

```
X=[1 -2 3];Y=[1 3 1];cross(X,Y)

    ans =

        -11     2     5
```

(b)

```
X=[1 0 3];Y=[1 -1 2];cross(X,Y)

    ans =

         3     1    -1
```

(c)

```
X=[1 2 -3];Y=[2 -1 2];cross(X,Y)

    ans =

         1    -8    -5
```

ML.2. (a)

```
X=[2 3 -1];Y=[2 3 1];cross(X,Y)

    ans =

         6    -4     0
```

(b)

```
X=[3 -1 1];Y=2*X;cross(X,Y)

    ans =

         0     0     0
```

Exercises 3.3

(c) X=[1 -2 1];Y=[3 1 -1];cross(X,Y)

```
ans =

     1     4     7
```

ML.5. Following Example 6 we proceed as follows in MATLAB.

```
X=[3 -2 1];Y=[1 2 3];Z=[2 -1 2];
vol=abs(dot(X,cross(Y,Z)))

vol =

     8
```

ML.6 We find the angle between the perpendicular perxy to plane P1 and the perpendicular pervw to plane P2.

```
X=[2 -1 2];Y=[3 -2 1];V=[1 3 1];W=[0 2 -1];
perxy=cross(X,Y)

perxy =

     3     4    -1

pervw=cross(V,W)

pervw =

    -5     1     2

angle=dot(perxy,pervw)/(norm(perxy)*norm(pervw))

angle =

   -0.4655

angdeg=(180/pi)*angle

angdeg =

  -26.6697
```

Exercises 3.4

1. Let $\mathbf{X} = \begin{bmatrix} x_1 \\ x_2 \end{bmatrix}$, $\mathbf{Y} = \begin{bmatrix} y_1 \\ y_2 \end{bmatrix}$, and $\mathbf{Z} = \begin{bmatrix} z_1 \\ z_2 \end{bmatrix}$ be any vectors in $\mathbf{R}^2$; let c and d be any scalars.

Property (α) $\mathbf{X} + \mathbf{Y} = \begin{bmatrix} x_1 \\ x_2 \end{bmatrix} + \begin{bmatrix} y_1 \\ y_2 \end{bmatrix} = \begin{bmatrix} x_1 + y_1 \\ x_2 + y_2 \end{bmatrix}$ which is in $\mathbf{R}^2$.

Thus $\mathbf{R}^2$ is closed under addition of vectors.

Property (a) $\mathbf{X} + \mathbf{Y} = \begin{bmatrix} x_1 \\ x_2 \end{bmatrix} + \begin{bmatrix} y_1 \\ y_2 \end{bmatrix} = \begin{bmatrix} x_1 + y_1 \\ x_2 + y_2 \end{bmatrix} = \begin{bmatrix} y_1 + x_1 \\ y_2 + x_2 \end{bmatrix}$

$$= \begin{bmatrix} y_1 \\ y_2 \end{bmatrix} + \begin{bmatrix} x_1 \\ x_2 \end{bmatrix} = \mathbf{Y} + \mathbf{X}.$$

Property (b) $\mathbf{X} + (\mathbf{Y} + \mathbf{Z}) = \begin{bmatrix} x_1 \\ x_2 \end{bmatrix} + \left[\begin{bmatrix} y_1 \\ y_2 \end{bmatrix} + \begin{bmatrix} z_1 \\ z_2 \end{bmatrix}\right]$

$$= \begin{bmatrix} x_1 \\ x_2 \end{bmatrix} + \begin{bmatrix} y_1 + z_1 \\ y_2 + z_2 \end{bmatrix} = \begin{bmatrix} x_1 + (y_1 + z_1) \\ x_2 + (y_2 + z_2) \end{bmatrix} = \begin{bmatrix} (x_1 + y_1) + z_1 \\ (x_2 + y_2) + z_2 \end{bmatrix}$$

$$= \begin{bmatrix} x_1 + y_1 \\ x_2 + y_2 \end{bmatrix} + \begin{bmatrix} z_1 \\ z_2 \end{bmatrix} = (\mathbf{X} + \mathbf{Y}) + \mathbf{Z}.$$

Property (c) Let $\mathbf{0} = \begin{bmatrix} 0 \\ 0 \end{bmatrix}$.

$$\mathbf{X} + \mathbf{0} = \begin{bmatrix} x_1 \\ x_2 \end{bmatrix} + \begin{bmatrix} 0 \\ 0 \end{bmatrix} = \begin{bmatrix} x_1 + 0 \\ x_2 + 0 \end{bmatrix} = \begin{bmatrix} x_1 \\ x_2 \end{bmatrix} = \mathbf{X}.$$

$$\mathbf{0} + \mathbf{X} = \begin{bmatrix} 0 \\ 0 \end{bmatrix} + \begin{bmatrix} x_1 \\ x_2 \end{bmatrix} = \begin{bmatrix} 0 + x_1 \\ 0 + x_2 \end{bmatrix} = \begin{bmatrix} x_1 \\ x_2 \end{bmatrix} = \mathbf{X}.$$

Property (d) Let $-\mathbf{X} = \begin{bmatrix} -x_1 \\ -x_2 \end{bmatrix}$.

$$\mathbf{X} + (-\mathbf{X}) = \begin{bmatrix} x_1 \\ x_2 \end{bmatrix} + \begin{bmatrix} -x_1 \\ -x_2 \end{bmatrix} = \begin{bmatrix} x_1 - x_1 \\ x_2 - x_2 \end{bmatrix} = \begin{bmatrix} 0 \\ 0 \end{bmatrix} = \mathbf{0}.$$

Property (β) $c\mathbf{X} = c\begin{bmatrix} x_1 \\ x_2 \end{bmatrix} = \begin{bmatrix} cx_1 \\ cx_2 \end{bmatrix}$ which is in $\mathbf{R}^2$. Thus $\mathbf{R}^2$ is closed under scalar multiplication.

Property (e) $c(\mathbf{X} + \mathbf{Y}) = c\begin{bmatrix} x_1 + y_1 \\ x_2 + y_2 \end{bmatrix} = \begin{bmatrix} c(x_1 + y_1) \\ c(x_2 + y_2) \end{bmatrix}$

$= \begin{bmatrix} cx_1 + cy_1 \\ cx_2 + cy_2 \end{bmatrix} = \begin{bmatrix} cx_1 \\ cx_2 \end{bmatrix} + \begin{bmatrix} cy_1 \\ cy_2 \end{bmatrix} = c\begin{bmatrix} x_1 \\ x_2 \end{bmatrix} + c\begin{bmatrix} y_1 \\ y_2 \end{bmatrix}$

$= c\mathbf{X} + c\mathbf{Y}$.

Property (f) $(c + d)\mathbf{X} = \begin{bmatrix} (c + d)x_1 \\ (c + d)x_2 \end{bmatrix} = \begin{bmatrix} cx_1 + dx_2 \\ cx_2 + dx_2 \end{bmatrix}$

$= \begin{bmatrix} cx_1 \\ cx_2 \end{bmatrix} + \begin{bmatrix} dx_1 \\ dx_2 \end{bmatrix} = c\mathbf{X} + d\mathbf{X}$.

Property (g) $c(d\mathbf{X}) = c\begin{bmatrix} dx_1 \\ dx_2 \end{bmatrix} = \begin{bmatrix} c(dx_1) \\ c(dx_2) \end{bmatrix} = \begin{bmatrix} (cd)x_1 \\ (cd)x_2 \end{bmatrix} = (cd)\mathbf{X}$.

Property (h) $1\mathbf{X} = 1\begin{bmatrix} x_1 \\ x_2 \end{bmatrix} = \begin{bmatrix} 1x_1 \\ 1x_2 \end{bmatrix} = \begin{bmatrix} x_1 \\ x_2 \end{bmatrix} = \mathbf{X}$.

3. Let **V** be the set of all ordered triples of real numbers of the form (x,y,0). Denote, $\mathbf{X} = (x,y,0)$, $\mathbf{Y} = (x',y',0)$, $\mathbf{Z} = (x'',y'',0)$, and let c and d be scalars. The operations $\oplus$ and $\circ$ are defined by

$\mathbf{X} \oplus \mathbf{Y} = (x,y,0) \oplus (x',y',0) = (x+x',y+y',0)$

$c \circ \mathbf{X} = c \circ (x,y,0) = (cx,cy,0)$.

Property (α) $\mathbf{X} \oplus \mathbf{Y} = (x+x',y+y',0)$, which is an ordered triple of the correct form. Thus $\mathbf{X} \oplus \mathbf{Y}$ is in **V**. That is, **V** is closed under the operation $\oplus$.

Property (a) $\mathbf{X} \oplus \mathbf{Y} = (x+x',y+y',0) = (x'+x,y'+y,0) = \mathbf{Y} \oplus \mathbf{X}$.

Property (b) $\mathbf{X} \oplus (\mathbf{Y} \oplus \mathbf{Z}) = (x,y,0) \oplus (x'+x'',y'+y'',0)$
$= (x+x'+x'',y+y'+y'',0)$
$(\mathbf{X} \oplus \mathbf{Y}) \oplus \mathbf{Z} = (x+x',y+y',0) \oplus (x'',y'',0)$
$= (x+x'+x'',y+y'+y'',0)$
Thus, $\mathbf{X} \oplus (\mathbf{Y} \oplus \mathbf{Z}) = (\mathbf{X} \oplus \mathbf{Y}) \oplus \mathbf{Z}$.

Property (c) Let **0** be the element in **V** of the form (0,0,0).
$\mathbf{X} \oplus \mathbf{0} = (x+0,y+0,0) = (x,y,0) = \mathbf{X}$
$\mathbf{0} \oplus \mathbf{X} = (0+x,0+y,0) = (x,y,0) = \mathbf{X}$
Thus, $\mathbf{X} \oplus \mathbf{0} = \mathbf{0} \oplus \mathbf{X} = \mathbf{X}$.

Property (d) Let $-\mathbf{X}$ be defined to be the triple (-x,-y,0). For any **X** in **V**, $-\mathbf{X}$ is also in **V**.
$\mathbf{X} \oplus -\mathbf{X} = (x+(-x),y+(-y),0) = (0,0,0) = \mathbf{0}$

Property (β) $c \circ \mathbf{X} = (cx,cy,0)$, which is an ordered triple of the correct form. Thus $c \circ \mathbf{X}$ is in $\mathbf{V}$. That is, $\mathbf{V}$ is closed under the operation $\circ$.

Property (e) $c \circ (\mathbf{X} \oplus \mathbf{Y}) = c \circ (x+x',y+y',0)$
$= (c(x+x'),c(y+y'),0) = (cx+cx',cy+cy',0)$
$= (cx,cy,0) \oplus (cx',cy',0)$
$= c \circ (x,y,0) \oplus c \circ (x',y',0)$
$= c \circ \mathbf{X} \oplus c \circ \mathbf{Y}$.

Property (f) $(c+d) \circ \mathbf{X} = ((c+d)x,(c+d)y,0) = (cx+dx,cy+dy,0)$
$= (cx,cy,0) \oplus (dx,dy,0)$
$= c \circ (x,y,0) \oplus d \circ (x,y,0)$
$= c \circ \mathbf{X} \oplus d \circ \mathbf{X}$.

Property (g) $c \circ (d \circ \mathbf{X}) = c \circ (dx,dy,0) = (cdx,cdy,0)$
$= (cd) \circ (x,y,0) = (cd) \circ \mathbf{X}$.

Property (h) $1 \circ \mathbf{X} = (1x,1y,0) = (x,y,0) = \mathbf{X}$.
Thus $\mathbf{V}$ with operations $\oplus$ and $\circ$ is a vector space.

5. Let $\mathbf{V}$ be the set of all real valued functions defined on $\mathbf{R}^1$. Let $\mathbf{X} = f$, $\mathbf{Y} = g$, $\mathbf{Z} = h$, denote real valued functions and let c and d be scalars. The operations $\oplus$ and $\circ$ are defined by

$$\mathbf{X} \oplus \mathbf{Y} = (f \oplus g)(t) = f(t) + g(t)$$
$$c \circ \mathbf{X} = (c \circ f)(t) = cf(t).$$

Here, $\oplus$ is the addition of functions which is defined to be the point-by-point sum of their evaluations at t. Thus, $f(t) + g(t)$ is the addition of real numbers.

Property (α) $\mathbf{X} \oplus \mathbf{Y} = (f \oplus g)(t) = f(t) + g(t)$.
For each real number t, the value of the function $(f \oplus g)$ at t is the real number $f(t) + g(t)$. Hence, $(f \oplus g)$ is a real valued function on $\mathbf{R}^1$. Thus $\mathbf{X} \oplus \mathbf{Y}$ is in $\mathbf{V}$. That is, $\mathbf{V}$ is closed under the operation $\oplus$.

Property (a) $\mathbf{X} \oplus \mathbf{Y} = (f \oplus g)(t) = f(t) + g(t) = g(t) + f(t)$
$= (g \oplus f)(t) = \mathbf{Y} \oplus \mathbf{X}$.

Property (b) $\mathbf{X} \oplus (\mathbf{Y} \oplus \mathbf{Z}) = (f \oplus (g \oplus h))(t) = f(t) + (g \oplus h)(t)$
$= f(t) + (g(t) + h(t)) = (f(t) + g(t)) + h(t)$
$= (f \oplus g)(t) + h(t) = ((f \oplus g) \oplus h)(t)$
$= (\mathbf{X} \oplus \mathbf{Y}) \oplus \mathbf{Z}$.

Property (c) Let $\mathbf{0}$ be the element in $\mathbf{V}$ such that $\mathbf{0}(t) = 0$ for all real values t.
$\mathbf{X} \oplus \mathbf{0} = (f \oplus \mathbf{0})(t) = f(t) + \mathbf{0}(t) = f(t) + 0$
$= f(t)$. Thus function $\mathbf{X} \oplus \mathbf{0} = \mathbf{X}$.
$\mathbf{0} \oplus \mathbf{X} = (\mathbf{0} \oplus f)(t) = \mathbf{0}(t) + f(t) = 0 + f(t)$
$= f(t)$. Thus function $\mathbf{0} \oplus \mathbf{X} = \mathbf{X}$.
Hence, $\mathbf{X} \oplus \mathbf{0} = \mathbf{0} \oplus \mathbf{X} = \mathbf{X}$.

Property (d) Let $-\mathbf{X} = -f$ be defined to be the function which when evaluated at t gives the real nunber $-f(t)$. For any $\mathbf{X}$ in $\mathbf{V}$, $-\mathbf{X}$ is also in $\mathbf{V}$. $\mathbf{X} \oplus -\mathbf{X} = (f \oplus -f)(t) = f(t) + (-f(t)) = 0$. Thus for any real number t, function $\mathbf{X} \oplus -\mathbf{X}$ gives the value zero. From property (c), there is only one such function, $\mathbf{0}$. Hence, $\mathbf{X} \oplus -\mathbf{X} = \mathbf{0}$.

Property (β) $c \circ \mathbf{X} = (c \circ f)(t) = cf(t)$. For any real number t, function $c \circ f$ gives the real value $cf(t)$, hence $c \circ f$ is a real valued function on $\mathbf{R}^1$. Thus $c \circ \mathbf{X}$ is in $\mathbf{V}$. That is, $\mathbf{V}$ is closed under the operation $\circ$.

Property (e) $c \circ (\mathbf{X} \oplus \mathbf{Y}) = c \circ (f \oplus g)(t) = c(f(t) + g(t))$
$= cf(t) + cg(t) = (c \circ f)(t) + (c \circ g)(t)$
$= c \circ f \oplus c \circ g = c \circ \mathbf{X} \oplus c \circ \mathbf{Y}$.

Property (f) $(c+d) \circ \mathbf{X} = ((c+d) \circ f)(t) = (c+d)f(t)$
$= cf(t) + df(t)$
$= (c \circ f)(t) \oplus (d \circ f)(t)$
$= c \circ f \oplus d \circ f = c \circ \mathbf{X} \oplus d \circ \mathbf{X}$.

Property (g) $c \circ (d \circ \mathbf{X}) = c \circ (d \circ f)(t) = c(df(t))$
$= (cd)f(t) = ((cd) \circ f)(t)$
$= (cd) \circ f = (cd) \circ \mathbf{X}$.

Property (h) $1 \circ \mathbf{X} = (1 \circ f) = (1 \circ f)(t) = 1f(t) = f(t)$
$= f = \mathbf{X}$.

Thus $\mathbf{V}$ with operations $\oplus$ and $\circ$ is a vector space.

7. $\mathbf{V}$ is the set of ordered triples with operations

$$(x,y,z) \oplus (x',y',z') = (x',y+y',z')$$
$$c \circ (x,y,z) = (cx,cy,cz).$$

Let $\mathbf{X} = (x,y,z)$, $\mathbf{Y} = (x',y',z')$, $\mathbf{Z} = (x'',y'',z'')$, and c and d be scalars.

Property (α) $\mathbf{X} \oplus \mathbf{Y} = (x',y+y',z')$ which is an ordered triple of real numbers. Thus $\mathbf{V}$ is closed with respect to $\oplus$.

Property (a) $\mathbf{X} \oplus \mathbf{Y} = (x',y+y',z')$
$\mathbf{Y} \oplus \mathbf{X} = (x,y'+y,z)$
Thus $\mathbf{X} \oplus \mathbf{Y} \neq \mathbf{Y} \oplus \mathbf{X}$, hence $\mathbf{V}$ with the operations of $\oplus$ and $\circ$ is not a vector space.

Property (b) $\mathbf{X} \oplus (\mathbf{Y} \oplus \mathbf{Z}) = (x,y,z) \oplus (x'',y'+y'',z'')$
$= (x'',y+y'+y'',z'')$
$(\mathbf{X} \oplus \mathbf{Y}) \oplus \mathbf{Z} = (x',y+y',z') \oplus (x'',y'',z'')$
$= (x'',y+y'+y'',z'')$
Thus $\mathbf{X} \oplus (\mathbf{Y} \oplus \mathbf{Z}) = (\mathbf{X} \oplus \mathbf{Y}) \oplus \mathbf{Z}$.

Property (c) Let $\mathbf{0} = (0,0,0)$.
$\mathbf{X} \oplus \mathbf{0} = (0,y+0,0) \neq \mathbf{X}$
This property fails to hold.

Property (d) Let $-\mathbf{X}$ denote $(-x,-y,-z)$.
$\mathbf{X} \oplus -\mathbf{X} = (-x,y-y,-z) = (-x,0,-z) \neq \mathbf{0}$
This property fails to hold.

Property (β) $c \circ \mathbf{X} = (cx,cy,cz)$ which is an ordered triple of real numbers. Thus $\mathbf{V}$ is closed with respect to $\circ$.

Property (e) $c \circ (\mathbf{X} \oplus \mathbf{Y}) = c \circ (x',y+y',z') = (cx',cy+cy',cz')$
$c \circ \mathbf{X} \oplus c \circ \mathbf{Y} = (cx,cy,cz) \oplus (cx',cy',cz')$
$= (cx',cy+cy',cz')$
Thus $c \circ (\mathbf{X} \oplus \mathbf{Y}) = c \circ \mathbf{X} \oplus c \circ \mathbf{Y}$.

Property (f) $(c + d) \circ \mathbf{X} = (cx+dx,cy+dy,cz+dz)$
$c \circ \mathbf{X} \oplus d \circ \mathbf{X} = (cx,cy,cz) + (dx,dy,dz)$
$= (dx,cy+dy,z)$
Thus this property fails to hold.

Property (g) $c \circ (d \circ \mathbf{X}) = c \circ (dx,dy,dz) = (cdx,cdy,cdz)$
$(cd) \circ \mathbf{X} = (cdx,cdy,cdz)$
Thus $c \circ (d \circ \mathbf{X}) = (cd) \circ \mathbf{X}$.

Property (h) $1 \circ \mathbf{X} = (x,y,z) = \mathbf{X}$.
In summary: Properties (a), (c), (d), and (f) fail to hold.

9. $\mathbf{V}$ is the set of ordered triples of the form $(0,0,z)$ with operations

$$(0,0,z) \oplus (0,0,z') = (0,0,z+z')$$
$$c \circ (0,0,z) = (0,0,cz).$$

Let $\mathbf{X} = (0,0,z)$, $\mathbf{Y} = (0,0,z')$, $\mathbf{Z} = (0,0,z")$, and c and d be scalars.

Property (α) $\mathbf{X} \oplus \mathbf{Y} = (0,0,z+z')$ which is an ordered triple of the correct form to belong to $\mathbf{V}$. Thus $\mathbf{V}$ is closed with respect to $\oplus$.

Property (a) $\mathbf{X} \oplus \mathbf{Y} = (0,0,z+z') = (0,0,z'+z) = \mathbf{Y} \oplus \mathbf{X}$
This property holds.

Property (b) $\mathbf{X} \oplus (\mathbf{Y} \oplus \mathbf{Z}) = (0,0,z) \oplus (0,0,z'+z")$
$= (0,0,z+z'+z")$
$(\mathbf{X} \oplus \mathbf{Y}) \oplus \mathbf{Z} = (0,0,z+z') \oplus (0,0,z")$
$= (0,0,z+z'+z")$
This property holds.

Property (c) Let $\mathbf{0} = (0,0,0)$.
$\mathbf{X} \oplus \mathbf{0} = (0,0,z+0) = (0,0,z) = \mathbf{X}$
$\mathbf{0} \oplus \mathbf{X} = (0,0,0+z) = (0,0,z) = \mathbf{X}$
This property holds.

Property (d) Let $-\mathbf{X}$ denote $(0,0,-z)$.
$\mathbf{X} \oplus -\mathbf{X} = (0,0,z-z) = (0,0,0) = \mathbf{0}$
This property holds.

Property (β) $c \circ \mathbf{X} = (0,0,cz)$ which is an ordered triple of of the correct form to belong to **V**. Thus **V** is closed with respect to $\circ$.

Property (e) $c \circ (\mathbf{X} \oplus \mathbf{Y}) = c \circ (0,0,z+z') = (0,0,cz+cz')$
$c \circ \mathbf{X} \oplus c \circ \mathbf{Y} = (0,0,cz) \oplus (0,0,cz')$
$= (0,0,cz+cz')$
This property holds.

Property (f) $(c + d) \circ \mathbf{X} = (0,0,cz+dz)$
$c \circ \mathbf{X} \oplus d \circ \mathbf{X} = (0,0,cz) \oplus (0,0,dz) = (0,0,cz+dz)$
This property holds.

Property (g) $c \circ (d \circ \mathbf{X}) = c \circ (0,0,dz) = (0,0,cdz)$
$(cd) \circ \mathbf{X} = (0,0,cdz)$
This property holds.

Property (h) $1 \circ \mathbf{X} = (0,0,z) = \mathbf{X}$.
This property holds.

In summary: **V** is a vector space with operations $\oplus$ and $\circ$.

11. Let **V** be the set of all ordered pairs (x,y), where $x \le 0$ with operations

$$(x,y) \oplus (x',y') = (x+x',y+y')$$
$$c \circ (x,y) = (cx,cy).$$

Let $\mathbf{X} = (x,y)$, $\mathbf{Y} = (x',y')$, $\mathbf{Z} = (x'',y'')$, and c and d be scalars.

Property (α) $\mathbf{X} \oplus \mathbf{Y} = (x+x',y+y')$. Since x and $x' \le 0$, $x + x' \le 0$. Then $\mathbf{X} \oplus \mathbf{Y}$ is an ordered pair with first entry ≤ 0. Thus **V** is closed with respect to $\oplus$.

Property (a) $\mathbf{X} \oplus \mathbf{Y} = (x+x',y+y') = (x'+x,y'+y) = \mathbf{Y} \oplus \mathbf{X}$
This property holds.

Property (b) $\mathbf{X} \oplus (\mathbf{Y} \oplus \mathbf{Z}) = (x,y) \oplus (x'+x'',y'+y'')$
$= (x+x'+x'',y+y'+y'')$
$(\mathbf{X} \oplus \mathbf{Y}) \oplus \mathbf{Z} = (x+x',y+y') \oplus (x'',y'')$
$= (x+x'+x'',y+y'+y'')$
This property holds.

Property (c) Let $\mathbf{0} = (0,0)$. Since its first entry is ≤ 0, $\mathbf{0}$ is in **V**.
$\mathbf{X} \oplus \mathbf{0} = (x+0,y+0) = (x,y) = \mathbf{X}$
$\mathbf{0} \oplus \mathbf{X} = (0+x,0+y) = (x,y) = \mathbf{X}$
This property holds.

Property (d) Let $-\mathbf{X}$ denote $(-x,-y)$. Since $x \le 0$, $-x \ge 0$. Hence $-\mathbf{X}$ is not in **V** for some **X** in **V**. Thus not every element in **V** has an additive inverse. It follows that **V** is not a vector space.

Property (β) $c \circ \mathbf{X} = (cx,cy)$. If $c < 0$ and $x \neq 0$, then $c \circ \mathbf{X}$ has its first component > 0 and hence is not in $\mathbf{V}$. Thus $\mathbf{V}$ is not closed under $\circ$.

Property (e) $c \circ (\mathbf{X} \oplus \mathbf{Y}) = c \circ (x+x',y+y') = (cx+cx',cy+cy')$ which may not be in $\mathbf{V}$. Hence this property may fail.

Property (f) $(c + d) \circ \mathbf{X} = (cx+dx,cy+dy)$ which may not be in $\mathbf{V}$. Hence this property may fail.

Property (g) $c \circ (d \circ \mathbf{X}) = c \circ (dx,dy) = (cdx,cdy)$ which may not be in $\mathbf{V}$. Hence this property may fail.

Property (h) $1 \circ \mathbf{X} = (1x,1y) = \mathbf{X}$.
This property holds.

In summary: Properties (d), (β), (e), (f), and (g) fail to hold.

13. $\mathbf{V}$ is the set of all positive real numbers with operations

$$\mathbf{X} \oplus \mathbf{Y} = \mathbf{XY}$$
$$c \circ \mathbf{X} = \mathbf{X}^c.$$

Let $\mathbf{X}$, $\mathbf{Y}$, and $\mathbf{Z}$ be real numbers, and c and d be scalars.

Property (α) $\mathbf{X} \oplus \mathbf{Y} = \mathbf{XY}$ which is a positive real number.
Thus $\mathbf{V}$ is closed with respect to $\oplus$.

Property (a) $\mathbf{X} \oplus \mathbf{Y} = \mathbf{XY} = \mathbf{YX} = \mathbf{Y} \oplus \mathbf{X}$
This property holds.

Property (b) $\mathbf{X} \oplus (\mathbf{Y} \oplus \mathbf{Z}) = \mathbf{X} \oplus (\mathbf{YZ}) = \mathbf{X}(\mathbf{YZ}) = (\mathbf{XY})\mathbf{Z}$
$= (\mathbf{X} \oplus \mathbf{Y}) \oplus \mathbf{Z}$
This property holds.

Property (c) Let $\mathbf{0} = 1$. Then $\mathbf{0}$ is in $\mathbf{V}$.
$\mathbf{X} \oplus \mathbf{0} = \mathbf{X}1 = \mathbf{X}$
$\mathbf{0} \oplus \mathbf{X} = 1\mathbf{X} = \mathbf{X}$
This property holds.

Property (d) Let $-\mathbf{X}$ denote $1/\mathbf{X}$. Then $-\mathbf{X}$ is in $\mathbf{V}$.
$\mathbf{X} \oplus -\mathbf{X} = \mathbf{X}(1/\mathbf{X}) = 1 = \mathbf{0}$
This property holds.

Property (β) $c \circ \mathbf{X} = \mathbf{X}^c$ which is a positive real number.
Thus $\mathbf{V}$ is closed under $\circ$.

Property (e) $c \circ (\mathbf{X} \oplus \mathbf{Y}) = c \circ (\mathbf{XY}) = (\mathbf{XY})^c$
$c \circ \mathbf{X} \oplus c \circ \mathbf{Y} = \mathbf{X}^c \oplus \mathbf{Y}^c = \mathbf{X}^c\mathbf{Y}^c = (\mathbf{XY})^c$
This property holds.

Property (f) $(c + d) \circ \mathbf{X} = \mathbf{X}^{c+d}$
$c \circ \mathbf{X} \oplus d \circ \mathbf{X} = \mathbf{X}^c \oplus \mathbf{X}^d = \mathbf{X}^c\mathbf{X}^d = \mathbf{X}^{c+d}$
This property holds.

Property (g) $c \circ (d \circ \mathbf{X}) = c \circ \mathbf{X}^d = (\mathbf{X}^d)^c = \mathbf{X}^{cd}$
$(cd) \circ \mathbf{X} = \mathbf{X}^{cd}$
This property holds.

Property (h) $1 \circ \mathbf{X} = \mathbf{X}^1 = \mathbf{X}$.
This property holds.
In summary: **V** with operations $\oplus$ and $\circ$ is a vector space.

<<**Strategy:** In Exercises 15 through 25 we use Theorem 3.8. That is, if both closure properties hold, then the subset is a subspace. If either closure property fails, then the subset is not a subspace.>>

15. **V** is the vector space $\mathbf{R}^3$.
(a) $\mathbf{W} = \{(a,b,2) \mid a \text{ and } b \text{ arbitrary}\}$.
Property (α) $(a,b,2) + (a',b',2) = (a+a',b+b',4)$ which is not in **W**.
Thus **W** is not a subspace of $\mathbf{R}^3$.

(b) $\mathbf{W} = \{(a,b,c) \mid a + b = c\}$.
Property (α) Suppose (a,b,c) and (a',b',c') are in **W**, then $a + b = c$ and $a' + b' = c'$.
$(a,b,c) + (a',b',c') = (a+a',b+b',c+c')$
This is in **W** if the sum of the first two entries gives the third entry.
$(a+a') + (b+b') = (a+b) + (a'+b') = c + c'$
It follows that **W** is closed under $\oplus$.

Property (β) Let d be a scalar and (a,b,c) be in **W**. Then, $d \circ (a,b,c) = (da,db,dc)$.
This is in **W** if the sum of the first two entries gives the third entry.
$da + db = d(a+b) = dc$
It follows that **W** is closed under $\circ$.
Thus **W** is a subspace of $\mathbf{R}^3$.

(c) $\mathbf{W} = \{(a,b,c) \mid c > 0\}$.
Property (α) Suppose (a,b,c) and (a',b',c') are in **W**, then $c > 0$ and $c' > 0$.
$(a,b,c) + (a',b',c') = (a+a',b+b',c+c')$
This is in **W** if the third entry is positive. Since both c and c′ are positive, their sum is positive. It follows that **W** is closed under $\oplus$.

Property (β) Let d be a scalar and (a,b,c) be in **W**. Then, $d \circ (a,b,c) = (da,db,dc)$.
This is in **W** if dc is positive. If $d < 0$, then dc is not positive. Thus **W** is not closed under $\circ$.
W is not a subspace of $\mathbf{R}^3$.

17. **V** is the vector space $\mathbf{R}^4$.

(a) $\mathbf{W} = \{(a,b,c,d) \mid a - b = 2\} = \{(a,a-2,c,d)$.

Property (α) $(a,a-2,c,d) + (a',a'-2,c',d')$
$= (a+a',a+a'-4,c+c',d+d')$ which is not in the proper form to belong to **W**.

W is not a subspace of $\mathbf{R}^4$.

(b) $\mathbf{W} = \{(a,b,c,d) \mid c = a + 2b \text{ and } d = a - 3b\}$
$= \{(a,b,a+2b,a-3b)\}$.

Property (α) $(a,b,a+2b,a-3b) + (a',b',a'+2b',a'-3b')$
$= (a+a',b+b',a+a'+2(b+b'),a+a'-3(b+b'))$
which is of the proper form to belong to **W**.

Property (β) $e(a,b,a+2b,a-3b) = (ea,eb,ea+2eb,ea-3eb)$
which is of the proper form to belong to **W**.

W is a subspace of $\mathbf{R}^4$.

(c) $\mathbf{W} = \{(a,b,c,d) \mid a = 0 \text{ and } b = -d\} = \{(0,b,c,-b)$.

Property (α) $(0,b,c,-b) + (0,b',c',-b')$
$= (0,b+b',c+c',-(b+b'))$ which is of the proper form to belong to **W**.

Property (β) $e(0,b,c,-b) = (0,eb,ec,-eb)$ which is of the proper form to belong to **W**.

W is a subspace of $\mathbf{R}^4$.

19. $\mathbf{P}_2$ is the set of all polynomials of degree 2 or less. That is, $\mathbf{P}_2 = \{a_0t^2 + a_1t + a_2\}$.

Property (α) $a_0t^2 + a_1t + a_2 + b_0t^2 + b_1t + b_2$
$= (a_0+b_0)t^2 + (a_1+b_1)t + (a_2+b_2)$ which is a polynomial of degree 2 or less, hence is in **W**.

Property (β) $c \cdot (a_0t^2 + a_1t + a_2) = ca_0{}^2 + ca_1t + ca_2$
which is a polynomial of degree 2 or less, hence is in **W**.

$\mathbf{P}_2$ is a subspace of **P**.

21. **V** is the vector space $\mathbf{R}^3$. Let $\mathbf{X} = (1,2,-3)$ and $\mathbf{Y} = (-2,3,0)$.
$\mathbf{W} = \{a\mathbf{X} + b\mathbf{Y} \mid$ for any real numbers a and b$\}$.

Property (α) $(a\mathbf{X} + b\mathbf{Y}) + (a'\mathbf{X} + b'\mathbf{Y}) = (a+a')\mathbf{X} + (b+b')\mathbf{Y}$
which is of the proper form to belong to **W**.

Property (β) $c(a\mathbf{X} + b\mathbf{Y}) = ca\mathbf{X} + cb\mathbf{Y}$ which is of the proper form to belong to **W**.

W is a subspace of $\mathbf{R}^3$.

Exercises 3.4

23. **V** is the vector space of all 2×3 matrices defined in Example 4.

(a) **W** = the set of all matrices of the form $\begin{bmatrix} a & b & c \\ d & 0 & 0 \end{bmatrix}$,

where $b = a + c$

$$= \left\{ \begin{bmatrix} a & a+c & c \\ d & 0 & 0 \end{bmatrix} \right\}.$$

Property (α) $\begin{bmatrix} a & a+c & c \\ d & 0 & 0 \end{bmatrix} + \begin{bmatrix} a' & a'+c' & c' \\ d' & 0 & 0 \end{bmatrix}$

$$= \begin{bmatrix} a+a' & (a+a')+(c+c') & c+c' \\ d+d' & 0 & 0 \end{bmatrix}$$

which is of the proper form to belong to **W**.

Property (β) $e\begin{bmatrix} a & a+c & c \\ d & 0 & 0 \end{bmatrix} = \begin{bmatrix} ea & ea+ec & ec \\ ed & 0 & 0 \end{bmatrix}$

which is of the proper form to belong to **W**.

W is a subspace of the vector space of 2×3 matrices.

(b) **W** = set of all matrices of the form $\begin{bmatrix} a & b & c \\ d & 0 & 0 \end{bmatrix}$, where $c > 0$.

Property (α) Let c and c′ be positive. Then $c + c' > 0$.

$$\begin{bmatrix} a & b & c \\ d & 0 & 0 \end{bmatrix} + \begin{bmatrix} a' & b' & c' \\ d' & 0 & 0 \end{bmatrix} = \begin{bmatrix} a+a' & b+b' & c+c' \\ d+d' & 0 & 0 \end{bmatrix}$$

which is of the proper form to belong to **W**.

Property (β) Let c be positive.

$$e\begin{bmatrix} a & b & c \\ d & 0 & 0 \end{bmatrix} = \begin{bmatrix} ea & eb & ec \\ ed & 0 & 0 \end{bmatrix}$$

If e is negative, then ec is negative and it follows that **W** is not closed under scalar multiplication.

W is not a subspace of the vector space of 2×3 matrices.

(c) **W** = the set of all matrices of the form $\begin{bmatrix} a & b & c \\ d & e & f \end{bmatrix}$, where

$a = -2c$ and $f = 2e + d$

$$= \left\{ \begin{bmatrix} -2c & b & c \\ d & e & 2e+d \end{bmatrix} \right\}.$$

Property (α) $\begin{bmatrix} -2c & b & c \\ d & e & 2e+d \end{bmatrix} + \begin{bmatrix} -2c' & b' & c' \\ d' & e' & 2e'+d' \end{bmatrix}$

$$= \begin{bmatrix} -2(c+c') & b+b' & c+c' \\ d+d' & e+e' & 2(e+e')+(d+d') \end{bmatrix}$$

which is of the proper form to belong to **W**.

Property (β) $k\begin{bmatrix} -2c & b & c \\ d & e & 2e+d \end{bmatrix}$

$= \begin{bmatrix} -2kc & kb & kc \\ kd & ke & 2ke+kd \end{bmatrix}$ which is of the proper form to belong to **W**.

W is a subspace of the vector space of 2 $\times$ 3 matrices.

25. Apply Theorem 3.8 to each case.

(a) Let **A** and **B** be two n $\times$ n symmetric matrices. Then from Section 1.3 we have $\mathbf{A}^T = \mathbf{A}$ and $\mathbf{B}^T = \mathbf{B}$.

Property (α) **A** + **B** will be symmetric provided $(\mathbf{A} + \mathbf{B})^T = \mathbf{A} + \mathbf{B}$. We have

$$\begin{aligned} (\mathbf{A} + \mathbf{B})^T &= \mathbf{A}^T + \mathbf{B}^T && \text{(by Theorem 1.4(b))} \\ &= \mathbf{A} + \mathbf{B} && \text{(since } \mathbf{A} \text{ and } \mathbf{B} \text{ are symmetric)} \end{aligned}$$

Hence we have closure for +.

Property (β) k**A** will be symmetric provided $(k\mathbf{A})^T = k\mathbf{A}$. We have

$$\begin{aligned} (k\mathbf{A})^T &= k\mathbf{A}^T && \text{(by Theorem 1.4(d))} \\ &= k\mathbf{A} && \text{(since A is symmetric)} \end{aligned}$$

Hence we have closure for $\cdot$.

Thus the set of symmetric matrices is a subspace of M_{nn}.

(b) Let **A** and **B** be nonsingular n $\times$ n matrices. Property (α) asks if the sum of two nonsingular matrices gives a nonsingular matrix. The answer is not necessarily. We know that $\mathbf{I}_n$ and $-\mathbf{I}_n$ are both nonsingular, yet their sum is the zero matrix which is singular. Hence it follows that the set of nonsingular n $\times$ n matrices is not a subspace of M_{nn}. (See also Exercise 22 in Section 1.5.)

(c) Let **A** and **B** be $n \times n$ diagonal matrices. Property (α) asks if **A** + **B** is diagonal. The answer is yes. This follows immediately from the definition of addition of matrices. (See also Exercise T2 Section 1.2.) Property (β) follows from the definition of scalar multiplication. Thus the set of diagonal matrices is a subspace of M_{nn}.

27. **V** is the vector space $\mathbf{C}(-\infty,\infty)$ of Example 14.

(a) **W** = all nonnegative functions $= \{f(t) \mid f(t) \geq 0 \text{ for all } t\}$.

Property (α) Let f(t) and g(t) be in **W**. Then $f(t) + g(t) \geq 0$. Thus **W** is closed with respect to addition.

Property (β) For a negative scalar c, $c \cdot f(t) \leq 0$. Thus **W** is not closed with respect to scalar multiplication.

W is not a subspace of vector space $\mathbf{C}(-\infty,\infty)$.

(b) **W** = all constant functions $= \{f(t) \mid f(t) = k, \text{ a constant, for all } t\}$.

Property (α) Let $f(t) = k$ and $g(t) = p$ be in **W**. Then $f(t) + g(t) = k + p$, which is a constant. Thus **W** is closed with respect to addition.

Property (β) For $f(t) = k$ in **W** and scalar c, $c \cdot f(t) = ck$, which is a constant. Thus **W** is closed with respect to scalar multiplication.

W is a subspace of vector space $\mathbf{C}(-\infty,\infty)$.

(c) **W** = $\{f(t) \mid f(0) = 0\}$.

Property (α) Let f(t) and g(t) be in **W**. Then $f(t) + g(t)$ evaluated at $t = 0$ gives $f(0) + g(0) = 0 + 0 = 0$. Thus $f(t) + g(t)$ belongs to **W**.

Property (β) Let f(t) be in **W** and c be a scalar. Then $c \cdot f(t)$ evaluated at $t = 0$ gives $cf(0) = c \cdot 0 = 0$. Thus $c \cdot f(t)$ belongs to **W**.

W is a subspace of vector space $\mathbf{C}(-\infty,\infty)$.

(d) **W** = $\{f(t) \mid f(0) = 5\}$.

Property (α) Let f(t) and g(t) be in **W**. Then $f(t) + g(t)$ evaluated at $t = 0$ gives $f(0) + g(0) = 5 + 5 = 10 \neq 5$. Hence **W** is not closed with respect to addition.

W is not a subspace of vector space $\mathbf{C}(-\infty,\infty)$.

(e) **W** = all differentiable functions.

Property (α) Let f(t) and g(t) be in **W**. Then from calculus we have

$$\frac{d}{dt}(f(t) + g(t)) = \frac{d}{dt} f(t) + \frac{d}{dt} g(t).$$

That is, the sum of differentiable functions is a differentiable function. Thus f(t) + g(t) belongs to **W**.

Property (β) Let f(t) be in **W** and c be a scalar. Then from calculus we have

$$\frac{d}{dt}(c \cdot f(t)) = c \frac{d}{dt} f(t).$$

That is, the product of a differentiable function and a scalar is a differentiable function. Thus c·f(t) belongs to **W**.

W is a subspace of the vector space $\mathbf{C}(-\infty,\infty)$.

29. (a) The line segment shown is in the second quadrant and stops at the origin. Hence if (a,b) is a point on the line segment $a \le 0$ and $b \ge 0$. Thus vector $\begin{bmatrix} a \\ b \end{bmatrix}$ is in the set of vectors represented by the line segment. However, $(-1)\begin{bmatrix} a \\ b \end{bmatrix}$ is not, since point (-a,-b) is not in the second quadrant and it follows that Property (β) is not satisfied. Hence this is not a subspace of R^2.

(b) The line shown goes through the origin. If (a,b) is any point on the line, the vector $\begin{bmatrix} a \\ b \end{bmatrix}$ is in the set of vectors represented by the line. One representation for the equation of the line is found using points (a,b) and (0,0). We have

$$y = (b/a)x \quad \Longleftrightarrow \quad ay = bx$$

If (c,d) is any other point on the line the cy = dx is another representation for the equation of the line. Thus we have that vector $\begin{bmatrix} r \\ s \end{bmatrix}$ is in the set represented by the line provided ry = sx.

Exercises 3.4

Property (α) Let $\begin{bmatrix} a \\ b \end{bmatrix}$ and $\begin{bmatrix} c \\ d \end{bmatrix}$ be any vectors in the set represented by the line. Then $\begin{bmatrix} a \\ b \end{bmatrix} + \begin{bmatrix} c \\ d \end{bmatrix}$ $= \begin{bmatrix} a+c \\ b+d \end{bmatrix}$ and this vector is in the set represented by the line since

$$(a+c)y = ay + cy = bx + dx = (b+d)x$$

Property (β) Let $\begin{bmatrix} a \\ b \end{bmatrix}$ be any vector in the set represented by the line and k be any scalar. Then $k\begin{bmatrix} a \\ b \end{bmatrix}$ $= \begin{bmatrix} ka \\ kb \end{bmatrix}$ and this vector is in the set represented by the line since

$$(ka)y = k(ay) = k(bx) = kb(x)$$

Thus this line represented a subspace of R^2.

T.1. Prove $c\mathbf{0} = \mathbf{0}$ for every scalar c.
Case c = 0: $c\mathbf{0} = 0\mathbf{0} = \mathbf{0}$ by Theorem 3.7(a).
We show that $c\mathbf{0}$ added to any vector $\mathbf{X}$ gives $\mathbf{X}$.
Case $c \neq 0$: $c\mathbf{0} + \mathbf{X} = c\mathbf{0} + c(1/c)\mathbf{X} = c(\mathbf{0} + (1/c)\mathbf{X}) = c(1/c)\mathbf{X}$ by parts (e) and (c) of Definition 1.
Thus $c\mathbf{0}$ must be the additive identity. It follows from part (c) of Definition 1 that the additive identity is unique, hence $c\mathbf{0} = \mathbf{0}$.

T.3. Prove that $-(-\mathbf{X}) = \mathbf{X}$.
That is, show that the additive inverse of $-\mathbf{X}$ is $\mathbf{X}$.
Clearly $(-\mathbf{X}) + [-(-\mathbf{X})] = \mathbf{0}$. Thus $-(-\mathbf{X})$ is the additive inverse of $-\mathbf{X}$. But $\mathbf{X} + (-\mathbf{X}) = \mathbf{0}$, hence $\mathbf{X}$ is the additive inverse of $-\mathbf{X}$. By part (d) of Definition 1, the additive inverse of a vector is unique, hence $-(-\mathbf{X}) = \mathbf{X}$.

T.5. Prove that if $\mathbf{X} + \mathbf{Y} = \mathbf{X} + \mathbf{Z}$, then $\mathbf{Y} = \mathbf{Z}$.

$$\begin{aligned} (-\mathbf{X}) + (\mathbf{X} + \mathbf{Y}) &= (-\mathbf{X}) + (\mathbf{X} + \mathbf{Z}) \\ (-\mathbf{X} + \mathbf{X}) + \mathbf{Y} &= (-\mathbf{X} + \mathbf{X}) + \mathbf{Z} \\ \mathbf{0} + \mathbf{Y} &= \mathbf{0} + \mathbf{Z} \\ \mathbf{Y} &= \mathbf{Z} \end{aligned}$$

T.7. If **W** is a subspace, then for **X**, **Y** in **W**, **X** + **Y** and c**X** are in **W** by properties (α) and (β) of Definition 1.

Conversely, assume (α) and (β) of Theorem 3.8 hold. We must show that properties (a) through (h) in Definition 1 hold.

Property (a) Since **X**, **Y** are in **W**, they are in **V**. Since **V** is a vector space, property (a) holds for **X** and **Y**.

Property (b) Since **X**, **Y**, and **Z** are in **W**, they are in **V**. Since **V** is a vector space, property (b) holds for **X**, **Y**, and **Z**.

Property (c) By (β), for c = 0, **0** = 0**X** lies in **W**. **X** + **0** = **0** + **X** = **0** since property (c) holds for **V**.

Property (d) By (β), for c= -1, -**X** = (-1)**X** lies in **W**. **X** + (-**X**) = **0** since property (d) holds for V.

Properties (e), (f), (g), and (h) follow for W because those properties hold for any scalars and any vectors in **V**.

T.9. **W** must be closed under vector addition and under scalar multiplication. Thus, along with $\mathbf{X}_1, \mathbf{X}_2, \ldots, \mathbf{X}_k$, **W** must contain $\sum_{i=1}^{k} a_i\mathbf{X}_i$ for any set of coefficients $a_1, a_2, \ldots, a_k$. Thus **W** contains span S.

Exercises 3.5

1. Let $\mathbf{X}_1 = (4,2,-3)$, $\mathbf{X}_2 = (2,1,-2)$, and $\mathbf{X}_3 = (-2,-1,0)$. Following Examples 1 and 2, we construct a linear system from the expression $c_1\mathbf{X}_1 + c_2\mathbf{X}_2 + c_3\mathbf{X}_3 = \mathbf{X}$. If the corresponding system is consistent, then $\mathbf{X}$ is a linear combination of $\{\mathbf{X}_1, \mathbf{X}_2, \mathbf{X}_3\}$, otherwise $\mathbf{X}$ is not a linear combination of $\{\mathbf{X}_1, \mathbf{X}_2, \mathbf{X}_3\}$.

   (a) $c_1\mathbf{X}_1 + c_2\mathbf{X}_2 + c_3\mathbf{X}_3 = (1,1,1)$
   $c_1(4,2,-3) + c_2(2,1,-2) + c_3(-2,-1,0) = (1,1,1)$
   Expanding and equating corresponding components leads to the linear system

$$\begin{aligned} 4c_1 + 2c_2 - 2c_3 &= 1 \\ 2c_1 + c_2 - c_3 &= 1 \\ -3c_1 - 2c_2 \quad\quad &= 1. \end{aligned}$$

   We solve this system by Gauss-Jordan reduction obtaining the reduced row echelon form of the associated augmented matrix, which is

$$\left[\begin{array}{ccc|c} 1 & 0 & -2 & 0 \\ 0 & 1 & 3 & 0 \\ 0 & 0 & 0 & 1 \end{array}\right].$$

   The system is inconsistent, hence (1,1,1) is not a linear combination of $\{\mathbf{X}_1, \mathbf{X}_2, \mathbf{X}_3\}$.

   (b) $c_1\mathbf{X}_1 + c_2\mathbf{X}_2 + c_3\mathbf{X}_3 = (4,2,-6)$
   $c_1(4,2,-3) + c_2(2,1,-2) + c_3(-2,-1,0) = (4,2,-6)$
   Expanding and equating corresponding components leads to the linear system

$$\begin{aligned} 4c_1 + 2c_2 - 2c_3 &= 4 \\ 2c_1 + c_2 - c_3 &= 2 \\ -3c_1 - 2c_2 \quad\quad &= -6. \end{aligned}$$

   We solve this system by Gauss-Jordan reduction obtaining the reduced row echelon form of the associated augmented matrix, which is

$$\left[\begin{array}{ccc|c} 1 & 0 & -2 & -2 \\ 0 & 1 & 3 & 6 \\ 0 & 0 & 0 & 0 \end{array}\right].$$

   We have $c_1 = -2 + 2r$, $c_2 = 6 - 3r$, $c_3 = r$, any real number. Since the system is consistent, (4,2,-6) is a linear combination of $\{\mathbf{X}_1, \mathbf{X}_2, \mathbf{X}_3\}$.

   (c) $c_1\mathbf{X}_1 + c_2\mathbf{X}_2 + c_3\mathbf{X}_3 = (-2,-1,1)$
   $c_1(4,2,-3) + c_2(2,1,-2) + c_3(-2,-1,0) = (-2,-1,1)$
   Expanding and equating corresponding components leads to the linear system

$$\begin{aligned} 4c_1 + 2c_2 - 2c_3 &= -2 \\ 2c_1 + c_2 - c_3 &= -1 \\ -3c_1 - 2c_2 \quad\quad &= 1. \end{aligned}$$

We solve this system by Gauss-Jordan reduction obtaining the reduced row echelon form of the associated augmented matrix, which is

$$\left[\begin{array}{ccc|c} 1 & 0 & -2 & -1 \\ 0 & 1 & 3 & 1 \\ 0 & 0 & 0 & 0 \end{array}\right].$$

We have $c_1 = 2r - 1$, $c_2 = -3r + 1$, $c_3 = r$, any real number. Since the system is consistent, $(-2,-1,1)$ is a linear combination of $\{\mathbf{X}_1, \mathbf{X}_2, \mathbf{X}_3\}$.

(d) $c_1\mathbf{X}_1 + c_2\mathbf{X}_2 + c_3\mathbf{X}_3 = (-1,2,3)$
$c_1(4,2,-3) + c_2(2,1,-2) + c_3(-2,-1,0) = (-1,2,3)$
Expanding and equating corresponding components leads to the linear system

$$\begin{aligned} 4c_1 + 2c_2 - 2c_3 &= -1 \\ 2c_1 + c_2 - c_3 &= 2 \\ -3c_1 - 2c_2 &= 3. \end{aligned}$$

We solve this system by Gauss-Jordan reduction obtaining the reduced row echelon form of the associated augmented matrix, which is

$$\left[\begin{array}{ccc|c} 1 & 0 & -2 & 0 \\ 0 & 1 & 3 & 0 \\ 0 & 0 & 0 & 1 \end{array}\right].$$

The system is inconsistent, hence $(-1,2,3)$ is not a linear combination of $\{\mathbf{X}_1, \mathbf{X}_2, \mathbf{X}_3\}$.

3. Let $p_1(t) = t^2 + 2t + 1$, $p_2(t) = t^2 + 3$, and $p_3(t) = t - 1$. To determine if a vector $p(t)$ is a linear combination of $\{p_1(t), p_2(t), p_3(t)\}$, we form the expression

$$ap_1(t) + bp_2(t) + cp_3(t) = p(t),$$

expand and equate coefficients of like powers of t to form a linear system. Construct the associated augmented matrix and compute its reduced row echelon form. The vector $p(t)$ is a linear combination of $\{p_1(t), p_2(t), p_3(t)\}$ if the linear system is consistent. In this case we have

$$a(t^2 + 2t + 1) + b(t^2 + 3) + c(t - 1) = p(t)$$
$$(a + b)t^2 + (2a + c)t + (a + 3b - c) = p(t).$$

(a) $p(t) = t^2 + t + 2$, hence the corresponding linear system is

$$\begin{aligned} a + b &= 1 \\ 2a + c &= 1 \\ a + 3b - c &= 2. \end{aligned}$$

The reduced row echelon form of the augmented matrix is

$$\left[\begin{array}{ccc|c} 1 & 0 & 1/2 & 1/2 \\ 0 & 1 & -1/2 & 1/2 \\ 0 & 0 & 0 & 0 \end{array}\right].$$

It follows that this system is consistent with solution $a = .5 - .5r$, $b = .5 + .5r$, $c = r$, any real number. Thus $t^2 + t + 2$ is a linear combination of $\{p_1(t), p_2(t), p_3(t)\}$.

(b) $p(t) = 2t^2 + 2t + 3$, hence the corresponding linear system is

$$\begin{aligned} a + b \quad\quad &= 2 \\ 2a \quad\quad + c &= 2 \\ a + 3b - c &= 3. \end{aligned}$$

The reduced row echelon form of the augmented matrix is

$$\left[\begin{array}{ccc|c} 1 & 0 & 1/2 & 1 \\ 0 & 1 & -1/2 & 1 \\ 0 & 0 & 0 & 1 \end{array}\right].$$

It follows that this system is inconsistent. Thus $2t^2 + 2t + 3$ is not a linear combination of $\{p_1(t), p_2(t), p_3(t)\}$.

(c) $p(t) = -t^2 + t - 4$, hence the corresponding linear system is

$$\begin{aligned} a + b \quad\quad &= -1 \\ 2a \quad\quad + c &= 1 \\ a + 3b - c &= -4. \end{aligned}$$

The reduced row echelon form of the augmented matrix is

$$\left[\begin{array}{ccc|c} 1 & 0 & 1/2 & 1/2 \\ 0 & 1 & -1/2 & -3/2 \\ 0 & 0 & 0 & 0 \end{array}\right].$$

It follows that this system is consistent with solution $a = .5 - .5r$, $b = -1.5 + .5r$, $c = r$, any real number. Thus $-t^2 + t - 4$ is a linear combination of $\{p_1(t), p_2(t), p_3(t)\}$.

(d) $p(t) = -2t^2 + 3t + 1$, hence the corresponding linear system is

$$\begin{aligned} a + b \quad\quad &= -2 \\ 2a \quad\quad + c &= 3 \\ a + 3b - c &= 1. \end{aligned}$$

The reduced row echelon form of the augmented matrix is

$$\left[\begin{array}{ccc|c} 1 & 0 & 1/2 & 3/2 \\ 0 & 1 & -1/2 & -7/2 \\ 0 & 0 & 0 & 1 \end{array}\right].$$

It follows that this system is inconsistent. Thus $-2t^2 + 3t + 1$ is not a linear combination of $\{p_1(t), p_2(t), p_3(t)\}$.

5. A vector $\mathbf{X}$ belongs to span$\{\mathbf{X}_1, \mathbf{X}_2, \mathbf{X}_3\}$ provided $\mathbf{X}$ can be expressed as a linear combination of $\mathbf{X}_1$, $\mathbf{X}_2$, $\mathbf{X}_3$. Following the procedure in Example 3, we form the expression

$$c_1\mathbf{X}_1 + c_2\mathbf{X}_2 + c_3\mathbf{X}_3 = \mathbf{X},$$

and determine whether there exist values c_1, c_2, and c_3 for which this is true. For $\mathbf{X}_1 = (1,0,0,1)$, $\mathbf{X}_2 = (1,-1,0,0)$, $\mathbf{X}_3 = (0,1,2,1)$ we have that

$$(c_1 + c_2,\ -c_2 + c_3,\ 2c_3,\ c_1 + c_3) = \mathbf{X}.$$

We equate corresponding components to obtain a linear system. Construct the associated augmented matrix and compute its reduced row echelon form. If the system is consistent, then $\mathbf{X}$ belongs to span$\{\mathbf{X}_1, \mathbf{X}_2, \mathbf{X}_3\}$.

(a) For $\mathbf{X} = (-1,4,2,2)$ the linear system is

$$\begin{aligned} c_1 + c_2 &= -1 \\ -c_2 + c_3 &= 4 \\ 2c_3 &= 2 \\ c_1 + c_3 &= 2 \end{aligned}$$

and the reduced row echelon form of the augmented matrix is $\left[\begin{array}{ccc|c} 1 & 0 & 0 & 0 \\ 0 & 1 & 0 & 0 \\ 0 & 0 & 1 & 0 \\ 0 & 0 & 0 & 1 \end{array}\right]$. The system is inconsistent. Hence $\mathbf{X}$ is not in span$\{\mathbf{X}_1, \mathbf{X}_2, \mathbf{X}_3\}$.

(b) For $\mathbf{X} = (1,2,0,1)$ the linear system is

$$\begin{aligned} c_1 + c_2 &= 1 \\ -c_2 + c_3 &= 2 \\ 2c_3 &= 0 \\ c_1 + c_3 &= 1 \end{aligned}$$

and the reduced row echelon form of the augmented matrix is $\left[\begin{array}{ccc|c} 1 & 0 & 0 & 0 \\ 0 & 1 & 0 & 0 \\ 0 & 0 & 1 & 0 \\ 0 & 0 & 0 & 1 \end{array}\right]$. The system is inconsistent. Hence $\mathbf{X}$ is not in span$\{\mathbf{X}_1, \mathbf{X}_2, \mathbf{X}_3\}$.

(c) For $\mathbf{X} = (-1,1,4,3)$ the linear system is

$$\begin{aligned} c_1 + c_2 &= -1 \\ -c_2 + c_3 &= 1 \\ 2c_3 &= 4 \\ c_1 + c_3 &= 3 \end{aligned}$$

and the reduced row echelon form of the augmented matrix is $\left[\begin{array}{ccc|c} 1 & 0 & 0 & 0 \\ 0 & 1 & 0 & 0 \\ 0 & 0 & 1 & 0 \\ 0 & 0 & 0 & 1 \end{array}\right]$. The system is inconsistent. Hence $\mathbf{X}$ is not in span$\{\mathbf{X}_1, \mathbf{X}_2, \mathbf{X}_3\}$.

(d) For $\mathbf{X} = (0,1,1,0)$ the linear system is

$$\begin{aligned} c_1 + c_2 \quad\quad &= 0 \\ - c_2 + c_3 &= 1 \\ 2c_3 &= 1 \\ c_1 \quad + c_3 &= 0 \end{aligned}$$

and the reduced row echelon form of the augmented matrix

is $\left[\begin{array}{ccc|c} 1 & 0 & 0 & 0 \\ 0 & 1 & 0 & 0 \\ 0 & 0 & 1 & 0 \\ 0 & 0 & 0 & 1 \end{array}\right]$. The system is inconsistent. Hence $\mathbf{X}$ is not in span$\{\mathbf{X}_1, \mathbf{X}_2, \mathbf{X}_3\}$.

7. We follow the method of Example 4. Let $\mathbf{X} = (a,b)$.

(a) Determine whether constants c_1 and c_2 exist such that
$$c_1(1,2) + c_2(-1,1) = \mathbf{X} = (a,b).$$
The corresponding linear system is
$$\begin{aligned} c_1 - c_2 &= a \\ 2c_1 + c_2 &= b \end{aligned}$$
which has solution $c_1 = (a + b)/3$ and $c_2 = (-2a+b)/3$. Since there is a solution for every choice of a and b, $\{(1,2), (-1,1)\}$ spans $\mathbf{R}^2$.

(b) Determine whether constants c_1, c_2, and c_3 exist such that
$$c_1(0,0) + c_2(1,1) + c_3(-2,-2) = \mathbf{X} = (a,b).$$
The corresponding linear system is
$$\begin{aligned} 0c_1 + c_2 - 2c_3 &= a \\ 0c_1 + c_2 - 2c_3 &= b. \end{aligned}$$

The associated augmented matrix is $\left[\begin{array}{ccc|c} 0 & 1 & -2 & a \\ 0 & 1 & -2 & b \end{array}\right]$, which

is row equivalent to $\left[\begin{array}{ccc|c} 0 & 1 & -2 & a \\ 0 & 0 & 0 & b-a \end{array}\right]$. We see that this system is consistent only if $a = b$. Thus, not every vector (a,b) can be written as a linear combination of $\{(0,0), (1,1), (-2,-2)\}$. Hence these vectors do not span $\mathbf{R}^2$.

(c) Determine whether constants c_1, c_2, and c_3 exist such that
$$c_1(1,3) + c_2(2,-3) + c_3(0,2) = \mathbf{X} = (a,b).$$

The corresponding linear system is

$$1c_1 + 2c_2 + 0c_3 = a$$
$$3c_1 - 3c_2 + 2c_3 = b.$$

The associated augmented matrix is $\left[\begin{array}{ccc|c} 1 & 2 & 0 & a \\ 3 & -3 & 2 & b \end{array}\right]$, which is row equivalent to

$$\left[\begin{array}{ccc|c} 1 & 0 & \frac{4}{9} & \frac{3a + 2b}{9} \\ 0 & 1 & \frac{-2}{9} & \frac{3a-b}{9} \end{array}\right].$$ This system is consistent for any choices of a and b. Thus {(1,3), (2,-3), (0,2)} spans $\mathbf{R}^2$.

(d) Determine whether constants c_1 and c_2 exist such that

$$c_1(2,4) + c_2(-1,2) = \mathbf{X} = (a,b).$$

The corresponding linear system is

$$2c_1 - c_2 = a$$
$$4c_1 + 2c_2 = b$$

which has solution $c_1 = (2a + b)/8$ and $c_2 = (b - 2a)/4$. Since there is a solution for every choice of a and b, {(2,4), (-1,2)} spans $\mathbf{R}^2$.

9. Following the technique of Example 4, let $\mathbf{X} = (a,b,c,d)$.

(a) Determine whether constants c_1, c_2, c_3, and c_4 exist such that

$$c_1(1,0,0,1) + c_2(0,1,0,0) + c_3(1,1,1,1) + c_4(1,1,1,0) = \mathbf{X} = (a,b,c,d).$$

The corresponding linear system is

$$\begin{array}{l} c_1 + c_3 + c_4 = a \\ c_2 + c_3 + c_4 = b \\ c_3 + c_4 = c \\ c_1 + c_3 = d \end{array}$$

which has augmented matrix $\left[\begin{array}{cccc|c} 1 & 0 & 1 & 1 & a \\ 0 & 1 & 1 & 1 & b \\ 0 & 0 & 1 & 1 & c \\ 1 & 0 & 1 & 0 & d \end{array}\right]$. The reduced row echelon form is $\left[\begin{array}{cccc|c} 1 & 0 & 0 & 0 & a-c \\ 0 & 1 & 0 & 0 & b-c \\ 0 & 0 & 1 & 0 & c+d-a \\ 0 & 0 & 0 & 1 & a-d \end{array}\right]$.

The system is consistent for any choices of a, b, c, and d. Hence {(1,0,0,1), (0,1,0,0), (1,1,1,1), (1,1,1,0)} spans $\mathbf{R}^4$.

Exercises 3.5

(b) Determine whether constants c_1, c_2, and c_3 exist such that
$c_1(1,2,1,0) + c_2(1,1,-1,0) + c_3(0,0,0,1) = \mathbf{X} = (a,b,c,d)$.
The corresponding linear system is

$$\begin{aligned} c_1 + c_2 \quad &= a \\ 2c_1 + c_2 \quad &= b \\ c_1 - c_2 \quad &= c \\ c_3 &= d \end{aligned}$$

which has augmented matrix $\left[\begin{array}{ccc|c} 1 & 1 & 0 & a \\ 2 & 1 & 0 & b \\ 1 & -1 & 0 & c \\ 0 & 0 & 1 & d \end{array}\right]$. Row operations $-2\mathbf{r}_1+\mathbf{r}_2$, $-\mathbf{r}_1+\mathbf{r}_3$, $-2\mathbf{r}_2+\mathbf{r}_3$ give

$\left[\begin{array}{ccc|c} 1 & 1 & 0 & a \\ 0 & -1 & 0 & b-2a \\ 0 & 0 & 0 & 3a-2b+c \\ 0 & 0 & 1 & d \end{array}\right]$. We see that the system is inconsistent unless $3a-2b+c = 0$. Thus there is not a solution for all choices of a, b, c, and d and it follows that $\{(1,2,1,0), (1,1,-1,0), (0,0,0,1)\}$ does not span $\mathbf{R}^4$.

(c) Determine whether constants c_1, c_2, c_3, c_4, and c_5 exist such that
$c_1(6,4,-2,4) + c_2(2,0,0,1) + c_3(3,2,-1,2)$
$+ c_4(5,6,-3,2) + c_5(0,4,-2,-1) = \mathbf{X} = (a,b,c,d)$.
The corresponding linear system is

$$\begin{aligned} 6c_1 + 2c_2 + 3c_3 + 5c_4 \qquad\quad &= a \\ 4c_1 \qquad + 2c_3 + 6c_4 + 4c_5 &= b \\ -2c_1 \qquad - c_3 - 3c_4 - 2c_5 &= c \\ 4c_1 + c_2 + 2c_3 + 2c_4 - c_5 &= d \end{aligned}$$

which has augmented matrix $\left[\begin{array}{ccccc|c} 6 & 2 & 3 & 5 & 0 & a \\ 4 & 0 & 2 & 6 & 4 & b \\ -2 & 0 & -1 & -3 & -2 & c \\ 4 & 1 & 2 & 2 & -1 & d \end{array}\right]$.

We see that rows 2 and 3 are proportional except in the augmented column. Thus applying row operation $2\mathbf{r}_3+\mathbf{r}_2$ gives

$\left[\begin{array}{ccccc|c} 6 & 2 & 3 & 5 & 0 & a \\ 0 & 0 & 0 & 0 & 0 & b+2c \\ -2 & 0 & -1 & -3 & -2 & c \\ 4 & 1 & 2 & 2 & -1 & d \end{array}\right]$. It follows that the system is inconsistent unless $b+2c = 0$. Thus there is not a solution for all choices of a, b, c, and d. Hence $\{(6,4,-2,4), (2,0,0,1), (3,2,-1,2), (5,6,-3,2), (0,4,-2,-1)\}$ does not span $\mathbf{R}^4$.

(d) Determine whether constants c_1, c_2, c_3, and c_4 exist such that

$$c_1(1,1,0,0) + c_2(1,2,-1,1) + c_3(0,0,1,1) + c_4(2,1,2,1) = \mathbf{X} = (a,b,c,d).$$

The corresponding linear system is

$$\begin{aligned} c_1 + c_2 + 2c_4 &= a \\ c_1 + 2c_2 + c_4 &= b \\ -c_2 + c_3 + 2c_4 &= c \\ c_2 + c_3 + c_4 &= d \end{aligned}$$

which has augmented matrix $\left[\begin{array}{cccc|c} 1 & 1 & 0 & 2 & a \\ 1 & 2 & 0 & 1 & b \\ 0 & -1 & 1 & 2 & c \\ 0 & 1 & 1 & 1 & d \end{array}\right]$.

The row operations $-\mathbf{r}_1+\mathbf{r}_2$, $-\mathbf{r}_2+\mathbf{r}_1$, $\mathbf{r}_2+\mathbf{r}_3$, $-\mathbf{r}_2+\mathbf{r}_4$, $-\mathbf{r}_3+\mathbf{r}_4$, $-\mathbf{r}_4+\mathbf{r}_3$, $\mathbf{r}_4+\mathbf{r}_2$, $-3\mathbf{r}_4+\mathbf{r}_1$ give an equivalent linear system with solution $c_1 = -4a+5b+3c-3d$, $c_2 = a-b-c+d$, $c_3 = -3a+3b+2c-d$, $c_4 = 2a-2b-c+d$. It follows that the system is consistent for all choices of a, b, c, and d. Hence $\{(1,1,0,0), (1,2,-1,1), (0,0,1,1), (2,1,2,1)\}$ spans $\mathbf{R}^4$.

11. Following the technique of Example 5, let $p(t) = at^3 + bt^2 + ct + d$. Determine whether constants c_1, c_2, c_3, and c_4 exist such that

$$c_1(t^3 + 2t + 1) + c_2(t^2 - t + 2) + c_3(t^3 + 2) + c_4(-t^3 + t^2 - 5t + 2) = p(t).$$

Expanding, grouping like terms, and equating coefficients of like powers of t gives the linear system

$$\begin{aligned} c_1 + c_3 - c_4 &= a \\ c_2 + c_4 &= b \\ 2c_1 - c_2 - 5c_4 &= c \\ c_1 + 2c_2 + 2c_3 + 2c_4 &= d \end{aligned}$$

which has augmented matrix $\left[\begin{array}{cccc|c} 1 & 0 & 1 & -1 & a \\ 0 & 1 & 0 & 1 & b \\ 2 & -1 & 0 & -5 & c \\ 1 & 2 & 2 & 2 & d \end{array}\right]$. Row operations $-2\mathbf{r}_1+\mathbf{r}_3$, $-\mathbf{r}_1+\mathbf{r}_4$, $\mathbf{r}_2+\mathbf{r}_3$, $-2\mathbf{r}_2+\mathbf{r}_4$, $2\mathbf{r}_4+\mathbf{r}_3$ give an equivalent system which contains a row of the form

$$[0 \quad 0 \quad 0 \quad 0 \quad 0 \mid -4a-3b+c+2d] .$$

It follows that the system does not have a solution for all values of a, b, c, and d. Hence $\{t^3 + 2t + 1, t^2 - t + 2, t^3 + 2, -t^3 + t^2 - 5t + 2\}$ does not span $\mathbf{P}_3$.

Exercises 3.5

13. Following the technique in Example 8 we find the reduced row echelon form of the augmented matrix $[\mathbf{A}|\mathbf{0}]$. We obtain

$$\left[\begin{array}{cccc|c} 1 & 0 & 0 & -1 & 0 \\ 0 & 1 & 2 & 0 & 0 \\ 0 & 0 & 0 & 0 & 0 \\ 0 & 0 & 0 & 0 & 0 \end{array}\right]$$

The general solution is then given by

$$x_4 = r,\ x_3 = s,\ x_2 = -2s,\ x_1 = r$$

where r and s are any real numbers. In matrix form we have that any member of the solution space is given by

$$\mathbf{X} = r\begin{bmatrix} 1 \\ 0 \\ 0 \\ 1 \end{bmatrix} + s\begin{bmatrix} 0 \\ 1 \\ -2 \\ 0 \end{bmatrix}$$

Hence $\left\{\begin{bmatrix} 1 \\ 0 \\ 0 \\ 1 \end{bmatrix}, \begin{bmatrix} 0 \\ 1 \\ -2 \\ 0 \end{bmatrix}\right\}$ spans the null space of **A**.

15. Following the technique used in Example 9, we form a linear combination of the vectors with arbitrary coefficients and set it equal to the zero vector. Expanding, adding the vectors, and equating corresponding components leads to a homogeneous system of equations. If this system of equations has a nontrivial solution then the vectors are linearly dependent. If the only solution is the trivial one, then the vectors are linearly independent.

(a) $c_1(1,1,2,1) + c_2(1,0,0,2) + c_3(4,6,8,6)$
$+ c_4(0,3,2,1) = (0,0,0,0)$

The corresponding homogeneous system is

$$\begin{aligned} c_1 + c_2 + 4c_3 \qquad &= 0 \\ c_1 \qquad + 6c_3 + 3c_4 &= 0 \\ 2c_1 \qquad + 8c_3 + 2c_4 &= 0 \\ c_1 + 2c_2 + 6c_3 + c_4 &= 0. \end{aligned}$$

The row reduced echelon form of the coefficient matrix is

$$\begin{bmatrix} 1 & 0 & 0 & -3 \\ 0 & 1 & 0 & -1 \\ 0 & 0 & 1 & 1 \\ 0 & 0 & 0 & 0 \end{bmatrix}$$. Hence the solution of this homogeneous

system is $c_1 = 3r$, $c_2 = r$, $c_3 = -r$, $c_4 = r$, any real number. Thus the vectors are linearly dependent. Let $r = 1$, then we have $3\mathbf{X}_1 + \mathbf{X}_2 - \mathbf{X}_3 + \mathbf{X}_4 = \mathbf{0}$. Hence, $\mathbf{X}_3 = 3\mathbf{X}_1 + \mathbf{X}_2 + \mathbf{X}_4$.

(b) $c_1(1,-2,3,-1) + c_2(-2,4,-6,2) = (0,0,0,0)$
The corresponding homogeneous system is

$$\begin{aligned} c_1 - 2c_2 &= 0 \\ -2c_1 + 4c_2 &= 0 \\ 3c_1 - 6c_2 &= 0 \\ -c_1 + 2c_2 &= 0. \end{aligned}$$

The reduced row echelon form of the coefficient matrix is

$$\begin{bmatrix} 1 & -2 \\ 0 & 0 \\ 0 & 0 \\ 0 & 0 \end{bmatrix}$$. Hence the general solution of this homogeneous

system is $c_1 = 2r$, $c_2 = r$. Thus the vectors are linearly dependent.

(c) $c_1(1,1,1,1) + c_2(2,3,1,2) + c_3(3,1,2,1) + c_4(2,2,1,1) = (0,0,0,0)$
The corresponding homogeneous system is

$$\begin{aligned} c_1 + 2c_2 + 3c_3 + 2c_4 &= 0 \\ c_1 + 3c_2 + c_3 + 2c_4 &= 0 \\ c_1 \quad c_2 + 2c_3 + c_4 &= 0 \\ c_1 + 2c_2 + c_3 + c_4 &= 0. \end{aligned}$$

The reduced row echelon form of the coefficient matrix is $\mathbf{I}_4$. Hence the only solution of this homogeneous system is the trivial solution $c_1 = c_2 = c_3 = c_4 = 0$. Thus the vectors are linearly independent.

(d) $c_1(4,2,-1,3) + c_2(6,5,-5,1) + c_3(2,-1,3,5) = (0,0,0,0)$
The corresponding homogeneous system is

$$\begin{aligned} 4c_1 + 6c_2 + 2c_3 &= 0 \\ 2c_1 + 5c_2 - c_3 &= 0 \\ -c_1 - 5c_2 + 3c_3 &= 0 \\ 3c_1 + c_2 + 5c_3 &= 0. \end{aligned}$$

The reduced row echelon form of the coefficient matrix is

$\begin{bmatrix} 1 & 0 & 2 \\ 0 & 1 & 1 \\ 0 & 0 & 0 \\ 0 & 0 & 0 \end{bmatrix}$. Hence the solution of this homogeneous system is $c_1 = -2r$, $c_2 = -r$, $c_3 = r$, any real number. Thus the vectors are linearly dependent. Let $r = 1$, then we have $-2\mathbf{X}_1 - \mathbf{X}_2 + \mathbf{X}_3 = \mathbf{0}$. Hence, $\mathbf{X}_3 = 2\mathbf{X}_1 + \mathbf{X}_2$.

17. To determine if $\{\mathbf{X}_1, \mathbf{X}_2, \mathbf{X}_3\}$ is linearly independent or not we form the equation

$$a_1\mathbf{X}_1 + a_2\mathbf{X}_2 + a_3\mathbf{X}_3 = \mathbf{0}$$

which leads to the homogeneous linear system with augmented matrix $[\mathbf{X}_1\ \mathbf{X}_2\ \mathbf{X}_3 \mid \mathbf{0}]$. The reduced row echelon form of this matrix is

$$\left[\begin{array}{ccc|c} 1 & 0 & 3/2 & 0 \\ 0 & 1 & -1/2 & 0 \\ 0 & 0 & 0 & 0 \end{array}\right]$$

There are two equations with three unknowns, hence there is a nontrivial solution. Thus $\{\mathbf{X}_1, \mathbf{X}_2, \mathbf{X}_3\}$ is linearly dependent.

19. Let $c_1(t + 3) + c_2(2t + \lambda^2 + 2) = \mathbf{0}$. Then we have $t(c_1 + 2c_2) + (3c_1 + c_2(\lambda^2 + 2)) = \mathbf{0}$. The corresponding linear system is

$$\begin{aligned} c_1 + \qquad 2c_2 &= 0 \\ 3c_1 + (\lambda^2 + 2)c_2 &= 0. \end{aligned}$$

The vectors are linearly independent provided the only solution of this linear system is the trivial solution. A homogeneous linear system has only the trivial solution if and only if the coefficient matrix is nonsingular. Thus we require that λ be chosen so that

$$\det\begin{bmatrix} 1 & 2 \\ 3 & \lambda^2+2 \end{bmatrix} = \lambda^2 + 2 - 6 = \lambda^2 - 4 \neq 0.$$

Thus the vectors are linearly independent for all values of λ except $\lambda = 2$ or -2.

T.1. If $c_1\mathbf{E}_1 + c_2\mathbf{E}_2 + \cdots + c_n\mathbf{E}_n = (c_1, c_2, \ldots, c_n) = (0, 0, \ldots, 0) = \mathbf{0}$ in $\mathbf{R}^n$, then the only solution is the trivial solution $c_1 = c_2 = \cdots = c_n = 0$. Thus vectors $\mathbf{E}_1, \mathbf{E}_2, \ldots, \mathbf{E}_n$ are linearly independent.

T.3. Assume that $S = \{\mathbf{X}_1, \mathbf{X}_2, \ldots, \mathbf{X}_k\}$ is linearly dependent. Then there are constants c_i, not all zero, such that

$$c_1\mathbf{X}_1 + c_2\mathbf{X}_2 + \cdots + c_k\mathbf{X}_k = \mathbf{0}.$$

Supose c_j is not zero. Then, solving for $\mathbf{X}_j$ we have

$$\mathbf{X}_j = -\frac{c_1}{c_j}\mathbf{X}_1 - \frac{c_2}{c_j}\mathbf{X}_2 - \cdots - \frac{c_{j-1}}{c_j}\mathbf{X}_{j-1} - \frac{c_{j+1}}{c_j}\mathbf{X}_{j+1} - \cdots - \frac{c_k}{c_j}\mathbf{X}_k.$$

Thus one of the vectors is a linear combination of the others.

Conversely, suppose one vector, $\mathbf{X}_j$, is a linear combination of the others:

$$\mathbf{X}_j = d_1\mathbf{X}_1 + d_2\mathbf{X}_2 + \cdots + d_{j-1}\mathbf{X}_{j-1} + d_{j+1}\mathbf{X}_{j+1} + \cdots + d_k\mathbf{X}_k.$$

Then we have

$$d_1\mathbf{X}_1 + d_2\mathbf{X}_2 + \cdots + d_{j-1}\mathbf{X}_{j-1} + (-1)\mathbf{X}_j + d_{j+1}\mathbf{X}_{j+1} + \cdots + d_k\mathbf{X}_k = \mathbf{0}.$$

This is a linear combination of the vectors in S, in which not all the coefficients are zero, that gives the zero vector. It follows that S is a linearly dependent set.

T.5. Form the linear combination

$$c_1\mathbf{Y}_1 + c_2\mathbf{Y}_2 + c_3\mathbf{Y}_3 = \mathbf{0}$$

which gives

$$c_1(\mathbf{X}_1 + \mathbf{X}_2) + c_2(\mathbf{X}_1 + \mathbf{X}_3) + c_3(\mathbf{X}_2 + \mathbf{X}_3)$$

$$= (c_1 + c_2)\mathbf{X}_1 + (c_1 + c_3)\mathbf{X}_2 + (c_2 + c_3)\mathbf{X}_3 = \mathbf{0}$$

Since S is linearly independent a linear combination of its members is the zero vector only when each coefficient is zero. Hence we obtain the linear system

$$\begin{aligned} c_1 + c_2 &= 0 \\ c_1 + c_3 &= 0 \\ c_2 + c_3 &= 0 \end{aligned}$$

whose augmented matrix is

$$\left[\begin{array}{ccc|c} 1 & 1 & 0 & 0 \\ 1 & 0 & 1 & 0 \\ 0 & 1 & 1 & 0 \end{array}\right]$$

The reduced row echelon form is

$$\left[\begin{array}{ccc|c} 1 & 0 & 0 & 0 \\ 0 & 1 & 0 & 0 \\ 0 & 0 & 1 & 0 \end{array}\right]$$

thus $c_1 = c_2 = c_3 = 0$ which implies that set $\{\mathbf{Y}_1, \mathbf{Y}_2, \mathbf{Y}_3\}$ is a linearly independent set.

T.7. $\{\mathbf{X}_1, \mathbf{X}_2\}$ is linearly independent. Then by T.3 $\mathbf{X}_2$ is not a linear combination of $\mathbf{X}_1$. Suppose vector $\mathbf{X}_3$ does not belong to span$\{\mathbf{X}_1, \mathbf{X}_2\}$. Then $\mathbf{X}_3$ is not a linear combination of $\mathbf{X}_1$ and $\mathbf{X}_2$. Hence no vector in set $\{\mathbf{X}_1, \mathbf{X}_2, \mathbf{X}_3\}$ is a linear combination of preceding vectors. Thus it follows from Theorem 3.9 that $\{\mathbf{X}_1, \mathbf{X}_2, \mathbf{X}_3\}$ is a linearly independent set.

T.9. Let $\mathbf{Y}_i = \sum_{j=1}^{k} a_{ij}\mathbf{X}_j$ for $i = 1,2,\ldots,m$. Then

$$\mathbf{Z} = \sum_{i=1}^{m} b_i\mathbf{Y}_i = \sum_{i=1}^{m} b_i\sum_{j=1}^{k} a_{ij}\mathbf{X}_j = \sum_{j=1}^{k}\left[\sum_{i=1}^{m} b_i a_{ij}\right]\mathbf{X}_j.$$

Since $\left[\sum_{i=1}^{m} b_i a_{ij}\right]$ is a constant for each j, $\mathbf{Z}$ is a linear combination of the vectors in S.

T.11. Let $\mathbf{V} = \mathbf{R}^2$ and $\mathbf{S}_2 = \{(0,0), (1,0), (0,1)\}$. Since $\mathbf{S}_2$ contains the zero vector for $\mathbf{R}^2$ it is linearly dependent. If $\mathbf{S}_1 = \{(0,0), (1,0)\}$ it is linearly dependent for the same reason. If $\mathbf{S}_1 = \{(1,0), (0,1)\}$, then

$$c_1(1,0) + c_2(0,1) = (0,0)$$

only if $c_1 = c_2 = 0$. Thus in this case $\mathbf{S}_1$ is linearly independent.

T.13. If $\{\mathbf{X}, \mathbf{Y}\}$ is linearly dependent, then there exist scalars c_1 and c_2, not both zero such that $c_1\mathbf{X} + c_2\mathbf{Y} = \mathbf{0}$. In fact since neither $\mathbf{X}$ nor $\mathbf{Y}$ is the zero vector, it follows that both c_1 and c_2 must be nonzero for $c_1\mathbf{X} + c_2\mathbf{Y} = \mathbf{0}$. Hence we have $\mathbf{Y} = -c_1/c_2\mathbf{X}$.

Alternatively, if $\mathbf{Y} = k\mathbf{X}$, then $k \neq 0$ since $\mathbf{Y} \neq \mathbf{0}$. Hence we have $\mathbf{Y} - k\mathbf{X} = \mathbf{0}$ which implies that $\{\mathbf{X}, \mathbf{Y}\}$ is linearly dependent.

ML.1. (a) Following Example 1, we construct the augmented matrix that results from the expression $c_1\mathbf{X}_1 + c_2\mathbf{X}_2 + c_3\mathbf{X}_3 = \mathbf{X}$. Note that since the vectors are rows we need to convert them to columns to form this matrix. Next we obtain the reduced row echelon form of the associated linear system.

```
X1=[1 0 0 1];X2=[0 1 1 0];X3=[1 1 1 1];X=[0 1 1 1];
rref([X1' X2' X3' X'])
```

```
ans =

     1     0     1     0
     0     1     1     0
     0     0     0     1
     0     0     0     0
```

Since this represents an augmented matrix, the system is inconsistent and hence has no solution. Thus **X** is not a linear combination of $\{\mathbf{X}_1, \mathbf{X}_2, \mathbf{X}_3\}$.

(b) Here the strategy is similar to that in part a except that the vectors are already columns. We use the transpose operator to conveniently enter the vectors.

```
X1=[1 2 -1]';X2=[2 -1 0]';X3=[-1 8 -3]';X=[0 5 -2]';
rref([X1 X2 X3 X])

ans =

     1     0     3     2
     0     1    -2    -1
     0     0     0     0
```

Since this matrix represents an augmented matrix, the system is consistent. It follows that **X** is a linear combination of $\{\mathbf{X}_1, \mathbf{X}_2, \mathbf{X}_3\}$.

ML.2. (a) Apply the procedure in ML.1(a).

```
X1=[1 2 1];X2=[3 0 1];X3=[1 8 3];X=[-2 14 4];
rref([X1' X2' X3' X'])

ans =

     1     0     4     7
     0     1    -1    -3
     0     0     0     0
```

This system is consistent so **X** is a linear combination of $\{\mathbf{X}_1, \mathbf{X}_2, \mathbf{X}_3\}$. In the general solution if we set $c_3 = 0$, then $c_1 = 7$ and $c_2 = -3$. Hence $7\mathbf{X}_1 - 3\mathbf{X}_2 = \mathbf{X}$. There are many other linear combinations that work.

(b) After entering the 2×2 matrices into MATLAB we associate a column with each one by 'reshaping' it into a 4×1 matrix. The linear system obtained from the linear combination of the reshaped vectors is the same as that obtained when we use the 2×2 matrices in $c_1\mathbf{X}_1 + c_2\mathbf{X}_2 + c_3\mathbf{X}_3 = \mathbf{X}$.

```
X1=[1 2;1 0];X2=[2 -1;1 2];X3=[-3 1;0 1];X=eye(2);
rref([reshape(X1,4,1) reshape(X2,4,1) ...
      reshape(X3,4,1) reshape(X,4,1)])
```

```
ans =

     1     0     0     0
     0     1     0     0
     0     0     1     0
     0     0     0     1
```

The system is inconsistent, hence **X** is not a linear combination of $\{\mathbf{X}_1, \mathbf{X}_2, \mathbf{X}_3\}$.

ML.3. (a) Follow the procedure in ML.1(b).

```
X1=[1 2 1 0 1]';X2=[0 1 2 -1 1]';
X3=[2 1 0 0 -1]';X4=[-2 1 1 1 1]';
X=[0 -1 1 -2 1]';
rref([X1 X2 X3 X4 X])

ans =

     1     0     0     0     0
     0     1     0     0     1
     0     0     1     0    -1
     0     0     0     1    -1
     0     0     0     0     0
```

The system is consistent and it follows that $0\mathbf{X}_1 + \mathbf{X}_2 - \mathbf{X}_3 - \mathbf{X}_4 = \mathbf{X}$.

(b) Associate a column vector of coefficients with each polynomial, then follow the method in part a.

```
V1=[2 -1 1]';V2=[1 0 -2]';V3=[0 1 -1]';X=[4 1 -5]';
rref([V1 V2 V3 X])

ans =

     1     0     0     1
     0     1     0     2
     0     0     1     2
```

Since the system is consistent, we have that $p_1(t) + 2p_2(t) + 2p_3(t) = p(t)$.

ML.4. Follow the method in ML.2(a).

```
X1=[1 1 0 1];X2=[1 -1 0 1];X3=[0 1 2 1];
```

(a)

```
X=[2 3 2 3];
rref([X1' X2' X3' X'])
```

```
ans =

     1     0     0     2
     0     1     0     0
     0     0     1     1
     0     0     0     0
```

Since the system is consistent, **X** is in span S. In fact, $\mathbf{X} = 2\mathbf{X}_1 + \mathbf{X}_3$.

(b)

```
X=[2 -3 -2 3];
rref([X1' X2' X3' X'])

ans =

     1     0     0     0
     0     1     0     0
     0     0     1     0
     0     0     0     1
```

The system is inconsistent hence, **X** is not in span S.

(c)

```
X=[0 1 2 3];
rref([X1' X2' X3' X'])

ans =

     1     0     0     0
     0     1     0     0
     0     0     1     0
     0     0     0     1
```

The system is inconsistent hence, **X** is not in span S.

ML.5. Associate a column vector with each polynomial as in ML.3(b).

```
V1=[0 1 -1]';V2=[0 1 1]';V3=[1 1 1]';
```

(a)

```
X=[1 2 4]';
rref([V1 V2 V3 X])

ans =

     1     0     0    -1
     0     1     0     2
     0     0     1     1
```

Since the system is consistent, p(t) is in span S.

(b)

```
X=[2 1 -2]';
rref([V1 V2 V3 X])

ans =

     1     0     0     1
     0     1     0    -2
     0     0     1     2
```

Since the system is consistent, p(t) is in span S.

(c)

```
X=[-2 0 1]';
rref([V1 V2 V3 X])

ans =

    1.0000         0         0   -0.5000
         0    1.0000         0    2.5000
         0         0    1.0000   -2.0000
```

Since the system is consistent, p(t) is in span S.

ML.6. In each case we form a linear combination of the vectors in S, set it equal to the zero vector, derive the associated linear system and find its reduced row echelon form.

(a)

```
X1=[1 0 0 1];X2=[0 1 1 0];X3=[1 1 1 1];
rref([X1' X2' X3' zeros(4,1)])

ans =

     1     0     1     0
     0     1     1     0
     0     0     0     0
     0     0     0     0
```

This represents a homogeneous system with 2 equations in 3 unknowns, hence there is a nontrivial solution. Thus S is linearly dependent.

(b)

```
X1=[1 2;1 0];X2=[2 -1;1 2];X3=[-3 1;0 1];
rref([reshape(X1,4,1) reshape(X2,4,1) reshape(X3,4,1)..
     zeros(4,1)])

ans =

     1     0     0     0
     0     1     0     0
     0     0     1     0
     0     0     0     0
```

The homogeneous system has only the trivial solution hence, S is linearly independent.

(c)

```
V1=[0 1 -1]';V2=[0 1 1]';V3=[1 1 1]';
rref([V1 V2 V3 zeros(3,1)])

ans =

     1     0     0     0
     0     1     0     0
     0     0     1     0
```

The homogeneous system has only the trivial solution hence, S is linearly independent.

ML.7. Form the augmented matrix [A|0] and row reduce it.

```
A=[1 2 0 1;1 1 1 2;2 -1 5 7;0 2 -2 -2];
rref([A zeros(4,1)])

ans =

     1     0     2     3     0
     0     1    -1    -1     0
     0     0     0     0     0
     0     0     0     0     0
```

The general solution is $x_4 = s$, $x_3 = t$, $x_2 = t + s$, $x_1 = -2t - 3s$. Hence

X = [-2t-3s t+s t s]′ = t[-2 1 1 0]′ + s[-3 1 0 1]′

and it follows that [-2 1 1 0]′ and [-3 1 0 1]′ span the solution space.

Exercises 3.6

1. Following the method of Example 2, we determine whether the sets are linearly independent and span $\mathbf{R}^2$.

(a) Let $\mathbf{X}_1 = (1,3)$ and $\mathbf{X}_2 = (1,-1)$. Form the linear combination $c_1\mathbf{X}_1 + c_2\mathbf{X}_2 = \mathbf{0} = (0,0)$. Adding vectors then equating corresponding components gives the linear system

$$\begin{aligned} c_1 + c_2 &= 0 \\ 3c_1 - c_2 &= 0. \end{aligned}$$

The coefficient matrix $\begin{bmatrix} 1 & 1 \\ 3 & -1 \end{bmatrix}$ has reduced row echelon form $\mathbf{I}_2$. Hence the only solution is the trivial solution $c_1 = c_2 = 0$ and $\{\mathbf{X}_1, \mathbf{X}_2\}$ is linearly independent. Let $\mathbf{X} = (a,b)$ where a and b are any real numbers. Form the linear combination $k_1\mathbf{X}_1 + k_2\mathbf{X}_2 = \mathbf{X}$. Adding vectors and equating corresponding components gives a linear system with augmented matrix $\left[\begin{array}{cc|c} 1 & 1 & a \\ 3 & -1 & b \end{array}\right]$ which has reduced row echelon form $\left[\begin{array}{cc|c} 1 & 0 & \frac{a+b}{4} \\ 0 & 1 & \frac{3a-b}{4} \end{array}\right]$. It follows that there is a solution k_1, k_2 for any choices of a and b. Hence $\{\mathbf{X}_1, \mathbf{X}_2\}$ spans $\mathbf{R}^2$. Thus $\{\mathbf{X}_1, \mathbf{X}_2\}$ is a basis for $\mathbf{R}^2$.

(b) Let $\mathbf{X}_1 = (0,0)$, $\mathbf{X}_2 = (1,2)$, and $\mathbf{X}_3 = (2,4)$. Form the linear combination $c_1\mathbf{X}_1 + c_2\mathbf{X}_2 + c_3\mathbf{X}_3 = \mathbf{0} = (0,0)$. Adding vectors then equating corresponding components gives the linear system

$$\begin{aligned} 0c_1 + c_2 + 2c_3 &= 0 \\ 0c_1 + 2c_2 + 4c_3 &= 0. \end{aligned}$$

The coefficient matrix $\begin{bmatrix} 0 & 1 & 2 \\ 0 & 2 & 4 \end{bmatrix}$ has reduced row echelon form $\begin{bmatrix} 0 & 1 & 2 \\ 0 & 0 & 0 \end{bmatrix}$. Thus we have $c_1 = r$, any real number, $c_2 = -c_3$, $c_3 = s$, any real number. Hence there are nontrivial solutions and $\{\mathbf{X}_1, \mathbf{X}_2, \mathbf{X}_3\}$ is linearly dependent. This set is not a basis for $\mathbf{R}^2$.

(c) Let $\mathbf{X}_1 = (1,2)$, $\mathbf{X}_2 = (2,-3)$, and $\mathbf{X}_3 = (3,2)$. Form the linear combination $c_1\mathbf{X}_1 + c_2\mathbf{X}_2 + c_3\mathbf{X}_3 = \mathbf{0} = (0,0)$. Adding vectors then equating corresponding components gives the linear system

$$\begin{aligned} c_1 + 2c_2 + 3c_3 &= 0 \\ 2c_1 - 3c_2 + 2c_3 &= 0. \end{aligned}$$

The coefficient matrix $\begin{bmatrix} 1 & 2 & 3 \\ 2 & -3 & 2 \end{bmatrix}$ has reduced row echelon form $\begin{bmatrix} 1 & 0 & \frac{13}{7} \\ 0 & 1 & \frac{4}{7} \end{bmatrix}$. Hence $c_1 = -(13/7)c_3$, $c_2 = -(4/7)c_3$, $c_3 =$ an arbitrary real number. Thus there are nontrivial solutions so the set is linearly dependent and is not basis for $\mathbf{R}^2$.

(d) Let $\mathbf{X}_1 = (1,3)$ and $\mathbf{X}_2 = (-2,6)$. Form the linear combination $c_1\mathbf{X}_1 + c_2\mathbf{X}_2 = \mathbf{0} = (0,0)$. Adding vectors then equating corresponding components gives the linear system

$$\begin{aligned} c_1 - 2c_2 &= 0 \\ 3c_1 + 6c_2 &= 0. \end{aligned}$$

The coefficient matrix $\begin{bmatrix} 1 & -2 \\ 3 & 6 \end{bmatrix}$ has reduced row echelon form $\mathbf{I}_2$. Hence the only solution is the trivial solution $c_1 = c_2 = 0$ and $\{\mathbf{X}_1, \mathbf{X}_2\}$ is linearly independent. Let $\mathbf{X} = (a,b)$ where a and b are any real numbers. Form the linear combination $k_1\mathbf{X}_1 + k_2\mathbf{X}_2 = \mathbf{X}$. Adding vectors and equating corresponding components gives a linear system with augmented matrix $\left[\begin{array}{cc|c} 1 & -2 & a \\ 3 & 6 & b \end{array}\right]$ which has reduced row echelon form $\left[\begin{array}{cc|c} 1 & 0 & \frac{3a+b}{6} \\ 0 & 1 & \frac{b-3a}{12} \end{array}\right]$. It follows that there is a solution k_1, k_2 for any choices of a and b. Hence $\{\mathbf{X}_1, \mathbf{X}_2\}$ spans $\mathbf{R}^2$. Thus $\{\mathbf{X}_1, \mathbf{X}_2\}$ is a basis for $\mathbf{R}^2$.

3. Following the method of Example 2, we determine whether the sets are linearly independent and span $\mathbf{R}^4$.

(a) Let $\mathbf{X}_1 = (1,0,0,1)$, $\mathbf{X}_2 = (0,1,0,0)$, $\mathbf{X}_3 = (1,1,1,1)$, and $\mathbf{X}_4 = (0,1,1,1)$. Form the linear combination $c_1\mathbf{X}_1 + c_2\mathbf{X}_2 + c_3\mathbf{X}_3 + c_4\mathbf{X}_4 = \mathbf{0} = (0,0,0,0)$. Adding vectors then equating corresponding components gives the linear system

$$\begin{aligned} c_1 \quad\;\; + c_3 \quad\;\; &= 0 \\ c_2 + c_3 + c_4 &= 0 \\ c_3 + c_4 &= 0 \\ c_1 \quad\;\; + c_3 + c_4 &= 0. \end{aligned}$$

The coefficient matrix $\begin{bmatrix} 1 & 0 & 1 & 0 \\ 0 & 1 & 1 & 1 \\ 0 & 0 & 1 & 1 \\ 1 & 0 & 1 & 1 \end{bmatrix}$ has reduced row echelon form $\mathbf{I}_4$. Hence the only solution is the trivial solution $c_1 = c_2 = c_3 = c_4 = 0$ and $\{\mathbf{X}_1, \mathbf{X}_2, \mathbf{X}_3, \mathbf{X}_4\}$ is linearly independent. Let $\mathbf{X} = (a,b,c,d)$ where a, b, c, d are any real numbers. Form the linear combination

$$k_1\mathbf{X}_1 + k_2\mathbf{X}_2 + k_3\mathbf{X}_3 + k_4\mathbf{X}_4 = \mathbf{X}.$$

Adding vectors and equating corresponding components gives a linear system with augmented matrix $\left[\begin{array}{cccc|c} 1 & 0 & 1 & 0 & a \\ 0 & 1 & 1 & 1 & b \\ 0 & 0 & 1 & 1 & c \\ 1 & 0 & 1 & 1 & d \end{array}\right]$ which has reduced row echelon form $[\mathbf{I}_4|\mathbf{Q}]$ where the components of $\mathbf{Q}$ are linear combinations of a, b, c, d. It follows that there is a solution for any choice of a, b, c, d. Hence $\{\mathbf{X}_1, \mathbf{X}_2, \mathbf{X}_3, \mathbf{X}_4\}$ spans $\mathbf{R}^4$. Thus it follows that $\{\mathbf{X}_1, \mathbf{X}_2, \mathbf{X}_3, \mathbf{X}_4\}$ is a basis for $\mathbf{R}^4$.

(b) Let $\mathbf{X}_1 = (1,-1,0,2)$, $\mathbf{X}_2 = (3,-1,2,1)$, and $\mathbf{X}_3 = (1,0,0,1)$. Form the linear combination $c_1\mathbf{X}_1 + c_2\mathbf{X}_2 + c_3\mathbf{X}_3 = \mathbf{0} = (0,0,0,0)$. Adding vectors then equating corresponding components gives the linear system

$$\begin{aligned} c_1 + 3c_2 + c_3 &= 0 \\ -c_1 - c_2 &= 0 \\ 2c_2 &= 0 \\ 2c_1 + c_2 + c_3 &= 0. \end{aligned}$$

The coefficient matrix $\begin{bmatrix} 1 & 3 & 1 \\ -1 & -1 & 0 \\ 0 & 2 & 0 \\ 2 & 1 & 1 \end{bmatrix}$ has reduced row echelon form $\begin{bmatrix} 1 & 0 & 0 \\ 0 & 1 & 0 \\ 0 & 0 & 1 \\ 0 & 0 & 0 \end{bmatrix}$. Hence the only solution is the trivial solution $c_1 = c_2 = c_3 = 0$ and $\{\mathbf{X}_1, \mathbf{X}_2, \mathbf{X}_3\}$ is linearly independent. Let $\mathbf{X} = (a,b,c,d)$ where a, b, c, d are any real numbers. Form the linear combination

$$k_1\mathbf{X}_1 + k_2\mathbf{X}_2 + k_3\mathbf{X}_3 = \mathbf{X}.$$

Adding vectors and equating corresponding components gives

a linear system with augmented matrix $\begin{bmatrix} 1 & 3 & 1 & | & a \\ -1 & -1 & 0 & | & b \\ 0 & 2 & 0 & | & c \\ 2 & 1 & 1 & | & d \end{bmatrix}$

which has reduced row echelon form in which the last row is [0 0 0 | -2a+2b+3c+2d]. It follows that there is not a solution k_1, k_2, k_3 for all choices of a, b, c, d. Hence $\{\mathbf{X}_1, \mathbf{X}_2, \mathbf{X}_3\}$ does not span $\mathbf{R}^4$. Thus $\{\mathbf{X}_1, \mathbf{X}_2, \mathbf{X}_3\}$ is not a basis for $\mathbf{R}^4$.

(c) Let $\mathbf{X}_1 = (-2,4,6,4)$, $\mathbf{X}_2 = (0,1,2,0)$, $\mathbf{X}_3 = (-1,2,3,2)$, $\mathbf{X}_4 = (-3,2,5,6)$, and $\mathbf{X}_5 = (-2,-1,0,4)$. Form the linear combination $c_1\mathbf{X}_1 + c_2\mathbf{X}_2 + c_3\mathbf{X}_3 + c_4\mathbf{X}_4 + c_5\mathbf{X}_5 = \mathbf{0}$ $= (0,0,0,0)$. Adding vectors then equating corresponding components gives the linear system

$$\begin{aligned} -2c_1 - c_3 - 3c_4 - 2c_5 &= 0 \\ 4c_1 + c_2 + 2c_3 + 2c_4 - c_5 &= 0 \\ 6c_1 + 2c_2 + 3c_3 + 5c_4 &= 0 \\ 4c_1 + 2c_3 + 6c_4 + 4c_5 &= 0. \end{aligned}$$

By Theorem 1.7, this homogeneous system of 4 equations in 5 unknowns has a nontrivial solution. Thus $\{\mathbf{X}_1, \mathbf{X}_2, \mathbf{X}_3, \mathbf{X}_4, \mathbf{X}_5\}$ is linearly dependent and is not a basis for $\mathbf{R}^4$.

(d) Let $\mathbf{X}_1 = (0,0,1,1)$, $\mathbf{X}_2 = (-1,1,1,2)$, $\mathbf{X}_3 = (1,1,0,0)$, and $\mathbf{X}_4 = (2,1,2,1)$. Form the linear combination $c_1\mathbf{X}_1 + c_2\mathbf{X}_2 + + c_3\mathbf{X}_3 + c_4\mathbf{X}_4 = \mathbf{0} = (0,0,0,0)$. Adding vectors then equating corresponding components gives the linear system

$$\begin{aligned} -c_2 + c_3 + 2c_4 &= 0 \\ c_2 + c_3 + c_4 &= 0 \\ c_1 + c_2 + 2c_4 &= 0 \\ c_1 + 2c_2 + c_4 &= 0. \end{aligned}$$

The coefficient matrix $\begin{bmatrix} 0 & -1 & 1 & 2 \\ 0 & 1 & 1 & 1 \\ 1 & 1 & 0 & 2 \\ 1 & 2 & 0 & 1 \end{bmatrix}$ has reduced row

echelon form $\mathbf{I}_4$. Hence the only solution is the trivial solution $c_1 = c_2 = c_3 = c_4 = 0$ and $\{\mathbf{X}_1, \mathbf{X}_2, \mathbf{X}_3, \mathbf{X}_4\}$ is linearly independent. Let $\mathbf{X} = (a,b,c,d)$ where a, b, c, d are any real numbers. Form the linear combination

$$k_1\mathbf{X}_1 + k_2\mathbf{X}_2 + k_3\mathbf{X}_3 + k_4\mathbf{X}_4 = \mathbf{X}.$$

Adding vectors and equating corresponding components gives

a linear system with augmented matrix $\begin{bmatrix} 0 & -1 & 1 & 2 & | & a \\ 0 & 1 & 1 & 1 & | & b \\ 1 & 1 & 0 & 2 & | & c \\ 1 & 2 & 0 & 1 & | & d \end{bmatrix}$

which has reduced row echelon form $[\mathbf{I}_4 | \mathbf{Q}]$ where $\mathbf{Q}$ is a linear combination of a, b, c, d. It follows that there is a solution k_1, k_2, k_3, k_4 for any choices of a, b, c, d. Hence $\{\mathbf{X}_1, \mathbf{X}_2, \mathbf{X}_3, \mathbf{X}_4\}$ spans $\mathbf{R}^4$ and is a basis for $\mathbf{R}^4$.

5. We determine whether the sets are linearly independent and span $\mathbf{P}_3$.
   (a) Let $p_1(t) = t^3 + 2t^2 + 3t$, $p_2(t) = 2t^3 + 1$, $p_3(t) = 6t^3 + 8t^2 + 6t + 4$, and $p_4(t) = t^3 + 2t^2 + t + 1$. Form the linear combination $c_1p_1(t) + c_2p_2(t) + c_3p_3(t) + c_4p_4(t) = \mathbf{0} = 0t^3 + 0t^2 + 0t + 0$. Expanding and combining terms with like powers of t we obtain

$$(c_1+2c_2+6c_3+c_4)t^3 + (2c_1+8c_3+2c_4)t^2 + (3c_1+6c_3+c_4)t + (c_2+4c_3+c_4) = \mathbf{0}.$$

   Equating coefficients of like powers of t from each side of the equation we obtain the linear system

$$\begin{aligned} c_1 + 2c_2 + 6c_3 + c_4 &= 0 \\ 2c_1 \qquad + 8c_3 + 2c_4 &= 0 \\ 3c_1 \qquad + 6c_3 + c_4 &= 0 \\ c_2 + 4c_3 + c_4 &= 0. \end{aligned}$$

   The coefficient matrix of this homogeneous system is

$$\begin{bmatrix} 1 & 2 & 6 & 1 \\ 2 & 0 & 8 & 2 \\ 3 & 0 & 6 & 1 \\ 0 & 1 & 4 & 1 \end{bmatrix}.$$ Applying row operations $-2\mathbf{r}_1+\mathbf{r}_2$, $-3\mathbf{r}_1+\mathbf{r}_3$, $(-1/4)\mathbf{r}_2$, $6\mathbf{r}_2+\mathbf{r}_3$, $-\mathbf{r}_2+\mathbf{r}_3$, $(-1/6)\mathbf{r}_3$, $-\mathbf{r}_3+\mathbf{r}_1$, $-4\mathbf{r}_3+\mathbf{r}_1$, $-3\mathbf{r}_3+\mathbf{r}_4$ gives the reduced row echelon form

$$\begin{bmatrix} 1 & 0 & 0 & \frac{-1}{3} \\ 0 & 1 & 0 & \frac{-1}{3} \\ 0 & 0 & 1 & \frac{1}{3} \\ 0 & 0 & 0 & 0 \end{bmatrix}.$$ Thus the general solution is $c_1 = (1/3)r$, $c_2 = (1/3)r$, $c_3 = (-1/3)r$, $c_4 = r$, any real number. Hence the vectors are linearly dependent and are not a basis for $\mathbf{P}_3$.

   (b) Let $p_1(t) = t^3 + t^2 + 1$, $p_2(t) = t^3 - 1$, and $p_3(t) = t^3 + t^2 + t$. Form the linear combination $c_1p_1(t) + c_2p_2(t) + c_3p_3(t) = \mathbf{0} = 0t^3 + 0t^2 + 0t + 0$. Expanding and combining terms with like powers of t we obtain

$$(c_1+c_2+c_3)t^3 + (c_1+c_3)t^2 + c_3t + (c_1-c_2) = \mathbf{0}.$$

   Equating coefficients of like powers of t from each side of the equation we obtain the linear system

$$\begin{aligned} c_1 + c_2 + c_3 &= 0 \\ c_1 \qquad + c_3 &= 0 \\ c_3 &= 0 \\ c_1 - c_2 \qquad &= 0. \end{aligned}$$

The coefficient matrix of this homogeneous system is

$\begin{bmatrix} 1 & 1 & 1 \\ 1 & 0 & 1 \\ 0 & 0 & 1 \\ 1 & -1 & 0 \end{bmatrix}$. Applying row operations $-\mathbf{r}_1+\mathbf{r}_2$, $-\mathbf{r}_1+\mathbf{r}_4$, $(-1)\mathbf{r}_2$, $-\mathbf{r}_2+\mathbf{r}_1$, $2\mathbf{r}_2+\mathbf{r}_4$, $-\mathbf{r}_3+\mathbf{r}_1$, $\mathbf{r}_3+\mathbf{r}_4$ gives the reduced row echelon form $\begin{bmatrix} 1 & 0 & 0 \\ 0 & 1 & 0 \\ 0 & 0 & 1 \\ 0 & 0 & 0 \end{bmatrix}$. Hence the only solution is the trivial solution and the vectors are linearly independent. Let $p(t) = at^3 + bt^2 + ct + d$ where a, b, c, and d are arbitrary real numbers. Form the linear combination $k_1p_1(t) + k_2p_2(t) + k_3p_3(t) = p(t)$. Expanding, combining terms with like powers of t, and equating coefficients of like powers from both sides of the equation we obtain the linear system

$$\begin{aligned} k_1 + k_2 + k_3 &= a \\ k_1 \qquad + k_3 &= b \\ k_3 &= c \\ k_1 - k_2 \qquad &= d. \end{aligned}$$

Applying the row operations given above, the reduced row echelon form is

$\left[\begin{array}{ccc|c} 1 & 0 & 0 & b-c \\ 0 & 1 & 0 & a-b \\ 0 & 0 & 1 & c \\ 0 & 0 & 0 & a-2b+c+d \end{array}\right]$. It follows that the system is inconsistent for certain choices of a, b, c, d. Hence the vectors do not span $\mathbf{P}_3$ and are not a basis for $\mathbf{P}_3$.

(c) Let $p_1(t) = t^3 + t^2 + t + 1$, $p_2(t) = t^3 + 2t^2 + t + 3$, and $p_3(t) = 2t^3 + t^2 + 3t + 2$, and $p_4(t) = t^3 + t^2 + 2t + 2$. Form the linear combination $c_1p_1(t) + c_2p_2(t) + c_3p_3(t) + c_4p_4(t) = \mathbf{0} = 0t^3 + 0t^2 + 0t + 0$. Expanding and combining terms with like powers of t we obtain

$$(c_1+c_2+2c_3+c_4)t^3 + (c_1+2c_2+c_3+c_4)t^2 + (c_1+c_2+3c_3+2c_4)t + (c_1+3c_2+2c_3+2c_4) = \mathbf{0}.$$

Equating coefficients of like powers of t from each side of the equation we obtain the linear system

$$\begin{aligned} c_1 + c_2 + 2c_3 + c_4 &= 0 \\ c_1 + 2c_2 + c_3 + c_4 &= 0 \\ c_1 + c_2 + 3c_3 + 2c_4 &= 0 \\ c_1 + 3c_2 + 2c_3 + 2c_4 &= 0. \end{aligned}$$

The coefficient matrix of this homogeneous system has reduced row echelon form $\mathbf{I}_4$. Hence the only solution is the trivial solution and the vectors are linearly independent. Let $p(t) = at^3 + bt^2 + ct + d$ where a, b, c, and d are arbitrary real numbers. Form the linear combination $k_1p_1(t) + k_2p_2(t) + k_3p_3(t) + k_4p_4(t) = p(t)$. Expanding, combining terms with like powers of t, and equating coefficients of like powers from both sides of the equation we obtain the linear system

$$\begin{aligned} k_1 + k_2 + 2k_3 + k_4 &= a \\ k_1 + 2k_2 + k_3 + k_4 &= b \\ k_1 + k_2 + 3k_3 + 2k_4 &= c \\ k_1 + 3k_2 + 2k_3 + 2k_4 &= d. \end{aligned}$$

The reduced row echelon form of the coefficient matrix has the form $[\mathbf{I}_4 | \mathbf{Q}]$, where the entries of $\mathbf{Q}$ are linear combinations of a, b, c, and d. Hence the system is consistent for all choices of a, b, c, and d. Thus the vectors span $\mathbf{P}_3$. Since the vectors are both linearly independent and span, they are a basis for $\mathbf{P}_3$.

(d) Let $p_1(t) = t^3 - t$, $p_2(t) = t^3 + t^2 + 1$, and $p_3(t) = t - 1$. Form the linear combination $c_1p_1(t) + c_2p_2(t) + c_3p_3(t) = \mathbf{0} = 0t^3 + 0t^2 + 0t + 0$. Expanding and combining terms with like powers of t we obtain

$$(c_1+c_2)t^3 + c_2t^2 + (-c_1+c_3)t + (c_2-c_3) = \mathbf{0}.$$

Equating coefficients of like powers of t from both sides of the equation we obtain the linear system

$$\begin{aligned} c_1 + c_2 \quad\quad &= 0 \\ c_2 \quad\quad &= 0 \\ -c_1 \quad + c_3 &= 0 \\ c_2 - c_3 &= 0. \end{aligned}$$

The coefficient matrix of this homogeneous system is $\begin{bmatrix} 1 & 1 & 0 \\ 0 & 1 & 0 \\ -1 & 0 & 1 \\ 0 & 1 & -1 \end{bmatrix}$. Applying row operations $\mathbf{r}_1+\mathbf{r}_3$, $-\mathbf{r}_2+\mathbf{r}_1$, $-\mathbf{r}_2+\mathbf{r}_3$, $-\mathbf{r}_2+\mathbf{r}_4$, $\mathbf{r}_3+\mathbf{r}_4$ gives the reduced row echelon form $\begin{bmatrix} 1 & 0 & 0 \\ 0 & 1 & 0 \\ 0 & 0 & 1 \\ 0 & 0 & 0 \end{bmatrix}$.

Hence the only solution is the trivial solution and the vectors are linearly independent. Let $p(t) = at^3 + bt^2 + ct + d$ where a, b, c, and d are arbitrary real numbers. Form the linear combination $k_1p_1(t) + k_2p_2(t) + k_3p_3(t) = p(t)$. Expanding, combining terms with like powers of t, and equating coefficients of like powers from both sides of the equation we obtain the linear system

$$\begin{aligned} k_1 + k_2 \quad\quad &= a \\ k_2 \quad\quad &= b \\ -k_1 \quad + k_3 &= c \\ k_2 - k_3 &= d. \end{aligned}$$

Applying the row operations given above we obtain the reduced row echelon form of the associated augmented matrix as

$$\left[\begin{array}{ccc|c} 1 & 0 & 0 & a-b \\ 0 & 1 & 0 & b \\ 0 & 0 & 1 & a-b+c \\ 0 & 0 & 0 & a-2b+c+d \end{array}\right].$$ It follows that there is not a solution for all choices of a, b, c, and d. Hence the vectors do not span $\mathbf{P}_3$ and are not a basis for $\mathbf{P}_3$.

7. We follow the technique of Example 2.

(a) Let $\mathbf{X}_1 = (1,1,1)$, $\mathbf{X}_2 = (1,2,3)$, $\mathbf{X}_3 = (0,1,0)$. Form the linear combination $c_1\mathbf{X}_1 + c_2\mathbf{X}_2 + c_3\mathbf{X}_3 = \mathbf{0} = (0,0,0)$. Expanding, adding vectors, and equating entries in corresponding positions gives the linear system

$$\begin{aligned} c_1 + c_2 \quad\quad &= 0 \\ c_1 + 2c_2 + c_3 &= 0 \\ c_1 + 3c_2 \quad\quad &= 0. \end{aligned}$$

The corresponding coefficient matrix is $\begin{bmatrix} 1 & 1 & 0 \\ 1 & 2 & 1 \\ 1 & 3 & 0 \end{bmatrix}$. Applying row operations $-\mathbf{r}_1+\mathbf{r}_2$, $-\mathbf{r}_1+\mathbf{r}_3$, $\mathbf{r}_2\leftrightarrow\mathbf{r}_3$, $(1/2)\mathbf{r}_2$, $-1\mathbf{r}_2+\mathbf{r}_1$, $-\mathbf{r}_2+\mathbf{r}_3$ gives the reduced row echelon form which is $\mathbf{I}_3$. Hence the only solution is the trivial solution and the vectors are linearly independent. Let $\mathbf{X} = (a,b,c)$ where a, b, and c are any real numbers. Form the linear combination

$$k_1\mathbf{X}_1 + k_2\mathbf{X}_2 + k_3\mathbf{X}_3 = \mathbf{X}.$$

Expanding, adding vectors and equating corresponding entries gives the linear system

$$\begin{aligned} k_1 + k_2 \quad &= a \\ k_1 + 2k_2 + k_3 &= b \\ k_1 + 3k_2 \quad &= c. \end{aligned}$$

Forming the augmented matrix and using the same row operations as above we obtain the reduced row echelon form

$$\left[\begin{array}{ccc|c} 1 & 0 & 0 & \frac{3a-c}{2} \\ 0 & 1 & 0 & \frac{c-a}{2} \\ 0 & 0 & 1 & \frac{-a+2b-c}{2} \end{array}\right].$$

The system is consistent for all choices of a, b, and c. Thus $\{\mathbf{X}_1, \mathbf{X}_2, \mathbf{X}_3\}$ spans $\mathbf{R}^3$. It follows that $\{\mathbf{X}_1, \mathbf{X}_2, \mathbf{X}_3\}$ is a basis for $\mathbf{R}^3$. Let a = 2, b= 1, c = 3, then from the preceding matrix we have $k_1 = 3/2$, $k_2 = 1/2$, $k_3 = -3/2$. Thus $\mathbf{X} = (2,1,3) = (3/2)\mathbf{X}_1 + (1/2)\mathbf{X}_2 + (-3/2)\mathbf{X}_3$.

(b) Let $\mathbf{X}_1 = (1,2,2)$, $\mathbf{X}_2 = (2,1,3)$, $\mathbf{X}_3 = (0,0,0)$. Form the linear combination $c_1\mathbf{X}_1 + c_2\mathbf{X}_2 + c_3\mathbf{X}_3 = \mathbf{0} = (0,0,0)$. Expanding, adding vectors, and equating entries in corresponding positions gives the linear system

$$\begin{aligned} c_1 + 2c_2 &= 0 \\ 2c_1 + c_2 &= 0 \\ 2c_1 + 3c_2 &= 0. \end{aligned}$$

The corresponding coefficient matrix is $\begin{bmatrix} 1 & 2 \\ 2 & 1 \\ 2 & 3 \end{bmatrix}$. Applying row operations $-2\mathbf{r}_1+\mathbf{r}_2$, $-2\mathbf{r}_1+\mathbf{r}_3$, $(-1/3)\mathbf{r}_2$, $-2\mathbf{r}_2+\mathbf{r}_1$, $\mathbf{r}_2+\mathbf{r}_3$ gives the reduced row echelon form $\begin{bmatrix} 1 & 0 & 0 \\ 0 & 1 & 0 \\ 0 & 0 & 0 \end{bmatrix}$. Hence the general solution is $c_1 = c_2 = 0$, $c_3 = r$, any real number. Thus there exist nontrivial solutions. Hence the vectors are linearly dependent and are not a basis for $\mathbf{R}^3$.

(c) Let $\mathbf{X}_1 = (-2,1,3)$, $\mathbf{X}_2 = (-1,2,3)$, $\mathbf{X}_3 = (-1,-4,-3)$. Form the linear combination $c_1\mathbf{X}_1 + c_2\mathbf{X}_2 + c_3\mathbf{X}_3 = \mathbf{0} = (0,0,0)$. Expanding, adding vectors, and equating entries in corresponding positions gives the linear system

$$\begin{aligned} -2c_1 - c_2 - c_3 &= 0 \\ c_1 + 2c_2 - 4c_3 &= 0 \\ 3c_1 + 3c_2 - 3c_3 &= 0. \end{aligned}$$

The corresponding coefficient matrix is $\begin{bmatrix} -2 & -1 & -1 \\ 1 & 2 & -4 \\ 3 & 3 & -3 \end{bmatrix}$.

Applying row operations $\mathbf{r}_1 \leftrightarrow \mathbf{r}_2$, $2\mathbf{r}_1+\mathbf{r}_2$, $-3\mathbf{r}_1+\mathbf{r}_3$, $\mathbf{r}_2+\mathbf{r}_3$, gives the equivalent system

$$\begin{bmatrix} 1 & 2 & -4 \\ 0 & 3 & -9 \\ 0 & 0 & 0 \end{bmatrix}.$$ There are two equations but three unknowns, so there exist nontrivial solutions. The set $\{\mathbf{X}_1, \mathbf{X}_2, \mathbf{X}_3\}$ is linearly dependent and is not a basis for $\mathbf{R}^3$.

9. We follow the technique given in Example 5. Let $S = \{\mathbf{X}_1, \mathbf{X}_2, \mathbf{X}_3, \mathbf{X}_4\}$ where $\mathbf{X}_1 = (1,2,2)$, $\mathbf{X}_2 = (3,2,1)$, $\mathbf{X}_3 = (11,10,7)$, $\mathbf{X}_4 = (7,6,4)$. Form the linear combination
$$c_1\mathbf{X}_1 + c_2\mathbf{X}_2 + c_3\mathbf{X}_3 + c_4\mathbf{X}_4 = \mathbf{0} = (0,0,0).$$
Expanding, adding vectors, and equating corresponding components gives the linear system
$$\begin{aligned} c_1 + 3c_2 + 11c_3 + 7c_4 &= 0 \\ 2c_1 + 2c_2 + 10c_3 + 6c_4 &= 0 \\ 2c_1 + c_2 + 7c_3 + 4c_4 &= 0. \end{aligned}$$
The coefficient matrix of this homogeneous system is

$$\begin{bmatrix} 1 & 3 & 11 & 7 \\ 2 & 2 & 10 & 6 \\ 2 & 1 & 7 & 4 \end{bmatrix}.$$ Applying row operations $-2\mathbf{r}_1+\mathbf{r}_2$, $-2\mathbf{r}_1+\mathbf{r}_3$, $(-1/4)\mathbf{r}_2$, $-3\mathbf{r}_2+\mathbf{r}_1$, $-5\mathbf{r}_2+\mathbf{r}_3$ gives reduced row echelon form

$$\begin{bmatrix} 1 & 0 & 2 & 1 \\ 0 & 1 & 3 & 2 \\ 0 & 0 & 0 & 0 \end{bmatrix}.$$ The leading 1's appear in columns 1 and 2, so $\{\mathbf{X}_1, \mathbf{X}_2\}$ is a basis for $W = \text{span } S$. Hence $\dim W = 2$.

11. We follow the technique given in Example 5. Let $S = \{p_1(t), p_2(t), p_3(t), p_4(t)\}$ where $p_1(t) = t^3+t^2-2t+1$, $p_2(t) = t^2+1$, $p_3(t) = t^3-2t$, $p_4(t) = 2t^3+3t^2-4t+3\}$. Form the linear combination
$$c_1p_1(t) + c_2p_2(t) + c_3p_3(t) + c_4p_4(t) = \mathbf{0}.$$
Expanding and adding like terms gives
$$(c_1+c_3+2c_4)t^3 + (c_1+c_2+3c_4)t^2 + (-2c_1-2c_3-4c_4)t + (c_1+c_2+3c_4) = \mathbf{0}.$$
Equating coefficients of like powers of t on each side of the equation gives the linear system
$$\begin{aligned} c_1 + c_3 + 2c_4 &= 0 \\ c_1 + c_2 + 3c_4 &= 0 \\ -2c_1 - 2c_3 - 4c_4 &= 0 \\ c_1 + c_2 + 3c_4 &= 0. \end{aligned}$$

Exercises 3.6

The coefficient matrix of this homogeneous system is

$$\begin{bmatrix} 1 & 0 & 1 & 2 \\ 1 & 1 & 0 & 3 \\ -2 & 0 & -2 & -4 \\ 1 & 1 & 0 & 3 \end{bmatrix}.$$ Applying row operations $-\mathbf{r}_1+\mathbf{r}_2$, $2\mathbf{r}_1+\mathbf{r}_3$, $-\mathbf{r}_1+\mathbf{r}_4$, $-\mathbf{r}_2+\mathbf{r}_4$, gives reduced row echelon form

$$\begin{bmatrix} 1 & 0 & 1 & 2 \\ 0 & 1 & -1 & 1 \\ 0 & 0 & 0 & 0 \\ 0 & 0 & 0 & 0 \end{bmatrix}.$$ The leading 1's appear in columns 1 and 2, so $\{p_1(t), p_2(t)\}$ is a basis for W = span S. Hence dim W = 2.

13. (a) Let W be the subspace of $\mathbf{R}^3$ of all vectors of the form (a,b,c) where b = a + c. Then

$$\begin{aligned}(a,b,c) &= (a,a+c,c) = (a,a,0) + (0,c,c)\\ &= a(1,1,0) + c(0,1,1).\end{aligned}$$

Let $\mathbf{X}_1 = (1,1,0)$ and $\mathbf{X}_2 = (0,1,1)$. From the preceding equation $\{\mathbf{X}_1, \mathbf{X}_2\}$ spans W. Form the linear combination $c_1\mathbf{X}_1 + c_2\mathbf{X}_2 = \mathbf{0} = (0,0,0)$. Expanding, adding vectors, and equating corresponding components gives the linear system

$$\begin{aligned} c_1 \quad\quad &= 0\\ c_1 + c_2 &= 0\\ c_2 &= 0.\end{aligned}$$

The only solution is the trivial solution, hence $\mathbf{X}_1$ and $\mathbf{X}_2$ are linearly independent. It follows that $\{\mathbf{X}_1, \mathbf{X}_2\}$ is a basis for W.

(b) Let W be the subspace of $\mathbf{R}^3$ of all vectors of the form (a,b,c) where b = a. Then

$$\begin{aligned}(a,b,c) &= (a,a,c) = (a,a,0) + (0,0,c)\\ &= a(1,1,0) + c(0,0,1).\end{aligned}$$

Let $\mathbf{X}_1 = (1,1,0)$ and $\mathbf{X}_2 = (0,0,1)$. From the preceding equation $\{\mathbf{X}_1, \mathbf{X}_2\}$ spans W. Form the linear combination $c_1\mathbf{X}_1 + c_2\mathbf{X}_2 = \mathbf{0} = (0,0,0)$. Expanding, adding vectors, and equating corresponding components gives the linear system

$$\begin{aligned} c_1 \quad\quad &= 0\\ c_1 \quad\quad &= 0\\ c_2 &= 0.\end{aligned}$$

The only solution is the trivial solution, hence $\mathbf{X}_1$ and $\mathbf{X}_2$ are linearly independent. It follows that $\{\mathbf{X}_1, \mathbf{X}_2\}$ is a basis for W.

(c) Let W be the subspace of $\mathbf{R}^3$ of all vectors of the form (a,b,c) where a =0. Then

$$\begin{aligned}(a,b,c) &= (0,b,c) = (0,b,0) + (0,0,c)\\ &= b(0,1,0) + c(0,0,1).\end{aligned}$$

Let $\mathbf{X}_1 = (0,1,0)$ and $\mathbf{X}_2 = (0,0,1)$. From the preceding equation $\{\mathbf{X}_1, \mathbf{X}_2\}$ spans W. Form the linear combination $c_1\mathbf{X}_1 + c_2\mathbf{X}_2 = \mathbf{0} = (0,0,0)$. Expanding, adding vectors, and equating corresponding components gives the linear system

$$\begin{aligned} c_1 \qquad &= 0 \\ c_2 &= 0. \end{aligned}$$

The only solution is the trivial solution, hence $\mathbf{X}_1$ and $\mathbf{X}_2$ are linearly independent. It follows that $\{\mathbf{X}_1, \mathbf{X}_2\}$ is a basis for W.

(d) Let W be the subspace of $\mathbf{R}^3$ of all vectors of the form $(a-b,b+c,2a-b+c)$. Then

$$\begin{aligned}(a-b,b+c,2a-b+c) &= (a,0,2a) + (-b,b,-b) + (0,c,c) \\ &= a(1,0,2) + b(-1,1,-1) + c(0,1,1).\end{aligned}$$

Let $\mathbf{X}_1 = (1,0,2)$, $\mathbf{X}_2 = (-1,1,-1)$, and $\mathbf{X}_3 = (0,1,1)$. The preceding equation implies that $\{\mathbf{X}_1, \mathbf{X}_2, \mathbf{X}_3\}$ spans W. Form the linear combination $c_1\mathbf{X}_1 + c_2\mathbf{X}_2 + c_3\mathbf{X}_3 = \mathbf{0} = (0,0,0)$. Expanding, adding vectors, and equating corresponding components gives the linear system

$$\begin{aligned} c_1 - c_2 \qquad &= 0 \\ c_2 + c_3 &= 0 \\ 2c_1 - c_2 + c_3 &= 0. \end{aligned}$$

The coefficient matrix of this homogeneous system is

$\begin{bmatrix} 1 & -1 & 0 \\ 0 & 1 & 1 \\ 2 & -1 & 1 \end{bmatrix}$. Applying row operations $-2\mathbf{r}_1+\mathbf{r}_3$, $-\mathbf{r}_2+\mathbf{r}_1$, $-\mathbf{r}_2+\mathbf{r}_3$ gives reduced row echelon form $\begin{bmatrix} 1 & 0 & 1 \\ 0 & 1 & 1 \\ 0 & 0 & 0 \end{bmatrix}$. The leading 1's appear in columns 1 and 2, so $\{\mathbf{X}_1, \mathbf{X}_2\}$ is a basis for W.

15. (a) In Exercise 1(a) it is shown that $\{(1,3), (1,-1)\}$ is a basis for $\mathbf{R}^2$. Thus $W = \text{span}\ \{(1,3), (1,-1)\} = \mathbf{R}^2$ and $\dim W = 2$.

(b) Let $\mathbf{X}_1 = (0,0)$, $\mathbf{X}_2 = (1,2)$, $\mathbf{X}_3 = (2,4)$, $S = \{\mathbf{X}_1, \mathbf{X}_2, \mathbf{X}_3\}$, and W = span S. Following the technique in Example 5, we determine a subset of S which is a basis for W. Form the linear combination

$$c_1\mathbf{X}_1 + c_2\mathbf{X}_2 + c_3\mathbf{X}_3 = \mathbf{0} = (0,0).$$

Expanding, adding vectors, and equating corresponding components gives the linear system

$$\begin{aligned} 0c_1 + c_2 + 2c_3 &= 0 \\ 0c_1 + 2c_2 + 4c_3 &= 0. \end{aligned}$$

The coefficient matrix of this homogeneous system is

$\begin{bmatrix} 0 & 1 & 2 \\ 0 & 2 & 4 \end{bmatrix}$. Applying row operation $-2\mathbf{r}_1 + \mathbf{r}_2$ gives reduced row echelon form $\begin{bmatrix} 0 & 1 & 2 \\ 0 & 0 & 0 \end{bmatrix}$. The leading 1 appears in column 2, so $\{\mathbf{X}_2\}$ is a basis for W. Thus dim W = 1.

(c) Let $\mathbf{X}_1 = (1,2)$, $\mathbf{X}_2 = (2,-3)$, $\mathbf{X}_3 = (3,2)$, $S = \{\mathbf{X}_1, \mathbf{X}_2, \mathbf{X}_3\}$, and W = span S. Following the technique in Example 5, we determine a subset of S which is a basis for W. Form the linear combination

$$c_1\mathbf{X}_1 + c_2\mathbf{X}_2 + c_3\mathbf{X}_3 = \mathbf{0} = (0,0).$$

Expanding, adding vectors, and equating corresponding components gives the linear system

$$\begin{aligned} c_1 + 2c_2 + 3c_3 &= 0 \\ 2c_1 - 3c_2 + 2c_3 &= 0. \end{aligned}$$

The coefficient matrix of this homogeneous system is

$\begin{bmatrix} 1 & 2 & 3 \\ 2 & -3 & 2 \end{bmatrix}$. Applying row operation $-2\mathbf{r}_1 + \mathbf{r}_2$, $(-1/7)\mathbf{r}_2$, $-2\mathbf{r}_2 + \mathbf{r}_1$ gives reduced row echelon form

$\begin{bmatrix} 1 & 0 & 13/7 \\ 0 & 1 & 1/7 \end{bmatrix}$. The leading 1's appears in columns 1 and 2, so $\{\mathbf{X}_1, \mathbf{X}_2\}$ is a basis for W. Thus dim W = 2.

(d) In Exercise 1(d) it is shown that $\{(1,3), (-2,6)\}$ is a basis for $\mathbf{R}^2$. Thus W = span $\{(1,3), (-2,6)\} = \mathbf{R}^2$ and dim W = 2.

17. (a) Let $\mathbf{X}_1 = (1,0,0,1)$, $\mathbf{X}_2 = (0,1,0,0)$, $\mathbf{X}_3 = (1,1,1,1)$, $\mathbf{X}_4 = (0,1,1,1)$, $S = \{\mathbf{X}_1, \mathbf{X}_2, \mathbf{X}_3, \mathbf{X}_4\}$, and W = span S. In Exercise 3(a) it is shown that S is linearly independent. Thus dim W = 4.

(b) Let $\mathbf{X}_1 = (1,-1,0,2)$, $\mathbf{X}_2 = (3,-1,2,1)$, $\mathbf{X}_3 = (1,0,0,1)$, $S = \{\mathbf{X}_1, \mathbf{X}_2, \mathbf{X}_3\}$, and W = span S. In Exercise 3(b) it is shown that S is linearly independent. Thus dim W = 3.

(c) Let $\mathbf{X}_1 = (-2,4,6,4)$, $\mathbf{X}_2 = (0,1,2,0)$, $\mathbf{X}_3 = (-1,2,3,2)$, $\mathbf{X}_4 = (-3,2,5,6)$, $\mathbf{X}_5 = (-2,-1,0,4)$, $S = \{\mathbf{X}_1, \mathbf{X}_2, \mathbf{X}_3, \mathbf{X}_4, \mathbf{X}_5\}$, and W = span S. Following the technique in Example 5, we determine a subset of S which is a basis for W. Form the linear combination

$$c_1\mathbf{X}_1 + c_2\mathbf{X}_2 + c_3\mathbf{X}_3 + c_4\mathbf{X}_4 + c_5\mathbf{X}_5 = \mathbf{0} = (0,0,0,0).$$

Expanding, adding vectors, and equating corresponding components gives the linear system

$$\begin{aligned} -2c_1 + 0c_2 - c_3 - 3c_4 - 2c_5 &= 0 \\ 4c_1 + c_2 + 2c_3 + 2c_4 - c_5 &= 0 \\ 6c_1 + 2c_2 + 3c_3 + 5c_4 + 0c_5 &= 0 \\ 4c_1 + 0c_2 + 2c_3 + 6c_4 + 4c_5 &= 0. \end{aligned}$$

The coefficient matrix of this homogeneous system is

$$\begin{bmatrix} -2 & 0 & -1 & -3 & -2 \\ 4 & 1 & 2 & 2 & -1 \\ 6 & 2 & 3 & 5 & 0 \\ 4 & 0 & 2 & 6 & 4 \end{bmatrix}$$. The reduced row echelon form is

$$\begin{bmatrix} 1 & 0 & 1/2 & 0 & -1/2 \\ 0 & 1 & 0 & 0 & -1 \\ 0 & 0 & 0 & 1 & 1 \\ 0 & 0 & 0 & 0 & 0 \end{bmatrix}$$. The leading 1's appears in

columns 1, 2, and 4 so $\{\mathbf{X}_1, \mathbf{X}_2, \mathbf{X}_4\}$ is a basis for W. Thus dim W = 3.

(d) Let $\mathbf{X}_1 = (0,0,1,1)$, $\mathbf{X}_2 = (-1,1,1,2)$, $\mathbf{X}_3 = (1,1,0,0)$, $\mathbf{X}_4 = (2,1,2,1)$, $S = \{\mathbf{X}_1, \mathbf{X}_2, \mathbf{X}_3, \mathbf{X}_4\}$, and W = span S. In Exercise 3(d) it is shown that S is linearly independent. Thus dim W = 4.

19. We follow the technique given in Example 9. Let $\mathbf{E}_1 = (1,0,0,0)$, $\mathbf{E}_2 = (0,1,0,0)$, $\mathbf{E}_3 = (0,0,1,0)$, $\mathbf{E}_4 = (0,0,0,1)$, $\mathbf{X}_1 = (1,0,1,0)$, $\mathbf{X}_2 = (0,1,-1,0)$, and $S = \{\mathbf{X}_1, \mathbf{X}_2, \mathbf{E}_1, \mathbf{E}_2, \mathbf{E}_3, \mathbf{E}_4\}$. S spans $\mathbf{R}^4$ since it contains a basis. We form the expression

$$c_1\mathbf{X}_1 + c_2\mathbf{X}_2 + c_3\mathbf{E}_1 + c_4\mathbf{E}_2 + c_5\mathbf{E}_3 + c_6\mathbf{E}_4 = (0, 0, 0, 0)$$

which leads to the homogeneous linear system with augmented matrix

$$\left[\begin{array}{cccccc|c} 1 & 0 & 1 & 0 & 0 & 0 & 0 \\ 0 & 1 & 0 & 1 & 0 & 0 & 0 \\ 1 & -1 & 0 & 0 & 1 & 0 & 0 \\ 0 & 0 & 0 & 0 & 0 & 1 & 0 \end{array}\right]$$

Transforming the augmented matrix to reduced row echelon form gives

$$\left[\begin{array}{cccccc|c} 1 & 0 & 0 & 1 & 1 & 0 & 0 \\ 0 & 1 & 0 & 1 & 0 & 0 & 0 \\ 0 & 0 & 1 & -1 & -1 & 0 & 0 \\ 0 & 0 & 0 & 0 & 0 & 1 & 0 \end{array}\right]$$

Since the leading 1's appear in columns 1, 2, 3, and 6, we conclude that $\{\mathbf{X}_1, \mathbf{X}_2, \mathbf{E}_1, \mathbf{E}_4\}$ is a basis for R^4 containing $\mathbf{X}_1$ and $\mathbf{X}_2$.

Exercises 3.6

21. Using Theorem 3.14(a) we find all values of a so that

$$S = \{\mathbf{X}_1, \mathbf{X}_2, \mathbf{X}_3\} = \{(a^2, 0, 1), (0, a, 2), (1, 0, 1)\}$$

is a linearly independent set. Form the expression

$$c_1(a^2, 0, 1) + c_2(0, a, 2) + c_3(1, 0, 1) = (0, 0, 0)$$

and combine the terms on the left. Equate corresponding entries from each side to obtain the linear system

$$\begin{aligned} a^2c_1 \qquad\quad + c_3 &= 0 \\ ac_2 \qquad &= 0 \\ c_1 + 2c_2 + c_3 &= 0 \end{aligned}$$

Form the augmented matrix and reduce it:

$$\left[\begin{array}{ccc|c} a^2 & 0 & 1 & 0 \\ 0 & a & 0 & 0 \\ 1 & 2 & 1 & 0 \end{array}\right]_{R_1 \leftrightarrow R_3} = \left[\begin{array}{ccc|c} 1 & 2 & 1 & 0 \\ 0 & a & 0 & 0 \\ a^2 & 0 & 1 & 0 \end{array}\right]_{-a^2R_1+R_3}$$

$$= \left[\begin{array}{ccc|c} 1 & 2 & 1 & 0 \\ 0 & a & 0 & 0 \\ 0 & -2a^2 & 1-a^2 & 0 \end{array}\right]$$

Note that if $a = 0$, then there will be nontrivial solutions. Hence for S to be linearly independent a must not be zero. Thus we can perform the row operation $(1/a)R_2$ to give

$$\left[\begin{array}{ccc|c} 1 & 2 & 1 & 0 \\ 0 & 1 & 0 & 0 \\ 0 & -2a^2 & 1-a^2 & 0 \end{array}\right]_{\substack{-2R_2+R_1 \\ 2a^2R_2+R_3}}$$

$$= \left[\begin{array}{ccc|c} 1 & 0 & 1 & 0 \\ 0 & 1 & 0 & 0 \\ 0 & 0 & 1-a^2 & 0 \end{array}\right]$$

Note that $1-a^2$ must not be zero, otherwise there will be nontrivial solutions and S would be linearly dependent. Thus it follows that $a \neq 1$ or -1. Hence S will be linearly independent for all values of a except -1, 1, and 0.

23. By observation

$$\begin{bmatrix} a & 0 & 0 \\ 0 & b & 0 \\ 0 & 0 & c \end{bmatrix} = a\begin{bmatrix} 1 & 0 & 0 \\ 0 & 0 & 0 \\ 0 & 0 & 0 \end{bmatrix} + b\begin{bmatrix} 0 & 0 & 0 \\ 0 & 1 & 0 \\ 0 & 0 & 0 \end{bmatrix} + c\begin{bmatrix} 0 & 0 & 0 \\ 0 & 0 & 0 \\ 0 & 0 & 1 \end{bmatrix}$$

Hence $S = \left\{ \begin{bmatrix} 1 & 0 & 0 \\ 0 & 0 & 0 \\ 0 & 0 & 0 \end{bmatrix}, \begin{bmatrix} 0 & 0 & 0 \\ 0 & 1 & 0 \\ 0 & 0 & 0 \end{bmatrix}, \begin{bmatrix} 0 & 0 & 0 \\ 0 & 0 & 0 \\ 0 & 0 & 1 \end{bmatrix} \right\}$ spans W.

Again by observation, dim W = 3, so Theorem 3.14(b) implies set S is a basis.

T.1. Let $\mathbf{X}_j = \sum_{k=1}^{j-1} c_k \mathbf{X}_k$. Also let $\mathbf{X}$ be any vector in V. Since S spans V, there exist constants $a_1, a_2, \ldots, a_n$ such that $\mathbf{X} = \sum_{i=1}^{n} a_i \mathbf{X}_i$. Then

$$\begin{aligned} \mathbf{X} &= a_1\mathbf{X}_1 + \cdots + a_{j-1}\mathbf{X}_{j-1} + a_j \sum_{k=1}^{j-1} c_k \mathbf{X}_k \\ &\quad + a_{j+1}\mathbf{X}_{j+1} + \cdots + a_n \mathbf{X}_n \\ &= \sum_{i=1}^{j-1} (a_i + c_i)\mathbf{X}_i + \sum_{i=j+1}^{n} a_i \mathbf{X}_i. \end{aligned}$$

Thus every vector in V is a linear combination of the vectors in S_1; that is, span S_1 = V.

T.3. Let dim V = n. First note that any set of vectors in W that is linearly independent in W is linearly independent in V. If W = {**0**}, then dim W = 0 and we are done. Suppose now that W is a nonzero subspace of V. Then W contains a nonzero vector $\mathbf{X}_1$, so $\{\mathbf{X}_1\}$ is linearly independent in W (and in V). If span $\{\mathbf{X}_1\}$ = W, then dim W = 1 and we are done. If span $\{\mathbf{X}_1\} \neq$ W, then there exists a vector $\mathbf{X}_2$ in W which is not in span $\{\mathbf{X}_1\}$. Then $\{\mathbf{X}_1, \mathbf{X}_2\}$ is linearly independent in W (and in V). Since dim V = n, no linearly independent set in V can have more than n vectors. Hence, no linearly independent set in W can have more than n vectors. Continuing the above process we find a basis for W containing at most n vectors. Hence W is finite dimensional and dim W $\leq$ dim V.

T.5. Supppose a set S of n-1 vectors in V spans V. By Theorem 3.11, some subset of S would be a basis for V. Thus dim V $\leq$ n-1. Contradiction that dim V = n.

T.7. (a) By Theorem 3.13, there is a basis T for V which contains $S = \{\mathbf{X}_1, \mathbf{X}_2, \ldots, \mathbf{X}_n\}$. Since dim V = n, T cannot have more vectors than S. Thus T = S.

(b) By Theorem 3.11, some subset T of S is a basis for V. Since dim V = n, T has n elements. Thus T = S.

Exercises 3.6

T.9. Let $V = R^3$. The trivial subspaces of any vector space are $\{\mathbf{0}\}$ and V. Hence $\{\mathbf{0}\}$ and R^3 are subspaces of R^3.

A line ℓ_0 passing through the origin in R^3 parallel to vector $\mathbf{v} = \begin{bmatrix} a \\ b \\ c \end{bmatrix}$ is the set of all points P(x,y,z) whose position vector $\mathbf{u} = \begin{bmatrix} x \\ y \\ z \end{bmatrix}$ is of the form $\mathbf{u} = t\mathbf{v}$, where t is any real scalar. Let $\mathbf{u}_1 = r\mathbf{v}$ and $\mathbf{u}_2 = s\mathbf{v}$ be the position vectors for two points on line ℓ_0. Then $\mathbf{u}_1 + \mathbf{u}_2 = (r + s)\mathbf{v}$ and hence is a position vector for a point on ℓ_0. Similarly, $c\mathbf{u}_1 = (cr)\mathbf{v}$ is a position vector for a point on ℓ_0. Thus ℓ_0 is a subspace of R^3.

Any plane π in R^3 through the origin has an equation of the form

$$ax + by + cz = 0$$

Sums and scalar multiples of any point on π will also satisfy this equation, hence π is a subspace of R^3.

To show that $\{\mathbf{0}\}$, V, lines, and planes are the only subspaces of R^3 we argue as follows. Let W be any subspace of R^3. Hence W contains the zero vector $\mathbf{0}$. If $W \neq \{\mathbf{0}\}$, then it contains a nonzero vector $\mathbf{Y} = [a \;\; b \;\; c]^T$ where at least one of a, b, or c is not zero. Since W is a subspace it contains span $\{\mathbf{Y}\}$. If W = span $\{\mathbf{Y}\}$ then W is a line in R^3 through the origin. Otherwise, there exists a vector $\mathbf{X}$ in W which is not in span $\{\mathbf{Y}\}$. Hence $\{\mathbf{Y}, \mathbf{X}\}$ is a linearly independent set. But then W contains span $\{\mathbf{Y}, \mathbf{X}\}$. If W = span $\{\mathbf{Y}, \mathbf{X}\}$ then W is a plane through the origin. Otherwise there is a vector $\mathbf{Z}$ in W that is not in span $\{\mathbf{Y}, \mathbf{X}\}$. Hence $\{\mathbf{Y}, \mathbf{X}, \mathbf{Z}\}$ is a linearly independent set in W and W contains span $\{\mathbf{Y}, \mathbf{X}, \mathbf{Z}\}$. But $\{\mathbf{Y}, \mathbf{X}, \mathbf{Z}\}$ is a linearly independent set in R^3, hence a basis for R^3. It follows in this case that $W = R^3$.

T.11. The set $T = \{\mathbf{Y}_1, \mathbf{Y}_2, \mathbf{Y}_3\}$ is a set of three vectors in a three dimensional vector space V. By Theorem 3.14(b) we need only show that T spans V. One may solve for the $\mathbf{X}$'s in terms of the $\mathbf{Y}$'s:

$$\mathbf{X}_3 = \mathbf{Y}_3$$

$$\mathbf{X}_2 = \mathbf{Y}_2 - \mathbf{X}_3 = \mathbf{Y}_2 - \mathbf{Y}_3$$

$$\mathbf{X}_1 = \mathbf{Y}_1 - \mathbf{X}_2 - \mathbf{X}_3 = \mathbf{Y}_1 - (\mathbf{Y}_2 - \mathbf{Y}_3) - \mathbf{Y}_3 = \mathbf{Y}_1 - \mathbf{Y}_2$$

Thus $S = \{\mathbf{X}_1, \mathbf{X}_2, \mathbf{X}_3\}$ is contained in span T and so V = span S is contained in span T. Hence T is a basis for V.

T.13. If $\mathbf{A}$ is nonsingular then linear system $\mathbf{AY} = \mathbf{0}$ has only the trivial solution $\mathbf{Y} = \mathbf{0}$. Let

$$c_1\mathbf{AX}_1 + c_2\mathbf{AX}_2 + \cdots + c_n\mathbf{AX}_n = \mathbf{0}$$

Then $\mathbf{A}\,[c_1\mathbf{X}_1 + \cdots + c_n\mathbf{X}_n] = \mathbf{0}$ and by the opening remark it must be that

$$c_1\mathbf{X}_1 + c_2\mathbf{X}_2 + \cdots + c_n\mathbf{X}_n = \mathbf{0}$$

However since $\{\mathbf{X}_1, \mathbf{X}_2, \ldots, \mathbf{X}_n\}$ is linearly independent it follows that $c_1 = c_2 = \cdots = c_n = 0$. Hence $\{\mathbf{AX}_1, \mathbf{AX}_2, \ldots, \mathbf{AX}_n\}$ is linearly independent.

ML.1. Follow the procedure in Exercise ML.1 in Section 3.5.

```
X1=[1 2 1]';X2=[2 1 1]';X3=[2 2 1]';
rref([X1 X2 X3 zeros(X1)])

ans =

     1     0     0     0
     0     1     0     0
     0     0     1     0
```

It follows that the only solution is the trivial solution so S is linearly independent and since dim V = 3, S is a basis for V.

ML.2. Follow the procedure in Exercise ML.3(b) in Section 3.5.

```
X1=[0 2 -2]';X2=[1 -3 1]';X3=[2 -8 4]';
rref([X1 X2 X3 zeros(X1)])

ans =

     1     0    -1     0
     0     1     2     0
     0     0     0     0
```

It follows that there is a nontrivial solution so S is linearly dependent and cannot be a basis for V.

ML.3. Proceed as in ML.1.

```
X1=[1 1 0 0]';X2=[2 1 1 -1]';X3=[0 0 1 1]';X4=[1 2 1 2]';
rref([X1 X2 X3 X4 zeros(X1)])
```

```
ans =

     1     0     0     0     0
     0     1     0     0     0
     0     0     1     0     0
     0     0     0     1     0
```

It follows that S is linearly independent and since dim V = 4, S is a basis for V.

ML.4. Here we do not know dim(span S), but dim(span S) = the number of linearly independent vectors in S. We proceed as we did in ML.1.

```
X1=[1 2 1 0]';X2=[2 1 3 1]';X3=[2 -2 4 2]';
rref([X1 X2 X3 zeros(X1)])

ans =
     1     0    -2     0
     0     1     2     0
     0     0     0     0
     0     0     0     0
```

The leading 1's imply that $\mathbf{X}_1$ and $\mathbf{X}_2$ are a linearly independent subset of S, hence dim(span S) = 2 and S is not a basis for V.

ML.5. Here we do not know dim(span S), but dim(span S) = the number of linearly independent vectors in S. We proceed as we did in ML.1.

```
X1=[1 2 1 0]';X2=[2 1 3 1]';X3=[2 2 1 2]';
rref([X1 X2 X3 zeros(X1)])

ans =

     1     0     0     0
     0     1     0     0
     0     0     1     0
     0     0     0     0
```

The leading 1's imply that S is a linearly independent set hence dim(span S) = 3 and S is a basis for V.

ML.6. Any vector in V has the form

$$(a, b, c) = (a, 2a-c, c) = a(1, 2, 0) + c(0, -1, 1)$$

It follows that T={(1, 2, 0), (0, -1, 1)} spans V and since the members of T are not multiples of one another T is a linearly independent subset of V. Thus dim V = 2. We need only determine if S is a linearly independent subset of V. Let

```
X1=[0 1 -1]';X2=[1 1 1]';
```

then

```
rref([X1 X2 zeros(X1)])

ans =

     1     0     0
     0     1     0
     0     0     0
```

It follows that S is linearly independent and so Theorem 3.14 implies that S is a basis for V.

In Exercises ML.7 through ML.9 we use the technique involving leading 1's as in Example 5.

ML.7.
```
X1=[1 1 0 0]';X2=[-2 -2 0 0]';X3=[1 0 2 1]';X4=[2 1 2 1]';
X5=[0 1 1 1]';
rref([X1 X2 X3 X4 X5 zeros(X1)])

ans =

     1    -2     0     1     0     0
     0     0     1     1     0     0
     0     0     0     0     1     0
     0     0     0     0     0     0
```

The leading 1's point to vectors $\mathbf{X}_1$, $\mathbf{X}_3$, and $\mathbf{X}_5$ and hence these vectors are a linearly independent set which also spans S. Thus $T = \{\mathbf{X}_1, \mathbf{X}_3, \mathbf{X}_5\}$ is a basis for span S. We have dim(span S) = 3 and span S $\neq R^4$.

ML.8. Associate a column with each 2 x 2 matrix as in Exercise ML.2(b) in Section 3.5.

```
X1=[1 2;1 2];X2=[1 0;1 1];X3=[0 2 ;0 1];X4=[2 4;2 4];
X5=[1 0;0 1];
rref([reshape(X1,4,1) reshape(X2,4,1) reshape(X3,4,1)...
     reshape(X4,4,1) reshape(X5,4,1) zeros(4,1)])

ans =

     1     0     1     2     0     0
     0     1    -1     0     0     0
     0     0     0     0     1     0
     0     0     0     0     0     0
```

Exercises 3.6

The leading 1's point to $\mathbf{X}_1$, $\mathbf{X}_2$, and $\mathbf{X}_5$ which form a basis for span S. We have dim(span S) = 3 and span S $\neq R_{22}$.

ML.9. Proceed as in ML.2.

```
X1=[0 1 -2]';X2=[0 2 -1]';X3=[0 4 -2]';X4=[1 -1 1]';
X5=[1 2 1]';
rref([X1 X2 X3 X4 X5 zeros(X1)])

ans =

     1     0     0     0    -1     0
     0     1     2     0     2     0
     0     0     0     1     1     0
```

It follows that T = {$\mathbf{X}_1$, $\mathbf{X}_2$, $\mathbf{X}_4$} is a basis for span S. We have dim(span S) = 3 and it follows that span S = P_2.

ML.10.

```
X1=[1 1 0 0]';X2=[1 0 1 0]';
rref([X1 X2 eye(4) zeros(X1)])

ans =

     1     0     0     1     0     0     0
     0     1     0     0     1     0     0
     0     0     1    -1    -1     0     0
     0     0     0     0     0     1     0
```

It follows that {$\mathbf{X}_1$, $\mathbf{X}_2$, $\mathbf{E}_1$ = [1 0 0 0]', $\mathbf{E}_4$ = [0 0 0 1]'} is a basis for V which contains S.

ML.11.

```
X1=[1 0 -1 1]';X2=[1 0 0 2]';
rref([X1 X2 eye(4) zeros(X1)])

ans =

1.0000         0         0         0   -1.0000         0         0
     0    1.0000         0         0    0.5000    0.5000         0
     0         0    1.0000         0    0.5000   -0.5000         0
     0         0         0    1.0000         0         0         0
```

It follows that {$\mathbf{X}_1$, $\mathbf{X}_2$, $\mathbf{E}_3$ = [0 0 1 0]', $\mathbf{E}_4$ = [0 0 0 1]'} is a basis for R^4. Hence the basis for P_3 is {$t^3 - t + 1$, $t^3 + 2$, t, 1}.

ML.12. Any vector in V has the form (a, 2d+e, a, d, e). It follows that

(a, 2d+e, a, d, e) = a(1, 0, 1 ,0 ,0) + d(0, 2, 0, 1, 0) + e(0, 1, 0, 0, 1) and T = {(1, 0, 1, 0, 0), (0, 2, 0, 1, 0), (0, 1, 0, 0, 1)} is a basis for V. Hence let

```
X1=[0 3 0 2 -1]';T1=[1 0 1 0 0]';T2=[0 2 0 1 0]';
T3=[0 1 0 0 1]';
```

then

```
rref([X1 T1 T2 T3 zeros(X1)])

ans =

     1     0     0    -1     0
     0     1     0     0     0
     0     0     1     2     0
     0     0     0     0     0
     0     0     0     0     0
```

Thus $\{\mathbf{X}_1, \mathbf{T}_1, \mathbf{T}_2\}$ is a basis for V containing S.

Exercises 3.7

<<**Strategy:** In Exercises 1-8 we form the augmented matrix associated with the homogeneous linear system, find its reduced row echelon form, and write out the solution assigning arbitary constants as needed. Following the method in Example 1 we determine a basis for the solution space and compute the dimension of the solution space.>>

1. Let [**A** | **0**] be the augmented matrix associated with the homogeneous linear system. We have

$$[\mathbf{A} \mid \mathbf{0}] = \left[\begin{array}{cccc|c} 1 & 1 & 1 & 1 & 0 \\ 2 & 1 & -1 & 1 & 0 \end{array}\right]$$

Using row operations we find that the reduced row echelon form of [**A** | **0**] is

$$\left[\begin{array}{cccc|c} 1 & 0 & -2 & 0 & 0 \\ 0 & 1 & 3 & 1 & 0 \end{array}\right]$$

It follows that every solution is of the form

$$\mathbf{X} = \begin{bmatrix} 2t \\ -s-3t \\ t \\ s \end{bmatrix} = s\begin{bmatrix} 0 \\ -1 \\ 0 \\ 1 \end{bmatrix} + t\begin{bmatrix} 2 \\ -3 \\ 1 \\ 0 \end{bmatrix}$$

Thus $\left\{\begin{bmatrix} 0 \\ -1 \\ 0 \\ 1 \end{bmatrix}, \begin{bmatrix} 2 \\ -3 \\ 1 \\ 0 \end{bmatrix}\right\}$ is a basis for the solution space and it has dimension 2.

3. Let [**A** | **0**] be the augmented matrix associated with the homogeneous linear system. We have

$$[\mathbf{A} \mid \mathbf{0}] = \left[\begin{array}{cccc|c} 1 & 2 & -1 & 3 & 0 \\ 2 & 2 & -1 & 2 & 0 \\ 1 & 0 & 3 & 3 & 0 \end{array}\right]$$

Using row operations we find that the reduced row echelon form of [**A** | **0**] is

$$\left[\begin{array}{cccc|c} 1 & 0 & 0 & -1 & 0 \\ 0 & 1 & 0 & 8/3 & 0 \\ 0 & 0 & 1 & 4/3 & 0 \end{array}\right]$$

It follows that every solution is of the form

$$\mathbf{X} = \begin{bmatrix} t \\ (-8/3)t \\ (-4/3)t \\ t \end{bmatrix} = t\begin{bmatrix} 1 \\ -8/3 \\ -4/3 \\ 1 \end{bmatrix}$$

Thus $\left\{\begin{bmatrix} 1 \\ -8/3 \\ -4/3 \\ 1 \end{bmatrix}\right\}$ is a basis for the solution space and it has dimension 1.

5. Let $[\mathbf{A} \mid \mathbf{0}]$ be the augmented matrix associated with the homogeneous linear system. We have

$$[\mathbf{A} \mid \mathbf{0}] = \left[\begin{array}{ccccc|c} 1 & 2 & 1 & 2 & 1 & 0 \\ 1 & 2 & 2 & 1 & 2 & 0 \\ 2 & 4 & 3 & 3 & 3 & 0 \\ 0 & 0 & 1 & -1 & -1 & 0 \end{array}\right]$$

Using row operations we find that the reduced row echelon form of $[\mathbf{A} \mid \mathbf{0}]$ is

$$\left[\begin{array}{ccccc|c} 1 & 2 & 0 & 3 & 0 & 0 \\ 0 & 0 & 1 & -1 & 0 & 0 \\ 0 & 0 & 0 & 0 & 1 & 0 \\ 0 & 0 & 0 & 0 & 0 & 0 \end{array}\right]$$

It follows that every solution is of the form

$$\mathbf{X} = \begin{bmatrix} -2s-3t \\ s \\ t \\ t \\ 0 \end{bmatrix} = s\begin{bmatrix} -2 \\ 1 \\ 0 \\ 0 \\ 0 \end{bmatrix} + t\begin{bmatrix} -3 \\ 0 \\ 1 \\ 1 \\ 0 \end{bmatrix}$$

Thus $\left\{\begin{bmatrix} -2 \\ 1 \\ 0 \\ 0 \\ 0 \end{bmatrix}, \begin{bmatrix} -3 \\ 0 \\ 1 \\ 1 \\ 0 \end{bmatrix}\right\}$ is a basis for the solution space and it has dimension 2.

Exercises 3.7

7. Let $[\mathbf{A} \mid \mathbf{0}]$ be the augmented matrix associated with the homogeneous linear system. We have

$$[\mathbf{A} \mid \mathbf{0}] = \left[\begin{array}{rrrrr|r} 1 & 2 & 2 & -1 & 1 & 0 \\ 0 & 2 & 2 & -2 & -1 & 0 \\ 2 & 6 & 2 & -4 & 1 & 0 \\ 1 & 4 & 0 & -3 & 0 & 0 \end{array}\right]$$

Using row operations we find that the reduced row echelon form of $[\mathbf{A} \mid \mathbf{0}]$ is

$$\left[\begin{array}{rrrrr|r} 1 & 0 & 0 & -1 & 2 & 0 \\ 0 & 1 & 0 & -1 & -1/2 & 0 \\ 0 & 0 & 1 & 0 & 0 & 0 \\ 0 & 0 & 0 & 0 & 0 & 0 \end{array}\right]$$

It follows that every solution is of the form

$$\mathbf{X} = \begin{bmatrix} -2s+t \\ (1/2)s+t \\ 0 \\ t \\ s \end{bmatrix} = s\begin{bmatrix} -2 \\ 1/2 \\ 0 \\ 0 \\ 1 \end{bmatrix} + t\begin{bmatrix} 1 \\ 1 \\ 0 \\ 1 \\ 0 \end{bmatrix}$$

Thus $\left\{\begin{bmatrix} -2 \\ 1/2 \\ 0 \\ 0 \\ 1 \end{bmatrix}, \begin{bmatrix} 1 \\ 1 \\ 0 \\ 1 \\ 0 \end{bmatrix}\right\}$ is a basis for the solution space and it has dimension 2.

9. The null space of matrix $\mathbf{A}$ is the same as the solution space of $\mathbf{AX} = \mathbf{0}$. Hence we follow the procedure used in Exercises 1-8. Let $[\mathbf{A} \mid \mathbf{0}]$ be the augmented matrix associated with the homogeneous linear system. We have

$$[\mathbf{A} \mid \mathbf{0}] = \left[\begin{array}{rrrr|r} 1 & 2 & 3 & -1 & 0 \\ 2 & 3 & 2 & 0 & 0 \\ 3 & 4 & 1 & 1 & 0 \\ 1 & 1 & -1 & 1 & 0 \end{array}\right]$$

Using row operations we find that the reduced row echelon form of $[\mathbf{A} \mid \mathbf{0}]$ is

$$\left[\begin{array}{rrrr|r} 1 & 0 & -5 & 3 & 0 \\ 0 & 1 & 4 & -2 & 0 \\ 0 & 0 & 0 & 0 & 0 \\ 0 & 0 & 0 & 0 & 0 \end{array}\right]$$

It follows that every solution is of the form

$$\mathbf{X} = \begin{bmatrix} -3s+5t \\ 2s-4t \\ t \\ s \end{bmatrix} = s\begin{bmatrix} -3 \\ 2 \\ 0 \\ 1 \end{bmatrix} + t\begin{bmatrix} 5 \\ -4 \\ 1 \\ 0 \end{bmatrix}$$

Thus $\left\{ \begin{bmatrix} -3 \\ 2 \\ 0 \\ 1 \end{bmatrix}, \begin{bmatrix} 5 \\ -4 \\ 1 \\ 0 \end{bmatrix} \right\}$ is a basis for the null space of **A**.

11. Following Example 2, we form the matrix $1\mathbf{I}_2 - \mathbf{A} = \begin{bmatrix} -2 & -2 \\ -1 & -1 \end{bmatrix}$.

Then the reduced row echelon form of matrix $\left[\begin{array}{cc|c} -2 & -2 & 0 \\ -1 & -1 & 0 \end{array}\right]$ is

$\left[\begin{array}{cc|c} 1 & 1 & 0 \\ 0 & 0 & 0 \end{array}\right]$. Hence every solution of the homogeneous system

$(1\mathbf{I}_2 - \mathbf{A})\mathbf{X} = \mathbf{0}$ is of the form $\mathbf{X} = \begin{bmatrix} -t \\ t \end{bmatrix}$, where t is any real number. It follows that a basis for the solution space of

$(1\mathbf{I}_2 - \mathbf{A})\mathbf{X} = \mathbf{0}$ is $\left\{ \begin{bmatrix} -1 \\ 1 \end{bmatrix} \right\}$.

13. Following Example 2, we form the matrix $1\mathbf{I}_3 - \mathbf{A} = \begin{bmatrix} 1 & 0 & -1 \\ -1 & 1 & 3 \\ 0 & -1 & -2 \end{bmatrix}$.

Then the reduced row echelon form of matrix $\left[\begin{array}{ccc|c} 1 & 0 & -1 & 0 \\ -1 & 1 & 3 & 0 \\ 0 & -1 & -2 & 0 \end{array}\right]$

is $\left[\begin{array}{ccc|c} 1 & 0 & -1 & 0 \\ 0 & 1 & 2 & 0 \\ 0 & 0 & 0 & 0 \end{array}\right]$. Hence every solution of the homogeneous system $(1\mathbf{I}_3 - \mathbf{A})\mathbf{X} = \mathbf{0}$ is of the form $\mathbf{X} = \begin{bmatrix} t \\ -2t \\ t \end{bmatrix}$, where t is any real number. It follows that a basis for the solution space of $(1\mathbf{I}_3 - \mathbf{A})\mathbf{X} = \mathbf{0}$ is $\left\{\begin{bmatrix} 1 \\ -2 \\ 1 \end{bmatrix}\right\}$.

15. Follow the procedure in Example 3.

$$\lambda\mathbf{I}_2 - \mathbf{A} = \lambda\begin{bmatrix} 1 & 0 \\ 0 & 1 \end{bmatrix} - \begin{bmatrix} 2 & 3 \\ 2 & -3 \end{bmatrix} = \begin{bmatrix} \lambda-2 & -3 \\ -2 & \lambda+3 \end{bmatrix}$$

The homogeneous system $(\lambda\mathbf{I}_2 - \mathbf{A})\mathbf{X} = \mathbf{0}$ has a nontrivial solution if and only if $\det(\lambda\mathbf{I}_2 - \mathbf{A}) = 0$. We have

$$\det(\lambda\mathbf{I}_2 - \mathbf{A}) = \det\left(\begin{bmatrix} \lambda-2 & -3 \\ -2 & \lambda+3 \end{bmatrix}\right) = (\lambda-2)(\lambda+3)-(-2)(-3)$$

$$= \lambda^2 + \lambda - 12 = (\lambda - 3)(\lambda + 4) = 0$$

only if $\lambda = 3$ or -4. It follows that $(\lambda\mathbf{I}_2 - \mathbf{A})\mathbf{X} = \mathbf{0}$ has a nontrivial solution when $\lambda = 3$ or -4.

17. Follow the procedure in Example 3.

$$\lambda\mathbf{I}_3 - \mathbf{A} = \lambda\begin{bmatrix} 1 & 0 & 0 \\ 0 & 1 & 0 \\ 0 & 0 & 1 \end{bmatrix} - \begin{bmatrix} 0 & 0 & 0 \\ 0 & 1 & -1 \\ 1 & 0 & 0 \end{bmatrix} = \begin{bmatrix} \lambda & 0 & 0 \\ 0 & \lambda-1 & 1 \\ -1 & 0 & \lambda \end{bmatrix}$$

The homogeneous system $(\lambda\mathbf{I}_3 - \mathbf{A})\mathbf{X} = \mathbf{0}$ has a nontrivial solution if and only if $\det(\lambda\mathbf{I}_3 - \mathbf{A}) = 0$. We have

$$\det(\lambda\mathbf{I}_3 - \mathbf{A}) = \det\left(\begin{bmatrix} \lambda & 0 & 0 \\ 0 & \lambda-1 & 1 \\ -1 & 0 & \lambda \end{bmatrix}\right) = \lambda^2(\lambda-1) = 0$$

only if $\lambda = 0$ or 1. It follows that $(\lambda\mathbf{I}_3 - \mathbf{A})\mathbf{X} = \mathbf{0}$ has a nontrivial solution when $\lambda = 0$ or 1.

T.1. Since each vector in S is a solution of $\mathbf{AX} = \mathbf{0}$, we have $\mathbf{AX}_i = \mathbf{0}$ for $i = 1, 2, \ldots, n$. The span of S consists of all possible linear combinations of the vectors in S, hence

$$\mathbf{Y} = c_1\mathbf{X}_1 + c_2\mathbf{X}_2 + \cdots + c_k\mathbf{X}_k$$

represents an arbitrary member of span S. We have

$$\begin{aligned} \mathbf{AY} &= c_1\mathbf{AX}_1 + c_2\mathbf{AX}_2 + \cdots + c_k\mathbf{AX}_k \\ &= c_1\mathbf{0} + c_2\mathbf{0} + \cdots + c_k\mathbf{0} \\ &= 0 \end{aligned}$$

Thus **Y** is a solution of **AX** = **0** and it follows that every member of span S is a solution of **AX** = **0**.

T.3. If **A** has a row of zeros, then det **A** = 0 and it follows that matrix **A** is singular. A homogeneous system with a singular coefficient matrix is guaranteed to have a nontrivial solution.

T.5. Since the reduced row echelon forms of matrices **A** and **B** are the same it follows that the solutions of the linear systems **AX** = **0** and **BX** = **0** are the same set of vectors. Hence the nullspaces of **A** and **B** are the same.

ML.1. Enter A into MATLAB and we find that

```
rref(A)

ans =

     1     0     2     1     2
     0     1     0     1    -1
     0     0     0     0     0
```

Write out the solution to the linear system **AX** = **0** as

$$\mathbf{X} = r\begin{bmatrix} -2 \\ 0 \\ 1 \\ 0 \\ 0 \end{bmatrix} + s\begin{bmatrix} -1 \\ -1 \\ 0 \\ 1 \\ 0 \end{bmatrix} + t\begin{bmatrix} -2 \\ 1 \\ 0 \\ 0 \\ 1 \end{bmatrix}$$

A basis for the null space of **A** consists of the three vectors above. We can compute such a basis directly using command **homsoln** as shown next.

```
homsoln(A)

ans =

    -2    -1    -2
     0    -1     1
     1     0     0
     0     1     0
     0     0     1
```

ML.2. Enter **A** into MATLAB and we find that

```
rref(A)
```

```
ans =

     1     0     0
     0     1     0
     0     0     1
     0     0     0
     0     0     0
```

The homogeneous system **AX** = **0** has only the trivial solution.

ML.3. Enter **A** into MATLAB and we find that

```
rref(A)

ans =

    1.0000         0   -1.0000   -1.3333
         0    1.0000    2.0000    0.3333
         0         0         0         0

rat(ans,'s')

ans =

     1          0         -1         -4/3
     0          1          2          1/3
     0          0          0           0
```

Write out the solution to the linear system **AX** = **0** as

$$\mathbf{X} = r\begin{bmatrix} 1 \\ -2 \\ 1 \\ 0 \end{bmatrix} + s\begin{bmatrix} 4/3 \\ -1/3 \\ 0 \\ 1 \end{bmatrix}$$

A basis for the null space of **A** consists of the two vectors above. We can compute such a basis directly using command **homsoln** as shown next.

```
homsoln(A)

ans =

     1.0000    1.3333
    -2.0000   -0.3333
     1.0000         0
          0    1.0000

rat(ans,'s')
```

```
ans =

        1           4/3
       -2          -1/3
        1             0
        0             1
```

ML.4. Form the matrix $3\mathbf{I}_2 - \mathbf{A}$ in MATLAB as follows.

```
C = 3*eye(2)-[1 2;2 1]

C =

     2    -2
    -2     2

rref(C)

ans =

     1    -1
     0     0
```

The solution is $\mathbf{X} = \begin{bmatrix} t \\ t \end{bmatrix}$, for t any real number. Just choose $t \neq 0$ to obtain a nontrivial solution.

ML.5. Form the matrix $6\mathbf{I}_3 - \mathbf{A}$ in MATLAB as follows.

```
C = 6*eye(3)-[1 2 3;3 2 1;2 1 3]

C =

     5    -2    -3
    -3     4    -1
    -2    -1     3

rref(C)

ans =

     1     0    -1
     0     1    -1
     0     0     0
```

The solution is $\mathbf{X} = \begin{bmatrix} t \\ t \\ t \end{bmatrix}$, for t any nonzero real number.

Just choose $t \neq 0$ to obtain a nontrivial solution.

Exercises 3.8

1. We follow the procedure used in Example 1. Let $S = \{\mathbf{X}_1, \mathbf{X}_2, \mathbf{X}_3, \mathbf{X}_4, \mathbf{X}_5\}$, where $\mathbf{X}_1 = (1,2,3)$, $\mathbf{X}_2 = (2,1,4)$, $\mathbf{X}_3 = (-1,-1,2)$, $\mathbf{X}_4 = (0,1,2)$, $\mathbf{X}_5 = (1,1,1)$. To find a basis for V = span S, we form a matrix **A** with rows $\mathbf{X}_1$, $\mathbf{X}_2$, $\mathbf{X}_3$, $\mathbf{X}_4$, $\mathbf{X}_5$ and transform it to reduced row echelon form. The nonzero rows of the reduced row echelon form are a basis for V.

$$\mathbf{A} = \begin{bmatrix} 1 & 2 & 3 \\ 2 & 1 & 4 \\ -1 & -1 & 2 \\ 0 & 1 & 2 \\ 1 & 1 & 1 \end{bmatrix}$$ and applying row operations we obtain the

reduced row echelon form $\begin{bmatrix} 1 & 0 & 0 \\ 0 & 1 & 0 \\ 0 & 0 & 1 \\ 0 & 0 & 0 \\ 0 & 0 & 0 \end{bmatrix}$. Thus {(1,0,0), (0,1,0), (0,0,1)} is a basis for V. (Hence, in this case $V = \mathbf{R}^3$.)

3. We follow the procedure used in Example 3. Let $S = \{\mathbf{X}_1, \mathbf{X}_2, \mathbf{X}_3, \mathbf{X}_4, \mathbf{X}_5\}$, where $\mathbf{X}_1 = \begin{bmatrix} 1 \\ 2 \\ 1 \\ 2 \end{bmatrix}$, $\mathbf{X}_2 = \begin{bmatrix} 2 \\ 1 \\ 2 \\ 1 \end{bmatrix}$, $\mathbf{X}_3 = \begin{bmatrix} 3 \\ 2 \\ 3 \\ 2 \end{bmatrix}$, $\mathbf{X}_4 = \begin{bmatrix} 3 \\ 3 \\ 3 \\ 3 \end{bmatrix}$,

$\mathbf{X}_5 = \begin{bmatrix} 5 \\ 3 \\ 5 \\ 3 \end{bmatrix}$. To find a basis for V = span S, we form a matrix $\mathbf{A}^T$

whose columns are the vectors $\mathbf{X}_1$, $\mathbf{X}_2$, $\mathbf{X}_3$, $\mathbf{X}_4$, $\mathbf{X}_5$ written in row form and transform it to reduced row echelon form. The nonzero rows of the reduced row echelon form written as columns are a basis for V.

$$\mathbf{A}^T = \begin{bmatrix} 1 & 2 & 1 & 2 \\ 2 & 1 & 2 & 1 \\ 3 & 2 & 3 & 2 \\ 3 & 3 & 3 & 3 \\ 5 & 3 & 5 & 3 \end{bmatrix}$$ applying row operations we obtain the reduced

row echelon form $\begin{bmatrix} 1 & 0 & 1 & 0 \\ 0 & 1 & 0 & 1 \\ 0 & 0 & 0 & 0 \\ 0 & 0 & 0 & 0 \\ 0 & 0 & 0 & 0 \end{bmatrix}$. Vectors $\left\{ \begin{bmatrix} 1 \\ 0 \\ 1 \\ 0 \end{bmatrix}, \begin{bmatrix} 0 \\ 1 \\ 0 \\ 1 \end{bmatrix} \right\}$ are a basis for V.

Notation: We use the symbol **rref(A)** to denote the reduced row echelon form of matrix **A**.

5. (a) To find a basis for the row space of **A** that are not row vectors of **A** we compute rref(**A**) and select the nonzero rows as in Example 1.

$$\text{rref}(\mathbf{A}) = \begin{bmatrix} 1 & 0 & -1 \\ 0 & 1 & 0 \\ 0 & 0 & 0 \\ 0 & 0 & 0 \end{bmatrix}$$

Thus {[1 0 -1], [0 1 0]} is a basis for for the row space of **A**.

(b) To find a basis for the row space of **A** that are row vectors of **A** we use the method described in Example 2. Compute rref(A^T) and then use the leading 1's to point to the rows of **A** that form a basis for the row space of **A**.

$$\text{rref}(\mathbf{A}^T) = \begin{bmatrix} 1 & 0 & -5 & -3 \\ 0 & 1 & 2 & 1 \\ 0 & 0 & 0 & 0 \end{bmatrix}$$

The leading 1's are in columns 1 and 2, hence rows 1 and 2 of **A** form a basis for the row space.

7. (a) To find a basis for the column space of **A** that does not consist of the columns of **A** we follow the procedure in Example 3. Compute rref($\mathbf{A}^T$) and then take the transposes of the nonzero rows to get the desired basis.

$$\text{rref}(\mathbf{A}^T) = \begin{bmatrix} 1 & 0 & 0 & 0 \\ 0 & 1 & 0 & 1/5 \\ 0 & 0 & 1 & 3/5 \\ 0 & 0 & 0 & 0 \end{bmatrix}$$

Thus $S = \left\{ \begin{bmatrix} 1 \\ 0 \\ 0 \\ 0 \end{bmatrix}, \begin{bmatrix} 0 \\ 1 \\ 0 \\ 1/5 \end{bmatrix}, \begin{bmatrix} 0 \\ 0 \\ 1 \\ 3/5 \end{bmatrix} \right\}$ is a basis for the column space of **A**.

(b) To find a basis for the column space of **A** that consists of columns of **A** we follow the procedure in the second solution in Example 3. Compute rref(**A**) and use the leading 1's to point to the columns of **A** that form the desired basis.

$$\text{rref}(\mathbf{A}) = \begin{bmatrix} 1 & 0 & 1 & 0 \\ 0 & 1 & -3 & 0 \\ 0 & 0 & 0 & 1 \\ 0 & 0 & 0 & 0 \end{bmatrix}$$

The leading 1's are in columns 1, 2, and 4 hence those columns of **A** form a basis for the column space of **A**.

9. The row rank of **A** is the number of nonzero rows in the row reduced echelon form of **A**. For $\mathbf{A} = \begin{bmatrix} 1 & 2 & 3 & 2 & 1 \\ 3 & 1 & -5 & -2 & 1 \\ 7 & 8 & -1 & 2 & 5 \end{bmatrix}$ the reduced row echelon form is $\begin{bmatrix} 1 & 0 & 0 & 0 & 0 \\ 0 & 1 & 0 & \frac{4}{13} & \frac{8}{13} \\ 0 & 0 & 1 & \frac{6}{13} & \frac{-1}{13} \end{bmatrix}$ and hence **A** has row rank 3. The column rank of A is the number of nonzero rows in the reduced row echelon form of $\mathbf{A}^T$. For matrix **A** this is $\begin{bmatrix} 1 & 0 & 0 \\ 0 & 1 & 0 \\ 0 & 0 & 1 \\ 0 & 0 & 0 \\ 0 & 0 & 0 \end{bmatrix}$. Thus the column rank of **A** is 3.

<<**Strategy:** In Exercises 11, 13, and 15 we compute the reduced row echelon form and count the number of nonzero rows to determine the rank. >>

11. $\text{rref}(\mathbf{A}) = \begin{bmatrix} 1 & 0 & -3 & \frac{-13}{3} \\ 0 & 1 & 2 & \frac{11}{3} \\ 0 & 0 & 0 & 0 \\ 0 & 0 & 0 & 0 \end{bmatrix}$ and rank **A** = 2. Since **A** represents the coefficients of a homogeneous system of four equations in four unknowns and rref(**A**) implies that the system is equivalent to two equations in four unknowns, it follows that two of the variables could be chosen arbitrarily hence the nullity of **A** is 2. Hence rank **A** + nullity **A** = 2 + 2 = n = 4.

13. $\text{rref}(\mathbf{A}) = \begin{bmatrix} 1 & 0 & 0 \\ 0 & 1 & 0 \\ 0 & 0 & 1 \end{bmatrix}$ and rank **A** = 3. Since **A** the represents the coefficients of a homogeneous system of three equations in three unknowns and rref(**A**) also represents the a system of 3 equations in three unknowns, it follows that no variable can be chosen arbitrarily hence the nullity of **A** is 0. Hence rank **A** + nullity **A** = 3 + 0 = n = 3.

15. $\text{rref}(\mathbf{A}) = \begin{bmatrix} 1 & 0 & \frac{7}{3} \\ 0 & 1 & \frac{5}{3} \\ 0 & 0 & 0 \\ 0 & 0 & 0 \end{bmatrix}$ and rank **A** = 2. Since **A** represents the coefficients of a homogeneous system of four equations in three unknowns and rref(**A**) implies that the system is equivalent to two equations in three unknowns, it follows that one of the variables could be chosen arbitrarily hence the nullity of **A** is 1. Hence rank **A** + nullity **A** = 2 + 1 = n = 3.

Exercises 3.8

17. For a 4×6 matrix the largest possible rank is 4, since there can be at most 4 linearly independent rows. However, row rank = column rank; there can be at most 4 linearly independent columns. Thus the columns are linearly dependent.

19. Since $\mathbf{A}$ is 7×3 with rank 3 it follows that there are 3 linearly independent rows (rank = row rank); hence the rows of $\mathbf{A}$ are linearly dependent.

<<**Strategy:** In Exercises 21 and 23 we compute the reduced row echelon form (rref) in order to compute the rank.>>

21. $\text{rref}(\mathbf{A}) = \begin{bmatrix} 1 & 0 & -3 \\ 0 & 1 & 0 \\ 0 & 0 & 0 \end{bmatrix}$ thus rank $\mathbf{A} = 2$. By Theorem 3.18, $\mathbf{A}$ is singular.

23. $\text{rref}(\mathbf{A}) = \mathbf{I}_4$, thus rank $\mathbf{A} = 4$. By Theorem 3.18 $\mathbf{A}$ is nonsingular.

25. $\text{rref}(\mathbf{A}) = \mathbf{I}_3$, thus rank $\mathbf{A} = 3$. Hence Corollary 3.3 implies that $\mathbf{AX} = \mathbf{B}$ has a unique solution for every 3×1 matrix $\mathbf{B}$.

27. Form the matrix $\mathbf{A} = \begin{bmatrix} 4 & 1 & 2 \\ 2 & 5 & -5 \\ 2 & -1 & 3 \end{bmatrix}$. Since $|\mathbf{A}| = 0$, by Corollary 3.4 set S is linearly dependent in $\mathbf{R}^3$.

<<**Strategy:** In Exercise 29 we compute the reduced row echelon form (rref) of $\mathbf{A}$ in order to find rank $\mathbf{A}$.>>

29. $\text{rref}(\mathbf{A}) = \begin{bmatrix} 1 & 0 & 3 \\ 0 & 1 & 0 \\ 0 & 0 & 0 \end{bmatrix}$, thus rank $\mathbf{A} = 2$. By Corollary 3.5 the homogeneous linear system $\mathbf{AX} = \mathbf{0}$ has nontrivial solutions.

<<**Strategy:** In Exercises 31 and 33 we compute the reduced row echelon form (rref) of the coefficient matrix **A** and the reduced row echelon form of the augmented matrix [**A**|**B**] to obtain the ranks needed. We follow the procedures given in Examples 7 and 8.>>

31. $\text{rref}(\mathbf{A}) = \begin{bmatrix} 1 & 0 & 0 & \frac{22}{83} \\ 0 & 1 & 0 & \frac{56}{83} \\ 0 & 0 & 1 & \frac{-60}{83} \end{bmatrix}$ and rref([**A**|**B**]) is the same but with a column of zeros attached. Thus rank **A** = rank [**A**|**B**], hence the linear system has a solution.

33. $\text{rref}(\mathbf{A}) = \begin{bmatrix} 1 & 0 & -1 & \frac{8}{7} \\ 0 & 1 & 1 & \frac{-10}{7} \\ 0 & 0 & 0 & 0 \end{bmatrix}$ and

$\text{rref}([\mathbf{A}|\mathbf{B}]) = \left[\begin{array}{cccc|c} 1 & 0 & -1 & \frac{8}{7} & 0 \\ 0 & 1 & 1 & \frac{-10}{7} & 0 \\ 0 & 0 & 0 & 0 & 1 \end{array}\right]$. Hence rank **A** ≠ rank [**A**|**B**]. Thus the linear system has no solution.

T.1. The row space of an $m \times n$ matrix **A** is subspace of $\mathbf{R}^n$ consisting of the set of all possible linear combinations of the rows of **A**. The row operations used to obtain the reduced row echelon form of **A**, which we denote as **B**, replace the original rows of **A** by a set of vectors that span the same subspace of $\mathbf{R}^n$. If **B** has r nonzero rows $\mathbf{B}_1, \mathbf{B}_2, \ldots, \mathbf{B}_r$, then span $\{\mathbf{B}_1, \mathbf{B}_2, \ldots, \mathbf{B}_r\}$ is the row space of **A**. Suppose next that some linear combination $\sum_{k=1}^{r} c_k\mathbf{B}_k = \mathbf{0}$. If the ith row of **B** has its leading entry in the jth column, then since all other rows have entry 0 in the jth column, the jth coordinate of $\Sigma c_k B_k$ is $c_i \cdot 1$, which must equal 0. Thus $c_i = 0$ for $i = 1,2,\ldots,r$. Hence the r nonzero rows are linearly independent, and hence constitute a basis for the row space of **A**.

T.3. Suppose that $\mathbf{AX} = \mathbf{B}$ has a unuque solution for every $n \times 1$ matrix **B**. Then the n linear systems $\mathbf{AX} = \mathbf{E}_1$, $\mathbf{AX} = \mathbf{E}_2$, ..., $\mathbf{AX} = \mathbf{E}_n$, where $\mathbf{E}_1$, $\mathbf{E}_2$, ..., $\mathbf{E}_n$ are the columns of $\mathbf{I}_n$, have unique solutions which we denote as $\mathbf{X}_1$, $\mathbf{X}_2$, ..., $\mathbf{X}_n$ respectively. Let $\mathbf{C} = [\mathbf{X}_1\ \mathbf{X}_2\ \ldots\ \mathbf{X}_n]$. Then

$$\mathbf{AC} = [\mathbf{AX}_1\ \mathbf{AX}_2\ \ldots\ \mathbf{AX}_n] = [\mathbf{E}_1\ \mathbf{E}_2\ \ldots\ \mathbf{E}_n] = \mathbf{I}_n$$

Hence, $\mathbf{C} = \mathbf{A}^{-1}$, so **A** is nonsingular and Theorem 3.18 implies that rank $\mathbf{A} = n$.

Suppose that rank $\mathbf{A} = n$. Then Theorem 3.18 implies that **A** is nonsingular, so $\mathbf{X} = \mathbf{A}^{-1}\mathbf{B}$ is a solution of the linear system $\mathbf{AX} = \mathbf{B}$. Suppose that both $\mathbf{X}_1$ and $\mathbf{X}_2$ are solutions of $\mathbf{AX} = \mathbf{B}$, then $\mathbf{AX}_1 = \mathbf{B}$ and $\mathbf{AX}_2 = \mathbf{B}$ thus $\mathbf{AX}_1 = \mathbf{AX}_2$. Multiplying both sides by $\mathbf{A}^{-1}$ gives $\mathbf{X}_1 = \mathbf{X}_2$. Thus, $\mathbf{AX} = \mathbf{B}$ has a unique solution.

T.5. If $\mathbf{AX} = \mathbf{0}$ has a nontrivial solution, then by Corollary 3.5 rank $\mathbf{A} < n$. Hence column rank of $\mathbf{A} < n$ and it follows that the columns of **A** are linearly dependent.

If the columns of **A** are linearly dependent, then by Corollary 3.4 $\det(\mathbf{A}) = 0$. It follows by Corollary 3.2 that rank $\mathbf{A} < n$ and then Corollary 3.5 implies that the linear system $\mathbf{AX} = \mathbf{0}$ has a nontrivial solution.

T.7. If the rows of **A** are linearly independent, then by Corollary 3.4 $\det(\mathbf{A}) \neq 0$. It follows by Corollary 3.2 that rank $\mathbf{A} = n$. Since column rank = rank we have that **A** has n linearly independent columns. By Theorem 3.14(a) the columns of **A** are a basis for R^n so they do indeed span R^n.

The preceding argument is reversible and shows that if the columns of **A** span R^n then the rows of **A** are linearly independent.

T.9. Suppose that the columns of **A** are linearly independent. Then rank $\mathbf{A} = n$, so by Theorem 3.17, nullity $\mathbf{A} = 0$. Hence, the homogeneous system $\mathbf{AX} = \mathbf{0}$ has only the trivial solution.

Suppose that the homogeneous system $\mathbf{AX} = \mathbf{0}$ has only the trivial solution. It follows that nullity $\mathbf{A} = 0$, so by Theorem 3.17 rank $\mathbf{A} = n$. Thus we have

column rank **A** = dimension of column space of **A** = n

Since **A** has n columns and these span its column space (which is of dimension n), it follows that the columns of **A** are linearly independent.

T.11. Since the rank of a matrix is the same as the row rank and column rank, the number of linearly independent rows of a matrix is the same as the number of linearly independent columns. It follows that the largest the rank can be is min{m,n}. Since $m \neq n$, it must be that either the rows or columns are linearly dependent.

ML.2. (a) One basis for the row space of **A** is the nonzero rows of rref(**A**).

```
A=[1 3 1;2 5 0;4 11 2;6 9 1];

rref(A)

ans =

     1     0     0
     0     1     0
     0     0     1
     0     0     0
```

Another basis is found using the leading 1's of rref($\mathbf{A}^T$) to point to rows of **A** that form a basis for the row space of **A**.

```
rref(A')

ans =

     1     0     2     0
     0     1     1     0
     0     0     0     1
```

It follows that rows 1, 2, and 4 of **A** are a basis for the row space of **A**.

(b) Follow the same procedure as in part a.

```
A = [2 1 2 0;0 0 0 0;1 2 2 1;4 5 6 2;3 3 4 1];

rref(A)

ans =

    1.0000         0    0.6667   -0.3333
         0    1.0000    0.6667    0.6667
         0         0         0         0
         0         0         0         0
         0         0         0         0
```

```
rat(ans,'s')

ans =

       1              0              2/3            -1/3
       0              1              2/3             2/3
       0              0               0               0
       0              0               0               0
       0              0               0               0

rref(A')
ans =

       1      0      0      1      1
       0      0      1      2      1
       0      0      0      0      0
       0      0      0      0      0
```

It follows that rows 1 and 2 of **A** are a basis for the row space of **A**.

ML.3. (a) The transposes of the nonzero rows of rref($\mathbf{A}^T$) give us one basis for the column space of **A**.

```
A=[1 3 1;2 5 0;4 11 2;6 9 1];

rref(A')

ans =

       1      0      2      0
       0      1      1      0
       0      0      0      1
```

The leading ones of rref(**A**) point to the columns of **A** that form a basis for the column space of **A**.

```
rref(A)

ans =

       1      0      0
       0      1      0
       0      0      1
       0      0      0
```

Thus columns 1, 2, and 3 of **A** are a basis for the column space of **A**.

(b) Follow the same procedure as in part a.

```
A=[2 1 2 0;0 0 0 0;1 2 2 1;4 5 6 2;3 3 4 1];
```

```
rref(A')

ans =

     1     0     0     1     1
     0     0     1     2     1
     0     0     0     0     0
     0     0     0     0     0

rref(A)

ans =
    1.0000         0    0.6667   -0.3333
         0    1.0000    0.6667    0.6667
         0         0         0         0
         0         0         0         0
         0         0         0         0
```

Thus columns 1 and 2 of **A** are a basis for the column space of **A**.

ML.4. (a) A=[3 2 1;1 2 -1;2 1 3];

```
rank(A)

ans =

     3
```

The nullity of A is 0.

(b) A=[1 2 1 2 1;2 1 0 0 2;1 -1 -1 -2 1;3 0 -1 -2 3];

```
rank(A)

ans =

     2
```

The nullity of A = 5 - rank(A) = 3

ML.5. Compare the rank of the coefficient matrix with the rank of the augmented matrix as in Theorem 3.19.

(a) A=[1 2 4 -1;0 1 2 0;3 1 1 -2];b=[21 8 16]';
rank(A),rank([A b])

```
ans =

     3

ans =

     3
```

Exercises 3.8

The system is consistent.

(b)
```
A=[1 2 1;1 1 0;2 1 -1];b=[3 3 3]';
rank(A),rank([A b])

ans =

      2

ans =

      3
```

The system is inconsistent.

(c)
```
A=[1 2;2 0;2 1;-1 2];b=[3 2 3 2]';
rank(A),rank([A b])

ans =

     2

ans =

     3
```

The system is inconsistent.

Exercises 3.9

1. Since S is the natural basis for R^2 we have

$$\mathbf{x} = \begin{bmatrix} 3 \\ -2 \end{bmatrix} = 3\begin{bmatrix} 1 \\ 0 \end{bmatrix} - 2\begin{bmatrix} 0 \\ 1 \end{bmatrix} \text{ and hence } \left[\mathbf{x}\right]_S = \begin{bmatrix} 3 \\ -2 \end{bmatrix}.$$

3. Following the procedure in Example 3(b), we solve for a_1 and a_2 where $a_1(t + 1) + a_2(t - 2) = t + 4$. Expanding the left side and collecting terms we have

$$(a_1 + a_2)t + (a_1 - 2a_2) = t + 4$$

Equating the coefficients of like terms we obtain the system of equations

$$\begin{aligned} a_1 + a_2 &= 1 \\ a_1 - 2a_2 &= 4 \end{aligned}$$

Form the augmented matrix $\left[\begin{array}{cc|c} 1 & 1 & 1 \\ 1 & -2 & 4 \end{array}\right]$. Its reduced row echelon form is $\left[\begin{array}{cc|c} 1 & 0 & 2 \\ 0 & 1 & -1 \end{array}\right]$, hence $\left[\mathbf{x}\right]_S = \begin{bmatrix} 2 \\ -1 \end{bmatrix}$.

5. Form the linear combination

$$a_1\begin{bmatrix} 1 & 0 \\ 0 & 0 \end{bmatrix} + a_2\begin{bmatrix} 0 & 0 \\ 1 & 0 \end{bmatrix} + a_3\begin{bmatrix} 0 & 1 \\ 0 & 0 \end{bmatrix} + a_4\begin{bmatrix} 0 & 0 \\ 0 & 1 \end{bmatrix} = \begin{bmatrix} 1 & 0 \\ -1 & 2 \end{bmatrix}$$

Combining terms on the left side gives

$$\begin{bmatrix} a_1 & a_3 \\ a_2 & a_4 \end{bmatrix} = \begin{bmatrix} 1 & 0 \\ -1 & 2 \end{bmatrix}$$

It follows that $\left[\mathbf{x}\right]_S = \begin{bmatrix} 1 \\ -1 \\ 0 \\ 2 \end{bmatrix}$.

Exercises 3.9

<<**Strategy:** In Exercises 7-12 we are given the coordinates

$[\mathbf{X}]_S = \begin{bmatrix} a_1 \\ a_2 \\ \cdot \\ \cdot \\ a_k \end{bmatrix}$ of a vector **X** with respect to to a basis

$S = \{\mathbf{X}_1, \mathbf{X}_2, \ldots, \mathbf{X}_k\}$ and asked to determine **X**. We use that $\mathbf{X} = a_1\mathbf{X}_1 + a_2\mathbf{X}_2 + \cdots + a_k\mathbf{X}_k$. >>

7. $\mathbf{X} = 1\begin{bmatrix} 2 \\ 1 \end{bmatrix} + 2\begin{bmatrix} -1 \\ 1 \end{bmatrix} = \begin{bmatrix} 0 \\ 3 \end{bmatrix}$

9. $\mathbf{X} = -2t + 3(2t - 1) = 4t - 3$

11. $\mathbf{X} = 2\begin{bmatrix} -1 & 0 \\ 1 & 0 \end{bmatrix} + 1\begin{bmatrix} 2 & 2 \\ 0 & 1 \end{bmatrix} - 1\begin{bmatrix} 1 & 2 \\ -1 & 3 \end{bmatrix} + 3\begin{bmatrix} 0 & 0 \\ 2 & 3 \end{bmatrix} = \begin{bmatrix} -1 & 0 \\ 9 & 7 \end{bmatrix}$

13. Let $S = \left\{\begin{bmatrix} 1 & 2 \end{bmatrix}, \begin{bmatrix} 0 & 1 \end{bmatrix}\right\}$ and $T = \left\{\begin{bmatrix} 1 & 1 \end{bmatrix}, \begin{bmatrix} 2 & 3 \end{bmatrix}\right\}$ be ordered bases of R^2 and let $\mathbf{X} = \begin{bmatrix} 1 & 5 \end{bmatrix}$ and $\mathbf{Y} = \begin{bmatrix} 5 & 4 \end{bmatrix}$ be vectors in R^2.

(a) Find $[\mathbf{X}]_T$ and $[\mathbf{Y}]_T$. We express **X** and **Y** in terms of the vectors in the T-basis.

$$a_1\begin{bmatrix} 1 & 1 \end{bmatrix} + a_2\begin{bmatrix} 2 & 3 \end{bmatrix} = \mathbf{X} = \begin{bmatrix} 1 & 5 \end{bmatrix}$$

leads to the linear system with augmented matrix

$\left[\begin{array}{cc|c} 1 & 2 & 1 \\ 1 & 3 & 5 \end{array}\right]$. Applying row operation $-\mathbf{r}_1 + \mathbf{r}_2$ gives

$\left[\begin{array}{cc|c} 1 & 2 & 1 \\ 0 & 1 & 4 \end{array}\right]$ and applying row operation $-2\mathbf{r}_2 + \mathbf{r}_1$ gives

the reduced row echelon form $\left[\begin{array}{cc|c} 1 & 0 & -7 \\ 0 & 1 & 4 \end{array}\right]$. Hence $a_1 = -7$

and $a_2 = 4$. Thus $[\mathbf{X}]_T = \begin{bmatrix} -7 \\ 4 \end{bmatrix}$. Similarly,

$$a_1[\ 1 \quad 1] + a_2[\ 2 \quad 3] = \mathbf{Y} = [\ 5 \quad 4]$$

leads to the linear system with augmented matrix

$\left[\begin{array}{cc|c} 1 & 2 & 5 \\ 1 & 3 & 4 \end{array}\right]$. Applying row operation $-\mathbf{r}_1 + \mathbf{r}_2$ gives

$\left[\begin{array}{cc|c} 1 & 2 & 5 \\ 0 & 1 & -1 \end{array}\right]$ and applying row operation $-2\mathbf{r}_2 + \mathbf{r}_1$ gives

the reduced row echelon form $\left[\begin{array}{cc|c} 1 & 0 & 7 \\ 0 & 1 & -1 \end{array}\right]$. Hence $a_1 = 7$

and $a_2 = -1$. Thus $[\mathbf{Y}]_T = \begin{bmatrix} 7 \\ -1 \end{bmatrix}$. (Note: In each of the

linear systems above the coefficient matrix was the same. Only the right-hand side of the linear system changed. Hence we could compute the coordinates of **X** and **Y** together by using the partitioned matrix

$\left[\begin{array}{cc|c|c} 1 & 2 & 1 & 5 \\ 1 & 3 & 5 & 4 \end{array}\right]$ with the same row operations. When

put into reduced row echelon form the coordinates appear in the last two columns. For efficiency we usually drop all but the first set of vertical separator bars. Hence

we would have $\left[\begin{array}{cc|cc} 1 & 2 & 1 & 5 \\ 1 & 3 & 5 & 4 \end{array}\right]$. This procedure

generalizes to vector spaces other than R^2 and to more than two vectors **X** and **Y**.

(b) Find the transition matrix from the T-basis to the S-basis. We first find the coordinates the vectors in the T-basis with respect to the S-basis. Hence we must solve the equations

$$a_1[\ 1 \quad 2] + a_2[\ 0 \quad 1] = [\ 1 \quad 1]$$

$$a_1[\ 1 \quad 2] + a_2[\ 0 \quad 1] = [\ 2 \quad 3]$$

Using the note in part a, we have the matrix

$$\left[\begin{array}{cc|cc} 1 & 0 & 1 & 2 \\ 2 & 1 & 1 & 3 \end{array}\right]$$

to row reduce. Apply row operation $-2\mathbf{r}_1 + \mathbf{r}_2$ and we obtain the reduced row echelon form of

$$\left[\begin{array}{cc|cc} 1 & 0 & 1 & 2 \\ 0 & 1 & -1 & -1 \end{array}\right]$$

Thus the transition matrix from the T-basis to S-basis is

$$\mathbf{P} = \begin{bmatrix} 1 & 2 \\ -1 & -1 \end{bmatrix}.$$

(c) We find the coordinates of **X** and **Y** with respect to the S-basis by multiplying their coordinates with respect to the T-basis by the transition matrix. Using the results from parts a and b we have

$$[\mathbf{X}]_S = \mathbf{P}[\mathbf{X}]_T = \begin{bmatrix} 1 & 2 \\ -1 & -1 \end{bmatrix}\begin{bmatrix} -7 \\ 4 \end{bmatrix} = \begin{bmatrix} 1 \\ 3 \end{bmatrix}$$

$$[\mathbf{Y}]_S = \mathbf{P}[\mathbf{Y}]_T = \begin{bmatrix} 1 & 2 \\ -1 & -1 \end{bmatrix}\begin{bmatrix} 7 \\ -1 \end{bmatrix} = \begin{bmatrix} 5 \\ -6 \end{bmatrix}$$

(d) To find the coordinate vectors of **X** and **Y** with respect to the S-basis directly we proceed as in part a. Here we find the coordinates using the partitioned matrix approach. Row reduce $\left[\begin{array}{cc|cc} 1 & 0 & 1 & 5 \\ 2 & 1 & 5 & 4 \end{array}\right]$. Applying row operation $-2\mathbf{r}_1 + \mathbf{r}_2$ gives the reduced row echelon form $\left[\begin{array}{cc|cc} 1 & 0 & 1 & 5 \\ 0 & 1 & 3 & -6 \end{array}\right]$. Hence $[\mathbf{X}]_S = \begin{bmatrix} 1 \\ 3 \end{bmatrix}$ and $[\mathbf{Y}]_S = \begin{bmatrix} 5 \\ -6 \end{bmatrix}$.

(e) To find the transition matrix **Q** from the S-basis to the T-basis we find the coordinates of the vectors in S with respect to the T-basis. This leads to the linear systems

$$a_1[\,1 \quad 1\,] + a_2[\,2 \quad 3\,] = [\,1 \quad 2\,]$$

$$a_1[\,1 \quad 1\,] + a_2[\,2 \quad 3\,] = [\,0 \quad 1\,]$$

Combining these systems into the partitioned form discussed above we row reduce the following matrix:

$$\left[\begin{array}{cc|cc} 1 & 2 & 1 & 0 \\ 1 & 3 & 2 & 1 \end{array}\right]{-1\mathbf{r}_1 + \mathbf{r}_2} \Longrightarrow \left[\begin{array}{cc|cc} 1 & 2 & 1 & 0 \\ 0 & 1 & 1 & 1 \end{array}\right]{-2\mathbf{r}_2 + \mathbf{r}_1}$$

$$\Longrightarrow \left[\begin{array}{cc|cc} 1 & 0 & -1 & -2 \\ 0 & 1 & 1 & 1 \end{array}\right] \quad \text{Hence } \mathbf{Q} = \begin{bmatrix} -1 & -2 \\ 1 & 1 \end{bmatrix}.$$

(f) From part d we have $[\mathbf{X}]_S = \begin{bmatrix} 1 \\ 3 \end{bmatrix}$ and $[\mathbf{Y}]_S = \begin{bmatrix} 5 \\ -6 \end{bmatrix}$. Then,

$$[\mathbf{X}]_T = Q[\mathbf{X}]_S = \begin{bmatrix} -1 & -2 \\ 1 & 1 \end{bmatrix}\begin{bmatrix} 1 \\ 3 \end{bmatrix} = \begin{bmatrix} -7 \\ 4 \end{bmatrix}$$

$$\text{and } [\mathbf{Y}]_T = Q[\mathbf{Y}]_S = \begin{bmatrix} -1 & -2 \\ 1 & 1 \end{bmatrix}\begin{bmatrix} 5 \\ -6 \end{bmatrix} = \begin{bmatrix} 7 \\ -1 \end{bmatrix}$$

15. Let $S = \{t^2 + 1, t - 2, t + 3\}$ and $T = \{2t^2 + t, t^2 + 3, t\}$ be ordered bases of P_2 and let $\mathbf{X} = 8t^2 - 4t + 6$ and $\mathbf{Y} = 7t^2 - t + 9$.

(a) Find $[\mathbf{X}]_T$ and $[\mathbf{Y}]_T$. We express **X** and **Y** in terms of the vectors in the T-basis.

$$a_1(2t^2 + t) + a_2(t^2 + 3) + a_3(t) = 8t^2 - 4t + 6$$

and

$$a_1(2t^2 + t) + a_2(t^2 + 3) + a_3(t) = 7t^2 - t + 9$$

Combining like terms on the left and then equating like powers of t on the left and right leads to systems of equations with the same coefficient matrix but different right-hand sides. We express the systems in the partition matrix form as used in Exercise 1 and row reduce it. We have,

$$\left[\begin{array}{ccc|cc} 2 & 1 & 0 & 8 & 7 \\ 1 & 0 & 1 & -4 & -1 \\ 0 & 3 & 0 & 6 & 9 \end{array}\right] \begin{array}{c} \mathbf{r}_1 \leftrightarrow \mathbf{r}_2 \\ -2\mathbf{r}_1 + \mathbf{r}_2 \end{array} \Longrightarrow \left[\begin{array}{ccc|cc} 1 & 0 & 1 & -4 & -1 \\ 0 & 1 & -2 & 16 & 9 \\ 0 & 3 & 0 & 6 & 9 \end{array}\right] -3\mathbf{r}_2 + \mathbf{r}_3$$

$$\Longrightarrow \left[\begin{array}{ccc|cc} 1 & 0 & 1 & -4 & -1 \\ 0 & 1 & -2 & 16 & 9 \\ 0 & 0 & 6 & -42 & -18 \end{array}\right] \begin{array}{c} (1/6)\mathbf{r}_3 \\ 2\mathbf{r}_3 + \mathbf{r}_2 \\ -1\mathbf{r}_3 + \mathbf{r}_1 \end{array} \Longrightarrow \left[\begin{array}{ccc|cc} 1 & 0 & 0 & 3 & 2 \\ 0 & 1 & 0 & 2 & 3 \\ 0 & 0 & 1 & -7 & -3 \end{array}\right]$$

$$\text{Thus } [\mathbf{X}]_T = \begin{bmatrix} 3 \\ 2 \\ -7 \end{bmatrix} \text{ and } [\mathbf{Y}]_T = \begin{bmatrix} 2 \\ 3 \\ -3 \end{bmatrix}.$$

(b) Find the transition matrix from the T-basis to the S-basis.

We first find the coordinates of the vectors in the T-basis with respect to the S-basis. Hence we must solve the equations

$$a_1(t^2 + 1) + a_2(t - 2) + a_3(t + 3) = 2t^2 + t, \text{ or } t^2 + 3, \text{ or } t$$

Exercises 3.9

Combining like terms on the left and then equating like powers of t on the left and right leads to systems of equations with the same coefficient matrix but different right-hand sides. We express the systems in the partition matrix form as used in Exercise 1 and row reduce it. We have,

$$\left[\begin{array}{ccc|ccc} 1 & 0 & 0 & 2 & 1 & 0 \\ 0 & 1 & 1 & 1 & 0 & 1 \\ 1 & -2 & 3 & 0 & 3 & 0 \end{array}\right] \begin{matrix} \\ \\ -1\mathbf{r}_1 + \mathbf{r}_3 \end{matrix} \Longrightarrow \left[\begin{array}{ccc|ccc} 1 & 0 & 0 & 2 & 1 & 0 \\ 0 & 1 & 1 & 1 & 0 & 1 \\ 0 & -2 & 3 & -2 & 2 & 0 \end{array}\right] \begin{matrix} \\ \\ 2\mathbf{r}_2 + \mathbf{r}_3 \end{matrix}$$

$$\Longrightarrow \left[\begin{array}{ccc|ccc} 1 & 0 & 0 & 2 & 1 & 0 \\ 0 & 1 & 1 & 1 & 0 & 1 \\ 0 & 0 & 5 & 0 & 2 & 2 \end{array}\right] \begin{matrix} \\ \\ (1/5)\mathbf{r}_3 \\ -1\mathbf{r}_3 + \mathbf{r}_2 \end{matrix}$$

$$\Longrightarrow \left[\begin{array}{ccc|ccc} 1 & 0 & 0 & 2 & 1 & 0 \\ 0 & 1 & 0 & 1 & -2/5 & 3/5 \\ 0 & 0 & 1 & 0 & 2/5 & 2/5 \end{array}\right]$$

$$\text{Thus, } P = \begin{bmatrix} 2 & 1 & 0 \\ 1 & -2/5 & 3/5 \\ 0 & 2/5 & 2/5 \end{bmatrix}.$$

(c) $$[\mathbf{X}]_S = P[\mathbf{X}]_T = \begin{bmatrix} 2 & 1 & 0 \\ 1 & -2/5 & 3/5 \\ 0 & 2/5 & 2/5 \end{bmatrix} \begin{bmatrix} 3 \\ 2 \\ -7 \end{bmatrix} = \begin{bmatrix} 8 \\ -2 \\ -2 \end{bmatrix}$$

$$[\mathbf{Y}]_S = \mathbf{P}[\mathbf{Y}]_T = \begin{bmatrix} 2 & 1 & 0 \\ 1 & -2/5 & 3/5 \\ 0 & 2/5 & 2/5 \end{bmatrix} \begin{bmatrix} 2 \\ 3 \\ -3 \end{bmatrix} = \begin{bmatrix} 7 \\ -1 \\ 0 \end{bmatrix}$$

(d) To find the coordinate vectors of **X** and **Y** with respect to the S-basis directly we proceed as in part a. Here we find the coordinates using the partitioned matrix approach. Row reduce the following matrix:

$$\left[\begin{array}{ccc|cc} 1 & 0 & 0 & 8 & 7 \\ 0 & 1 & 1 & -4 & -1 \\ 1 & -2 & 3 & 6 & 9 \end{array}\right] \begin{matrix} \\ \\ -1\mathbf{r}_1 + \mathbf{r}_3 \end{matrix} \Longrightarrow \left[\begin{array}{ccc|cc} 1 & 0 & 0 & 8 & 7 \\ 0 & 1 & 1 & -4 & -1 \\ 0 & -2 & 3 & -2 & 2 \end{array}\right] \begin{matrix} \\ \\ 2\mathbf{r}_2 + \mathbf{r}_3 \end{matrix}$$

$$\Longrightarrow \left[\begin{array}{ccc|cc} 1 & 0 & 0 & 8 & 7 \\ 0 & 1 & 1 & -4 & -1 \\ 0 & 0 & 5 & -10 & 0 \end{array}\right] \begin{matrix} \\ \\ (1/5)\mathbf{r}_3 \\ -1\mathbf{r}_3 + \mathbf{r}_2 \end{matrix} \Longrightarrow \left[\begin{array}{ccc|cc} 1 & 0 & 0 & 8 & 7 \\ 0 & 1 & 0 & -2 & -1 \\ 0 & 0 & 1 & -2 & 0 \end{array}\right]$$

Thus $[\mathbf{x}]_S = \begin{bmatrix} 8 \\ -2 \\ -2 \end{bmatrix}$ and $[\mathbf{y}]_S = \begin{bmatrix} 7 \\ -1 \\ 0 \end{bmatrix}$.

(e) To find the transition matrix Q from the S-basis to the T-basis we find the coordinates of the vectors in S with respect to the T-basis. This leads to the equations

$$a_1(2t^2 + t) + a_2(t^2 + 3) + a_3(t) = t^2 + 1, \text{ or } t - 2, \text{ or } t + 3$$

Combining like terms on the left and then equating like powers of t on the left and right leads to systems of equations with the same coefficient matrix but different right-hand sides. We express the systems in the partitioned matrix form as used in Exercise 1 and row reduce it. We have,

$$\left[\begin{array}{ccc|ccc} 2 & 1 & 0 & 1 & 0 & 0 \\ 1 & 0 & 1 & 0 & 1 & 1 \\ 0 & 3 & 0 & 1 & -2 & 3 \end{array}\right] \begin{array}{l} \mathbf{r}_1 \leftrightarrow \mathbf{r}_2 \\ -2\mathbf{r}_1 + \mathbf{r}_2 \end{array}$$

$$\Longrightarrow \left[\begin{array}{ccc|ccc} 1 & 0 & 1 & 0 & 1 & 1 \\ 0 & 1 & -2 & 1 & -2 & -2 \\ 0 & 3 & 0 & 1 & -2 & 3 \end{array}\right] -3\mathbf{r}_2 + \mathbf{r}_3$$

$$\Longrightarrow \left[\begin{array}{ccc|ccc} 1 & 0 & 1 & 0 & 1 & 1 \\ 0 & 1 & -2 & 1 & -2 & -2 \\ 0 & 0 & 6 & -2 & 4 & 9 \end{array}\right] \begin{array}{l} (1/6)\mathbf{r}_3 \\ 2\mathbf{r}_3 + \mathbf{r}_2 \\ -1\mathbf{r}_3 + \mathbf{r}_1 \end{array}$$

$$\Longrightarrow \left[\begin{array}{ccc|ccc} 1 & 0 & 0 & 1/3 & 1/3 & -1/2 \\ 0 & 1 & 0 & 1/3 & -2/3 & 1 \\ 0 & 0 & 1 & -1/3 & 2/3 & 3/2 \end{array}\right]$$

Thus $Q = \begin{bmatrix} 1/3 & 1/3 & -1/2 \\ 1/3 & -2/3 & 1 \\ -1/3 & 2/3 & 3/2 \end{bmatrix}$.

(f) $[\mathbf{x}]_T = Q[\mathbf{x}]_S = \begin{bmatrix} 1/3 & 1/3 & -1/2 \\ 1/3 & -2/3 & 1 \\ -1/3 & 2/3 & 3/2 \end{bmatrix} \begin{bmatrix} 8 \\ -2 \\ -2 \end{bmatrix} = \begin{bmatrix} 3 \\ 2 \\ -7 \end{bmatrix}$

$[\mathbf{y}]_T = Q[\mathbf{y}]_S = \begin{bmatrix} 1/3 & 1/3 & -1/2 \\ 1/3 & -2/3 & 1 \\ -1/3 & 2/3 & 3/2 \end{bmatrix} \begin{bmatrix} 7 \\ -1 \\ 0 \end{bmatrix} = \begin{bmatrix} 2 \\ 3 \\ -3 \end{bmatrix}$

17. Let

$$S = \{\mathbf{s_1}, \mathbf{s_2}, \mathbf{s_3}, \mathbf{s_4}\} = \left\{\begin{bmatrix} 1 & 0 \\ 0 & 0 \end{bmatrix}, \begin{bmatrix} 0 & 1 \\ 1 & 0 \end{bmatrix}, \begin{bmatrix} 0 & 2 \\ 0 & 1 \end{bmatrix}, \begin{bmatrix} 0 & 0 \\ 1 & 1 \end{bmatrix}\right\}$$

and

$$T = \{\mathbf{t_1}, \mathbf{t_2}, \mathbf{t_3}, \mathbf{t_4}\} = \left\{\begin{bmatrix} 1 & 1 \\ 0 & 0 \end{bmatrix}, \begin{bmatrix} 0 & 0 \\ 1 & 0 \end{bmatrix}, \begin{bmatrix} 0 & 0 \\ 0 & 1 \end{bmatrix}, \begin{bmatrix} 1 & 0 \\ 0 & 0 \end{bmatrix}\right\}$$

be ordered bases of M_{22} and let $\mathbf{X} = \begin{bmatrix} 1 & 1 \\ 1 & 1 \end{bmatrix}$ and

$$\mathbf{Y} = \begin{bmatrix} 1 & 2 \\ -2 & 1 \end{bmatrix}.$$

(a) Find $[\mathbf{X}]_T$ and $[\mathbf{Y}]_T$. We express **X** and **Y** in terms of the vectors in the T-basis. Equation

$$a_1\mathbf{t_1} + a_2\mathbf{t_2} + a_3\mathbf{t_3} + a_4\mathbf{t_4} = \mathbf{X}$$

leads to $\begin{bmatrix} a_1+a_4 & a_1 \\ a_2 & a_3 \end{bmatrix} = \begin{bmatrix} 1 & 1 \\ 1 & 1 \end{bmatrix}$ and equating corresponding elements gives a linear system whose augmented matrix is $\left[\begin{array}{cccc|c} 1 & 0 & 0 & 1 & 1 \\ 1 & 0 & 0 & 0 & 1 \\ 0 & 1 & 0 & 0 & 1 \\ 0 & 0 & 1 & 0 & 1 \end{array}\right]$. Applying row operation $-\mathbf{r}_1 + \mathbf{r}_2$ gives $\left[\begin{array}{cccc|c} 1 & 0 & 0 & 1 & 1 \\ 0 & 0 & 0 & -1 & 0 \\ 0 & 1 & 0 & 0 & 1 \\ 0 & 0 & 1 & 0 & 1 \end{array}\right]$. It follows that $a_1 = a_2 = a_3 = 1$ and $a_4 = 0$. Thus $[\mathbf{X}]_T = \begin{bmatrix} 1 \\ 1 \\ 1 \\ 0 \end{bmatrix}$. To determine $[\mathbf{Y}]_T$ we replace **X** in the previous steps by **Y**. The corresponding linear system is

$$\begin{bmatrix} a_1+a_4 & a_1 \\ a_2 & a_3 \end{bmatrix} = \begin{bmatrix} 1 & 2 \\ -2 & 1 \end{bmatrix}$$

Equating corresponding elements gives a linear system whose augmented matrix is $\left[\begin{array}{cccc|c} 1 & 0 & 0 & 1 & 1 \\ 1 & 0 & 0 & 0 & 2 \\ 0 & 1 & 0 & 0 & -2 \\ 0 & 0 & 1 & 0 & 1 \end{array}\right]$. Row reducing this systems gives $a_1 = 2$, $a_2 = -2$, $a_3 = 1$, $a_4 = -1$. Thus $[\mathbf{Y}]_T = \begin{bmatrix} 2 \\ -2 \\ 1 \\ -1 \end{bmatrix}$.

(b) Find the transition matrix from the T-basis to the S-basis.

We first find the coordinates the vectors in the T-basis with respect to the S-basis. Hence we must solve the equations

$$a_1\mathbf{s_1} + a_2\mathbf{s_2} + a_3\mathbf{s_3} + a_4\mathbf{s_4} = \mathbf{t_1} \mid \mathbf{t_2} \mid \mathbf{t_3} \mid \mathbf{t_4}$$

That is, there are really four equations to be solved. The right-hand side is the only thing that changes in each of the corresponding systems. Using the ideas from Exercise 13, we substitute in the vectors from M_{22}, equate corresponding components and obtain a set of linear systems with the same coefficient matrix but different right hand sides. For efficiency we express this as in Exercise 13 using a partitioned matrix. We obtain the following matrix which we row reduce.

$$\left[\begin{array}{cccc|cccc} 1 & 0 & 0 & 0 & 1 & 0 & 0 & 1 \\ 0 & 1 & 2 & 0 & 1 & 0 & 0 & 0 \\ 0 & 1 & 0 & 1 & 0 & 1 & 0 & 0 \\ 0 & 0 & 1 & 1 & 0 & 0 & 1 & 0 \end{array}\right] \begin{array}{l} \\ \\ -1\mathbf{r}_2 + \mathbf{r}_3 \\ \end{array} \Longrightarrow$$

$$\left[\begin{array}{cccc|cccc} 1 & 0 & 0 & 0 & 1 & 0 & 0 & 1 \\ 0 & 1 & 2 & 0 & 1 & 0 & 0 & 0 \\ 0 & 0 & -2 & 1 & -1 & 1 & 0 & 0 \\ 0 & 0 & 1 & 1 & 0 & 0 & 1 & 0 \end{array}\right] \mathbf{r}_3 \leftrightarrow \mathbf{r}_4 \Longrightarrow$$

$$\left[\begin{array}{cccc|cccc} 1 & 0 & 0 & 0 & 1 & 0 & 0 & 1 \\ 0 & 1 & 2 & 0 & 1 & 0 & 0 & 0 \\ 0 & 0 & 1 & 1 & 0 & 0 & 1 & 0 \\ 0 & 0 & -2 & 1 & -1 & 1 & 0 & 0 \end{array}\right] \begin{array}{l} \\ \\ -2\mathbf{r}_3 + \mathbf{r}_2 \\ 2\mathbf{r}_3 + \mathbf{r}_4 \end{array} \Longrightarrow$$

$$\left[\begin{array}{cccc|cccc} 1 & 0 & 0 & 0 & 1 & 0 & 0 & 1 \\ 0 & 1 & 0 & -2 & 1 & 0 & -2 & 0 \\ 0 & 0 & 1 & 1 & 0 & 0 & 1 & 0 \\ 0 & 0 & 0 & 3 & -1 & 1 & 2 & 0 \end{array}\right] \begin{array}{l} \\ \\ (-1/3)\mathbf{r}_4 + \mathbf{r}_3 \\ (2/3)\mathbf{r}_4 + \mathbf{r}_2 \\ (1/3)\mathbf{r}_4 \end{array} \Longrightarrow$$

$$\left[\begin{array}{cccc|cccc} 1 & 0 & 0 & 0 & 1 & 0 & 0 & 1 \\ 0 & 1 & 0 & 0 & 1/3 & 2/3 & -2/3 & 0 \\ 0 & 0 & 1 & 0 & 1/3 & -1/3 & 1/3 & 0 \\ 0 & 0 & 0 & 1 & -1/3 & 1/3 & 2/3 & 0 \end{array}\right]$$

The coordinates of $\mathbf{t}_j$, $j = 1,2,3,4$ are respectively the last four columns of the preceding matrix. Hence the transition matrix is the right hand 4×4 block of the preceding matrix.

(c) We find the coordinates of **X** and **Y** with respect to the S-basis by multiplying their coordinates respect to the T-basis by the transition matrix. Using the results from parts a and b we have

$$[\mathbf{X}]_S = \mathbf{P}[\mathbf{X}]_T = \begin{bmatrix} 1 & 0 & 0 & 1 \\ 1/3 & 2/3 & -2/3 & 0 \\ 1/3 & -1/3 & 1/3 & 0 \\ -1/3 & 1/3 & 2/3 & 0 \end{bmatrix} \begin{bmatrix} 1 \\ 1 \\ 1 \\ 0 \end{bmatrix} = \begin{bmatrix} 1 \\ 1/3 \\ 1/3 \\ 2/3 \end{bmatrix}$$

$$[\mathbf{Y}]_S = \mathbf{P}[\mathbf{Y}]_T = \begin{bmatrix} 1 & 0 & 0 & 1 \\ 1/3 & 2/3 & -2/3 & 0 \\ 1/3 & -1/3 & 1/3 & 0 \\ -1/3 & 1/3 & 2/3 & 0 \end{bmatrix} \begin{bmatrix} 2 \\ -2 \\ 1 \\ -1 \end{bmatrix} = \begin{bmatrix} 1 \\ -4/3 \\ 5/3 \\ -2/3 \end{bmatrix}$$

(d) To find the coordinate vectors of **X** and **Y** with respect to the S-basis directly we proceed as in part a. Here we find the coordinates using the partitioned matrix approach. From

$$a_1\mathbf{s_1} + a_2\mathbf{s_2} + a_3\mathbf{s_3} + a_4\mathbf{s}_4 = \mathbf{X} \mid \mathbf{Y}$$

we obtain linear systems whose augmented matrices written in partitioned form are represented by

$$\left[\begin{array}{cccc|cc} 1 & 0 & 0 & 0 & 1 & 1 \\ 0 & 1 & 2 & 0 & 1 & 2 \\ 0 & 1 & 0 & 1 & 1 & -2 \\ 0 & 0 & 1 & 1 & 1 & 1 \end{array}\right]$$

Performing the same row operations as in part (b) we obtain

$$\left[\begin{array}{cccc|cc} 1 & 0 & 0 & 0 & 1 & 1 \\ 0 & 1 & 0 & 0 & 1/3 & -4/3 \\ 0 & 0 & 1 & 0 & 1/3 & 5/3 \\ 0 & 0 & 0 & 1 & 2/3 & -2/3 \end{array}\right]$$

Thus $[\mathbf{X}]_S = \begin{bmatrix} 1 \\ 1/3 \\ 1/3 \\ 2/3 \end{bmatrix}$ and $[\mathbf{Y}]_S = \begin{bmatrix} 1 \\ -4/3 \\ 5/3 \\ -2/3 \end{bmatrix}$.

(e) We could proceed directly as in part (b) reversing the roles of the S and T bases. However, if we call $\mathbf{Q}$ the transition matrix from the S to the T-basis, then we have that $\mathbf{Q} = \mathbf{P}^{-1}$. We form the partitioned matrix $[\mathbf{P} \mid \mathbf{I}_4]$ and obtain its reduced row echelon form. The result is $[\mathbf{I}_4 \mid \mathbf{P}^{-1}]$. The result is

$$\mathbf{Q} = \mathbf{P}^{-1} = \begin{bmatrix} 0 & 1 & 2 & 0 \\ 0 & 1 & 0 & 1 \\ 0 & 0 & 1 & 1 \\ 1 & -1 & -2 & 0 \end{bmatrix}.$$

(f) From part (d) we have $[\mathbf{X}]_S$ and $[\mathbf{Y}]_S$ and from part (e) we have $\mathbf{Q}$. Then

$$[\mathbf{X}]_T = Q[\mathbf{X}]_S = \begin{bmatrix} 1 \\ 1 \\ 1 \\ 0 \end{bmatrix} \text{ and } [\mathbf{Y}]_T = Q[\mathbf{Y}]_S = \begin{bmatrix} 2 \\ -2 \\ 1 \\ -1 \end{bmatrix}$$

19. We find the transition matrix $\mathbf{P}$ from the T-basis to the S-basis and then $[\mathbf{X}]_S = \mathbf{P}[\mathbf{X}]_T$.

To find $\mathbf{P}$ we express the basis vectors in T in terms of those in S. The equations

$$a_1[1\ \ -1] + a_2[2\ \ 1] = [3\ \ 0]$$
$$a_1[1\ \ -1] + a_2[2\ \ 1] = [4\ \ -1]$$

lead to the partioned matrix

$$\left[\begin{array}{cc|cc} 1 & 2 & 3 & 4 \\ -1 & 1 & 0 & -1 \end{array}\right]$$

which we row reduce. We obtain

$$\left[\begin{array}{cc|cc} 1 & 0 & 1 & 2 \\ 0 & 1 & 1 & 1 \end{array}\right]$$

Thus it follows that $\mathbf{P} = \begin{bmatrix} 1 & 2 \\ 1 & 1 \end{bmatrix}$ and $[\mathbf{x}]_S = \mathbf{P}[\mathbf{x}]_T = \begin{bmatrix} 5 \\ 3 \end{bmatrix}$.

21. Following the procedure in Exercise 19 we are led to the partitioned matrix

$$\left[\begin{array}{ccc|ccc} -1 & 0 & 0 & -1 & 0 & -2 \\ 1 & 1 & 1 & 2 & 1 & 2 \\ 0 & 0 & 1 & 1 & 1 & 1 \end{array}\right]$$

Its reduced row echelon form is

$$\left[\begin{array}{ccc|ccc} 1 & 0 & 0 & 1 & 0 & 2 \\ 0 & 1 & 0 & 0 & 0 & -1 \\ 0 & 0 & 1 & 1 & 1 & 1 \end{array}\right]$$

and it follows that $\mathbf{Q}$, the transition matrix from S to T, is

$$\mathbf{Q} = \begin{bmatrix} 1 & 0 & 2 \\ 0 & 0 & -1 \\ 1 & 1 & 1 \end{bmatrix}$$

Thus $[\mathbf{x}]_T = \begin{bmatrix} 4 \\ -1 \\ 3 \end{bmatrix}$.

23. Denote the transition matrix from T to S by $\mathbf{P}$. Then from Equation (5) we have that the columns of $\mathbf{P}$ are the coordinates of the T-basis vectors with respect to the S-basis. Hence

$$[\mathbf{Y}_1]_S = \begin{bmatrix} 1 \\ 2 \\ -1 \end{bmatrix} \implies \mathbf{Y}_1 = 1\mathbf{X}_1 + 2\mathbf{X}_2 - \mathbf{X}_3 = \begin{bmatrix} 3 \\ 2 \\ 0 \end{bmatrix}$$

$$[\mathbf{Y}_2]_S = \begin{bmatrix} 1 \\ 1 \\ -1 \end{bmatrix} \implies \mathbf{Y}_2 = 1\mathbf{X}_1 + 1\mathbf{X}_2 - \mathbf{X}_3 = \begin{bmatrix} 2 \\ 1 \\ 0 \end{bmatrix}$$

$$[\mathbf{Y}_3]_S = \begin{bmatrix} 2 \\ 1 \\ 1 \end{bmatrix} \implies \mathbf{Y}_3 = 2\mathbf{X}_1 + 1\mathbf{X}_2 + 1\mathbf{X}_3 = \begin{bmatrix} 3 \\ 1 \\ 3 \end{bmatrix}$$

25. Following the procedure used in Exercise 23, we have

$$[\mathbf{Y}_1]_S = \begin{bmatrix} 2 \\ 1 \end{bmatrix} \implies \mathbf{Y}_1 = 2\mathbf{X}_1 + 1\mathbf{X}_2 = \begin{bmatrix} 2 \\ 5 \end{bmatrix}$$

$$[\mathbf{Y}_2]_S = \begin{bmatrix} 1 \\ 1 \end{bmatrix} \implies \mathbf{Y}_2 = 1\mathbf{X}_1 + 1\mathbf{X}_2 = \begin{bmatrix} 1 \\ 3 \end{bmatrix}$$

T.1. Let $\mathbf{X} = \mathbf{Y}$. The coordinates of a vector with respect to basis S are the coefficients used to express the vector in terms of the members of S. A vector has a unique expression in terms of the vectors of a basis, hence it follows that $[\mathbf{X}]_S$ must equal $[\mathbf{Y}]_S$.

Let $[\mathbf{X}]_S = [\mathbf{Y}]_S = \begin{bmatrix} a_1 \\ a_2 \\ \cdot \\ \cdot \\ a_n \end{bmatrix}$. Then $\mathbf{X} = a_1\mathbf{X}_1 + a_2\mathbf{X}_2 + \cdots + a_n\mathbf{X}_n$ and $\mathbf{Y} = a_1\mathbf{X}_1 + a_2\mathbf{X}_2 + \cdots + a_n\mathbf{X}_n$. Hence $\mathbf{X} = \mathbf{Y}$.

T.3. Suppose that $\left\{[\mathbf{Y}_1]_S, [\mathbf{Y}_2]_S, \ldots, [\mathbf{Y}_k]_S\right\}$ is linearly dependent. Then there exist scalars, a_i, $i = 1,2,\ldots,k$, that are not all zero such that

$$a_1[\mathbf{Y}_1]_S + a_2[\mathbf{Y}_2]_S + \cdots + a_k[\mathbf{Y}_k]_S = [\mathbf{0}_V]_S$$

Using Exercise T.2 we find that the preceding equation is equivalent to

$$[a_1\mathbf{Y}_1 + a_2\mathbf{Y}_2 + \cdots + a_k\mathbf{Y}_k]_S = [\mathbf{0}_V]_S$$

By Exercise T.1 we have that

$$a_1\mathbf{Y}_1 + a_2\mathbf{Y}_2 + \cdots + a_n\mathbf{Y}_n = \mathbf{0}_V$$

Since the $\mathbf{Y}$'s are linearly independent the preceding equation is only true when all $a_i = 0$. Hence we have a contradiction and our assumption that the $[\mathbf{Y}_i]_S$'s are linearly dependent must be false. It follows that $\left\{[Y_1]_S, [Y_2]_S, \ldots, [Y_k]_S\right\}$ is linearly independent.

ML.1. Since S is a subset of 3 vectors in a 3-dimensional vector space, we can show that S is a basis by verifying that the vectors in S are linearly independent. It follows that if the reduced row echelon form of the three columns is I_3, they are linearly independent.

```
A=[1 2 1;2 1 0;1 0 2];
rref(A)

ans =

     1     0     0
     0     1     0
     0     0     1
```

To find the coordinates of **X** we solve the system **Ac** = **X**. We can do all three parts simultaneously as follows. Put the three columns whose coordinates we want to find into a matrix **B**.

```
B=[8 2 4;4 0 3;7 -3 3];
rref([A B])

ans =

     1     0     0     1    -1     1
     0     1     0     2     2     1
     0     0     1     3    -1     1
```

The coordinates appear in the last three columns of the matrix above.

ML.2. Proceed as in ML.1 by making each of the vectors in S a column in matrix **A**.

```
A=[1 0 1 1;1 2 1 3;0 2 1 1;0 1 0 0]'
rref(A)

ans =

     1     0     0     0
     0     1     0     0
     0     0     1     0
     0     0     0     1
```

To find the coordinates of **X** we solve a linear system. We can do all three parts simultaneously as follows. Associate with each vector **X** a column. Form a matrix **B** from these columns.

```
B=[4 12 8 14;1/2 0 0 0;1 1 1 7/3]';
rref([A B])

ans =

1.0000       0       0       0   1.0000   0.5000   0.3333
     0  1.0000       0       0   3.0000        0   0.6667
     0       0  1.0000       0   4.0000  -0.5000        0
     0       0       0  1.0000  -2.0000   1.0000  -0.3333
```

The coordinates are the last three columns of the preceding matrix.

ML.3. Associate a column with each matrix and proceed as in ML.2.

```
A=[1 1 2 2;0 1 2 0;3 -1 1 0;-1 0 0 0]';
rref(A)

ans =

     1     0     0     0
     0     1     0     0
     0     0     1     0
     0     0     0     1

B=[1 0 0 1;2 7/6 10/3 2;1 1 1 1]';
rref([A B])

ans =

1.0000       0       0       0   0.5000   1.0000   0.5000
     0  1.0000       0       0  -0.5000   0.5000   0.1667
     0       0  1.0000       0        0   0.3333  -0.3333
     0       0       0  1.0000  -0.5000        0  -1.5000
```

The last three columns of the preceding matrix are the coordinates.

ML.4.

```
A=[1 0 1;1 1 0;0 1 1];
B=[2 1 1;1 2 1;1 1 2];
rref([A B])

ans =

     1     0     0     1     1     0
     0     1     0     0     1     1
     0     0     1     1     0     1
```

The transition matrix from the T-basis to the S-basis is P = ans(:,4:6)

Exercises 3.9

```
      P =

           1        1        0
           0        1        1
           1        0        1
```

ML.5.
```
A=[0 0 1 -1;0 0 1 1;0 1 1 0;1 0 -1 0]';
B=[0 1 0 0;0 0 -1 1;0 -1 0 2;1 1 0 0]';
rref([A B])

ans =

  Columns 1 through 6

    1.0000         0         0         0   -0.5000   -1.0000
         0    1.0000         0         0   -0.5000         0
         0         0    1.0000         0    1.0000         0
         0         0         0    1.0000         0         0

  Columns 7 and 8

   -0.5000         0
    1.5000         0
   -1.0000    1.0000
         0    1.0000
```

The transition matrix P is found in columns 5 through 8 of the preceding matrix.

ML.6.
```
A=[1 2 3 0;0 1 2 3;3 0 1 2;2 3 0 1]';
B=eye(4);
rref([A B])

ans =

  Columns 1 through 6

    1.0000         0         0         0    0.0417    0.0417
         0    1.0000         0         0   -0.2083    0.0417
         0         0    1.0000         0    0.2917   -0.2083
         0         0         0    1.0000    0.0417    0.2917

  Columns 7 and 8

    0.2917   -0.2083
    0.0417    0.2917
    0.0417    0.0417
   -0.2083    0.0417
```

The transition matrix P is found in columns 5 through 8 of the preceding matrix.

ML.7. We put basis S into matrix A, T into B, and U into C.

```
A=[1 1 0;1 2 1;1 1 1];
B=[1 1 0;0 1 1;1 0 2];
C=[2 -1 1;1 2 -2;1 1 1];
```

(a) The transition matrix from U to T will be the last 3 columns of rref([B C]).

```
rref([B C])
ans =

   1.0000        0        0   1.0000  -1.6667   2.3333
        0   1.0000        0   1.0000   0.6667  -1.3333
        0        0   1.0000        0   1.3333  -0.6667

P=ans(:,4:6)

P =

   1.0000  -1.6667   2.3333
   1.0000   0.6667  -1.3333
        0   1.3333  -0.6667
```

(b) The transition matrix from T to S will be the last 3 columns of rref([A B]).

```
rref([A B])

ans =

     1     0     0     2     0     1
     0     1     0    -1     1    -1
     0     0     1     0    -1     2

Q=ans(:,4:6)

Q =

     2     0     1
    -1     1    -1
     0    -1     2
```

(c) The transition matrix from U to S will be the last 3 columns of rref([A C]).

```
rref([A C])

ans =

     1     0     0     2    -2     4
     0     1     0     0     1    -3
     0     0     1    -1     2     0
```

Exercises 3.9

```
Z=ans(:,4:6)

Z =

     2    -2     4
     0     1    -3
    -1     2     0
```

(d) **Q*P** gives **Z**.

Exercises 3.10

1. If the dot product of each pair of vectors is zero, then the set of vectors is orthogonal.
   (a) $(1,-1,2)\cdot(0,2,-1) = -4$
   $(1,-1,2)\cdot(-1,1,1) = 0$
   $(0,2,-1)\cdot(-1,1,1) = 1$ Thus the set is not orthogonal.

   (b) $(1,2,-1,1)\cdot(0,-1,-2,0) = 0$
   $(1,2,-1,1)\cdot(1,0,0,-1) = 0$
   $(0,-1,-2,0)\cdot(1,0,0,-1) = 0$ Thus the set is orthogonal.

   (c) $(0,1,0,-1)\cdot(1,0,1,1) = -1$
   $(0,1,0,-1)\cdot(-1,1,-1,2) = -1$
   $(1,0,1,1)\cdot(-1,1,-1,2) = 0$ Thus the set is not orthogonal.

3. $\mathbf{X}\cdot\mathbf{Y} = a - 1 - 4 = 0$ only if $a = 5$.

<<**Strategy:** In Exercises 5, 7, and 9 we use the technique developed in Example 5 to transform a basis for a subspace into an orthonormal basis.>>

5. Let $\mathbf{X}_1 = (1,-1,0)$ and $\mathbf{X}_2 = (2,0,1)$. Define $\mathbf{Y}_1 = \mathbf{X}_1$. Compute

$$\mathbf{Y}_2 = \mathbf{X}_2 - \left[\frac{\mathbf{X}_2\cdot\mathbf{Y}_1}{\mathbf{Y}_1\cdot\mathbf{Y}_1}\right]\mathbf{Y}_1 = (2,0,1) - \frac{2}{2}(1,-1,0) = (1,1,1).$$

   Then $\{\mathbf{Y}_1, \mathbf{Y}_2\}$ is an orthogonal basis for W. Let

$$\mathbf{Z}_1 = \frac{\mathbf{Y}_1}{\|\mathbf{Y}_1\|} = 1/\sqrt{2}\,(1,-1,0)$$

   and let

$$\mathbf{Z}_2 = \frac{\mathbf{Y}_2}{\|\mathbf{Y}_2\|} = 1/\sqrt{3}\,(1,1,1).$$

   Then $\{\mathbf{Z}_1, \mathbf{Z}_2\}$ is an orthonormal basis for W.

7. Let $\mathbf{X}_1 = (1,-1,0,1)$, $\mathbf{X}_2 = (2,0,0,-1)$, and $\mathbf{X}_3 = (0,0,1,0)$. Define $\mathbf{Y}_1 = \mathbf{X}_1$. Compute

$$\mathbf{Y}_2 = \mathbf{X}_2 - \left[\frac{\mathbf{X}_2\cdot\mathbf{Y}_1}{\mathbf{Y}_1\cdot\mathbf{Y}_1}\right]\mathbf{Y}_1 = (2,0,0,-1) - \frac{1}{3}(1,-1,0,1)$$
$$= (5/3,1/3,0,-4/3),$$

$$Y_3 = X_3 - \left[\frac{X_3 \cdot Y_1}{Y_1 \cdot Y_1}\right] Y_1 - \left[\frac{X_3 \cdot Y_2}{Y_2 \cdot Y_2}\right] Y_2$$
$$= (0,0,1,0) - 0\cdot(1,-1,0,1) - 0\cdot(5/3,1/3,0,-4/3)$$
$$= (0,0,1,0).$$

Then $\{Y_1, Y_2, Y_3\}$ is an orthogonal basis for W. Clearing fractions in Y_2, we find that $\{Y_1', Y_2', Y_3'\} = \{(1,-1,0,1), (5,1,0,-4), (0,0,1,0)\}$ is also an orthogonal basis for W. Let

$$Z_1 = \frac{Y_1'}{\|Y_1'\|} = 1/\sqrt{3}\ (1,-1,0,1),$$

$$Z_2 = \frac{Y_2'}{\|Y_2'\|} = 1/\sqrt{42}\ (5,1,0,-4),$$

$$Z_3 = \frac{Y_3'}{\|Y_3'\|} = (0,0,1,0).$$

Then $\{Z_1, Z_2, Z_3\}$ is an orthonormal basis for W.

9. Let $X_1 = (1,2)$ and $X_2 = (-3,4)$.
   (a) Define $Y_1 = X_1$. Compute

$$Y_2 = X_2 - \left[\frac{X_2 \cdot Y_1}{Y_1 \cdot Y_1}\right] Y_1 = (-3,4) - 1\cdot(1,2) = (-4,2).$$

   Then $\{Y_1, Y_2\}$ is an orthogonal basis for $\mathbf{R}^2$.

   (b) Let

$$Z_1 = \frac{Y_1}{\|Y_1\|} = 1/\sqrt{5}\ (1,2)$$

   and let

$$Z_2 = \frac{Y_2}{\|Y_2\|} = 1/2\sqrt{5}\ (-4,2).$$

   Then $\{Z_1, Z_2\}$ is an orthonormal basis for $\mathbf{R}^2$.

11. Let $X_1 = (2/3,-2/3,1/3)$ and $X_2 = (2/3,1/3,-2/3)$. We first find a basis for $\mathbf{R}^3$ containing X_1 and X_2 following the technique of Example 9 of Section 3.6. Then we use the Gram-Schmidt process to transform it to an orthonormal basis.

    Let $E_1 = (1,0,0)$, $E_2 = (0,1,0)$, $E_3 = (0,0,1)$, and $S = \{X_1, X_2, E_1, E_2, E_3\}$. S spans $\mathbf{R}^3$ since it contains the basis $\{E_1, E_2, E_3\}$. Form the equation

$$c_1X_1 + c_2X_2 + c_3E_1 + c_4E_2 + c_5E_3 = (0,0,0)$$

    which leads to the homogeneous system

$$\left[\begin{array}{ccccc|c} 2/3 & 2/3 & 1 & 0 & 0 & 0 \\ -2/3 & 1/3 & 0 & 1 & 0 & 0 \\ 1/3 & -2/3 & 0 & 0 & 1 & 0 \end{array}\right]$$

Transforming this augmented matrix to reduced row echelon form we obtain

$$\left[\begin{array}{ccccc|c} 1 & 0 & 0 & -2 & -1 & 0 \\ 0 & 1 & 0 & -1 & -2 & 0 \\ 0 & 0 & 1 & 2 & 2 & 0 \end{array}\right]$$

The leading ones indicate that $\{\mathbf{X}_1, \mathbf{X}_2, \mathbf{E}_1\}$ is a basis for R^3.

Next we use the Gram-Schmidt process. Let $\mathbf{Y}_1 = \mathbf{X}_1$. Compute

$$\mathbf{Y}_2 = \mathbf{X}_2 - \left(\frac{\mathbf{X}_2 \cdot \mathbf{Y}_1}{\mathbf{Y}_1 \cdot \mathbf{Y}_1}\right) \mathbf{Y}_1 = \mathbf{X}_2 - 0\ \mathbf{Y}_1 = (2/3,1/3,-2/3)$$

and

$$\mathbf{Y}_3 = \mathbf{E}_1 - \left(\frac{\mathbf{E}_1 \cdot \mathbf{Y}_1}{\mathbf{Y}_1 \cdot \mathbf{Y}_1}\right) \mathbf{Y}_1 - \left(\frac{\mathbf{E}_1 \cdot \mathbf{Y}_2}{\mathbf{Y}_2 \cdot \mathbf{Y}_2}\right) \mathbf{Y}_2$$

$$= \mathbf{E}_1 - (2/3)\ \mathbf{Y}_1 - (2/3)\ \mathbf{Y}_2 = (1/9,2/9,2/9).$$

The set $\{\mathbf{Y}_1, \mathbf{Y}_2, \mathbf{Y}_3\}$ is an orthogonal basis for $\mathbf{R}^3$. Then $\{\mathbf{Z}_1, \mathbf{Z}_2, \mathbf{Z}_3\}$ is an othonormal basis, where

$$\mathbf{Z}_1 = \frac{\mathbf{Y}_1}{\|\mathbf{Y}_1\|} = (2/3,-2/3,1/3),$$

$$\mathbf{Z}_2 = \frac{\mathbf{Y}_2}{\|\mathbf{Y}_2\|} = (2/3,1/3,-2/3),$$

$$\mathbf{Z}_3 = \frac{\mathbf{Y}_3}{\|\mathbf{Y}_3\|} = (1/3,2/3,2/3).$$

13. Let $\mathbf{X}_1 = (1,1,0,0)$, $\mathbf{X}_2 = (2,-1,0,1)$, $\mathbf{X}_3 = (3,-3,0,-2)$, $\mathbf{X}_4 = (1,-2,0,-3)$, and $S = \{\mathbf{X}_1, \mathbf{X}_2, \mathbf{X}_3, \mathbf{X}_4\}$. We find a basis for W = span S following the method in Example 5 of Section 3.6. To determine a basis for span S form the linear combination $c_1\mathbf{X}_1 + c_2\mathbf{X}_2 + c_3\mathbf{X}_3 + c_4\mathbf{X}_4 = \mathbf{0}$. Expanding, adding vectors, and equating corresponding components from each side of the equation we obtain a homogenous system with coefficient

matrix $\begin{bmatrix} 1 & 2 & +3 & 1 \\ 1 & -1 & -3 & -2 \\ 0 & 0 & 0 & 0 \\ 0 & 1 & -2 & -3 \end{bmatrix}$ which has reduced row echelon form

$$\begin{bmatrix} 1 & 0 & 0 & 0 \\ 0 & 1 & 0 & -1 \\ 0 & 0 & 1 & 1 \\ 0 & 0 & 0 & 0 \end{bmatrix}$$. The leading 1's are in columns 1, 2, and 3 so $S1 = \{\mathbf{X}_1, \mathbf{X}_2, \mathbf{X}_3\}$ is a basis for span S. Next we use the Gram-Schmidt procedure. Let $\mathbf{Y}_1 = \mathbf{X}_1$. Compute

$$\mathbf{Y}_2 = \mathbf{X}_2 - \left[\frac{\mathbf{X}_2 \cdot \mathbf{Y}_1}{\mathbf{Y}_1 \cdot \mathbf{Y}_1}\right] \mathbf{Y}_1 = (2,-1,0,1) - (1/2)\ (1,1,0,0),$$
$$= (3/2,-3/2,0,1),$$

$$\mathbf{Y}_3 = \mathbf{X}_3 - \left[\frac{\mathbf{X}_3 \cdot \mathbf{Y}_1}{\mathbf{Y}_1 \cdot \mathbf{Y}_1}\right] \mathbf{Y}_1 - \left[\frac{\mathbf{X}_3 \cdot \mathbf{Y}_2}{\mathbf{Y}_2 \cdot \mathbf{Y}_2}\right] \mathbf{Y}_2$$
$$= (3,-3,0,-2) - 0\ (1,1,0,0) - (14/11)\ (3/2,-3/2,0,1)$$
$$= (12/11,-12/11,0,-36/11).$$

Then $\{\mathbf{Y}_1, \mathbf{Y}_2, \mathbf{Y}_3\}$ is an orthogonal basis for W. Clearing fractions, we have $\{\mathbf{Y}_1', \mathbf{Y}_2', \mathbf{Y}_3'\} = \{(1,1,0,0), (3,-3,0,2), (12,-12,0,-36)\}$ is also an orthogonal basis for W. Let

$$\mathbf{Z}_1 = \frac{\mathbf{Y}_1'}{\|\mathbf{Y}_1'\|} = 1/\sqrt{2}\ (1,1,0,0),$$

$$\mathbf{Z}_2 = \frac{\mathbf{Y}_2'}{\|\mathbf{Y}_2'\|} = 1/\sqrt{22}\ (3,-3,0,2),$$

$$\mathbf{Z}_3 = \frac{\mathbf{Y}_3'}{\|\mathbf{Y}_3'\|} = 1/\sqrt{11}\ (1,-1,0,-3).$$

It follows that $\{\mathbf{Z}_1, \mathbf{Z}_2, \mathbf{Z}_3\}$ is an orthonormal basis for W.

15. Let W be the subspace of R^4 consisting of all vectors of the form (a, a+b, c, b+c). Since

$$(a, a+b, c, b+c) = a(1,1,0,0) + b(0,1,0,1) + c(0,0,1,1)$$

it follows that $S = \{\mathbf{X}_1, \mathbf{X}_2, \mathbf{X}_3\} = \{(1,1,0,0), (0,1,0,1), (0,0,1,1)\}$ spans W. To show S is a basis for W we show S is linearly independent. Form the expression

$$c_1\mathbf{X}_1 + c_2\mathbf{X}_2 + c_3\mathbf{X}_3 = (0,0,0,0)$$

which has augmented matrix

$$\left[\begin{array}{ccc|c} 1 & 0 & 0 & 0 \\ 1 & 1 & 0 & 0 \\ 0 & 0 & 1 & 0 \\ 0 & 1 & 1 & 0 \end{array}\right]$$

The reduced row echelon form of this matrix is

$$\left[\begin{array}{ccc|c} 1 & 0 & 0 & 0 \\ 0 & 1 & 0 & 0 \\ 0 & 0 & 1 & 0 \\ 0 & 0 & 0 & 0 \end{array}\right]$$

so $c_1 = c_2 = c_3 = 0$. Hence S is linearly independent. We next apply the Gram-Schmidt process to S. Let $\mathbf{Y}_1 = \mathbf{X}_1$ and

$$\mathbf{Y}_2 = \mathbf{X}_2 - \left(\frac{\mathbf{X}_2 \cdot \mathbf{Y}_1}{\mathbf{Y}_1 \cdot \mathbf{Y}_1}\right) \mathbf{Y}_1 = (0,1,0,1) - (1/2)(1,1,0,0)$$
$$= (-1/2,1/2,0,1)$$

$$\mathbf{Y}_3 = \mathbf{X}_3 - \left(\frac{\mathbf{X}_3 \cdot \mathbf{Y}_1}{\mathbf{Y}_1 \cdot \mathbf{Y}_1}\right) \mathbf{Y}_1 - \left(\frac{\mathbf{X}_3 \cdot \mathbf{Y}_2}{\mathbf{Y}_2 \cdot \mathbf{Y}_2}\right) \mathbf{Y}_2 = (0,0,1,1) - 0(1,1,0,0)$$
$$- (1/(3/2))(-1/2,1/2,0,1) = (1/3,-1/3,1,1/3)$$

$$\mathbf{Z}_1 = \frac{\mathbf{Y}_1}{\|\mathbf{Y}_1\|} = (1/\sqrt{2},1/\sqrt{2},0,0)$$

$$\mathbf{Z}_2 = \frac{\mathbf{Y}_2}{\|\mathbf{Y}_2\|} = (-1/\sqrt{6},1/\sqrt{6},0,2/\sqrt{6})$$

$$\mathbf{Z}_3 = \frac{\mathbf{Y}_3}{\|\mathbf{Y}_3\|} = (1/\sqrt{12},-1/\sqrt{12},3/\sqrt{12},1/\sqrt{12}).$$

17. Let W be the subspace of R^4 consisting of all vectors of the form (a, b, c, d) such that $a - b - 2c + d = 0$. We have $a = b + 2c - d$ so W is all vectors of the form

$$(b+2c-d,\ b,\ c,\ d) = b(1,1,0,0) + c(2,0,1,0) + d(-1,0,0,1)$$

It follows that $S = \{\mathbf{X}_1, \mathbf{X}_2, \mathbf{X}_3\} = \{(1,1,0,0), (2,0,1,0), (-1,0,0,1)\}$ spans W. To show S is a basis for W we show S is linearly independent. Form the expression

$$c_1\mathbf{X}_1 + c_2\mathbf{X}_2 + c_3\mathbf{X}_3 = (0,0,0,0)$$

which has augmented matrix

$$\left[\begin{array}{ccc|c} 1 & 2 & -1 & 0 \\ 1 & 0 & 0 & 0 \\ 0 & 1 & 0 & 0 \\ 0 & 0 & 1 & 0 \end{array}\right]$$

The reduced row echelon form of this matrix is

$$\left[\begin{array}{ccc|c} 1 & 0 & 0 & 0 \\ 0 & 1 & 0 & 0 \\ 0 & 0 & 1 & 0 \\ 0 & 0 & 0 & 0 \end{array}\right]$$

so $c_1 = c_2 = c_3 = 0$. Hence S is linearly independent. We next apply the Gram-Schmidt process to S. Let $\mathbf{Y}_1 = \mathbf{X}_1$ and

$$\mathbf{Y}_2 = \mathbf{X}_2 - \left[\frac{\mathbf{X}_2 \cdot \mathbf{Y}_1}{\mathbf{Y}_1 \cdot \mathbf{Y}_1}\right] \mathbf{Y}_1 = (2,0,1,0) - (2/2)(1,1,0,0)$$
$$= (1,-1,1,0)$$

$$\mathbf{Y}_3 = \mathbf{X}_3 - \left[\frac{\mathbf{X}_3 \cdot \mathbf{Y}_1}{\mathbf{Y}_1 \cdot \mathbf{Y}_1}\right] \mathbf{Y}_1 - \left[\frac{\mathbf{X}_3 \cdot \mathbf{Y}_2}{\mathbf{Y}_2 \cdot \mathbf{Y}_2}\right] \mathbf{Y}_2 = (-1,0,0,1)$$
$$- (-1/2)(1,1,0,0) - (1/3)(-1,0,0,1) = (-1/6,1/6,1/3,1)$$

$$\mathbf{Z}_1 = \frac{\mathbf{Y}_1}{\|\mathbf{Y}_1\|} = (1/\sqrt{2},1/\sqrt{2},0,0)$$

$$\mathbf{Z}_2 = \frac{\mathbf{Y}_2}{\|\mathbf{Y}_2\|} = (1/\sqrt{3},-1/\sqrt{3},1/\sqrt{3},0)$$

$$\mathbf{Z}_3 = \frac{\mathbf{Y}_3}{\|\mathbf{Y}_3\|} = (-1/\sqrt{42},1/\sqrt{42},2/\sqrt{42},6/\sqrt{42}).$$

19. Form the augmented matrix $\left[\begin{array}{ccc|c} 1 & 1 & -1 & 0 \\ 2 & 1 & 3 & 0 \\ 1 & 2 & -6 & 0 \end{array}\right]$ and find its reduced row echelon form. We obtain $\left[\begin{array}{ccc|c} 1 & 0 & 4 & 0 \\ 0 & 1 & -5 & 0 \\ 0 & 0 & 0 & 0 \end{array}\right]$ and it follows that the solution is $\begin{bmatrix} -4t \\ 5t \\ t \end{bmatrix}$ for any real number t.

Hence $\mathbf{X}_1 = \begin{bmatrix} -4 \\ 5 \\ 1 \end{bmatrix}$ is a basis for the solution space. To find an orthonormal basis we compute $\mathbf{X}_1/\|\mathbf{X}_1\| = (1/\sqrt{42})\begin{bmatrix}-4\\5\\1\end{bmatrix}$.

21. Let $\mathbf{Z}_1 = 1/\sqrt{5}(1,0,2)$, $\mathbf{Z}_2 = 1/\sqrt{5}(-2,0,1)$, $\mathbf{Z}_3 = (0,1,0)$, and $\mathbf{X} = (2,-3,1)$. Following the procedure in Example 5, we compute

$c_1 = \mathbf{X}\cdot\mathbf{Z}_1 = 4/\sqrt{5}$, $c_2 = \mathbf{X}\cdot\mathbf{Z}_2 = -3/\sqrt{5}$, $c_3 = \mathbf{X}\cdot\mathbf{Z}_3 = -3$.

Then $\mathbf{X} = 4/\sqrt{5}\ \mathbf{Z}_1 - 3/\sqrt{5}\ \mathbf{Z}_2 - 3\ \mathbf{Z}_3$.

23. Let $\mathbf{X}_1 = (1/\sqrt{2},0,0,-1/\sqrt{2})$, $\mathbf{X}_2 = (0,0,1,0)$, $\mathbf{X}_3 = (1/\sqrt{2},0,0,1/\sqrt{2})$, and $\mathbf{X} = (1,0,2,3)$. We follow the technique used in Example 6. Compute

$$\begin{aligned}\mathbf{Z} = \text{proj}_W\mathbf{X} &= (\mathbf{X}\cdot\mathbf{X}_1)\mathbf{X}_1 + (\mathbf{X}\cdot\mathbf{X}_2)\mathbf{X}_2 + (\mathbf{X}\cdot\mathbf{X}_3)\mathbf{X}_3\\ &= -2/\sqrt{2}\ \mathbf{X}_1 + 2\ \mathbf{X}_2 + 4/\sqrt{2}\ \mathbf{X}_3\\ &= (1,0,2,3)\end{aligned}$$

and

$$\mathbf{Y} = \mathbf{X} - \mathbf{Z} = (0,0,0,0).$$

This implies $\mathbf{X}$ is in W.

25. Let $\mathbf{X}_1 = (1/\sqrt{2},0,0,-1/\sqrt{2})$, $\mathbf{X}_2 = (0,0,1,0)$, $\mathbf{X}_3 = (1/\sqrt{2},0,0,1/\sqrt{2})$, and $\mathbf{X} = (1,2,-1,0)$. We follow the technique used in Example 6. Compute

$$\begin{aligned}\mathbf{Z} = \text{proj}_W\mathbf{X} &= (\mathbf{X}\cdot\mathbf{X}_1)\mathbf{X}_1 + (\mathbf{X}\cdot\mathbf{X}_2)\mathbf{X}_2 + (\mathbf{X}\cdot\mathbf{X}_3)\mathbf{X}_3\\ &= 1/\sqrt{2}\ \mathbf{X}_1 + (-1)\mathbf{X}_2 + 1/\sqrt{2}\ \mathbf{X}_3\\ &= (1,0,-1,0)\end{aligned}$$

and

$$\begin{aligned}\mathbf{Y} = \mathbf{X} - \mathbf{Z} &= (1,2,-1,0) - (1,0,-1,0)\\ &= (0,2,0,0).\end{aligned}$$

Then the distance from $\mathbf{X}$ to W is

$$\|\mathbf{Y}\| = 2.$$

T.1. $\mathbf{E}_i\cdot\mathbf{E}_j = \begin{cases}0 & \text{for } i \neq j\\ 1 & \text{for } i = j\end{cases}$.

T.3. By Corollary 3.6, an orthonormal set of n vectors is linearly independent. By Theorem 3.14(a) such a set is a basis for $\mathbf{R}^n$.

Exercises 3.10

T.5. If $\mathbf{X}$ is orthogonal to $S = \{\mathbf{Y}_1, \mathbf{Y}_2, \ldots, \mathbf{Y}_n\}$ then $\mathbf{X}\cdot\mathbf{Y}_j = 0$ for $j = 1, \ldots, n$. Let $\mathbf{Z}$ be in span S. Then $\mathbf{Z}$ is a linear combination of the vectors in S:

$$\mathbf{Z} = \sum_{j=1}^{n} c_j \mathbf{Y}_j.$$

Thus

$$\mathbf{X}\cdot\mathbf{Z} = \sum_{j=1}^{n} c_j (\mathbf{X}\cdot\mathbf{Y}_j) = \sum_{j=1}^{n} c_j \, 0 = 0.$$

Hence $\mathbf{X}$ is orthogonal to every vector in span S.

T.7. If $\mathbf{X}\cdot\mathbf{Y} = 0$, then $x_1y_1 + x_2y_2 + \cdots + x_ny_n = 0$. We have

$$\begin{aligned}(c\mathbf{X})\cdot\mathbf{Y} &= cx_1y_1 + cx_2y_2 + \cdots + cx_ny_n \\ &= c(x_1y_1 + x_2y_2 + \cdots + x_ny_n) = c(0) = 0.\end{aligned}$$

ML.1. Use the following MATLAB commands.

```
A = [1 1 0;1 0 1;0 0 1];
gschmidt(A)

ans =

    0.7071    0.7071         0
    0.7071   -0.7071         0
         0         0    1.0000
```

Write in the columns in terms of $\sqrt{2}$. Note that $\sqrt{2}/2 \approx 0.7071$.

ML.2. Use the following MATLAB commands.

```
A=[1 0 1 1;1 2 1 3;0 2 1 1;0 1 0 0]';
gschmidt(A)

ans =

    0.5774   -0.2582   -0.1690    0.7559
         0    0.7746    0.5071    0.3780
    0.5774   -0.2582    0.6761   -0.3780
    0.5774    0.5164   -0.5071   -0.3780
```

ML.3. To find the orthonormal basis we proceed as follows in MATLAB.

```
A=[0 -1 1;0 1 1;1 1 1]';
G=gschmidt(A)

G =

         0         0    1.0000
   -0.7071    0.7071         0
    0.7071    0.7071         0
```

To find the coordinates of each vector with respect to the orthonormal basis T, which is the columns of matrix G, we express each vector as a linear combination of the columns of G. It follows that $[\mathbf{X}]_T$ is the solution to the linear system **GZ** = **X**. We perform the solution of all three parts at the same time as follows.

```
coord = rref([G [1 2 0;1 1 1;-1 0 1]'])

coord =

1.0000         0         0   -1.4142         0    0.7071
     0    1.0000         0    1.4142    1.4142    0.7071
     0         0    1.0000    1.0000    1.0000   -1.0000
```

Columns 4, 5, and 6 are the solutions to parts a, b, and c respectively.

ML.4. We have that all vectors of the form (a,0,a+b,b+c) can be expressed as follows.

(a,0,a+b,b+c) = a(1,0,1,0) + b(0,0,1,1) + c(0,0,0,1)

By the same type of arguments used in Exercises 16-19 we find that

S = {X1, X2, X3} = {(1,0,1,0), (0,0,1,1), (0,0,0,1)}

is a basis for the subspace. Apply routine **gschmidt** to the vectors of S.

```
A=[1 0 1 0;0 0 1 1;0 0 0 1]';

gschmidt(A,1)

ans =

    1.0000   -0.5000    0.3333
         0         0         0
    1.0000    0.5000   -0.3333
         0    1.0000    0.3333
```

The columns are an orthogonal basis for the subspace.

ML.5. Proceed as follows in MATLAB.

```
A=[1 2 0;1 -1 1;0 0 0;1 0 1];
```

Find an orthonormal basis for W.

Exercises 3.10

```
G=gschmidt(A)

G =

   0.5774    0.7715   -0.2673
   0.5774   -0.6172   -0.5345
        0         0         0
   0.5774   -0.1543    0.8018
```

(a)

```
X=[0 0 1 1]';
proj=dot(G(:,1),X)*G(:,1) +
     dot(G(:,2),X)*G(:,2) + dot(G(:,3),X)*G(:,3)

proj =

   0.0000
        0
        0
   1.0000
```

(b) The distance from X to W is given by

```
norm(X-proj)

ans =

     1
```

Chapter 3 Supplementary Exercises

1. Let $\mathbf{X} = (1,-1,2,3)$ and $\mathbf{Y} = (2,3,1,-2)$.

(a) $\|\mathbf{X}\| = \sqrt{(1)^2 + (-1)^2 + (2)^2 + (3)^2} = \sqrt{15}$

(b) $\|\mathbf{Y}\| = \sqrt{(2)^2 + (3)^2 + (1)^2 + (-2)^2} = 3\sqrt{2}$

(c) $\|\mathbf{X} - \mathbf{Y}\| = \sqrt{(1-2)^2 + (-1-3)^2 + (2-1)^2 + (3-(-2))^2} = \sqrt{43}$

(d) $\mathbf{X}\cdot\mathbf{Y} = (1)(2) + (-1)(3) + (2)(1) + (3)(-2) = -5$

(e) $\cos(\theta) = \dfrac{\mathbf{X}\cdot\mathbf{Y}}{\|\mathbf{X}\|\ \|\mathbf{Y}\|} = \dfrac{-5}{3\sqrt{30}} = \dfrac{-\sqrt{5}}{3\sqrt{6}}$

3. $\|c(1,-2,2,0)\| = \sqrt{c^2(1^2 + (-2)^2 + 2^2 + 0^2)} = \sqrt{9c^2}$

$= 3|c| = 9$

Thus $|c| = 3$, hence $c = 3$ or $c = -3$.

5. Form the linear combination

$$c_1(1,2,1) + c_2(-1,1,2) + c_3(-3,-3,0) = (4,2,1).$$

Expanding, adding vectors, and equating corresponding components on both sides of the equation gives a

linear system with augmented matrix $\left[\begin{array}{ccc|c} 1 & -1 & -3 & 4 \\ 2 & 1 & -3 & 2 \\ 1 & 2 & 0 & 1 \end{array}\right]$.

Applying row operations $-2\mathbf{r}_1+\mathbf{r}_2$, $-\mathbf{r}_1+\mathbf{r}_3$, $-\mathbf{r}_2+\mathbf{r}_3$ gives an equivalent linear system with augmented matrix

$\left[\begin{array}{ccc|c} 1 & -1 & -3 & 4 \\ 0 & 3 & 3 & -6 \\ 0 & 0 & 0 & 3 \end{array}\right]$ which is inconsistent. Hence there are no

solutions for any choice of constants c_1, c_2, c_3; thus $(4,2,1)$ is not a linear combination of these vectors.

7. Let $S = \{\mathbf{X}_1, \mathbf{X}_2, \mathbf{X}_3\}$ where

$$\mathbf{X}_1 = t^2 + 2t + 2,\ \mathbf{X}_2 = 2t^2 + 3t + 1,\ \mathbf{X}_3 = -t - 3.$$

Form the linear combination $c_1\mathbf{X}_1 + c_2\mathbf{X}_2 + c_3\mathbf{X}_3 = \mathbf{0}$. Expanding, adding like terms, and equating corresponding coefficients of like powers of t from both sides of the equation we obtain a homogeneous linear system whose coefficient matrix is $\begin{bmatrix} 1 & 2 & 0 \\ 2 & 3 & -1 \\ 2 & 1 & -3 \end{bmatrix}$. Applying row operations $-2\mathbf{r}_1+\mathbf{r}_2$, $-2\mathbf{r}_1+\mathbf{r}_3$, $(-1)\mathbf{r}_2$, $3\mathbf{r}_2+\mathbf{r}_3$, $-2\mathbf{r}_2+\mathbf{r}_1$ gives the reduced row echelon form $\begin{bmatrix} 1 & 0 & -2 \\ 0 & 1 & 1 \\ 0 & 0 & 0 \end{bmatrix}$. Thus we have $c_1 = 2r$, $c_2 = -r$, $c_3 = r$, where r is any real number. It follows that S is a linearly dependent set. For r = 1, we have $2\mathbf{X}_1 - \mathbf{X}_2 + \mathbf{X}_3 = \mathbf{0}$ which gives $\mathbf{X}_3 = -2\mathbf{X}_1 + \mathbf{X}_2$.

9. Let W be the subspace of $\mathbf{R}^4$ of all vectors of the form

$$(a+b, b+c, a-b-2c, b+c).$$

Then

$$(a+b, b+c, a-b-2c, b+c) = a(1,0,1,0) + b(1,1,-1,1) + c(0,1,-2,1).$$

Let $\mathbf{X}_1 = (1,0,1,0)$, $\mathbf{X}_2 = (1,1,-1,1)$, $\mathbf{X}_3 = (0,1,-2,1)$, and $S = \{\mathbf{X}_1, \mathbf{X}_2, \mathbf{X}_3\}$. From the preceding equation we have that span S = W. Thus we need only determine if S is linearly independent. Form the linear combination $c_1\mathbf{X}_1 + c_2\mathbf{X}_2 + c_3\mathbf{X}_3 = \mathbf{0}$, expand, add vectors, and equate corresponding components from both sides of the equation to obtain a linear system with coefficient matrix $\begin{bmatrix} 1 & 1 & 0 \\ 0 & 1 & 1 \\ 1 & -1 & -2 \\ 0 & 1 & 1 \end{bmatrix}$. Applying row operations $-\mathbf{r}_1+\mathbf{r}_3$, $2\mathbf{r}_2+\mathbf{r}_3$, $-\mathbf{r}_2+\mathbf{r}_4$, $-\mathbf{r}_2+\mathbf{r}_1$ we have the reduced row echelon form $\begin{bmatrix} 1 & 0 & -1 \\ 0 & 1 & 1 \\ 0 & 0 & 0 \\ 0 & 0 & 0 \end{bmatrix}$. The solution is $c_1 = r$, $c_2 = -r$, $c_3 = r$, where r is any real number. Thus there are nontrivial solutions, hence S is linearly dependent. Let r = 1. Then $\mathbf{X}_1 - \mathbf{X}_2 + \mathbf{X}_3 = \mathbf{0}$ and it follows that $\mathbf{X}_3 = \mathbf{X}_2 - \mathbf{X}_1$. We drop $\mathbf{X}_3$ from S and hence $S_1 = \{\mathbf{X}_1, \mathbf{X}_2\}$ spans W. S_1 is linearly independent since $\mathbf{X}_1$ is not a multiple of $\mathbf{X}_2$. We have that S_1 is a basis for W and dim W = 2.

11. The coefficient matrix of the augmented matrix is
$\begin{bmatrix} 1 & 2 & -1 & 1 & 2 \\ 1 & 1 & 2 & -3 & 1 \end{bmatrix}$. Applying row operations $-\mathbf{r}_1+\mathbf{r}_2$,
$-1\mathbf{r}_2$, $-2\mathbf{r}_2+\mathbf{r}_1$, gives reduced row echelon form
$\begin{bmatrix} 1 & 0 & 5 & -7 & 0 \\ 0 & 1 & -3 & 4 & 1 \end{bmatrix}$. Thus the solution is given by
$x_1 = -5r + 7s$, $x_2 = 3r - 4s - t$, $x_3 = r$, $x_4 = s$, $x_5 = t$, where r, s, and t are any real numbers. Then the solution vector is

$$\mathbf{X} = \begin{bmatrix} -5r+7s \\ 3r-4s-t \\ r \\ s \\ t \end{bmatrix} = r\begin{bmatrix} -5 \\ 3 \\ 1 \\ 0 \\ 0 \end{bmatrix} + s\begin{bmatrix} 7 \\ -4 \\ 0 \\ 1 \\ 0 \end{bmatrix} + t\begin{bmatrix} 0 \\ -1 \\ 0 \\ 0 \\ 1 \end{bmatrix}.$$

It follows that $\left\{\begin{bmatrix} -5 \\ 3 \\ 1 \\ 0 \\ 0 \end{bmatrix}, \begin{bmatrix} 7 \\ -4 \\ 0 \\ 1 \\ 0 \end{bmatrix}, \begin{bmatrix} 0 \\ -1 \\ 0 \\ 0 \\ 1 \end{bmatrix}\right\}$ is a basis for the solution space which is of dimension 3.

13. Let $S = \{\mathbf{X}_1, \mathbf{X}_2\}$ where $\mathbf{X}_1 = t + 3$ and $\mathbf{X}_2 = 2t + \lambda^2 + 2$. Form the linear combination $c_1\mathbf{X}_1 + c_2\mathbf{X}_2 = \mathbf{0}$, expand, add like powered terms, and equate coefficients of like powers of t from both sides of the equation to obtain the homogeneous linear system with coefficient matrix $\begin{bmatrix} 1 & 2 \\ 3 & \lambda^2+2 \end{bmatrix}$. Applying row operation $-3\mathbf{r}_1+\mathbf{r}_2$ gives an equivalent linear system with coefficient matrix $\begin{bmatrix} 1 & 2 \\ 0 & \lambda^2-4 \end{bmatrix}$. This system has only the zero solution provided $\lambda^2 - 4 \neq 0$. Hence S will be linearly independent for all values of λ excepts $\lambda = 2$ or $\lambda = -2$.

15. Vector $(a^2,a,1)$ is in span $\{(1,2,3), (1,1,1), (0,1,2)\}$ provided there exist constants c_1, c_2, c_3 such that

$$c_1(1,2,3) + c_2(1,1,1) + c_3(0,1,2) = (a^2,a,1).$$

Expanding, adding vectors, and equating corresponding components gives the linear system

$$\begin{aligned} c_1 + c_2 \quad\quad &= a^2 \\ 2c_1 + c_2 + \ \ c_3 &= a \\ 3c_1 + c_2 + 2c_3 &= 1. \end{aligned}$$

Form the augmented matrix $\left[\begin{array}{ccc|c} 1 & 1 & 0 & a^2 \\ 2 & 1 & 1 & a \\ 3 & 1 & 2 & 1 \end{array}\right]$ and use row operations $-2\mathbf{r}_1+\mathbf{r}_2$, $-3\mathbf{r}_1+\mathbf{r}_3$, $-\mathbf{r}_2+\mathbf{r}_3$ to obtain an equivalent linear system represented by

$$\left[\begin{array}{ccc|c} 1 & 1 & 0 & a^2 \\ 0 & -1 & 1 & a-2a^2 \\ 0 & 0 & 0 & a^2-2a+1 \end{array}\right].$$

This system is consistent provided $a^2 - 2a + 1 = (a-1)^2 = 0$. That is, provided $a = 1$.

17. Let $\mathbf{Y}$ be in span $\{\mathbf{X}_1, \mathbf{X}_2, \mathbf{X}_3\}$. Then there exist constants c_1, c_2, c_3 such that

$$\mathbf{Y} = c_1\mathbf{X}_1 + c_2\mathbf{X}_2 + c_3\mathbf{X}_3.$$

$\mathbf{Y}$ is in span $\{\mathbf{X}_1+\mathbf{X}_2, \mathbf{X}_1-\mathbf{X}_3, \mathbf{X}+\mathbf{X}_3\}$ provided we can find constants k_1, k_2, k_3 such that

$$\begin{aligned} \mathbf{Y} = c_1\mathbf{X}_1 + c_2\mathbf{X}_2 + c_3\mathbf{X}_3 &= k_1(\mathbf{X}_1+\mathbf{X}_2) + k_2(\mathbf{X}_1-\mathbf{X}_3) + k_3(\mathbf{X}_1+\mathbf{X}_3) \\ &= (k_1+k_2+k_3)\mathbf{X}_1 + k_1\mathbf{X}_2 + (-k_2+k_3)\mathbf{X}_3. \end{aligned}$$

Equating coefficients of corresponding vectors gives the linear system

$$\begin{aligned} k_1 + k_2 + k_3 &= c_1 \\ k_1 \quad\quad\quad\quad &= c_2 \\ - k_2 + k_3 &= c_3. \end{aligned}$$

Form the augmented matrix

$$\left[\begin{array}{ccc|c} 1 & 1 & 1 & c_1 \\ 1 & 0 & 0 & c_2 \\ 0 & -1 & 1 & c_3 \end{array}\right].$$

and use row operations to obtain its reduced row echelon form which is

$$\left[\begin{array}{ccc|c} 1 & 0 & 0 & c_2 \\ 0 & 1 & 0 & (c_1-c_2-c_3)/2 \\ 0 & 0 & 1 & (c_1-c_2+c_3)/2 \end{array}\right].$$

Since the linear system is consistent, $\mathbf{Y}$ is in span $\{\mathbf{X}_1+\mathbf{X}_2, \mathbf{X}_1-\mathbf{X}_3, \mathbf{X}_1+\mathbf{X}_3\}$.

On the other hand, let $\mathbf{Z}$ be in span $\{\mathbf{X}_1+\mathbf{X}_2, \mathbf{X}_1-\mathbf{X}_3, \mathbf{X}_1+\mathbf{X}_3\}$. Then there exist constants k_1, k_2, k_3 such that

$$\mathbf{Z} = k_1(\mathbf{X}_1+\mathbf{X}_2) + k_2(\mathbf{X}_1-\mathbf{X}_3) + k_3(\mathbf{X}_1+\mathbf{X}_3).$$

However,

$$\mathbf{Z} = (k_1+k_2+k_3)\mathbf{X}_1 + k_1\mathbf{X}_2 + (-k_2+k_3)\mathbf{X}_3,$$

hence $\mathbf{Z}$ is in span$\{\mathbf{X}_1, \mathbf{X}_2, \mathbf{X}_3\}$.

19. We check closure of vector addition and scalar multiplication.
(a) Let $\mathbf{X}_1 = (x_1, mx_1 + b)$ and $\mathbf{X}_2 = (x_2, mx_2 + b)$. Then

$$\mathbf{X}_1 + \mathbf{X}_2 = (x_1+x_2, m(x_1+x_2) + 2b)$$

is of the form $(x, mx + b)$ provided $b = 0$. For any real scalar c,

$$c\mathbf{X}_1 = (cx_1, cmx_1 + cb) = (cx_1, cmx_1).$$

(Assuming $b = 0$.) Thus $c\mathbf{X}_1$ is of the form $(x, mx + b) = (x, mx)$ for all values of m. It follows that $\{(x, mx + b)\}$ is a subspace of $\mathbf{R}^2$ for $b = 0$ and m any real number.

(b) Let $\mathbf{X}_1 = (x_1, rx_1^2)$ and $\mathbf{X}_2 = (x_2, rx_2^2)$. Then

$$\mathbf{X}_1 + \mathbf{X}_2 = (x_1+x_2, r(x_1^2+x_2^2))$$

is of the form (x, rx^2) provided

$$r(x_1^2 + x_2^2) = r(x_1+x_2)^2 = r(x_1^2 + 2x_1x_2 + x_2^2).$$

Thus, $2rx_1x_2$ must be zero for all x_1 and x_2. Hence $r = 0$. For any real scalar c,

$$c\mathbf{X}_1 = (cx_1, crx_1^2) = (cx_1, 0)$$

has the appropriate form for $r = 0$. It follows that $\{(x, rx^2)\}$ is a subspace of $\mathbf{R}^2$ for $r = 0$.

21. Let $S = \{(1,0,0), (0,1,0), (0,0,1)\}$. If $W = \mathbf{R}^3$, then W contains S. On the other hand, if W contains S, then W contains span $S = \mathbf{R}^3$.

23. (a) The verification of Definition 1 in Section 3.4 follows from the properties of continuous functions and real numbers. In particular, in calculus it is shown that the sum of continuous functions is continuous and a real number times a continuous function is again a continuous function. This verifies (α) and (β) of Definition 1. We demonstrate that (a) and (e) hold and (b), (c), (d), (f), (g), (h) are shown in a similar way. To show (a), let f and g belong to C[a,b] and for t in [a,b]

$$(f \oplus g)(t) = f(t) + g(t) = g(t) + f(t) = (g \oplus f)(t)$$

since f(t) and g(t) are real numbers and the addition of real numbers is commutative. To show (e), let c be any real number. Then

$$\begin{aligned} c \odot (f \oplus g)(t) &= c(f(t) + g(t)) = cf(t) + cg(t) \\ &= c \odot f(t) + c \odot g(t) = (c\odot f \oplus c\odot g)(t) \end{aligned}$$

since c, f(t), and g(t) are real numbers and multiplication of real numbers distributes over addition of real numbers.

(b) Let f and g be in W(k) and c be any real scalar. Then

$$(f + g)(a) = f(a) + g(a) = k + k = 2k.$$

Hence (f + g)(a) = k provided k = 0. Also,,

$$(c \,.\, f)(a) = cf(a) = ck.$$

Thus (c .) f)(a) = k for arbitrary c only if k = 0. W(k) is a subspace of C[a,b] for k = 0.

(c) Let f and g have roots at t_i, $i = 1,2,\ldots,n$; that is, $f(t_i) = g(t_i) = 0$. It follows that f + g has roots at the t_i since $(f + g)(t_i) = f(t_i) + g(t_i) = 0 + 0 = 0$. Similarly k . f has roots at the t_i since $(k \,.\, f)(t_i) = kf(t_i) = k\cdot 0 = 0$.

25. The row reduced echelon form of the matrix is

$$\begin{bmatrix} 1 & 0 & 0 & 2/13 & 15/13 \\ 0 & 1 & 0 & 17/13 & -9/13 \\ 0 & 0 & 1 & -3/13 & -16/13 \\ 0 & 0 & 0 & 0 & 0 \\ 0 & 0 & 0 & 0 & 0 \end{bmatrix}.$$ Hence the rank is 3.

27. S = {(1,0,-1), (0,1,1), (0 0 1)} and T = {(1,0,0), (0,1,-1), (1,-1,2)} are bases for R^3. Let **X** = (2,3,5).

(a) To find $[\mathbf{x}]_T$ we solve a linear system obtained from the expression $c_1(1,0,0) + c_2(0,1,-1) + c_3(1,-1,2) = (2,3,5)$ by finding the reduced row echelon form of the associated augmented matrix. The reduced row echelon form is

$$\left[\begin{array}{ccc|c} 1 & 0 & 0 & -6 \\ 0 & 1 & 0 & 11 \\ 0 & 0 & 1 & 8 \end{array}\right]$$

It follows that $[\mathbf{x}]_T = \begin{bmatrix} -6 \\ 11 \\ 8 \end{bmatrix}$.

(b) To find $[\mathbf{x}]_S$ we solve a linear system obtained from the expression $c_1(1,0,-1) + c_2(0,1,1) + c_3(0,0,1) = (2,3,5)$ by finding the reduced row echelon form of the associated augmented matrix. The reduced row echelon form is

$$\left[\begin{array}{ccc|c} 1 & 0 & 0 & 2 \\ 0 & 1 & 0 & 3 \\ 0 & 0 & 1 & 4 \end{array}\right]$$

It follows that $[\mathbf{x}]_S = \begin{bmatrix} 2 \\ 3 \\ 4 \end{bmatrix}$.

(c) Following the procedure used in Exercise 13 in Section 3.9 we find the coordinates of the vectors of the members of T in terms of the vectors in S. Writing out the expressions leads to three systems of equations with the same coefficient matrix and the right sides are the vectors in T written as columns. We write these three systems in partioned form and find the reduced row echelon form of the associated matrix. Specifically,

rref of $\left[\begin{array}{ccc|ccc} 1 & 0 & 0 & 1 & 0 & 1 \\ 0 & 1 & 0 & 0 & 1 & -1 \\ -1 & 1 & 1 & 0 & -1 & 2 \end{array}\right]$ is

$$\left[\begin{array}{ccc|ccc} 1 & 0 & 0 & 1 & 0 & 1 \\ 0 & 1 & 0 & 0 & 1 & -1 \\ 0 & 0 & 1 & 1 & -2 & 4 \end{array}\right]$$

Hence $\mathbf{P} = \begin{bmatrix} 1 & 0 & 1 \\ 0 & 1 & -1 \\ 1 & -2 & 4 \end{bmatrix}$.

(d) $[\mathbf{X}]_S = \mathbf{P}[\mathbf{X}]_T = \begin{bmatrix} 2 \\ 3 \\ 4 \end{bmatrix}$

(e) Reversing the procedure in part c, we find

$$\mathbf{Q} = \mathbf{P}^{-1} = \begin{bmatrix} 2 & -2 & -1 \\ -1 & 3 & 1 \\ -1 & 2 & 1 \end{bmatrix}$$

(f) $[\mathbf{X}]_T = \mathbf{Q}[\mathbf{X}]_S = \begin{bmatrix} -6 \\ 11 \\ 8 \end{bmatrix}$

29. $\mathbf{X}\cdot\mathbf{Y} = 4a - 6 + 2a = 0$ only if $a = 1$.

31. Let $\mathbf{X} = (1,-2,1)$ and $\mathbf{Y} = (y_1, y_2, y_3)$. Then

$$\mathbf{X}\cdot\mathbf{Y} = y_1 - 2y_2 + y_3 = 0$$

gives one equation in three unknowns. Hence two unknowns can be chosen arbitrarily. Let $y_2 = r$ and $y_3 = s$ where r and s are any real numbers. Then the solution is $y_1 = 2r - s$, $y_2 = r$, $y_3 = s$. Thus

$$\mathbf{Y} = (2r-s,\ r,\ s) = r(2,1,0) + s(-1,0,1)$$

and $S = \{(2,1,0),\ (-1,0,1)\}$ is a basis for the subspace of vectors in $\mathbf{R}^3$ that are orthogonal to $\mathbf{X}$.

33. Let $S = \{\mathbf{X}_1, \mathbf{X}_2, \mathbf{X}_3\}$ where $\mathbf{X}_1 = 1/\sqrt{2}\ (1,0,-1)$, $\mathbf{X}_2 = (0,1,0)$, $\mathbf{X}_3 = 1/\sqrt{2}\ (1,0,1)$. Let $\mathbf{X} = (1,2,3)$. Then $\mathbf{X} = c_1\mathbf{X}_1 + c_2\mathbf{X}_2 + c_3\mathbf{X}_3$ where

$$c_1 = \mathbf{X}\cdot\mathbf{X}_1 = -\sqrt{2}, \quad c_2 = \mathbf{X}\cdot\mathbf{X}_2 = 2, \quad c_3 = \mathbf{X}\cdot\mathbf{X}_3 = 2\sqrt{2}.$$

Thus $\mathbf{X} = -\sqrt{2}\ \mathbf{X}_1 + 2\ \mathbf{X}_2 + 2\sqrt{2}\ \mathbf{X}_3$.

35. Let $W = \text{span}\{\mathbf{X}_1, \mathbf{X}_2\} = \text{span}\{(1,0,1),\ (0,1,0)\}$.

(a) Let V be the orthogonal complement of W. If $\mathbf{Y} = (y_1, y_2, y_3)$ is in V, then $\mathbf{Y}\cdot\mathbf{X}_1 = 0$ and $\mathbf{Y}\cdot\mathbf{X}_2 = 0$. Thus we have the linear system

$$\begin{aligned} y_1 \qquad + y_3 &= 0 \\ y_2 \qquad &= 0 \end{aligned}$$

which has solution $y_1 = -r$, $y_2 = 0$, $y_3 = r$, where r is any real number. Hence all vectors in V have the form

$(-r,0,r) = r(-1,0,1)$. It follows that $\{(-1,0,1)\}$ is a basis for V.

(b) We show that $T = \{(1,0,1), (0,1,0), (-1,0,1)\}$ is linearly independent; then Theorem 3.14 implies that T is a basis for $\mathbf{R}^3$. Form

$$c_1(1,0,1) + c_2(0,1,0) + c_3(-1,0,1) = (0,0,0)$$

and construct the coefficient matrix of the corresponding homogeneous linear system. We obtain

$$\begin{bmatrix} 1 & 0 & -1 \\ 0 & 1 & 0 \\ 1 & 0 & 1 \end{bmatrix}$$

which has reduced row echelon form $\mathbf{I}_3$. Hence the only solution is the trivial solution $c_1 = c_2 = c_3 = 0$. Thus T is linearly independent.

(c) Use Theorem 3.21. Let $S = (\mathbf{X}_1, \mathbf{X}_2\} = \{(1,0,1), (0,1,0)\}$. S is an orthogonal basis for W, so we normalize the vectors to obtain an orthonormal basis. Let

$$\mathbf{X}_1' = \mathbf{X}_1/\|\mathbf{X}_1\| = 1/\sqrt{2}\ \mathbf{X}_1 \text{ and } \mathbf{X}_2' = \mathbf{X}_2/\|\mathbf{X}_2\| = \mathbf{X}_2.$$

(i) Let $\mathbf{X} = (1,0,0)$. From Equation (7) in Section 3.8

$$\begin{aligned}\mathbf{Z} &= (\mathbf{X}\cdot\mathbf{X}_1')\mathbf{X}_1' + (\mathbf{X}\cdot\mathbf{X}_2')\mathbf{X}_2' = 1/\sqrt{2}\ \mathbf{X}_1' + 0\mathbf{X}_2' \\ &= (1/2)(1,0,1).\end{aligned}$$

Then

$$\begin{aligned}\mathbf{Y} &= \mathbf{X} - \mathbf{Z} = (1,0,0) - (1/2)(1,0,1) \\ &= (1/2,\ 0,\ -1/2).\end{aligned}$$

Hence

$$\mathbf{X} = (1,0,0) = (1/2,0,-1/2) + (1/2,0,1/2) = \mathbf{Y} + \mathbf{Z}.$$

(ii) Let $\mathbf{X} = (1,2,3)$. From Equation (7) in Section 3.8

$$\begin{aligned}\mathbf{Z} &= (\mathbf{X}\cdot\mathbf{X}_1')\mathbf{X}_1' + (\mathbf{X}\cdot\mathbf{X}_2')\mathbf{X}_2' = 4/\sqrt{2}\ \mathbf{X}_1' + \mathbf{2X}_2' \\ &= (2,2,2).\end{aligned}$$

Then

$$\mathbf{Y} = \mathbf{X} - \mathbf{Z} = (1,2,3) - (2,2,2) = (-1,0,1).$$

Hence

$$\mathbf{X} = (1,2,3) = (-1,0,1) + (2,2,2) = \mathbf{Y} + \mathbf{Z}.$$

T.1. If $\mathbf{A}$ is nonsingular then $\mathbf{AX} = \mathbf{0}$ has only the trivial solution. Thus the dimension of the solution space is zero. Conversely, if $\mathbf{AX} = \mathbf{0}$ has solution space of dimension zero, then $\mathbf{X} = \mathbf{0}$ is the only solution. Thus $\mathbf{A}$ is nonsingular.

T.3. If $\mathbf{X}\cdot\mathbf{Y} = 0$, then

$$\|\mathbf{X} + \mathbf{Y}\| = \sqrt{(\mathbf{X} + \mathbf{Y})\cdot(\mathbf{X} + \mathbf{Y})} = \sqrt{\mathbf{X}\cdot\mathbf{X} + 2\mathbf{X}\cdot\mathbf{Y} + \mathbf{Y}\cdot\mathbf{Y}}$$
$$= \sqrt{\mathbf{X}\cdot\mathbf{X} + \mathbf{Y}\cdot\mathbf{Y}}$$

and

$$\|\mathbf{X} - \mathbf{Y}\| = \sqrt{(\mathbf{X} - \mathbf{Y})\cdot(\mathbf{X} - \mathbf{Y})} = \sqrt{\mathbf{X}\cdot\mathbf{X} - 2\mathbf{X}\cdot\mathbf{Y} + \mathbf{Y}\cdot\mathbf{Y}}$$
$$= \sqrt{\mathbf{X}\cdot\mathbf{X} + \mathbf{Y}\cdot\mathbf{Y}}$$

hence $\|\mathbf{X} + \mathbf{Y}\| = \|\mathbf{X} - \mathbf{Y}\|$.

On the other hand, if

$$\|\mathbf{X} + \mathbf{Y}\| = \|\mathbf{X} - \mathbf{Y}\|,$$

then

$$\|\mathbf{X} + \mathbf{Y}\|^2 = \mathbf{X}\cdot\mathbf{X} + 2\mathbf{X}\cdot\mathbf{Y} + \mathbf{Y}\cdot\mathbf{Y} = \mathbf{X}\cdot\mathbf{X} - 2\mathbf{X}\cdot\mathbf{Y} + \mathbf{Y}\cdot\mathbf{Y} = \|\mathbf{X} - \mathbf{Y}\|^2.$$

Simplifying we have $2\mathbf{X}\cdot\mathbf{Y} = -2\mathbf{X}\cdot\mathbf{Y}$, hence $\mathbf{X}\cdot\mathbf{Y} = 0$.

T.5. rank $\mathbf{A}$ = row rank $\mathbf{A}$ = column rank $\mathbf{A}^T$ = rank $\mathbf{A}^T$. (See Theorem 3.16.)

T.7. (a) From the definition of a matrix product, the rows of $\mathbf{AB}$ are linear combinations of the rows of $\mathbf{B}$. Hence, the row space of $\mathbf{AB}$ is a subspace of the row space of $\mathbf{B}$ and it follows that rank $\mathbf{AB} \le$ rank $\mathbf{B}$. From Exercise T.5, rank $\mathbf{AB} \le$ rank $(\mathbf{AB})^T$ = rank $\mathbf{B}^T\mathbf{A}^T$. A similar argument shows rank $\mathbf{AB} \le$ rank $\mathbf{A}^T$ = rank $\mathbf{A}$. It follows that rank $\mathbf{AB} \le \min\{\text{rank } \mathbf{A}, \text{rank } \mathbf{B}\}$.

(b) One such pair of matrices is $\mathbf{A} = \begin{bmatrix} 1 & 0 \\ 0 & 0 \end{bmatrix}$ and

$$\mathbf{B} = \begin{bmatrix} 0 & 0 \\ 0 & 1 \end{bmatrix}.$$

(c) Since $\mathbf{A} = (\mathbf{AB})\mathbf{B}^{-1}$, by part (a), rank $\mathbf{A} \le$ rank $\mathbf{AB}$. But part (a) also implies that rank $\mathbf{AB} \le$ rank $\mathbf{A}$, so rank $\mathbf{AB}$ = rank $\mathbf{A}$.

(d) Since $\mathbf{B} = \mathbf{A}^{-1}(\mathbf{AB})$, by part (a), rank $\mathbf{B} \le$ rank $\mathbf{AB}$. But, part (a) also implies that rank $\mathbf{AB} \le$ rank $\mathbf{B}$, thus rank $\mathbf{AB}$ = rank $\mathbf{B}$.

(e) rank $\mathbf{PAQ}$ = rank $\mathbf{A}$

Exercises 4.1

<<**Strategy:** In Exercises 1 and 3, we follow the techniques in Examples 1 and 2 to determine if **L** is a linear transformation. >>

1. (a) Let $\mathbf{X} = (x_1,y_1)$ and $\mathbf{Y} = (x_2,y_2)$. Then

$$\mathbf{L}(\mathbf{X} + \mathbf{Y}) = \mathbf{L}((x_1,y_1) + (x_2,y_2))$$
$$= \mathbf{L}(x_1+x_2,y_1+y_2) = (x_1+x_2+1,y_1+y_2,x_1+x_2+y_1+y_2).$$

On the other hand

$$\mathbf{L}(\mathbf{X}) + \mathbf{L}(\mathbf{Y}) = \mathbf{L}(x_1,y_1) + \mathbf{L}(x_2,y_2)$$
$$= (x_1+1,y_1,x_1+y_1) + (x_2+1,y_2,x_2+y_2)$$
$$= (x_1+x_2+2,y_1+y_2,x_1+x_2+y_1+y_2).$$

However, $\mathbf{L}(\mathbf{X} + \mathbf{Y}) \neq \mathbf{L}(\mathbf{X}) + \mathbf{L}(\mathbf{Y})$. Thus **L** is not a linear transformation.

(b) Let $\mathbf{X} = \begin{bmatrix} x_1 \\ y_1 \\ z_1 \end{bmatrix}$ and $\mathbf{Y} = \begin{bmatrix} x_2 \\ y_2 \\ z_2 \end{bmatrix}$. Then

$$\mathbf{L}(\mathbf{X} + \mathbf{Y}) = \mathbf{L}\left(\begin{bmatrix} x_1+x_2 \\ y_1+y_2 \\ z_1+z_2 \end{bmatrix}\right) = \begin{bmatrix} x_1+x_2+y_1+y_2 \\ y_1+y_2 \\ x_1+x_2-z_1-z_2 \end{bmatrix}.$$ On the other hand

$$\mathbf{L}(\mathbf{X}) + \mathbf{L}(\mathbf{Y}) = \begin{bmatrix} x_1+y_1 \\ y_1 \\ x_1-z_1 \end{bmatrix} + \begin{bmatrix} x_2+y_2 \\ y_2 \\ x_2-z_2 \end{bmatrix} = \begin{bmatrix} x_1+x_2+y_1+y_2 \\ y_1+y_2 \\ x_1+x_2-z_1-z_2 \end{bmatrix}.$$ Thus

$\mathbf{L}(\mathbf{X} + \mathbf{Y}) = \mathbf{L}(\mathbf{X}) + \mathbf{L}(\mathbf{Y})$. Next, let c be any real number.

Then $$\mathbf{L}(c\mathbf{X}) = \mathbf{L}\left(\begin{bmatrix} cx_1 \\ cy_1 \\ cz_1 \end{bmatrix}\right) = \begin{bmatrix} cx_1+cy_1 \\ cy_1 \\ cx_1-cz_1 \end{bmatrix} = \begin{bmatrix} c(x_1+y_1) \\ cy_1 \\ c(x_1-z_1) \end{bmatrix} = c\begin{bmatrix} x_1+y_1 \\ y_1 \\ x_1-z_1 \end{bmatrix}$$

$= c\mathbf{L}(\mathbf{X})$. Thus **L** is a linear transformation.

(c) Let $\mathbf{X} = (x_1,y_1)$ and $\mathbf{Y} = (x_2,y_2)$. Then

$$\mathbf{L}(\mathbf{X} + \mathbf{Y}) = \mathbf{L}((x_1,y_1) + (x_2,y_2))$$
$$= \mathbf{L}(x_1+x_2,y_1+y_2) = ((x_1+x_2)^2+x_1+x_2,y_1+y_2-(y_1+y_2)^2)$$
$$= (x_1^2+2x_1x_2+x_2^2+x_1+x_2,y_1+y_2-y_1^2-2y_1y_2-y_2^2).$$

On the other hand

$$\mathbf{L}(\mathbf{X}) + \mathbf{L}(\mathbf{Y}) = \mathbf{L}(x_1,y_1) + \mathbf{L}(x_2,y_2)$$
$$= (x_1^2+x_1,y_1-y_1^2) + (x_2^2+x_2,y_2-y_2^2)$$
$$= (x_1^2+x_1+x_2^2+x_2,y_1-y_1^2+y_2-y_2^2).$$

However, $\mathbf{L}(\mathbf{X} + \mathbf{Y}) \neq \mathbf{L}(\mathbf{X}) + \mathbf{L}(\mathbf{Y})$. Thus $\mathbf{L}$ is not a linear transformation.

3. (a) Let $\mathbf{X} = (x_1,y_1,z_1)$ and $\mathbf{Y} = (x_2,y_2,z_2)$. Then

$$\mathbf{L}(\mathbf{X} + \mathbf{Y}) = \mathbf{L}((x_1,y_1,z_1) + (x_2,y_2,z_2))$$
$$= \mathbf{L}(x_1+x_2,y_1+y_2,z_1+z_2) = (x_1+x_2+y_1+y_2,0,2x_1+2x_2-z_1-z_2).$$

On the other hand

$$\mathbf{L}(\mathbf{X}) + \mathbf{L}(\mathbf{Y}) = \mathbf{L}(x_1,y_1,z_1) + \mathbf{L}(x_2,y_2,z_2)$$
$$= (x_1+y_1,0,2x_1-z_1) + (x_2+y_2,0,2x_2-z_2)$$
$$= (x_1+x_2+y_1+y_2,0,2x_1+2x_2-z_1-z_2).$$

Thus $\mathbf{L}(\mathbf{X} + \mathbf{Y}) = \mathbf{L}(\mathbf{X}) + \mathbf{L}(\mathbf{Y})$. Next let c be any real number. Then

$$\mathbf{L}(c\mathbf{X}) = \mathbf{L}(cx_1,cy_1,cz_1) = (cx_1+cy_1,0,2cx_1-cz_1)$$
$$= (c(x_1+y_1),0,c(2x_1-z_1)) = c(x_1+y_1,0,2x_1-z_1) = c\mathbf{L}(\mathbf{X}).$$

Hence $\mathbf{L}$ is a linear transformation.

(b) Let $\mathbf{X} = \begin{bmatrix} x_1 \\ y_1 \end{bmatrix}$ and $\mathbf{Y} = \begin{bmatrix} x_2 \\ y_2 \end{bmatrix}$. Then

$$\mathbf{L}(\mathbf{X} + \mathbf{Y}) = \mathbf{L}\left(\begin{bmatrix} x_1+x_2 \\ y_1+y_2 \end{bmatrix}\right) = \begin{bmatrix} (x_1+x_2)^2-(y_1+y_2)^2 \\ (x_1+x_2)^2+(y_1+y_2)^2 \end{bmatrix}$$
$$= \begin{bmatrix} x_1^2+2x_1x_2+x_2^2-y_1^2-2y_1y_2-y_2^2 \\ x_1^2+2x_1x_2+x_2^2+y_1^2+2y_1y_2+y_2^2 \end{bmatrix}.$$

On the other hand

$$\mathbf{L}(\mathbf{X}) + \mathbf{L}(\mathbf{Y}) = \begin{bmatrix} x_1^2-y_1^2 \\ x_1^2+y_1^2 \end{bmatrix} + \begin{bmatrix} x_2^2-y_2^2 \\ x_2^2+y_2^2 \end{bmatrix} = \begin{bmatrix} x_1^2+x_2^2-y_1^2-y_2^2 \\ x_1^2+x_2^2+y_1^2+y_2^2 \end{bmatrix}.$$

However, $\mathbf{L}(\mathbf{X} + \mathbf{Y}) \neq \mathbf{L}(\mathbf{X}) + \mathbf{L}(\mathbf{Y})$. Thus $\mathbf{L}$ is not a linear transformation.

(c) Let $\mathbf{X} = (x_1,y_1)$ and $\mathbf{Y} = (x_2,y_2)$. Then

$$\mathbf{L}(\mathbf{X} + \mathbf{Y}) = \mathbf{L}((x_1,y_1) + (x_2,y_2))$$
$$= \mathbf{L}(x_1+x_2,\ y_1+y_2) = (x_1+x_2-(y_1+y_2),0,2(x_1+x_2)+3)$$
$$= (x_1+x_2-y_1-y_2,0,2x_1+2x_2+3).$$

On the other hand

$$\mathbf{L}(\mathbf{X}) + \mathbf{L}(\mathbf{Y}) = \mathbf{L}(x_1,y_1) + \mathbf{L}(x_2,y_2)$$
$$= (x_1-y_1,0,2x_1+3) + (x_2-y_2,0,2x_2+3)$$
$$=(x_1+x_2-y_1-y_2,0,2x_1+2x_2+6).$$

However, $\mathbf{L}(\mathbf{X} + \mathbf{Y}) \neq \mathbf{L}(\mathbf{X}) + \mathbf{L}(\mathbf{Y})$. Thus $\mathbf{L}$ is not a linear transformation.

5. Let $p(t) = a_1t^2 + b_1t + c_1$ and $q(t) = a_2t^2 + b_2t + c_2$.

(a) $$\mathbf{L}(p(t) + q(t)) = \mathbf{L}((a_1+a_2)t^2 + (b_1+b_2)t + (c_1+c_2))$$
$$= (a_1+a_2)t + (b_1+ b_2) + 1$$

$$\mathbf{L}(p(t)) + \mathbf{L}(q(t)) = (a_1t + b_1 + 1) + (a_2t + b_2 + 1)$$
$$= (a_1+a_2)t + (b_1+b_2) + 2$$

Hence $L(p(t) + q(t)) \neq \mathbf{L}(p(t)) + \mathbf{L}(q(t))$, so $\mathbf{L}$ is not a linear transformation. (It can also be shown that $\mathbf{L}(kp(t)) \neq k\mathbf{L}(p(t))$.)

(b) $$\mathbf{L}(p(t) + q(t)) = \mathbf{L}((a_1+a_2)t^2 + (b_1+b_2)t + (c_1+c_2))$$
$$= 2(a_1+a_2)t - (b_1+ b_2)$$
$$= (2a_1t - b_1) + (2a_2t - b_2)$$
$$= \mathbf{L}(p(t)) + \mathbf{L}(q(t))$$

$$\mathbf{L}(kp(t)) = \mathbf{L}(ka_1t^2 + kb_1t + kc_1) = 2ka_1t - kb_1$$
$$= k(2a_1t - b_1) = k\mathbf{L}(p(t)$$

Hence $\mathbf{L}$ is a linear transformation.

(c) For variety we start with the scalar multiple property here.

$$\mathbf{L}(kp(t)) = \mathbf{L}(ka_1t^2 + kb_1t + kc_1)$$
$$= (ka_1 + 2)t + kb_1 - ka_1$$
$$= ka_1t + 2t + kb_1 - ka_1$$

Exercises 4.1

$$kL(p(t)) = k[(a_1 + 2)t + (b_1 - a_1)]$$
$$= ka_1t + 2kt + kb_1 - ka_1$$

Hence $\mathbf{L}(kp(t)) \neq k\mathbf{L}(p(t))$, so $\mathbf{L}$ is not a linear transformation. (It is also the case that $\mathbf{L}(p(t) + q(t)) \neq \mathbf{L}(p(t)) + \mathbf{L}(q(t))$.)

7. Let $\mathbf{A}$ and $\mathbf{B}$ be in M_{nn}. We have

$$\mathbf{L}(\mathbf{A} + \mathbf{B}) = \mathbf{C}(\mathbf{A} + \mathbf{B}) = \mathbf{CA} + \mathbf{CB} = \mathbf{L}(\mathbf{A}) + \mathbf{L}(\mathbf{B})$$

and

$$\mathbf{L}(k\mathbf{A}) = \mathbf{C}(k\mathbf{A}) = k(\mathbf{CA}) = k\mathbf{L}(\mathbf{A}).$$

It follows that $\mathbf{L}$ is a linear transformation.

9. Let $\mathbf{X}_1 = \begin{bmatrix} 1 \\ 1 \end{bmatrix}$, $\mathbf{X}_2 = \begin{bmatrix} 0 \\ 1 \end{bmatrix}$, and $S = \{\mathbf{X}_1, \mathbf{X}_2\}$. If S is a basis for $\mathbf{R}^2$, then for any $\mathbf{X}$ in $\mathbf{R}^2$ we can use Theorem 4.3 to compute $\mathbf{L}(\mathbf{X})$. The two nonzero vectors are linearly independent since they are not scalar multiples of one another. Since the dimension of $\mathbf{R}^2$ is two, Theorem 3.14(a) implies that S is a basis for $\mathbf{R}^2$. Hence $\mathbf{L}(\mathbf{X})$ is completely determined by

$\mathbf{L}(\mathbf{X}_1) = \begin{bmatrix} 2 \\ -3 \end{bmatrix}$ and $\mathbf{L}(\mathbf{X}_2) = \begin{bmatrix} 1 \\ 2 \end{bmatrix}$.

(a) It is easy to show that $\begin{bmatrix} 3 \\ -2 \end{bmatrix} = 3\mathbf{X}_1 - 5\mathbf{X}_2$. Thus $\mathbf{L}\left(\begin{bmatrix} 3 \\ -2 \end{bmatrix}\right) =$

$$\mathbf{L}(3\mathbf{X}_1 - 5\mathbf{X}_2) = 3\mathbf{L}(\mathbf{X}_1) - 5\mathbf{L}(\mathbf{X}_2) = 3\begin{bmatrix} 2 \\ -3 \end{bmatrix} - 5\begin{bmatrix} 1 \\ 2 \end{bmatrix} = \begin{bmatrix} 1 \\ -19 \end{bmatrix}.$$

(b) It is easy to show that $\begin{bmatrix} a \\ b \end{bmatrix} = a\mathbf{X}_1 + (b-a)\mathbf{X}_2$. Thus $\mathbf{L}\left(\begin{bmatrix} a \\ b \end{bmatrix}\right) =$

$$\mathbf{L}(a\mathbf{X}_1 + (b-a)\mathbf{X}_2) = a\mathbf{L}(\mathbf{X}_1) + (b-a)\mathbf{L}(\mathbf{X}_2) = a\begin{bmatrix} 2 \\ -3 \end{bmatrix} + (b-a)\begin{bmatrix} 1 \\ 2 \end{bmatrix}$$

$$= \begin{bmatrix} a+b \\ -5a+2b \end{bmatrix}.$$

11. (a) Let **X** and **Y** be any vectors in $\mathbf{R}^3$ and k any real number. From Example 3, **L**(**X**) = r**X** where r is a real number. Then,

$$\mathbf{L}(\mathbf{X} + \mathbf{Y}) = r(\mathbf{X} + \mathbf{Y}) = r\mathbf{X} + r\mathbf{Y} = \mathbf{L}(\mathbf{X}) + \mathbf{L}(\mathbf{Y})$$

and

$$\mathbf{L}(k\mathbf{X}) = r(k\mathbf{X}) = k(r\mathbf{X}) = k\mathbf{L}(\mathbf{X}).$$

Hence **L** is a linear transformation.

(b) Let **X** and **Y** be any vectors in $\mathbf{R}^2$, k be any real number,

and $\mathbf{A} = \begin{bmatrix} 1 & 0 \\ 0 & 1 \\ 1 & -1 \end{bmatrix}$. From Example 4, **L**(**X**) = **AX**. Then,

$$\mathbf{L}(\mathbf{X} + \mathbf{Y}) = \mathbf{A}(\mathbf{X} + \mathbf{Y}) = \mathbf{AX} + \mathbf{AY} = \mathbf{L}(\mathbf{X}) + \mathbf{L}(\mathbf{Y})$$

and

$$\mathbf{L}(k\mathbf{X}) = \mathbf{A}(k\mathbf{X}) = k(\mathbf{AX}) = k\mathbf{L}(\mathbf{X}).$$

Thus **L** is a linear transformation.

(c) Let **X** = (x_1,y_1) and **Y** = (x_2,y_2). From Example 9, **L**(**X**) = $(x_1,-y_1)$. Then,

$$\mathbf{L}(\mathbf{X} + \mathbf{Y}) = \mathbf{L}(x_1+x_2,y_1+y_2) = (x_1+x_2,-y_1-y_2)$$

$$= (x_1,-y_1) + (x_2,-y_2) = \mathbf{L}(\mathbf{X}) + \mathbf{L}(\mathbf{Y})$$

and

$$\mathbf{L}(k\mathbf{X}) = \mathbf{L}(kx_1,ky_1) = (kx_1,-ky_1) = k(x_1,-y_1) = k\mathbf{L}(\mathbf{X}).$$

Thus **L** is a linear transformation.

13. Let f and g be differentiable functions and k be any real scalar. From Example 11 we have **L**(f) = f′, the derivative of f. We assume that the properties of derivatives from calculus are familiar. Then

$$\mathbf{L}(f + g) = (f + g)' = f' + g' = \mathbf{L}(f) + \mathbf{L}(g)$$

and

$$\mathbf{L}(kf) = (kf)' = kf' = k\mathbf{L}(f).$$

Thus **L** is a linear transformation.

Exercises 4.1

15. (a) Following the technique of Example 8, we determine the numbers associated with the letters in the message.

$$\begin{matrix} S & E & N & D & & H & I & M & & M & O & N & E & Y \\ \updownarrow & \updownarrow & \updownarrow & \updownarrow & & \updownarrow & \updownarrow & \updownarrow & & \updownarrow & \updownarrow & \updownarrow & \updownarrow & \updownarrow \\ 19 & 5 & 14 & 4 & & 8 & 9 & 13 & & 13 & 15 & 14 & 5 & 25 \end{matrix}$$

We break this string of numbers into four vectors in $\mathbf{R}^3$ and multiply each vector by the matrix $\mathbf{A} = \begin{bmatrix} 1 & 2 & 3 \\ 1 & 1 & 2 \\ 0 & 1 & 2 \end{bmatrix}$.

We obtain

$$\mathbf{A}\begin{bmatrix} 19 \\ 5 \\ 14 \end{bmatrix} = \begin{bmatrix} 71 \\ 52 \\ 33 \end{bmatrix}, \mathbf{A}\begin{bmatrix} 4 \\ 8 \\ 9 \end{bmatrix} = \begin{bmatrix} 47 \\ 30 \\ 26 \end{bmatrix}, \mathbf{A}\begin{bmatrix} 13 \\ 13 \\ 15 \end{bmatrix} = \begin{bmatrix} 84 \\ 56 \\ 43 \end{bmatrix}, \mathbf{A}\begin{bmatrix} 14 \\ 5 \\ 25 \end{bmatrix} = \begin{bmatrix} 99 \\ 69 \\ 55 \end{bmatrix}.$$

The final version of the code message is the string of numbers

71 52 33 47 30 26 84 56 43 99 69 55.

(b) We are given the coded message

67 44 41 49 39 19 113 76 62 104 69 55.

Breaking this into vectors in $\mathbf{R}^3$, the decoded vector is obtained by multiplying each of the vectors by

$\mathbf{A}^{-1} = \begin{bmatrix} 0 & 1 & -1 \\ 2 & -2 & -1 \\ -1 & 1 & 1 \end{bmatrix}$. We obtain $\mathbf{A}^{-1}\begin{bmatrix} 67 \\ 44 \\ 41 \end{bmatrix} = \begin{bmatrix} 3 \\ 5 \\ 18 \end{bmatrix}$,

$$\mathbf{A}^{-1}\begin{bmatrix} 49 \\ 39 \\ 19 \end{bmatrix} = \begin{bmatrix} 20 \\ 1 \\ 9 \end{bmatrix}, \quad \mathbf{A}^{-1}\begin{bmatrix} 113 \\ 76 \\ 62 \end{bmatrix} = \begin{bmatrix} 14 \\ 12 \\ 25 \end{bmatrix}, \quad \mathbf{A}^{-1}\begin{bmatrix} 104 \\ 69 \\ 55 \end{bmatrix} = \begin{bmatrix} 14 \\ 15 \\ 20 \end{bmatrix}.$$

Then the message is

$$\begin{matrix} 3 & 5 & 18 & 20 & 1 & 9 & 14 & 12 & 25 & 14 & 15 & 20 \\ \updownarrow & \updownarrow & \updownarrow & \updownarrow & \updownarrow & \updownarrow & \updownarrow & \updownarrow & \updownarrow & \updownarrow & \updownarrow & \updownarrow \\ C & E & R & T & A & I & N & L & Y & N & O & T \end{matrix}.$$

17. $\mathbf{L}:V \to R^n$ is the function that associates to vector in V its coordinates with respect to the S-basis; $\mathbf{L}(\mathbf{X}) = [\mathbf{X}]_S$. We find $[\mathbf{X}]_S$ by solving the linear equation

$$c_1\mathbf{Y}_1 + c_2\mathbf{Y}_2 + \cdots + c_n\mathbf{Y}_n = \mathbf{X}$$

Let $\mathbf{A} = [\mathbf{Y}_1\ \mathbf{Y}_2\ \ldots\ \mathbf{Y}_n]$ and $\mathbf{c} = [c_1\ c_2\ \ldots\ c_n]^T$, then the preceding equation can be expressed in matrix form as

$$\mathbf{Ac} = \mathbf{X}$$

and it follows that

$$\left[\mathbf{X}\right]_S = \mathbf{c} = \mathbf{A}^{-1}\mathbf{X} = \mathbf{L}(\mathbf{X})$$

Let $\mathbf{X}$ and $\mathbf{Y}$ be in V. Then

$$\begin{aligned}\mathbf{L}(\mathbf{X} + \mathbf{Y}) &= \mathbf{A}^{-1}(\mathbf{X} + \mathbf{Y}) = \mathbf{A}^{-1}\mathbf{X} + \mathbf{A}^{-1}\mathbf{Y} \\ &= \left[\mathbf{X}\right]_S + \left[\mathbf{Y}\right]_S = \mathbf{L}(\mathbf{X}) + \mathbf{L}(\mathbf{Y})\end{aligned}$$

$$\mathbf{L}(k\mathbf{X}) = \mathbf{A}^{-1}(k\mathbf{X}) = k(\mathbf{A}^{-1}\mathbf{X}) = k\mathbf{L}(\mathbf{X})$$

Hence $\mathbf{L}$ is a linear transformation.

T.1. $\mathbf{L}(\mathbf{X} + \mathbf{Y}) = \mathbf{A}(\mathbf{X} + \mathbf{Y}) = \mathbf{AX} + \mathbf{AY} = \mathbf{L}(\mathbf{X}) + \mathbf{L}(\mathbf{Y})$ and $\mathbf{L}(c\mathbf{X}) = \mathbf{A}(c\mathbf{X}) = c(\mathbf{AX}) = c\mathbf{L}(\mathbf{X})$.

T.3. $\mathbf{L}(\mathbf{X} - \mathbf{Y}) = \mathbf{L}(\mathbf{X} + (-1)\mathbf{Y}) = \mathbf{L}(\mathbf{X}) + \mathbf{L}((-1)\mathbf{Y}) = \mathbf{L}(\mathbf{X}) - \mathbf{L}(\mathbf{Y})$.

T.5. Let $\mathbf{A}$ and $\mathbf{B}$ be $n \times n$ matrices and be c any scalar. We have $\mathbf{L}(\mathbf{A}) = \sum_{i=1}^{n} a_{ii}$. Then

$$\mathbf{L}(\mathbf{A} + \mathbf{B}) = \sum_{i=1}^{n}(a_{ii} + b_{ii}) = \sum_{i=1}^{n} a_{ii} + \sum_{i=1}^{n} b_{ii}$$

$$= \mathbf{L}(\mathbf{A}) + \mathbf{L}(\mathbf{B})$$

and

$$\mathbf{L}(c\mathbf{A}) = \sum_{i=1}^{n}(ca_{ii}) = c\sum_{i=1}^{n} a_{ii} = c\mathbf{L}(\mathbf{A}).$$

Thus $\mathbf{L}$ is a linear transformation.

T.7. No. We show by example that $\mathbf{L}(\mathbf{A} + \mathbf{B})$ need not be equal to $\mathbf{L}(\mathbf{A}) + \mathbf{L}(\mathbf{B})$. Let $n = 2$, $\mathbf{A} = \begin{bmatrix} 1 & 0 \\ 0 & 0 \end{bmatrix}$, $\mathbf{B} = \begin{bmatrix} 0 & 0 \\ 0 & 1 \end{bmatrix}$. Then

$$\mathbf{L}(\mathbf{A} + \mathbf{B}) = \mathbf{L}(\mathbf{I}_2) = \mathbf{I}_2$$

but

$$\mathbf{L}(\mathbf{A}) + \mathbf{L}(\mathbf{B}) = \mathbf{0} + \mathbf{0} \neq \mathbf{I}_2.$$

Exercises 4.1

ML.1. (a)
```
A=[1 0;0 0];B=[0 0;0 1];
det(A+B)

ans =

      1

det(A)+det(B)

ans =

     0
```

(b)
```
A=eye(3);B=-ones(3);
det(A+B)

ans =

     -2

det(A)+det(B)

ans =

      1
```

ML.2. (a)
```
A=eye(2);B=eye(2);
rank(A+B)

ans =

      2

rank(A)+rank(B)

ans =

      4
```

(b) Repeat the preceding with A=eye(3) and B=eye(3).

ML.3. (a)
```
v=[1 2]';w=[0 3]';
norm(v+w)

ans =

      5.0990

norm(v)+norm(w)

ans =

      5.2361
```

(b)
```
v=[1 2 3]';w=[6 0 1]';
norm(v+w)

ans =

      8.3066

norm(v)+norm(w)

ans =

      9.8244
```

Exercises 4.2

1. Let $\mathbf{L}:\mathbf{R}^2 \rightarrow \mathbf{R}^2$ be the linear transformation defined by $\mathbf{L}(a_1,a_2) = (a_1,0)$.

   (a) $\mathbf{L}(0,2) = (0,0)$, thus $(0,2)$ is in ker $\mathbf{L}$.

   (b) $\mathbf{L}(2,2) = (2,0)$, thus $(2,2)$ is not in ker $\mathbf{L}$.

   (c) Since $\mathbf{L}(3,0) = (3,0)$, $(3,0)$ is in range $\mathbf{L}$.

   (d) Vector $(3,2)$ is not in range $\mathbf{L}$ since it is not of the form $(a_1,0)$.

   (e) Vector (x_1,x_2) is in ker $\mathbf{L}$ provided $\mathbf{L}(x_1,x_2) = (0,0)$. Since $\mathbf{L}(x_1,x_2) = (x_1,0)$ it follows that (x_1,x_2) is in ker $\mathbf{L}$ only if $x_1 = 0$. Thus ker $\mathbf{L} = \{(0,r) \mid r$ is any real number$\}$.

   (f) Range $\mathbf{L} = \{(r,0) \mid r$ is any real number$\}$.

3. Let $\mathbf{L}:\mathbf{R}^2 \rightarrow \mathbf{R}^3$ be defined by $\mathbf{L}(x,y) = (x,x+y,y)$.

   (a) ker $\mathbf{L}$ is the set of all vectors whose image is $(0,0,0)$. Thus we set $\mathbf{L}(x,y) = (x,x+y,y) = (0,0,0)$. Equating corresponding components gives $x = y = 0$, and it follows that ker $\mathbf{L} = \{(0,0)\}$.

   (b) From (a), ker $\mathbf{L} = \{\mathbf{0}\}$ hence Theorem 4.5 implies that $\mathbf{L}$ is one-to-one.

   (c) Let $\mathbf{Y} = (y_1,y_2,y_3)$. To determine if $\mathbf{L}$ is onto, we ask if there is a vector $\mathbf{X} = (x_1,x_2)$ such that $\mathbf{L}(\mathbf{X}) = \mathbf{Y}$ for an arbitrary vector $\mathbf{Y}$. We have $\mathbf{L}(\mathbf{X}) = (x_1,x_1+x_2,x_2) = (y_1,y_2,y_3)$. Equating corresponding components gives a

   linear system with augmented matrix $\left[\begin{array}{cc|c} 1 & 0 & y_1 \\ 1 & 1 & y_2 \\ 0 & 1 & y_3 \end{array}\right]$ which

   has reduced row echelon form

   $\left[\begin{array}{cc|c} 1 & 0 & y_1 \\ 0 & 1 & y_3 \\ 0 & 0 & y_2-y_1-y_3 \end{array}\right]$. The system is

   inconsistent unless $y_2-y_1-y_3 = 0$, hence $\mathbf{L}$ is not onto.

5. Let $\mathbf{A} = \begin{bmatrix} 1 & 0 & -1 & 3 & -1 \\ 1 & 0 & 0 & 2 & -1 \\ 2 & 0 & -1 & 5 & -1 \\ 0 & 0 & -1 & 1 & 0 \end{bmatrix}$ and $\mathbf{L}(\mathbf{X}) = \mathbf{AX}$ for $\mathbf{X}$ in $\mathbf{R}^5$.

(a) $\mathbf{X}$ is in ker $\mathbf{L}$ provided $\mathbf{AX} = \mathbf{0}$. Thus we have a homogeneous linear system with coefficient matrix $\mathbf{A}$. The reduced row echelon form of $\mathbf{A}$ is $\begin{bmatrix} 1 & 0 & 0 & 2 & 0 \\ 0 & 0 & 1 & -1 & 0 \\ 0 & 0 & 0 & 0 & 1 \\ 0 & 0 & 0 & 0 & 0 \end{bmatrix}$. Thus the solution is $x_1 = -2r$, $x_2 = s$, $x_3 = r$, $x_4 = r$, $x_5 = 0$, where r and s are any real numbers. In vector form the solution is

$$\begin{bmatrix} -2r \\ s \\ r \\ r \\ 0 \end{bmatrix} = r\begin{bmatrix} -2 \\ 0 \\ 1 \\ 1 \\ 0 \end{bmatrix} + s\begin{bmatrix} 0 \\ 1 \\ 0 \\ 0 \\ 0 \end{bmatrix}.$$

Thus $\left\{\begin{bmatrix} -2 \\ 0 \\ 1 \\ 1 \\ 0 \end{bmatrix}, \begin{bmatrix} 0 \\ 1 \\ 0 \\ 0 \\ 0 \end{bmatrix}\right\}$ is a basis for ker $\mathbf{L}$.

(b) Let $S = \{\mathbf{X}_1, \mathbf{X}_2, \mathbf{X}_3, \mathbf{X}_4, \mathbf{X}_5\}$ where $\mathbf{X}_j$ = column j of $\mathbf{A}$. Then span S = range $\mathbf{L}$. However S is linearly dependent since $\mathbf{X}_2$ is the zero vector. Thus we can drop $\mathbf{X}_2$ and $S_1 = \{\mathbf{X}_1, \mathbf{X}_3, \mathbf{X}_4, \mathbf{X}_5\}$ spans range $\mathbf{L}$. Following Example 4 of Section 3.7, we form a matrix with rows corresponding to the vectors in S_1:

$$\begin{bmatrix} 1 & 1 & 2 & 0 \\ -1 & 0 & -1 & -1 \\ 3 & 2 & 5 & 1 \\ -1 & -1 & -1 & 0 \end{bmatrix}.$$

The reduced row echelon form is

$$\begin{bmatrix} 1 & 0 & 0 & 1 \\ 0 & 1 & 0 & -1 \\ 0 & 0 & 1 & 0 \\ 0 & 0 & 0 & 0 \end{bmatrix}.$$

The nonzero rows written as columns are

$$\left\{\begin{bmatrix}1\\0\\0\\1\end{bmatrix}, \begin{bmatrix}0\\1\\0\\-1\end{bmatrix}, \begin{bmatrix}0\\0\\1\\0\end{bmatrix}\right\}$$ and these form a basis for range of **L**.

Alternatively, a linearly independent subset of S can be determined using the method of Example 5 section 3.6. The reduced row echelon form of A is $\begin{bmatrix}1 & 0 & 0 & 2 & 0\\0 & 0 & 1 & -1 & 0\\0 & 0 & 0 & 0 & 1\\0 & 0 & 0 & 0 & 0\end{bmatrix}$. Since the leading 1's are in columns 1, 3, and 5 $\{\mathbf{X}_1, \mathbf{X}_3, \mathbf{X}_5\}$ =

$$\left\{\begin{bmatrix}1\\-1\\3\\-1\end{bmatrix}, \begin{bmatrix}1\\0\\2\\-1\end{bmatrix}, \begin{bmatrix}0\\-1\\1\\0\end{bmatrix}\right\}$$ is a basis for span S.

(c) $\dim(\ker \mathbf{L}) = 2$, $\dim(\text{range } \mathbf{L}) = 3$, $\dim \mathbf{R}^5 = 5$, thus

$$\dim(\ker \mathbf{L}) + \dim(\text{range } \mathbf{L}) = \dim \mathbf{R}^5.$$

7. Let $\mathbf{L}:\mathbf{R}^4 \to \mathbf{R}^3$ be defined by $\mathbf{L}\left(\begin{bmatrix}x\\y\\z\\w\end{bmatrix}\right) = \begin{bmatrix}x+y\\y-z\\z-w\end{bmatrix}$.

(a) For arbitrary real numbers y_1, y_2, y_3 we determine if there is a vector in $\mathbf{R}^4$ whose image is $\begin{bmatrix}y_1\\y_2\\y_3\end{bmatrix}$. We have $\begin{bmatrix}x+y\\y-z\\z-w\end{bmatrix}$ $= \begin{bmatrix}y_1\\y_2\\y_3\end{bmatrix}$. Equating corresponding components gives a linear system with augmented matrix $\left[\begin{array}{cccc|c}1 & 1 & 0 & 0 & y_1\\0 & 1 & -1 & 0 & y_2\\0 & 0 & 1 & -1 & y_3\end{array}\right]$. The reduced row echelon form is

$$\left[\begin{array}{cccc|c} 1 & 0 & 0 & 1 & y_1-y_2-y_3 \\ 0 & 1 & 0 & -1 & y_2+y_3 \\ 0 & 0 & 1 & -1 & y_3 \end{array}\right].$$

The system is consistent for any values of y_1, y_2, y_3, hence **L** is onto.

(b) From (a) **L** is onto, thus range **L** = $\mathbf{R}^3$. Then from Theorem 4.7 we have dim(ker **L**) = dim $\mathbf{R}^4$ - dim(range **L**) = 4 - 3 = 1.

(c) To verify Theorem 4.7, we find dim(ker **L**) directly by finding a basis for ker **L**. Vector **X** in $\mathbf{R}^4$ is in ker **L**

provided $\mathbf{L}(\mathbf{X}) = \begin{bmatrix} x+y \\ y-z \\ z-w \end{bmatrix} = \begin{bmatrix} 0 \\ 0 \\ 0 \end{bmatrix} = \mathbf{0}$. Equating corresponding

components gives a homogeneous linear system with

coefficient matrix $\begin{bmatrix} 1 & 1 & 0 & 0 \\ 0 & 1 & -1 & 0 \\ 0 & 0 & 1 & -1 \end{bmatrix}$ whose reduced row echelon

form is $\begin{bmatrix} 1 & 0 & 0 & 1 \\ 0 & 1 & 0 & -1 \\ 0 & 0 & 1 & -1 \end{bmatrix}$. The solution is x = -r, y = r,

z = r, w = r, where r is any real number. Thus every

vector in ker **L** is of the form $\begin{bmatrix} -r \\ r \\ r \\ r \end{bmatrix} = r\begin{bmatrix} -1 \\ 1 \\ 1 \\ 1 \end{bmatrix}$ and dim(ker **L**)

= 1. Theorem 4.7 follows.

9. (a) Let **L**(x,y) = (x+y,y). Then $\mathbf{L}:\mathbf{R}^2 \to \mathbf{R}^2$. We determine ker **L** and range **L**. Set **L**(x,y) = (x+y,y) = (0,0). Equating corresponding components gives the homogeneous linear system

$$\begin{aligned} x + y &= 0 \\ y &= 0 \end{aligned}$$

whose only solution is x = y = 0. Thus ker **L** = {(0,0)} and dim(ker **L**) = 0. By Theorem 4.5, L is one-to-one and by Corollary 4.1, **L** is onto. Thus range **L** = $\mathbf{R}^2$ and dim(range **L**) = 2. Hence dim(ker **L**) + dim(range **L**) = 0 + 2 = dim($\mathbf{R}^2$).

Exercises 4.2

(b) Let $\mathbf{L}(\mathbf{X}) = \mathbf{L}\left(\begin{bmatrix} x \\ y \\ z \end{bmatrix}\right) = \mathbf{AX} = \begin{bmatrix} 4 & -1 & -1 \\ 2 & 2 & 3 \\ 2 & -3 & -4 \end{bmatrix}\begin{bmatrix} x \\ y \\ z \end{bmatrix}$. Then ker $\mathbf{L}$ is the solution space of $\mathbf{AX} = \mathbf{0}$. The reduced row echelon form of $\mathbf{A}$ is

$$\begin{bmatrix} 1 & 0 & 1/10 \\ 0 & 1 & 7/5 \\ 0 & 0 & 0 \end{bmatrix}$$

and it follows that the solution is $x = (-1/10)r$, $y = (-7/5)r$, $z = r$, where r is any real number. Hence ker $\mathbf{L}$ is the set of all vectors of the form $r\begin{bmatrix} -1/10 \\ -7/5 \\ 1 \end{bmatrix}$ and dim(ker $\mathbf{L}$) = 1. Next we determine a basis for range $\mathbf{L}$.

$$\mathbf{L}(\mathbf{X}) = x\begin{bmatrix} 4 \\ 2 \\ 2 \end{bmatrix} + y\begin{bmatrix} -1 \\ 2 \\ -3 \end{bmatrix} + z\begin{bmatrix} -1 \\ 3 \\ -4 \end{bmatrix}$$

Let $S = \left\{\begin{bmatrix} 4 \\ 2 \\ 2 \end{bmatrix}, \begin{bmatrix} -1 \\ 2 \\ -3 \end{bmatrix}, \begin{bmatrix} -1 \\ 3 \\ -4 \end{bmatrix}\right\}$. Then span S = range $\mathbf{L}$.

Following Example 4 of Section 3.7 we form the matrix with rows that are the vectors in S: $\begin{bmatrix} 4 & 2 & 2 \\ -1 & 2 & -3 \\ -1 & 3 & -4 \end{bmatrix}$. The reduced row echelon form of this matrix is $\begin{bmatrix} 1 & 0 & 1 \\ 0 & 1 & -1 \\ 0 & 0 & 0 \end{bmatrix}$. The nonzero rows written as columns form a basis for range $\mathbf{L}$, thus dim(range $\mathbf{L}$) = 2. Hence dim(ker $\mathbf{L}$) + dim(range $\mathbf{L}$) = 1 + 2 = 3 = dim($\mathbf{R}^3$).

(c) Let $\mathbf{L}(x,y,z) = (x+y-z,x+y,y+z)$. To find a basis for ker $\mathbf{L}$, set $\mathbf{L}(x,y,z) = (x+y-z,x+y,y+z) = (0,0,0)$ and equate corresponding components. The resulting homogeneous system is

$$\begin{aligned} x + y - z &= 0 \\ x + y \phantom{{}- z} &= 0 \\ y + z &= 0. \end{aligned}$$

The reduced row echelon form of the coefficient matrix is $\mathbf{I}_3$, hence the only solution is $x = y = z = 0$. Thus ker $\mathbf{L}$ = $\{\mathbf{0}\}$ and dim(ker $\mathbf{L}$) = 0. Then $\mathbf{L}$ is one-to-one by Theorem 4.5 and Corollary 4.1 implies that $\mathbf{L}$ is onto. It follows that range $\mathbf{L} = \mathbf{R}^3$ and dim(range $\mathbf{L}$) = 3. We have dim(ker $\mathbf{L}$) + dim(range $\mathbf{L}$) = 0 + 3 = 3 = dim($\mathbf{R}^3$).

11. Let $\mathbf{L}:\mathbf{P}_2 \to \mathbf{P}_2$ be defined by

$$\mathbf{L}(at^2 + bt + c) = (a + c)t^2 + (b + c)t.$$

(a) $\mathbf{L}(t^2 - t - 1) = 0t^2 - 2t \neq 0$, thus $t^2 - t - 1$ is not in ker $\mathbf{L}$.

(b) $\mathbf{L}(t^2 + t - 1) = 0t^2 + 0t = 0$, thus $t^2 + t - 1$ is in ker $\mathbf{L}$.

(c) Set $\mathbf{L}(at^2 + bt + c) = (a + c)t^2 + (b + c)t = 2t^2 - t$. Equating coefficients of like powers of t on both sides of the equation gives the linear system

$$\begin{aligned} a \phantom{{}+ b} + c &= 2 \\ b + c &= -1. \end{aligned}$$

Form the corresponding augmented matrix $\left[\begin{array}{ccc|c} 1 & 0 & 1 & 2 \\ 0 & 1 & 1 & -1 \end{array}\right]$.

It is already in reduced row echelon form, thus the system is consistent and $2t^2 - t$ is in range $\mathbf{L}$.

(d) Set $\mathbf{L}(at^2 + bt + c) = (a + c)t^2 + (b + c)t = t^2 - t - 2$. Equating coefficients of like powers of t on both sides of the equation gives the linear system

$$\begin{aligned} a \phantom{{}+ b} + c &= 1 \\ b + c &= -1 \\ 0 &= -2. \end{aligned}$$

The system is inconsistent, thus $t^2 - t - 2$ is not in range $\mathbf{L}$.

(e) Set $\mathbf{L}(at^2 + bt + c) = (a + c)t^2 + (b + c)t = 0t^2 + 0t + 0$. Equating coefficients of like powers of t on both sides of the equation gives the linear system

$$\begin{aligned} a \phantom{{}+ b} + c &= 0 \\ b + c &= 0 \end{aligned}$$

whose solution is $a = -r$, $b = -r$, $c = r$, where r is any real number. Thus ker **L** is the set of all elements in $\mathbf{P}_2$ of the form $-rt^2 - rt + r$ and $\{-t^2 - t + 1\}$ is a basis.

(f) Let $S = \{t^2, t\}$. From the definition of **L** we see that span S = range **L**. Since S is linearly independent, it is a basis for range **L**.

13. (a) From $\mathbf{L}\left(\begin{bmatrix} a & b \\ c & d \end{bmatrix}\right) = \begin{bmatrix} a+b & b+c \\ a+d & b+d \end{bmatrix} = \begin{bmatrix} 0 & 0 \\ 0 & 0 \end{bmatrix}$ it follows that

$$\begin{aligned} a + b &= 0 \\ b + c &= 0 \\ a + d &= 0 \\ b + d &= 0 \end{aligned}$$

This system of equations has augmented matrix

$$\left[\begin{array}{cccc|c} 1 & 1 & 0 & 0 & 0 \\ 0 & 1 & 1 & 0 & 0 \\ 1 & 0 & 0 & 1 & 0 \\ 0 & 1 & 0 & 1 & 0 \end{array}\right]$$

whose reduced row echelon form is $[\mathbf{I}_4 \mid \mathbf{0}]$. Thus $a = b = c = d = 0$ and hence the kernel consists only of the zero matrix.

(b) $$\mathbf{L}\left(\begin{bmatrix} a & b \\ c & d \end{bmatrix}\right) = \begin{bmatrix} a+b & b+c \\ a+d & b+d \end{bmatrix} = a\begin{bmatrix} 1 & 0 \\ 1 & 0 \end{bmatrix} + b\begin{bmatrix} 1 & 1 \\ 0 & 1 \end{bmatrix} + c\begin{bmatrix} 0 & 1 \\ 0 & 0 \end{bmatrix} + d\begin{bmatrix} 0 & 0 \\ 1 & 1 \end{bmatrix}$$

Hence range **L** is spanned by the four 2×2 matrices in the preceding expression. Since $\dim M_{22} = 4$, Theorem 3.14b implies these four matrices are linearly independent. Thus

$$\left\{ \begin{bmatrix} 1 & 0 \\ 1 & 0 \end{bmatrix}, \begin{bmatrix} 1 & 1 \\ 0 & 1 \end{bmatrix}, \begin{bmatrix} 0 & 1 \\ 0 & 0 \end{bmatrix}, \begin{bmatrix} 0 & 0 \\ 1 & 1 \end{bmatrix} \right\}$$

is a basis for range **L**.

15. Let $\mathbf{L}:\mathbf{P}_2 \to \mathbf{P}_1$ be defined by

$$\mathbf{L}(p(t)) = \mathbf{L}(at^2 + bt + c) = p'(t) = 2at + b.$$

(a) Set $\mathbf{L}(p(t)) = 2at + b = \mathbf{0}$. Equating coeffients of like powers of t from both sides of the equation gives $a = 0$, $b = 0$. Hence, c can be chosen arbitrarily. Let $c = r$, any real number. Then ker **L** is the set of all polynomials of the form $0t^2 + 0t + r$, that is, the set of all constant polynomials. Thus {1} is a basis for ker **L**.

(b) From the definition of **L** we have that span S = span {t,1} = range **L**. Since S is linearly independent, {t, 1} is a basis for range **L**.

17. Let $\mathbf{L}:\mathbf{R}^4 \to \mathbf{R}^6$ be a linear transformation.

(a) If dim(ker **L**) = 2, then from Theorem 4.7
dim(range **L**) = dim $\mathbf{R}^4$ - dim(ker **L**) = 4 - 2 = 2.

(b) If dim(range **L**) = 3, then from Theorem 4.7
dim(ker **L**) = dim $\mathbf{R}^4$ - dim(range **L**) = 4 - 3 = 1.

T.1. By Theorem 4.7 and the assumption that dim V = dim W, we have dim(ker **L**) + dim(range **L**) = dim W.

(a) If **L** is one-to-one, then ker **L** = {**0**}. Thus dim(ker **L**) = 0 and dim(range **L**) = dim W. Hence **L** is onto.

(b) If **L** is onto, then range **L** = W. Thus dim(ker **L**) = 0, so **L** is one-to-one.

T.3. By Theorem 4.5, **L** is one-to-one if and only if ker **L** = {**0**}. Using Theorem 4.7, ker **L** = {**0**} if and only if dim(range **L**) = dim V = dim $\mathbf{R}^n$ = n. Using Exercise T.2, dim(range **L**) = dim(column space **A**) = n. Since dim(column space **A**) = rank **A**, we have rank **A** = n. But an $n \times n$ matrix has rank n if and only if **A** is nonsingular. **A** is nonsingular if an only if $|\mathbf{A}| \neq 0$. In summary, **L** is one-to-one if and only if $|\mathbf{A}| \neq 0$.

T.5. (a) From Theorem 4.7, dim(ker **L**) + dim(range **L**) = dim V. However, dim(ker **L**) ≥ 0 hence dim(range **L**) $\leq$ dim V.

(b) If **L** is onto, then range **L** = W. Then from Theorem 4.7, dim(ker **L**) + dim W = dim V. Since dim(ker **L**) ≥ 0, dim W $\leq$ dim V.

T.7. **L** is one-to-one if and only if dim(ker **L**) = 0. Then from Theorem 4.7 we have that dim(range **L**) = dim V.

Exercises 4.2

T.9. The "only if" portion follows from Exercise T.8. If the image of a basis for V is a basis for W, then dim(range **L**) = dim W = dim V, and hence ker **L** has dimension 0 from Theorem 4.7. Thus ker **L** = {**0**} and Theorem 4.5 implies **L** is one-to-one.

For each of the following MATLAB exercises we use the observations made in Example 11. If **L**(**X**) = **AX**, then a basis for the kernel of **L** is obtained from the general solution of **AX** = **0** and a basis for range **L** is a basis for the column space of **A**. In MATLAB we use **rref(A)** and then form the general solution of **AX** = **0**. The nonzero columns of **rref(A')'** form a basis for range **L**.

ML.1.

```
A=[1 2 5 5;-2 -3 -8 -7];
rref(A)

ans =

     1     0     1    -1
     0     1     2     3
```

It follows that the general solution of **AX** = **0** is obtained from

$$\begin{aligned} x_1 \qquad + x_3 - x_4 &= 0 \\ x_2 + 2x_3 + 3x_4 &= 0 \end{aligned}$$

Let $x_3 = r$ and $x_4 = s$, then $x_2 = -2r - 3s$ and $x_1 = -r + s$. Thus

$$\mathbf{X} = \begin{bmatrix} -r + s \\ -2r - 3s \\ r \\ s \end{bmatrix} = r\begin{bmatrix} -1 \\ -2 \\ 1 \\ 0 \end{bmatrix} + s\begin{bmatrix} 1 \\ -3 \\ 0 \\ 1 \end{bmatrix}$$

and $\left\{\begin{bmatrix} -1 \\ -2 \\ 1 \\ 0 \end{bmatrix}, \begin{bmatrix} 1 \\ -3 \\ 0 \\ 1 \end{bmatrix}\right\}$ is a basis for ker **L**.

To find a basis for range **L** proceed as follows.

```
rref(A')'

ans =

     1     0     0     0
     0     1     0     0
```

Then $\left\{\begin{bmatrix}1\\0\end{bmatrix}, \begin{bmatrix}0\\1\end{bmatrix}\right\}$ is a basis for range **L**.

ML.2.
```
A=[-3 2 -7;2 -1 4;2 -2 6];
rref(A)

ans =

     1     0     1
     0     1    -2
     0     0     0
```

It follows that the general solution of **AX** = **0** is obtained from

$$\begin{aligned} x_1 \qquad\quad + x_3 &= 0 \\ x_2 - 2x_3 &= 0 \end{aligned}$$

Let $x_3 = r$, then $x_2 = 2r$ and $x_1 = -r$. Thus

$$\mathbf{X} = \begin{bmatrix}-r\\2r\\r\end{bmatrix} = r\begin{bmatrix}-1\\2\\1\end{bmatrix}$$

and $\begin{bmatrix}-1\\2\\1\end{bmatrix}$ is a basis for ker **L**. To find a basis for range **L** proceed as follows.

```
rref(A')'

ans =

     1     0     0
     0     1     0
    -2    -2     0
```

Then $\left\{\begin{bmatrix}1\\0\\-2\end{bmatrix}, \begin{bmatrix}0\\1\\-2\end{bmatrix}\right\}$ is a basis for range **L**.

ML.3.
```
A=[3 3 -3 1 11;-4 -4 7 -2 -19;2 2 -3 1 9];
rref(A)
```

Exercises 4.2

```
ans =
     1     1     0     0     2
     0     0     1     0    -1
     0     0     0     1     2
```

It follows that the general solution of $\mathbf{AX} = \mathbf{0}$ is obtained from

$$\begin{aligned} x_1 + x_2 \qquad\qquad + 2x_5 &= 0 \\ x_3 \qquad - x_5 &= 0 \\ x_4 + 2x_5 &= 0 \end{aligned}$$

Let $x_5 = r$ and $x_2 = s$, then $x_4 = -2r$ and $x_3 = r$, $x_1 = -s - 2r$. Thus

$$\mathbf{X} = \begin{bmatrix} -2r - s \\ s \\ r \\ -2r \\ r \end{bmatrix} = r\begin{bmatrix} -2 \\ 0 \\ 1 \\ -2 \\ 1 \end{bmatrix} + s\begin{bmatrix} -1 \\ 1 \\ 0 \\ 0 \\ 0 \end{bmatrix}$$

and $\left\{\begin{bmatrix} -2 \\ 0 \\ 1 \\ -2 \\ 1 \end{bmatrix}, \begin{bmatrix} -1 \\ 1 \\ 0 \\ 0 \\ 0 \end{bmatrix}\right\}$ is a basis for ker **L**. To find a basis for range **L** proceed as follows.

```
rref(A')'

ans =

     1     0     0     0     0
     0     1     0     0     0
     0     0     1     0     0
```

Thus the columns of $\mathbf{I}_3$ are a basis for range **L**.

Exercises 4.3

1. Let $\mathbf{L}: R^2 \to R^2$ be defined by $\mathbf{L}(x,y) = (x-2y, x+2y)$, $S = \{\mathbf{X}_1, \mathbf{X}_2\} = \{(1,-1), (0,1)\}$, and $T = \{\mathbf{Y}_1, \mathbf{Y}_2\} = \{(1,0), (0,1)\}$. We follow the steps given in the summary of the procedure in Theorem 4.8.

(a) $\mathbf{L}(\mathbf{X}_1) = \mathbf{L}((1,-1)) = (3,-1)$, $\mathbf{L}(\mathbf{X}_2) = \mathbf{L}((0,1)) = (-2,2)$

To find $\left[\mathbf{L}(\mathbf{X}_i)\right]_S$ we express $\mathbf{L}(\mathbf{X}_i)$ as linear combinations of the vectors in S. Thus we solve for c_1 and c_2 in

$$c_1\mathbf{X}_1 + c_2\mathbf{X}_2 = \mathbf{L}(\mathbf{X}_1)$$

and k_1 and k_2 in

$$k_1\mathbf{X}_1 + k_2\mathbf{X}_2 = \mathbf{L}(\mathbf{X}_2).$$

Substituting in the expression for the vectors and combining terms leads to the linear systems

$$\begin{aligned} c_1 &= 3 \\ -c_1 + c_2 &= -1 \end{aligned} \qquad \begin{aligned} k_1 &= -2 \\ -k_1 + k_2 &= 2 \end{aligned}$$

respectively. These systems have the same coefficient matrix, but different right hand sides. Hence form the partitioned matrix

$$\left[\begin{array}{rr|r|r} 1 & 0 & 3 & -2 \\ -1 & 1 & -1 & 2 \end{array}\right]$$

and find its reduced row echelon form which is

$$\left[\begin{array}{rr|r|r} 1 & 0 & 3 & -2 \\ 0 & 1 & 2 & 0 \end{array}\right]$$

It follows that $c_1 = 3$ and $c_2 = 2$, hence $\left[\mathbf{L}(\mathbf{X}_1)\right]_S = \begin{bmatrix} 3 \\ 2 \end{bmatrix}$.

Similarly, $k_1 = -2$ and $k_2 = 0$, so $\left[\mathbf{L}(\mathbf{X}_2)\right]_S = \begin{bmatrix} -2 \\ 0 \end{bmatrix}$. Hence the matrix of $\mathbf{L}$ with respect to S is

$$\left[\left[\mathbf{L}(\mathbf{X}_1)\right]_S \ \left[\mathbf{L}(\mathbf{X}_2)\right]_S\right] = \begin{bmatrix} 3 & -2 \\ 2 & 0 \end{bmatrix}.$$

(b) From part a, $\mathbf{L}(\mathbf{X}_1) = (3,-1)$, $\mathbf{L}(\mathbf{X}_2) = (-2,2)$. Here we must solve

$$c_1\mathbf{Y}_1 + c_2\mathbf{Y}_2 = \mathbf{L}(\mathbf{X}_1)$$

and

$$k_1\mathbf{Y}_1 + k_2\mathbf{Y}_2 = \mathbf{L}(\mathbf{X}_2).$$

Since T is the natural basis for R^2, we have $\left[\mathbf{L}(\mathbf{X}_i)\right]_T = (\mathbf{L}(\mathbf{X}_i))^T$, for i=1,2. Thus the matrix representing **L** with respect to S and T is

$$\left[\left[\mathbf{L}(\mathbf{X}_1)\right]_T \ \left[\mathbf{L}(\mathbf{X}_2)\right]_T\right] = \begin{bmatrix} 3 & -2 \\ -1 & 2 \end{bmatrix}.$$

(c) $\mathbf{L}(\mathbf{Y}_1) = (1,1)$, $\mathbf{L}(\mathbf{Y}_2) = (-2,2)$

To find $\left[\mathbf{L}(\mathbf{Y}_i)\right]_S$ we express $\mathbf{L}(\mathbf{Y}_i)$ as linear combinations of the vectors in S. Thus we solve for c_1 and c_2 in

$$c_1\mathbf{X}_1 + c_2\mathbf{X}_2 = \mathbf{L}(\mathbf{Y}_1)$$

and k_1 and k_2 in

$$k_1\mathbf{X}_1 + k_2\mathbf{X}_2 = \mathbf{L}(\mathbf{Y}_2).$$

Substituting in the expression for the vectors and combining terms leads to the linear systems

$$\begin{aligned} c_1 &= 1 \\ -c_1 + c_2 &= 1 \end{aligned} \qquad \begin{aligned} k_1 &= -2 \\ -k_1 + k_2 &= 2 \end{aligned}$$

respectively. These systems have the same coefficient matrix, but different right hand sides. Hence form the partitioned matrix

$$\left[\begin{array}{cc|c|c} 1 & 0 & 1 & -2 \\ -1 & 1 & 1 & 2 \end{array}\right]$$

and find its reduced row echelon form which is

$$\left[\begin{array}{cc|c|c} 1 & 0 & 1 & -2 \\ 0 & 1 & 2 & 0 \end{array}\right]$$

It follows that $c_1 = 1$ and $c_2 = 2$, hence $\left[\mathbf{L}(\mathbf{Y}_1)\right]_S = \begin{bmatrix} 1 \\ 2 \end{bmatrix}$.

Similarly, $k_1 = -2$ and $k_2 = 0$, so $\left[\mathbf{L}(\mathbf{Y}_2)\right]_S = \begin{bmatrix} -2 \\ 0 \end{bmatrix}$. Hence the matrix of **L** with respect to T and S is

$$\left[\left[\mathbf{L}(\mathbf{Y}_1)\right]_S \ \left[\mathbf{L}(\mathbf{Y}_2)\right]_S\right] = \begin{bmatrix} 1 & -2 \\ 2 & 0 \end{bmatrix}.$$

(d) From part c, $\mathbf{L}(\mathbf{Y}_1) = (1,1)$, $\mathbf{L}(\mathbf{Y}_2) = (-2,2)$. Here we must solve

$$c_1\mathbf{Y}_1 + c_2\mathbf{Y}_2 = \mathbf{L}(\mathbf{Y}_1)$$

and

$$k_1\mathbf{Y}_1 + k_2\mathbf{Y}_2 = \mathbf{L}(\mathbf{Y}_2).$$

Since T is the natural basis for R^2, we have $[\mathbf{L}(\mathbf{Y}_i)]_T = (\mathbf{L}(\mathbf{Y}_i))^T$, for i=1,2. Thus the matrix representing $\mathbf{L}$ with respect to T is

$$\left[[\mathbf{L}(\mathbf{Y}_1)]_T \ [\mathbf{L}(\mathbf{Y}_2)]_T\right] = \begin{bmatrix} 1 & -2 \\ 1 & 2 \end{bmatrix}.$$

(e) Let $\mathbf{Z} = (2,-1)$, then $\mathbf{L}(\mathbf{Z}) = \mathbf{L}((2,-1)) = (4,0)$. We find that $[\mathbf{Z}]_S = \begin{bmatrix} 2 \\ 1 \end{bmatrix}$ and $[\mathbf{Z}]_T = \begin{bmatrix} 2 \\ -1 \end{bmatrix}$. We proceed as follows.

From part a: $[\mathbf{L}(\mathbf{Z})]_S = \begin{bmatrix} 3 & -2 \\ 2 & 0 \end{bmatrix}[\mathbf{Z}]_S = \begin{bmatrix} 3 & -2 \\ 2 & 0 \end{bmatrix}\begin{bmatrix} 2 \\ 1 \end{bmatrix} = \begin{bmatrix} 4 \\ 4 \end{bmatrix}$

Thus $\mathbf{L}(\mathbf{Z}) = 4\mathbf{X}_1 + 4\mathbf{X}_2 = (4,0)$.

From part b: $[\mathbf{L}(\mathbf{Z})]_T = \begin{bmatrix} 3 & -2 \\ -1 & 2 \end{bmatrix}[\mathbf{Z}]_S = \begin{bmatrix} 3 & -2 \\ -1 & 2 \end{bmatrix}\begin{bmatrix} 2 \\ 1 \end{bmatrix} = \begin{bmatrix} 4 \\ 0 \end{bmatrix}$

Thus $\mathbf{L}(\mathbf{Z}) = 4\mathbf{Y}_1 + 0\mathbf{Y}_2 = (4,0)$.

From part c: $[\mathbf{L}(\mathbf{Z})]_S = \begin{bmatrix} 1 & -2 \\ 2 & 0 \end{bmatrix}[\mathbf{Z}]_T = \begin{bmatrix} 1 & -2 \\ 2 & 0 \end{bmatrix}\begin{bmatrix} 2 \\ -1 \end{bmatrix} = \begin{bmatrix} 4 \\ 4 \end{bmatrix}$

Thus $\mathbf{L}(\mathbf{Z}) = 4\mathbf{X}_1 + 4\mathbf{X}_2 = (4,0)$.

From part d: $[\mathbf{L}(\mathbf{Z})]_T = \begin{bmatrix} 1 & -2 \\ 1 & 2 \end{bmatrix}[\mathbf{Z}]_T = \begin{bmatrix} 1 & -2 \\ 1 & 2 \end{bmatrix}\begin{bmatrix} 2 \\ -1 \end{bmatrix} = \begin{bmatrix} 4 \\ 0 \end{bmatrix}$

Thus $\mathbf{L}(\mathbf{Z}) = 4\mathbf{Y}_1 + 0\mathbf{Y}_2 = (4,0)$.

3. Let $\mathbf{L}:\mathbf{R}^2 \to \mathbf{R}^3$ be defined by $\mathbf{L}\left(\begin{bmatrix} x \\ y \end{bmatrix}\right) = \begin{bmatrix} x - 2y \\ 2x + y \\ x + y \end{bmatrix}$, $S = \{\mathbf{X}_1, \mathbf{X}_2\}$ be the natural basis for $\mathbf{R}^2$, and $T = \{\mathbf{Y}_1, \mathbf{Y}_2, \mathbf{Y}_3\}$ be the natural basis for $\mathbf{R}^3$. Also, let $S' = \{\mathbf{X}_1', \mathbf{X}_2'\} = \left\{\begin{bmatrix} 1 \\ -1 \end{bmatrix}, \begin{bmatrix} 0 \\ 1 \end{bmatrix}\right\}$ be

another basis for $\mathbf{R}^2$ and $T' = \{\mathbf{Y}_1, \mathbf{Y}_2, \mathbf{Y}_3\} = \left\{ \begin{bmatrix} 1 \\ 1 \\ 0 \end{bmatrix}, \begin{bmatrix} 0 \\ 1 \\ 1 \end{bmatrix}, \begin{bmatrix} 1 \\ -1 \\ 1 \end{bmatrix} \right\}$

be another basis for $\mathbf{R}^3$. We follow the steps given in the summary of the procedure in Theorem 4.8.

(a) $$\mathbf{L}(\mathbf{X}_1) = \mathbf{L}\left(\begin{bmatrix} 1 \\ 0 \end{bmatrix}\right) = \begin{bmatrix} 1 \\ 2 \\ 1 \end{bmatrix}, \quad \mathbf{L}(\mathbf{X}_2) = \mathbf{L}\left(\begin{bmatrix} 0 \\ 1 \end{bmatrix}\right) = \begin{bmatrix} -2 \\ 1 \\ 1 \end{bmatrix}$$

Next we determine the coordinates of $\mathbf{L}(\mathbf{X}_1)$ and $\mathbf{L}(\mathbf{X}_2)$ with respect to T. Since T is the natural basis we have

$$[\mathbf{L}(\mathbf{X}_1)]_T = \begin{bmatrix} 1 \\ 2 \\ 1 \end{bmatrix}, \quad [\mathbf{L}(\mathbf{X}_2)]_T = \begin{bmatrix} -2 \\ 1 \\ 1 \end{bmatrix}.$$

The matrix $\mathbf{A}$ of $\mathbf{L}$ with respect to S and T is

$$\mathbf{A} = [\ [\mathbf{L}(\mathbf{X}_1)]_T \quad [\mathbf{L}(\mathbf{X}_2)]_T\ ] = \begin{bmatrix} 1 & -2 \\ 2 & 1 \\ 1 & 1 \end{bmatrix}.$$

(b) $$\mathbf{L}(\mathbf{X}_1{}') = \mathbf{L}\left(\begin{bmatrix} 1 \\ -1 \end{bmatrix}\right) = \begin{bmatrix} 3 \\ 1 \\ 0 \end{bmatrix}, \quad \mathbf{L}(\mathbf{X}_2{}') = \mathbf{L}\left(\begin{bmatrix} 0 \\ 1 \end{bmatrix}\right) = \begin{bmatrix} -2 \\ 1 \\ 1 \end{bmatrix}$$

Next we determine the coordinates of $\mathbf{L}(\mathbf{X}_1{}')$ and $\mathbf{L}(\mathbf{X}_2{}')$ with respect to T'. (since T' is not the natural basis we must solve a system of equations to obtain the coordinates.)

$$[\mathbf{L}(\mathbf{X}_1{}')] = \begin{bmatrix} 3 \\ 1 \\ 0 \end{bmatrix} = c_1\mathbf{Y}_1{}' + c_2\mathbf{Y}_2{}' + c_3\mathbf{Y}_3{}'$$

Adding vectors and equating corresponding components from both sides of the equation gives the linear system

$$\begin{bmatrix} 1 & 0 & 1 \\ 1 & 1 & -1 \\ 0 & 1 & 1 \end{bmatrix}\begin{bmatrix} c_1 \\ c_2 \\ c_3 \end{bmatrix} = \begin{bmatrix} 3 \\ 1 \\ 0 \end{bmatrix}.$$

Forming the augmented matrix and applying row operations we obtain the reduced row echelon form

$$\left[\begin{array}{ccc|c} 1 & 0 & 0 & 7/3 \\ 0 & 1 & 0 & -2/3 \\ 0 & 0 & 1 & 2/3 \end{array}\right].$$

Thus $c_1 = 7/3$, $c_2 = -2/3$, $c_3 = 2/3$ and

$$[\mathbf{L}(\mathbf{X}_1')]_{T'} = \begin{bmatrix} 7/3 \\ -2/3 \\ 2/3 \end{bmatrix}.$$

$$\mathbf{L}(\mathbf{X}_2') = \begin{bmatrix} -2 \\ 1 \\ 1 \end{bmatrix} = c_1\mathbf{Y}_1' + c_2\mathbf{Y}_2' + c_3\mathbf{Y}_3'$$

Adding vectors and equating corresponding components from both sides of the equation gives the linear system

$$\begin{bmatrix} 1 & 0 & 1 \\ 1 & 1 & -1 \\ 0 & 1 & 1 \end{bmatrix}\begin{bmatrix} c_1 \\ c_2 \\ c_3 \end{bmatrix} = \begin{bmatrix} -2 \\ 1 \\ 1 \end{bmatrix}.$$

Forming the augmented matrix and applying row operations we obtain the reduced row echelon form

$$\left[\begin{array}{ccc|c} 1 & 0 & 0 & -4/3 \\ 0 & 1 & 0 & 5/3 \\ 0 & 0 & 1 & -2/3 \end{array}\right].$$

Thus $c_1 = -4/3$, $c_2 = 5/3$, $c_3 = -2/3$ and

$$[\mathbf{L}(\mathbf{X}_2')]_{T'} = \begin{bmatrix} -4/3 \\ 5/3 \\ -2/3 \end{bmatrix}.$$

The matrix $\mathbf{A}$ of $\mathbf{L}$ with respect to S' and T' is

$$\mathbf{A} = [\ [\mathbf{L}(\mathbf{X}_1)]_{T'} \quad [\mathbf{L}(\mathbf{X}_2)]_{T'}] = \begin{bmatrix} 7/3 & -4/3 \\ -2/3 & 5/3 \\ 2/3 & -2/3 \end{bmatrix}.$$

(c) Let $\mathbf{X} = \begin{bmatrix} 1 \\ 2 \end{bmatrix}$ and $\mathbf{Y} = \mathbf{L}(\mathbf{X})$. Then from the definition of $\mathbf{L}$,

$\mathbf{Y} = \mathbf{L}(\mathbf{X}) = \mathbf{L}\left(\begin{bmatrix} 1 \\ 2 \end{bmatrix}\right) = \begin{bmatrix} -3 \\ 4 \\ 3 \end{bmatrix}$. Next we use Equation (2) in Theorem 4.8. We emphasize that Equation (2) gives the relationship between coordinates of vectors in V with respect to a basis for V and coordinates of vectors in W with respect to a basis for W. From (a)

$$[\mathbf{Y}]_T = \mathbf{A}[\mathbf{X}]_S = \begin{bmatrix} 1 & -2 \\ 2 & 1 \\ 1 & 1 \end{bmatrix}\begin{bmatrix} 1 \\ 2 \end{bmatrix}_S = \begin{bmatrix} 1 & -2 \\ 2 & 1 \\ 1 & 1 \end{bmatrix}\begin{bmatrix} 1 \\ 2 \end{bmatrix} = \begin{bmatrix} -3 \\ 4 \\ 3 \end{bmatrix}_T$$

hence $\mathbf{Y} = -3\mathbf{Y}_1 + 4\mathbf{Y}_2 + 3\mathbf{Y}_3 = \begin{bmatrix} -3 \\ 4 \\ 3 \end{bmatrix}$. From (b)

$$[\mathbf{Y}]_{T'} = \mathbf{A}[\mathbf{X}]_{S'} = \begin{bmatrix} 7/3 & -4/3 \\ -2/3 & 5/3 \\ 2/3 & -2/3 \end{bmatrix} \begin{bmatrix} 1 \\ 2 \end{bmatrix}_{S'} = \begin{bmatrix} 7/3 & -4/3 \\ -2/3 & 5/3 \\ 2/3 & -2/3 \end{bmatrix} \begin{bmatrix} 1 \\ 3 \end{bmatrix}$$

$= \begin{bmatrix} -5/3 \\ 13/3 \\ -4/3 \end{bmatrix}_{T'}$ where we have used $\mathbf{X} = 1\mathbf{X}_1' + 3\mathbf{X}_2'$. Hence

$$\mathbf{Y} = (-5/3)\mathbf{Y}_1' + (13/3)\mathbf{Y}_2' + (-4/3)\mathbf{Y}_3' = \begin{bmatrix} -3 \\ 4 \\ 3 \end{bmatrix}.$$

5. Let $\mathbf{L}:\mathbf{R}^3 \to \mathbf{R}^2$ be defined by $\mathbf{L}\left(\begin{bmatrix} x \\ y \\ z \end{bmatrix}\right) = \begin{bmatrix} x + y \\ y - z \end{bmatrix}$, $S = \{\mathbf{X}_1, \mathbf{X}_2, \mathbf{X}_3\}$ be the natural basis for $\mathbf{R}^3$, and $T = \{\mathbf{Y}_1, \mathbf{Y}_2\}$ be the natural basis for $\mathbf{R}^2$. Also, let $S' = \{\mathbf{X}_1', \mathbf{X}_2', \mathbf{X}_3'\} = \left\{\begin{bmatrix} 1 \\ 1 \\ 0 \end{bmatrix}, \begin{bmatrix} 0 \\ 1 \\ 0 \end{bmatrix}, \begin{bmatrix} -1 \\ 1 \\ 1 \end{bmatrix}\right\}$ be another basis for $\mathbf{R}^3$ and $T' = \{\mathbf{Y}_1, \mathbf{Y}_2\}$ $= \left\{\begin{bmatrix} -1 \\ 1 \end{bmatrix}, \begin{bmatrix} 1 \\ 2 \end{bmatrix}\right\}$ be another basis for $\mathbf{R}^2$. We follow the steps given in the summary of the procedure in Theorem 4.8.

(a) $\mathbf{L}(\mathbf{X}_1) = \mathbf{L}\left(\begin{bmatrix} 1 \\ 0 \\ 0 \end{bmatrix}\right) = \begin{bmatrix} 1 \\ 0 \end{bmatrix}$, $\mathbf{L}(\mathbf{X}_2) = \mathbf{L}\left(\begin{bmatrix} 0 \\ 1 \\ 0 \end{bmatrix}\right) = \begin{bmatrix} 1 \\ 1 \end{bmatrix}$,

$$\mathbf{L}(\mathbf{X}_3) = \mathbf{L}\left(\begin{bmatrix} 0 \\ 0 \\ 1 \end{bmatrix}\right) = \begin{bmatrix} 0 \\ -1 \end{bmatrix}$$

Next we determine the coordinates of $\mathbf{L}(\mathbf{X}_1)$, $\mathbf{L}(\mathbf{X}_2)$, and $\mathbf{L}(\mathbf{X}_3)$ with respect to T. Since T is the natural basis we have

$$[\mathbf{L}(\mathbf{X}_1)]_T = \begin{bmatrix} 1 \\ 0 \end{bmatrix}, \quad [\mathbf{L}(\mathbf{X}_2)]_T = \begin{bmatrix} 1 \\ 1 \end{bmatrix}, \quad [\mathbf{L}(\mathbf{X}_3)]_T = \begin{bmatrix} 0 \\ -1 \end{bmatrix}.$$

The matrix **A** of **L** with respect to S and T is

$$\mathbf{A} = [\ [\mathbf{L}(\mathbf{X}_1)]_T \quad [\mathbf{L}(\mathbf{X}_2)]_T \quad [\mathbf{L}(\mathbf{X}_3)]_T\ ] = \begin{bmatrix} 1 & 1 & 0 \\ 0 & 1 & -1 \end{bmatrix}.$$

(b) $$\mathbf{L}(\mathbf{X}_1') = \mathbf{L}\left(\begin{bmatrix} 1 \\ 1 \\ 0 \end{bmatrix}\right) = \begin{bmatrix} 2 \\ 1 \end{bmatrix}, \quad \mathbf{L}(\mathbf{X}_2') = \mathbf{L}\left(\begin{bmatrix} 0 \\ 1 \\ 0 \end{bmatrix}\right) = \begin{bmatrix} 1 \\ 1 \end{bmatrix},$$

$$\mathbf{L}(\mathbf{X}_3') = \mathbf{L}\left(\begin{bmatrix} -1 \\ 1 \\ 1 \end{bmatrix}\right) = \begin{bmatrix} 0 \\ 0 \end{bmatrix}$$

Next we determine the coordinates of $\mathbf{L}(\mathbf{X}_1')$, $\mathbf{L}(\mathbf{X}_2')$, and $\mathbf{L}(\mathbf{X}_3')$ with respect to T′. (since T′ is not the natural basis we must solve a system of equations to obtain the coordinates.)

$$\mathbf{L}(\mathbf{X}_1') = \begin{bmatrix} 2 \\ 1 \end{bmatrix} = c_1\mathbf{Y}_1' + c_2\mathbf{Y}_2'$$

Adding vectors and equating corresponding components from both sides of the equation gives the linear system

$$\begin{bmatrix} -1 & 1 \\ 1 & 2 \end{bmatrix}\begin{bmatrix} c_1 \\ c_2 \end{bmatrix} = \begin{bmatrix} 2 \\ 1 \end{bmatrix}.$$

Forming the augmented matrix and applying row operations we obtain the reduced row echelon form

$$\left[\begin{array}{cc|c} 1 & 0 & -1 \\ 0 & 1 & 1 \end{array}\right].$$

Thus $c_1 = -1$, $c_2 = 1$ and

$$[\mathbf{L}(\mathbf{X}_1')]_{T'} = \begin{bmatrix} -1 \\ 1 \end{bmatrix}.$$

$$\mathbf{L}(\mathbf{X}_2') = \begin{bmatrix} 1 \\ 1 \end{bmatrix} = c_1\mathbf{Y}_1' + c_2\mathbf{Y}_2'$$

Adding vectors and equating corresponding components from both sides of the equation gives the linear system

$$\begin{bmatrix} -1 & 1 \\ 1 & 2 \end{bmatrix}\begin{bmatrix} c_1 \\ c_2 \end{bmatrix} = \begin{bmatrix} 1 \\ 1 \end{bmatrix}.$$

Forming the augmented matrix and applying row operations we obtain the reduced row echelon form

$$\left[\begin{array}{cc|c} 1 & 0 & -1/3 \\ 0 & 1 & 2/3 \end{array}\right].$$

Thus $c_1 = -1/3$, $c_2 = 2/3$ and

$$[\mathbf{L}(\mathbf{X}_2')]_{T'} = \begin{bmatrix} -1/3 \\ 2/3 \end{bmatrix}.$$

$$\mathbf{L}(\mathbf{X}_3') = \begin{bmatrix} 0 \\ 0 \end{bmatrix} = c_1\mathbf{Y}_1' + c_2\mathbf{Y}_2'$$

Thus $c_1 = 0$ and $c_2 = 0$. Hence,

$$[\mathbf{L}(\mathbf{X}_3')]_{T'} = \begin{bmatrix} 0 \\ 0 \end{bmatrix}.$$

The matrix $\mathbf{A}$ of $\mathbf{L}$ with respect to S' and T' is

$$\mathbf{A} = [[\mathbf{L}(\mathbf{X}_1')]_{T'}\ [\mathbf{L}(\mathbf{X}_2')]_{T'}\ [\mathbf{L}(\mathbf{X}_3')]_{T'}] = \begin{bmatrix} -1 & -1/3 & 0 \\ 1 & 2/3 & 0 \end{bmatrix}.$$

(c) Let $\mathbf{X} = \begin{bmatrix} 1 \\ 2 \\ 3 \end{bmatrix}$ and $\mathbf{Y} = \mathbf{L}(\mathbf{X})$. Then from the definition of $\mathbf{L}$,

$\mathbf{Y} = \mathbf{L}(\mathbf{X}) = \mathbf{L}\left(\begin{bmatrix} 1 \\ 2 \\ 3 \end{bmatrix}\right) = \begin{bmatrix} 3 \\ -1 \end{bmatrix}$. Next we use Equation (2) from

Theorem 4.8 for (a) and (b). We emphasize that Equation (2)gives the relationship between coordinates of vectors in V with respect to a basis for V and coordinates of vectors in W with respect to a basis for W. From (a)

$$[\mathbf{Y}]_T = \mathbf{A}[\mathbf{X}]_S = \begin{bmatrix} 1 & 1 & 0 \\ 0 & 1 & -1 \end{bmatrix}\begin{bmatrix} 1 \\ 2 \\ 3 \end{bmatrix}_S = \begin{bmatrix} 1 & 1 & 0 \\ 0 & 1 & -1 \end{bmatrix}\begin{bmatrix} 1 \\ 2 \\ 3 \end{bmatrix}$$

$= \begin{bmatrix} 3 \\ -1 \end{bmatrix}_T$ hence $\mathbf{Y} = 3\mathbf{Y}_1 - 1\mathbf{Y}_2 = \begin{bmatrix} 3 \\ -1 \end{bmatrix}$. From (b)

$$[\mathbf{Y}]_{T'} = \mathbf{A}[\mathbf{X}]_{S'} = \begin{bmatrix} -1 & -1/3 & 0 \\ 1 & 2/3 & 0 \end{bmatrix} \begin{bmatrix} 1 \\ 2 \\ 3 \end{bmatrix}_{S'}$$

$$= \begin{bmatrix} -1 & -1/3 & 0 \\ 1 & 2/3 & 0 \end{bmatrix} \begin{bmatrix} 4 \\ -5 \\ 3 \end{bmatrix} = \begin{bmatrix} -7/3 \\ 2/3 \end{bmatrix}_{T'}$$

where we have used $\mathbf{X} = 4\mathbf{X}_1' - 5\mathbf{X}_2' + 3\mathbf{X}_3'$. Hence

$$\mathbf{Y} = (-7/3)\mathbf{Y}_1' + (2/3)\mathbf{Y}_2' = \begin{bmatrix} 3 \\ -1 \end{bmatrix}.$$

7. Let $\mathbf{L}:\mathbf{P}_1 \to \mathbf{P}_3$ be defined by $\mathbf{L}(p(t)) = t^2p(t)$. Let $S = \{p_1(t), p_2(t)\} = \{t, 1\}$ and $S' = \{p_1'(t), p_2'(t)\} = \{t, t + 1\}$ be bases for $\mathbf{P}_1$. Let $T = \{q_1(t), q_2(t), q_3(t), q_4(t)\} = \{t^3, t^2, t, 1\}$ and $T' = \{q_1'(t), q_2'(t), q_3'(t), q_4'(t)\} = \{t^3, t^2 - 1, t, t + 1\}$ be bases for $\mathbf{P}_3$. We proceed as in Example 7.

(a) $\mathbf{L}(p_1(t)) = t^2(t) = t^3$, $\mathbf{L}(p_2(t)) = t^2(1) = t^2$
Next we determine the coordinates of $\mathbf{L}(p_j(t))$ with respect to the T basis. Since T is the natural basis we have

$$[\mathbf{L}(p_1(t))]_T = \begin{bmatrix} 1 \\ 0 \\ 0 \\ 0 \end{bmatrix} \text{ and } [\mathbf{L}(p_2(t))]_T = \begin{bmatrix} 0 \\ 1 \\ 0 \\ 0 \end{bmatrix}.$$

Then the matrix of $\mathbf{L}$ with respect to S and T is

$$\mathbf{A} = [\ [\mathbf{L}(p_1(t))]_T \quad [\mathbf{L}(p_2(t))]_T\ ] = \begin{bmatrix} 1 & 0 \\ 0 & 1 \\ 0 & 0 \\ 0 & 0 \end{bmatrix}.$$

(b) $\mathbf{L}(p_1'(t)) = t^2(t) = t^3$, $\mathbf{L}(p_2'(t)) = t^2(t + 1) = t^3 + t^2$.
Next we determine the coordinates of $\mathbf{L}(p_j(t))$ with respect to the T′ basis. $\mathbf{L}(p_1'(t)) = 1q_1(t)$, thus

$$[\mathbf{L}(p_1'(t))]_{T'} = \begin{bmatrix} 1 \\ 0 \\ 0 \\ 0 \end{bmatrix}.$$

$\mathbf{L}(p_2'(t)) = 1q_1'(t) + 1q_2'(t) - 1q_3'(t) + 1q_4'(t)$, thus

$$[\mathbf{L}(p_2'(t))]_{T'} = \begin{bmatrix} 1 \\ 1 \\ -1 \\ 1 \end{bmatrix}.$$

Then the matrix of **L** with respect to S′ and T′ is

$$\mathbf{A} = [\ [\mathbf{L}(p_1'(t))]_{T'}\quad [\mathbf{L}(p_2'(t))]_{T'}\ ] = \begin{bmatrix} 1 & 1 \\ 0 & 1 \\ 0 & -1 \\ 0 & 1 \end{bmatrix}.$$

9. Let $\mathbf{L}:M_{22} \to M_{22}$ be defined by $\mathbf{L}(\mathbf{A}) = \mathbf{A}^T$, $S = \{\mathbf{X}_1, \mathbf{X}_2, \mathbf{X}_3, \mathbf{X}_4\}$ $= \left\{\begin{bmatrix} 1 & 0 \\ 0 & 0 \end{bmatrix}, \begin{bmatrix} 0 & 1 \\ 0 & 0 \end{bmatrix}, \begin{bmatrix} 0 & 0 \\ 1 & 0 \end{bmatrix}, \begin{bmatrix} 0 & 0 \\ 0 & 1 \end{bmatrix}\right\}$, and

$$T = \{\mathbf{Y}_1, \mathbf{Y}_2, \mathbf{Y}_3, \mathbf{Y}_4\} = \left\{\begin{bmatrix} 1 & 1 \\ 0 & 0 \end{bmatrix}, \begin{bmatrix} 0 & 1 \\ 0 & 0 \end{bmatrix}, \begin{bmatrix} 0 & 0 \\ 1 & 1 \end{bmatrix}, \begin{bmatrix} 1 & 0 \\ 0 & 1 \end{bmatrix}\right\}.$$

(a) $\mathbf{L}(\mathbf{X}_1) = \mathbf{X}_1$, $\mathbf{L}(\mathbf{X}_2) = \mathbf{X}_3$, $\mathbf{L}(\mathbf{X}_3) = \mathbf{X}_2$, $\mathbf{L}(\mathbf{X}_4) = \mathbf{X}_4$

Thus it follows that

$$[\mathbf{L}(\mathbf{X}_1)]_S = \begin{bmatrix} 1 \\ 0 \\ 0 \\ 0 \end{bmatrix}, \quad [\mathbf{L}(\mathbf{X}_2)]_S = \begin{bmatrix} 0 \\ 0 \\ 1 \\ 0 \end{bmatrix},$$

$$[\mathbf{L}(\mathbf{X}_3)]_S = \begin{bmatrix} 0 \\ 1 \\ 0 \\ 0 \end{bmatrix}, \quad [\mathbf{L}(\mathbf{X}_4)]_S = \begin{bmatrix} 0 \\ 0 \\ 0 \\ 1 \end{bmatrix}.$$

and hence the matrix of **L** with respect to S is

$$\left[[\mathbf{L}(\mathbf{X}_1)]_S\ [\mathbf{L}(\mathbf{X}_2)]_S\ [\mathbf{L}(\mathbf{X}_3)]_S\ [\mathbf{L}(\mathbf{X}_4)]_S\right] = \begin{bmatrix} 1 & 0 & 0 & 0 \\ 0 & 0 & 1 & 0 \\ 0 & 1 & 0 & 0 \\ 0 & 0 & 0 & 1 \end{bmatrix}.$$

(b) Use the computations of $\mathbf{L}(\mathbf{X}_i)$ $i = 1,2,3,4$ in part a. Here each $\mathbf{L}(\mathbf{X}_i)$ must be expressed in terms of the T-basis. Hence we consider four equations

$$c_1\mathbf{Y}_1 + c_2\mathbf{Y}_2 + c_3\mathbf{Y}_3 + c_4\mathbf{Y}_4 = \mathbf{L}(\mathbf{X}_i), \quad i=1,2,3,4. \qquad (*)$$

We form the linear system associated with (*). In each case the coefficient matrix is the same, but the right hand side is different. We illustrate the construction in the case that i = 1 and then show how it can be used to find $\left[\mathbf{L}(\mathbf{X}_i)\right]_T$ for i =1,2,3,4. Combining the terms on the left side of (*) gives

$$\begin{bmatrix} c_1 + c_4 & c_1 + c_2 \\ c_3 & c_3 + c_4 \end{bmatrix} = \mathbf{L}(\mathbf{X}_1) = \begin{bmatrix} 1 & 0 \\ 0 & 0 \end{bmatrix}$$

and equating corresponding entries gives the linear system

$$\begin{aligned} c_1 \quad\quad\quad\quad\quad + c_4 &= 1 \\ c_1 + c_2 \quad\quad\quad\quad &= 0 \\ c_3 \quad\quad &= 0 \\ c_3 + c_4 &= 0 \end{aligned}$$

In matrix form we have

$$\begin{bmatrix} 1 & 0 & 0 & 1 \\ 1 & 1 & 0 & 0 \\ 0 & 0 & 1 & 0 \\ 0 & 0 & 1 & 1 \end{bmatrix}\begin{bmatrix} c_1 \\ c_2 \\ c_3 \\ c_4 \end{bmatrix} = \begin{bmatrix} 1 \\ 0 \\ 0 \\ 0 \end{bmatrix}$$

For $\mathbf{L}(\mathbf{X}_2)$, $\mathbf{L}(\mathbf{X}_3)$, and $\mathbf{L}(\mathbf{X}_4)$ the right hand sides are respectively $\begin{bmatrix} 0 \\ 0 \\ 1 \\ 0 \end{bmatrix}$, $\begin{bmatrix} 0 \\ 1 \\ 0 \\ 0 \end{bmatrix}$, and $\begin{bmatrix} 0 \\ 0 \\ 0 \\ 1 \end{bmatrix}$. To solve the four systems efficiently we find the reduced row echelon form of the partitioned matrix

$$\left[\begin{array}{cccc|c|c|c|c} 1 & 0 & 0 & 1 & 1 & 0 & 0 & 0 \\ 1 & 1 & 0 & 0 & 0 & 0 & 1 & 0 \\ 0 & 0 & 1 & 0 & 0 & 1 & 0 & 0 \\ 0 & 0 & 1 & 1 & 0 & 0 & 0 & 1 \end{array}\right].$$

We obtain

$$\left[\begin{array}{cccc|c|c|c|c} 1 & 0 & 0 & 0 & 1 & 1 & 0 & -1 \\ 0 & 1 & 0 & 0 & -1 & -1 & 1 & 1 \\ 0 & 0 & 1 & 0 & 0 & 1 & 0 & 0 \\ 0 & 0 & 0 & 1 & 0 & -1 & 0 & 1 \end{array}\right] =$$

$$\left[\mathbf{I}_4 \;\middle|\; \left[\mathbf{L}(\mathbf{X}_1)\right]_T \;\middle|\; \left[\mathbf{L}(\mathbf{X}_2)\right]_T \;\middle|\; \left[\mathbf{L}(\mathbf{X}_3)\right]_T \;\middle|\; \left[\mathbf{L}(\mathbf{X}_3)\right]_T\right]$$

Thus it follows that the matrix of **L** with respect to S and T is

$$\begin{bmatrix} 1 & 1 & 0 & -1 \\ -1 & -1 & 1 & 1 \\ 0 & 1 & 0 & 0 \\ 0 & -1 & 0 & 1 \end{bmatrix}.$$

(c) Here we compute $\mathbf{L}(\mathbf{Y}_i)$ i=1,2,3,4 and then find $[\mathbf{L}(\mathbf{Y}_i)]_S$ to determine the columns of the matrix representing **L** with respect to T and S.

$$\mathbf{L}(\mathbf{Y}_1) = \begin{bmatrix} 1 & 0 \\ 1 & 0 \end{bmatrix},\ \mathbf{L}(\mathbf{Y}_2) = \begin{bmatrix} 0 & 0 \\ 1 & 0 \end{bmatrix},$$

$$\mathbf{L}(\mathbf{Y}_3) = \begin{bmatrix} 0 & 1 \\ 0 & 1 \end{bmatrix},\ \mathbf{L}(\mathbf{Y}_4) = \begin{bmatrix} 1 & 0 \\ 0 & 1 \end{bmatrix}$$

Following part b we consider the four equations

$$c_1\mathbf{X}_1 + c_2\mathbf{X}_2 + c_3\mathbf{X}_3 + c_4\mathbf{X}_4 = \mathbf{L}(\mathbf{Y}_i),\ i=1,2,3,4$$

Combining terms on the left gives

$$\begin{bmatrix} c_1 & c_2 \\ c_3 & c_4 \end{bmatrix} = \mathbf{L}(\mathbf{Y}_i),\ 1=1,2,3,4$$

It follows that we are led to the partitioned matrix

$$\left[\begin{array}{cccc|c|c|c|c} 1 & 0 & 0 & 0 & 1 & 0 & 0 & 1 \\ 0 & 1 & 0 & 0 & 0 & 0 & 1 & 0 \\ 0 & 0 & 1 & 0 & 1 & 1 & 0 & 0 \\ 0 & 0 & 0 & 1 & 0 & 0 & 1 & 1 \end{array}\right]$$

which is already in reduced row echelon form. Hence the matrix representing **L** with respect to T and S is

$$\begin{bmatrix} 1 & 0 & 0 & 1 \\ 0 & 0 & 1 & 0 \\ 1 & 1 & 0 & 0 \\ 0 & 0 & 1 & 1 \end{bmatrix}.$$

(d) Using the computations of $\mathbf{L}(\mathbf{Y}_i)$ from part c, we consider the four equations

$$c_1\mathbf{Y}_1 + c_2\mathbf{Y}_2 + c_3\mathbf{Y}_3 + c_4\mathbf{Y}_4 = \mathbf{L}(\mathbf{Y}_i),\ i=1,2,3,4$$

Combining the terms of the left gives

$$\begin{bmatrix} c_1+c_4 & c_1+c_2 \\ c_3 & c_4 \end{bmatrix} = \mathbf{L}(\mathbf{Y}_i),\ i=1,2,3,4.$$

It follows that we are led to the partitioned matrix

$$\left[\begin{array}{cccc|c|c|c|c} 1 & 0 & 0 & 1 & 1 & 0 & 0 & 1 \\ 1 & 1 & 0 & 0 & 0 & 0 & 1 & 0 \\ 0 & 0 & 1 & 0 & 1 & 1 & 0 & 0 \\ 0 & 0 & 0 & 1 & 0 & 0 & 1 & 1 \end{array}\right]$$

whose reduced row echelon form is

$$\left[\begin{array}{cccc|c|c|c|c} 1 & 0 & 0 & 0 & 1 & 0 & -1 & 0 \\ 0 & 1 & 0 & 0 & -1 & 0 & 2 & 0 \\ 0 & 0 & 1 & 0 & 1 & 1 & 0 & 0 \\ 0 & 0 & 0 & 1 & 0 & 0 & 1 & 1 \end{array}\right].$$

Hence the matrix of $\mathbf{L}$ with respect to T is

$$\begin{bmatrix} 1 & 0 & -1 & 0 \\ -1 & 0 & 2 & 0 \\ 1 & 1 & 0 & 0 \\ 0 & 0 & 1 & 1 \end{bmatrix}.$$

11. (a) By Theorem 4.8

$$[\mathbf{L}(\mathbf{X}_1)]_S = \mathbf{A}[\mathbf{X}_1]_S = \mathbf{A}\begin{bmatrix} 1 \\ 2 \end{bmatrix}_S = \begin{bmatrix} 2 & -3 \\ -1 & 4 \end{bmatrix}\begin{bmatrix} 1 \\ 0 \end{bmatrix} = \begin{bmatrix} 2 \\ -1 \end{bmatrix}$$

$$[\mathbf{L}(\mathbf{X}_2)]_S = \mathbf{A}[\mathbf{X}_2]_S = \mathbf{A}\begin{bmatrix} 1 \\ -1 \end{bmatrix}_S = \begin{bmatrix} 2 & -3 \\ -1 & 4 \end{bmatrix}\begin{bmatrix} 0 \\ 1 \end{bmatrix} = \begin{bmatrix} -3 \\ 4 \end{bmatrix}.$$

(b) Using the results of (a),

$$\mathbf{L}(\mathbf{X}_1) = 2\mathbf{X}_1 - \mathbf{X}_2 = \begin{bmatrix} 1 \\ 5 \end{bmatrix} \text{ and } \mathbf{L}(\mathbf{X}_2) = -3\mathbf{X}_1 + 4\mathbf{X}_2 = \begin{bmatrix} 1 \\ -10 \end{bmatrix}.$$

(c) We first compute the coordinates of $\mathbf{X} = \begin{bmatrix} -2 \\ 3 \end{bmatrix}$ with respect to S. Set $c_1\mathbf{X}_1 + c_2\mathbf{X}_2 = \mathbf{X}$. Adding vectors and equating corresponding components from both sides of the equation gives the linear system

$$\begin{bmatrix} 1 & 1 \\ 2 & -1 \end{bmatrix}\begin{bmatrix} c_1 \\ c_2 \end{bmatrix} = \begin{bmatrix} -2 \\ 3 \end{bmatrix}.$$

Form the augmented matrix and row reduce it to obtain $c_1 = 1/3$ and $c_2 = -7/3$. Thus

$$[\mathbf{X}]_S = \begin{bmatrix} -2 \\ 3 \end{bmatrix}_S = \begin{bmatrix} 1/3 \\ -7/3 \end{bmatrix}.$$

By Theorem 4.8,

$$\left[\mathbf{L}\left(\begin{bmatrix}-2\\3\end{bmatrix}\right)\right]_S = \mathbf{A}\begin{bmatrix}-2\\3\end{bmatrix}_S = \begin{bmatrix}2 & -3\\-1 & 4\end{bmatrix}\begin{bmatrix}1/3\\-7/3\end{bmatrix} = \begin{bmatrix}23/3\\-29/3\end{bmatrix}.$$

Hence

$$\mathbf{L}\left(\begin{bmatrix}-2\\3\end{bmatrix}\right) = (23/3)\mathbf{X}_1 - (29/3)\mathbf{X}_2 = \begin{bmatrix}-2\\25\end{bmatrix}.$$

13. Let $\mathbf{L}:P_1 \to P_2$ be represented by $\mathbf{A} = \begin{bmatrix}1 & 0\\2 & 1\\-1 & -2\end{bmatrix}$ with respect to bases $S = \{\mathbf{X}_1, \mathbf{X}_2\} = \{t+1, t-1\}$ and $T = \{\mathbf{Y}_1, \mathbf{Y}_2, \mathbf{Y}_3\} = \{t^2+1, t, t-1\}$.

(a) $[\mathbf{L}(\mathbf{X}_i)]_T = \mathbf{A}[\mathbf{X}_i]_S$ for i=1,2, so we first find $[\mathbf{X}_i]_S$. It follows that

$$[\mathbf{X}_1]_S = \begin{bmatrix}1\\0\end{bmatrix} \text{ and } [\mathbf{X}_2]_S = \begin{bmatrix}0\\1\end{bmatrix}.$$

Thus

$$[\mathbf{L}(\mathbf{X}_1)]_T = \begin{bmatrix}1 & 0\\2 & 1\\-1 & -2\end{bmatrix}\begin{bmatrix}1\\0\end{bmatrix} = \begin{bmatrix}1\\2\\-1\end{bmatrix} \text{ and}$$

$$[\mathbf{L}(\mathbf{X}_2)]_T = \begin{bmatrix}1 & 0\\2 & 1\\-1 & -2\end{bmatrix}\begin{bmatrix}0\\1\end{bmatrix} = \begin{bmatrix}0\\1\\-2\end{bmatrix}.$$

(b) $\mathbf{L}(\mathbf{X}_1) = 1\mathbf{Y}_1 + 2\mathbf{Y}_2 - 1\mathbf{Y}_3 = (t^2+1) + 2t - (t-1) = t^2 + t + 2$

$\mathbf{L}(\mathbf{X}_2) = 0\mathbf{Y}_1 + 1\mathbf{Y}_1 - 2\mathbf{Y}_3 = t - 2(t-1) = -t + 2$

(c) To compute $\mathbf{L}(2t + 1)$ we first compute $[2t + 1]_S$. Hence we solve

$$c_1\mathbf{X}_1 + c_2\mathbf{X}_2 = c_1(t+1) + c_2(t-1) = 2t + 1 \qquad (**)$$

We have

$$(c_1 + c_2)t + (c_1 - c_2) = 2t + 1$$

which gives linear system

$$\begin{aligned} c_1 + c_2 &= 2 \\ c_1 - c_2 &= 1 \end{aligned}$$

It follows that $c_1 = 3/2$ and $c_2 = 1/2$ so

$$[2t + 1]_S = \begin{bmatrix} 3/2 \\ 1/2 \end{bmatrix}.$$

Hence $[\mathbf{L}(2t + 1)]_T = \mathbf{A}[2t + 1]_S = \begin{bmatrix} 1 & 0 \\ 2 & 1 \\ -1 & -2 \end{bmatrix}\begin{bmatrix} 3/2 \\ 1/2 \end{bmatrix} = \begin{bmatrix} 3/2 \\ 7/2 \\ -5/2 \end{bmatrix}$

so

$$\begin{aligned} \mathbf{L}(2t + 1) &= (3/2)\mathbf{Y}_1 + (7/2)\mathbf{Y}_2 + (-5/2)\mathbf{Y}_3 \\ &= (3/2)(t^2+1) + (7/2)t - (5/2)(t-1) \\ &= (3/2)t^2 + t + 4. \end{aligned}$$

(d) In part c the right side of (**) is replaced by at + b so we are led to the linear system

$$\begin{aligned} c_1 + c_2 &= a \\ c_1 - c_2 &= b \end{aligned}$$

whose solution is $c_1 = (a+b)/2$, $c_2 = (a-b)/2$. Thus

$$[\mathbf{L}(at + b)]_T = \mathbf{A}\begin{bmatrix} (a+b)/2 \\ (a-b)/2 \end{bmatrix} = \begin{bmatrix} (a+b)/2 \\ (3a+b)/2 \\ (-3a+b)/2 \end{bmatrix}$$

and

$$\begin{aligned} \mathbf{L}(at + b) &= ((a+b)/2)\mathbf{Y}_1 + ((3a+b)/2)\mathbf{Y}_2 + ((-3a+b)/2)\mathbf{Y}_3 \\ &= ((a+b)/2)(t^2+1) + ((3a+b)/2)t \\ &\quad + ((-3a+b)/2)(t-1) \\ &= ((a+b)/2)t^2 + bt + 2a \end{aligned}$$

15. Let $\mathbf{L}:P_1 \to P_1$ where $\mathbf{L}(t+1) = t-1$ and $\mathbf{L}(t-1) = 2t+1$.

(a) $S = \{\mathbf{X}_1, \mathbf{X}_2\} = \{t+1, t-1\}$ is a basis for P_1. The matrix representing **L** with respect to S is

$$\left[[\mathbf{L}(\mathbf{X}_1)]_S \ [\mathbf{L}(\mathbf{X}_2)]_S \right].$$

Thus we must solve the two systems

$$c_1\mathbf{X}_1 + c_2\mathbf{X}_2 = \mathbf{L}(\mathbf{X}_i), \ i=1,2$$

We are led to linear systems where the corresponding partitioned matrix is

$$\left[\begin{array}{cc|c|c} 1 & 1 & 1 & 2 \\ 1 & -1 & -1 & 1 \end{array}\right]$$

The reduced row echelon form of this matrix is

$$\left[\begin{array}{cc|c|c} 1 & 0 & 0 & 3/2 \\ 0 & 1 & 1 & 1/2 \end{array}\right].$$

Thus the matrix representing **L** with respect to S is

$$\begin{bmatrix} 0 & 3/2 \\ 1 & 1/2 \end{bmatrix}.$$

(b) To compute **L**(2t+3) from the definition of **L** we first express 2t + 3 in terms of S. We have

$$\begin{aligned} 2t + 3 = k_1\mathbf{X}_1 + k_2\mathbf{X}_2 &= k_1(t+1) + k_2(t-1) \\ &= (k_1+k_2)t + (k_1-k_2) \end{aligned}$$

which leads to the system

$$\begin{aligned} k_1 + k_2 &= 2 \\ k_1 - k_2 &= 3 \end{aligned}$$

whose solution is $k_1 = 5/2$, $k_2 = -1/2$. Then since **L** is a linear transformation we have

$$\begin{aligned} \mathbf{L}(2t + 3) &= \mathbf{L}((5/2)\mathbf{X}_1 - (1/2)\mathbf{X}_2) \\ &= (5/2)\mathbf{L}(\mathbf{X}_1) - (1/2)\mathbf{L}(\mathbf{X}_2) \\ &= (5/2)(t-1) - (1/2)(2t+1) \\ &= (3/2)t - 3. \end{aligned}$$

To use the matrix from part a we use the fact that

$$\left[2t + 3\right]_S = \begin{bmatrix} 5/2 \\ -1/2 \end{bmatrix},$$

so

$$\left[\mathbf{L}(2t + 3)\right]_S = (-3/4)\mathbf{X}_1 + (9/4)\mathbf{X}_2 = (3/2)t - 3.$$

(c) Using part b, we replace 2t + 3 by at + b and we are led to the linear system

$$\begin{aligned} k_1 + k_2 &= a \\ k_1 - k_2 &= b \end{aligned}$$

whose solution is $k_1 = (a+b)/2$, $k_2 = (a-b)/2$. Then

$$[\mathbf{L}(at + b)]_S = \mathbf{A}\begin{bmatrix}(a+b)/2\\(a-b)/2\end{bmatrix} = \begin{bmatrix}0 & 3/2\\1 & 1/2\end{bmatrix}\begin{bmatrix}(a+b)/2\\(a-b)/2\end{bmatrix} = \begin{bmatrix}3(a-b)/4\\(3a+b)/4\end{bmatrix}$$

and it follows that

$$\begin{aligned}\mathbf{L}(at + b) &= (3(a-b)/4)(t+1) + ((3a+b)/4)(t-1)\\ &= ((3a-b)/2)t - b\end{aligned}$$

17. $\mathbf{L}:P_1 \to P_1$ is represented by $\mathbf{A} = \begin{bmatrix}2 & 3\\-1 & -2\end{bmatrix}$ with respect to basis $S = \{\mathbf{X}_1, \mathbf{X}_2\} = \{t+1, t-1\}$. Let $T = \{t, 1\}$ be the natural basis for P_1. To find the matrix representing $\mathbf{L}$ with respect to T we determine $\mathbf{P}$, the transition matrix matrix from T to S and then compute $\mathbf{P}^{-1}\mathbf{AP}$. We have that

$$\mathbf{P} = \begin{bmatrix}[t]_S & [1]_S\end{bmatrix}.$$

To find $[t]_S$ we solve $c_1\mathbf{X}_1 + c_2\mathbf{X}_2 = t$ and to find $[1]_S$ we solve $k_1\mathbf{X}_1 + k_2\mathbf{X}_2 = 1$. Each of these leads to a linear system with the same coefficient matrix, but a different right hand side. Hence we are led to the partitioned matrix

$$\left[\begin{array}{cc|c|c}1 & 1 & 1 & 0\\1 & -1 & 0 & 1\end{array}\right].$$

The reduced row echelon form is

$$\left[\begin{array}{cc|c|c}1 & 0 & 1/2 & 1/2\\0 & 1 & 1/2 & -1/2\end{array}\right].$$

Thus $\mathbf{P} = \begin{bmatrix}1/2 & 1/2\\1/2 & -1/2\end{bmatrix}$ and $\mathbf{P}^{-1}\mathbf{AP} = \begin{bmatrix}1 & 0\\4 & -1\end{bmatrix}$.

19. $\mathbf{L}:P_3 \to P_3$ is defined by $\mathbf{L}(p(t)) = p''(t) + p(0)$. Let $S = \{\mathbf{X}_1, \mathbf{X}_2, \mathbf{X}_3, \mathbf{X}_4\} = \{1, t, t^2, t^3\}$ and $T = \{\mathbf{Y}_1, \mathbf{Y}_2, \mathbf{Y}_3, \mathbf{Y}_4\} = \{t^3, t^2-1, t, 1\}$ be bases for P_3.

(a)
$$\begin{aligned}\mathbf{L}(\mathbf{X}_1) &= \mathbf{L}(1) = (1)'' + 1 = 1\\ \mathbf{L}(\mathbf{X}_2) &= \mathbf{L}(t) = (t)'' + 0 = 0\\ \mathbf{L}(\mathbf{X}_3) &= \mathbf{L}(t^2) = (t^2)'' + 0 = 2\\ \mathbf{L}(\mathbf{X}_4) &= \mathbf{L}(t^3) = (t^3)'' + 0 = 6t\end{aligned}$$

Since S is the natural basis for P_3, $[\mathbf{L}(\mathbf{X}_i)]_S$ for $i=1,2,3,4$ is easily computed:

$$[\mathbf{L}(\mathbf{X}_1)]_S = \begin{bmatrix}1\\0\\0\\0\end{bmatrix}, \quad [\mathbf{L}(\mathbf{X}_2)]_S = \begin{bmatrix}0\\0\\0\\0\end{bmatrix},$$

$$[\mathbf{L}(\mathbf{X}_3)]_S = \begin{bmatrix}2\\0\\0\\0\end{bmatrix}, \quad [\mathbf{L}(\mathbf{X}_4)]_S = \begin{bmatrix}0\\6\\0\\0\end{bmatrix}.$$

Hence the matrix representing **L** with respect to S is

$$\mathbf{A} = \begin{bmatrix}1 & 0 & 2 & 0\\0 & 0 & 0 & 6\\0 & 0 & 0 & 0\\0 & 0 & 0 & 0\end{bmatrix}.$$

(b) $\mathbf{L}(\mathbf{Y}_1) = \mathbf{L}(t^3) = (t^3)'' + 0 = 6t$
$\mathbf{L}(\mathbf{Y}_2) = \mathbf{L}(t^2-1) = (t^2-1)'' + (-1) = 1$
$\mathbf{L}(\mathbf{Y}_3) = \mathbf{L}(t) = (t)'' + 0 = 0$
$\mathbf{L}(\mathbf{Y}_4) = \mathbf{L}(1) = (1)'' + 1 = 1$

Next we express $\mathbf{L}(\mathbf{Y}_i)$ in terms of the vectors in T. This can be done here without solving any linear systems because of the simplicity of the vectors in T. We have

$$[\mathbf{L}(\mathbf{Y}_1)]_T = \begin{bmatrix}0\\0\\6\\0\end{bmatrix}, \quad [\mathbf{L}(\mathbf{Y}_2)]_T = \begin{bmatrix}0\\0\\0\\1\end{bmatrix},$$

$$[\mathbf{L}(\mathbf{Y}_3)]_T = \begin{bmatrix}0\\0\\0\\0\end{bmatrix}, \quad [\mathbf{L}(\mathbf{Y}_4)]_T = \begin{bmatrix}0\\0\\0\\1\end{bmatrix}.$$

It follows that the matrix representing **L** with respect to T is

$$\mathbf{B} = \begin{bmatrix}0 & 0 & 0 & 0\\0 & 0 & 0 & 0\\6 & 0 & 0 & 0\\0 & 1 & 0 & 1\end{bmatrix}.$$

(c) Given that **A** from part a represents **L** with respect to the S-basis we compute the matrix **L** with respect to the T-basis indirectly by determining the transition matrix **P** from T to S and proceeding as in Figure 4.7. From Section 3.8 we have that

$$\mathbf{P} = \left[[\mathbf{Y}_1]_S \ [\mathbf{Y}_2]_S \ [\mathbf{Y}_3]_S \ [\mathbf{Y}_4]_S \right]$$

Because S is the natural basis the $[\mathbf{Y}_i]_S$ are easy to determine. We have

$$\mathbf{P} = \begin{bmatrix} 0 & -1 & 0 & 1 \\ 0 & 0 & 1 & 0 \\ 0 & 1 & 0 & 0 \\ 1 & 0 & 0 & 0 \end{bmatrix}.$$

Computing $\mathbf{P}^{-1}$ we have that the matrix representing $\mathbf{L}$ with respect to T is

$$\mathbf{P}^{-1}\mathbf{AP} = \begin{bmatrix} 0 & 0 & 0 & 1 \\ 0 & 0 & 1 & 0 \\ 0 & 1 & 0 & 0 \\ 1 & 0 & 1 & 0 \end{bmatrix} \begin{bmatrix} 1 & 0 & 2 & 0 \\ 0 & 0 & 0 & 6 \\ 0 & 0 & 0 & 0 \\ 0 & 0 & 0 & 0 \end{bmatrix} \begin{bmatrix} 0 & -1 & 0 & 1 \\ 0 & 0 & 1 & 0 \\ 0 & 1 & 0 & 0 \\ 1 & 0 & 0 & 0 \end{bmatrix} = \mathbf{B}.$$

21. $\mathbf{L}(e^t) = (e^t)' = e^t \quad ===> \quad [\mathbf{L}(e^t)]_S = \begin{bmatrix} 1 \\ 0 \end{bmatrix}$

$\mathbf{L}(e^{-t}) = (e^{-t})' = -e^{-t} \quad ===> \quad [\mathbf{L}(e^{-t})]_S = \begin{bmatrix} 0 \\ -1 \end{bmatrix}$

Thus the matrix representing $\mathbf{L}$ with respect to S is $\begin{bmatrix} 1 & 0 \\ 0 & -1 \end{bmatrix}$.

T.1. If $\mathbf{Y} = \mathbf{L}(\mathbf{X})$ for some $\mathbf{X}$ in V, and if $\mathbf{X} = a_1\mathbf{X}_1 + \cdots + a_n\mathbf{X}_n$, then

$$\mathbf{Y} = \mathbf{L}(\mathbf{X}) = \mathbf{L}\left(\sum_{j=1}^{n} a_j\mathbf{X}_j \right) = \sum_{j=1}^{n} a_j\mathbf{L}(\mathbf{X}_j)$$

$$= a_1 \begin{bmatrix} c_{11} \\ c_{21} \\ \cdot \\ \cdot \\ c_{m1} \end{bmatrix} + a_2 \begin{bmatrix} c_{12} \\ c_{22} \\ \cdot \\ \cdot \\ c_{m2} \end{bmatrix} + \cdots + a_n \begin{bmatrix} c_{1n} \\ c2n \\ \cdot \\ \cdot \\ c_{mn} \end{bmatrix}$$

$$= \begin{bmatrix} c_{11} & c_{12} & \cdot & \cdot & \cdot & c_{1n} \\ c_{21} & c_{22} & \cdot & \cdot & \cdot & c_{2n} \\ \cdot & \cdot & & & & \cdot \\ \cdot & \cdot & & & & \cdot \\ c_{m1} & c_{m2} & \cdot & \cdot & \cdot & c_{mn} \end{bmatrix} \begin{bmatrix} a_1 \\ a_2 \\ \cdot \\ \cdot \\ a_n \end{bmatrix} = \mathbf{A}\ [\mathbf{X}]_S.$$

Exercises 4.3

Regarding the uniqueness of matrix **A**, for each $1 \le j \le n$, the jth column of **A** is uniquely determined, since for

$$\mathbf{X} = \mathbf{X}_j = 0\cdot\mathbf{X}_1 + \cdots + 1\cdot\mathbf{X}_j + \cdots + 0\cdot\mathbf{X}_n;$$

$$[\mathbf{Y}]_T = \mathbf{A}\cdot\mathbf{E}_j = \text{jth column of } \mathbf{A}.$$

T.3. Let $S = \{\mathbf{X}_1, \mathbf{X}_2, \ldots, \mathbf{X}_n\}$ be a basis for V, $T = \{\mathbf{Y}_1, \mathbf{Y}_2, \ldots, \mathbf{Y}_m\}$ a basis for W. Then

$$\mathbf{0}(\mathbf{X}_j) = \mathbf{0}_W = 0\cdot\mathbf{Y}_1 + \cdots + 0\cdot\mathbf{Y}_m.$$

If **A** is the matrix of the zero transformation with respect to these bases, then the jth column of **A** is **0**. Thus **A** is the $m \times n$ zero matrix.

ML.1. From the definition of **L**, note that we can compute images under **L** as a matrix multiply: **L**(**X**) = **CX** where

```
C=[2 - 1 0;1 1 -3]

C =

   2    -1     0
   1     1    -3
```

This observation makes it easy to compute **L**(**X**i) in MATLAB. Entering the vectors in set S and computing their images we have

```
X1=[1 1 1]';X2=[1 2 1]';X3=[0 1 -1]';
```

Denote the images as LXi:

```
LX1=C*X1

LX1 =

      1
     -1

LX2=C*X2

LX2 =

      0
      0

LX3=C*X3
```

```
LX3 =

     -1
      4
```

To find the coordinates of LXi relative to the T basis we solve the three systems involved all at once using the rref command.

```
rref([[1 2;2 1] LX1 LX2 LX3])

ans =

       1      0     -1      0      3
       0      1      1      0     -2
```

The last 3 columns give the matrix **A** representing **L** with respect to bases S and T.

```
A=ans(:,3:5)
A =

    -1      0      3
     1      0     -2
```

ML.2. Enter **C**, and the vectors from the S and T bases into MATLAB. Then compute the images of $\mathbf{X}_i$ as LXi = C*Xi.

```
C=[1 2 0;2 1 -1;3 1 0;-1 0 2]

C =

     1      2      0
     2      1     -1
     3      1      0
    -1      0      2

X1=[1 0 1]';X2=[2 0 1]';X3=[0 1 2]';
Y1=[1 1 1 2]';Y2=[1 1 1 0]';Y3=[0 1 1 -1]';Y4=[0 0 1 0]';
LX1=C*X1;LX2=C*X2;LX3=C*X3;
rref([Y1 Y2 Y3 Y4 LX1 LX2 LX3])

ans =

  1.0000         0         0         0    0.5000    0.5000    0.5000
       0    1.0000         0         0    0.5000    1.5000    1.5000
       0         0    1.0000         0         0    1.0000   -3.0000
       0         0         0    1.0000    2.0000    3.0000    2.0000
```

It follows that **A** consists of the last 3 columns of ans.

```
A=ans(:,5:7)

A =

    0.5000    0.5000    0.5000
    0.5000    1.5000    1.5000
         0    1.0000   -3.0000
    2.0000    3.0000    2.0000
```

ML.3. Note that images under **L** can be computed as a matrix multiply using matrix

```
C=[-1 2;3 -1]

C =

   -1     2
    3    -1
```

Enter each of the basis vectors into MATLAB

```
X1=[1 2]';X2=[-1 1]';
Y1=[-2 1]';Y2=[1 1]';
```

(a) Compute the images of Xi under **L**.

```
LX1=C*X1

LX1 =

     3
     1

LX2=C*X2

LX2 =

     3
    -4
```

To compute the matrix representing **L** with respect to S, compute

```
rref([X1 X2 LX1 LX2])

ans =

    1.0000         0    1.3333   -0.3333
         0    1.0000   -1.6667   -3.3333

A=ans(:,3:4)
```

```
A =

    1.3333   -0.3333
   -1.6667   -3.3333
```

(b) Compute the images of Yi under **L**.

```
LY1=C*Y1

LY1 =

     4
    -7

LY2=C*Y2

LY2 =

     1
     2
```

To compute the matrix representing **L** with respect to T, compute

```
rref([Y1 Y2 LY1 LY2])

ans =

    1.0000         0   -3.6667    0.3333
         0    1.0000   -3.3333    1.6667

B=ans(:,3:4)

B =

   -3.6667    0.3333
   -3.3333    1.6667
```

(c) To find the transition matrix from T to S we find the coordinates of the T-basis vectors in terms of S. This is done by solving 2 linear systems with coefficient matrix consisting of the S basis vectors and right hand sides the T-basis vectors. We use

```
rref([X1 X2 Y1 Y2])

ans =

    1.0000         0   -0.3333    0.6667
         0    1.0000    1.6667   -0.3333
```

Then the transition matrix **P** is

```
P=ans(:,3:4)

P =

   -0.3333    0.6667
    1.6667   -0.3333
```

(d) We have

```
inverse(P)*A*P

ans =

   -3.6667    0.3333
   -3.3333    1.6667
```

which is indeed **B**.

Exercises 4.4

1. (a) Let $\mathbf{L}(\mathbf{X}) = \mathbf{L}\left(\begin{bmatrix} x \\ y \end{bmatrix}\right) = \begin{bmatrix} -x \\ y \end{bmatrix}$. Then

$$\mathbf{L}\left(\begin{bmatrix} 1 \\ 0 \end{bmatrix}\right) = \begin{bmatrix} -1 \\ 0 \end{bmatrix} \text{ and } \mathbf{L}\left(\begin{bmatrix} 0 \\ 1 \end{bmatrix}\right) = \begin{bmatrix} 0 \\ 1 \end{bmatrix}.$$

Hence the matrix representation with respect to the natural basis is $\mathbf{A} = \begin{bmatrix} -1 & 0 \\ 0 & 1 \end{bmatrix}$.

(b) Let $\mathbf{L}(\mathbf{X}) = \mathbf{L}\left(\begin{bmatrix} x \\ y \end{bmatrix}\right) = \begin{bmatrix} y \\ x \end{bmatrix}$. Then

$$\mathbf{L}\left(\begin{bmatrix} 1 \\ 0 \end{bmatrix}\right) = \begin{bmatrix} 0 \\ 1 \end{bmatrix} \text{ and } \mathbf{L}\left(\begin{bmatrix} 0 \\ 1 \end{bmatrix}\right) = \begin{bmatrix} 1 \\ 0 \end{bmatrix}.$$

Hence the matrix representation with respect to the natural basis is $\mathbf{A} = \begin{bmatrix} 0 & 1 \\ 1 & 0 \end{bmatrix}$.

(c) Following Example 3, $\mathbf{A} = \begin{bmatrix} \cos 90° & -\sin 90° \\ \sin 90° & \cos 90° \end{bmatrix} = \begin{bmatrix} 0 & -1 \\ 1 & 0 \end{bmatrix}$.

(d) Following Example 3, $\mathbf{A} = \begin{bmatrix} \cos(-30°) & -\sin(-30°) \\ \sin(-30°) & \cos(-30°) \end{bmatrix}$

$$= \begin{bmatrix} \sqrt{3}/2 & 1/2 \\ -1/2 & \sqrt{3}/2 \end{bmatrix}.$$

3. (a) Let $S = \{\mathbf{X}_1, \mathbf{X}_2\} = \left\{\begin{bmatrix} 1 \\ 0 \end{bmatrix}, \begin{bmatrix} 0 \\ 1 \end{bmatrix}\right\}$ be the natural basis for R^2. Then $\mathbf{L}(\mathbf{X}_1) = \begin{bmatrix} 1 \\ k \end{bmatrix}$, $\mathbf{L}(\mathbf{X}_2) = \begin{bmatrix} 0 \\ 1 \end{bmatrix}$. It follows that the matrix representing $\mathbf{L}$ with respect to S is

$$\mathbf{A} = \left[[\mathbf{L}(\mathbf{X}_1)]_S \ [\mathbf{L}(\mathbf{X}_2)]_S \right] = \begin{bmatrix} 1 & 0 \\ k & 1 \end{bmatrix}.$$

(b) Let the vertices of rectangle R be denoted as vectors in R^2:

$$\mathbf{X}_1 = \begin{bmatrix} 1 \\ 1 \end{bmatrix}, \ \mathbf{X}_2 = \begin{bmatrix} 1 \\ 4 \end{bmatrix}, \ \mathbf{X}_3 = \begin{bmatrix} 3 \\ 1 \end{bmatrix}, \ \mathbf{X}_4 = \begin{bmatrix} 3 \\ 4 \end{bmatrix}.$$

From part a we have $\mathbf{A} = \begin{bmatrix} 1 & 0 \\ -2 & 1 \end{bmatrix}$ and since S is the natural basis $[\mathbf{X}_i]_S = \mathbf{X}_i$ and $\mathbf{L}(\mathbf{X}_i) = \mathbf{A}\mathbf{X}_i$. Hence

$$\mathbf{L}(\mathbf{X}_1) = \begin{bmatrix} 1 \\ -1 \end{bmatrix}, \ \mathbf{L}(\mathbf{X}_2) = \begin{bmatrix} 1 \\ 2 \end{bmatrix}, \ \mathbf{L}(\mathbf{X}_3) = \begin{bmatrix} 3 \\ -5 \end{bmatrix}, \ \mathbf{L}(\mathbf{X}_4) = \begin{bmatrix} 3 \\ -2 \end{bmatrix}.$$

Graphically we have

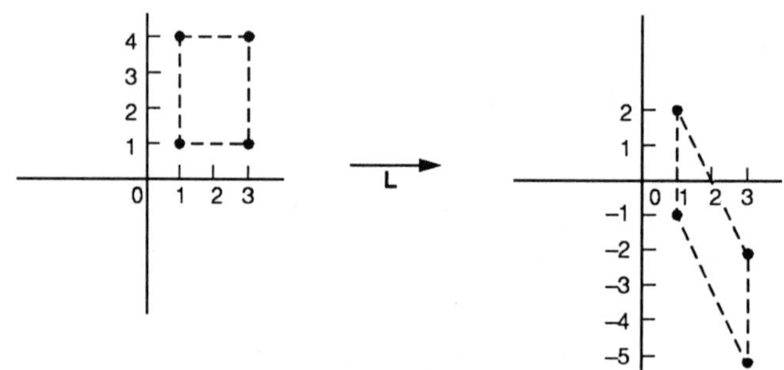

5. Let $\mathbf{L}(\mathbf{X}) = \mathbf{L}\left(\begin{bmatrix} x \\ y \end{bmatrix}\right) = \begin{bmatrix} x \\ ry \end{bmatrix}$. Then $\mathbf{L}\left(\begin{bmatrix} 1 \\ 0 \end{bmatrix}\right) = \begin{bmatrix} 1 \\ 0 \end{bmatrix}$ and $\mathbf{L}\left(\begin{bmatrix} 0 \\ 1 \end{bmatrix}\right) = \begin{bmatrix} 0 \\ r \end{bmatrix}$.

Hence the matrix representation with respect to the natural basis is $\mathbf{A} = \begin{bmatrix} 1 & 0 \\ 0 & r \end{bmatrix}$.

7. (a) Following the technique of Example 5,

$$\text{area} = \frac{1}{2}\left| \begin{vmatrix} 1 & 1 & 1 \\ -3 & -3 & 1 \\ 2 & -1 & 1 \end{vmatrix} \right| = \frac{1}{2}\,|12| = 6.$$

(b) $\mathbf{L}\left(\begin{bmatrix}1\\1\end{bmatrix}\right) = \mathbf{A}\begin{bmatrix}1\\1\end{bmatrix} = \begin{bmatrix}1\\-2\end{bmatrix}$, $\mathbf{L}\left(\begin{bmatrix}-3\\-3\end{bmatrix}\right) = \mathbf{A}\begin{bmatrix}-3\\-3\end{bmatrix} = \begin{bmatrix}-3\\6\end{bmatrix}$,

$$\mathbf{L}\left(\begin{bmatrix}2\\-1\end{bmatrix}\right) = \mathbf{A}\begin{bmatrix}2\\-1\end{bmatrix} = \begin{bmatrix}11\\-10\end{bmatrix}.$$

(c) $\text{area} = \frac{1}{2}\left|\,\begin{vmatrix}1 & -2 & 1\\-3 & 6 & 1\\11 & -10 & 1\end{vmatrix}\,\right| = \frac{1}{2}\,|48| = 24.$

T.1. Let $X(x_1,y_1)$, $Y(x_2,y_2)$, $Z(x_3,y_3)$ be the vertices of a triangle. Then from Example 5, we have

$$\text{area of } XYZ = \frac{1}{2}\left|\,\begin{vmatrix}x_1 & y_1 & 1\\x_2 & y_2 & 1\\x_3 & y_3 & 1\end{vmatrix}\,\right| = \frac{1}{2}\,|B|$$

$$= \frac{1}{2}|x_1y_2 + y_1x_3 + x_2y_3 - x_3y_2 - y_3x_1 - x_2y_1|.$$

Let $\mathbf{A}$ be the matrix representing a rotation through an angle ϕ. Thus

$$\mathbf{A} = \begin{bmatrix}\cos\phi & -\sin\phi\\ \sin\phi & \cos\phi\end{bmatrix}.$$

Define $X' = \mathbf{A}\begin{bmatrix}x_1\\y_1\end{bmatrix} = \begin{bmatrix}x_1\cos\phi - y_1\sin\phi\\x_1\sin\phi + y_1\cos\phi\end{bmatrix}$,

$$Y' = \mathbf{A}\begin{bmatrix}x_2\\y_2\end{bmatrix} = \begin{bmatrix}x_2\cos\phi - y_2\sin\phi\\x_2\sin\phi + y_2\cos\phi\end{bmatrix},$$

$$Z' = \mathbf{A}\begin{bmatrix}x_3\\y_3\end{bmatrix} = \begin{bmatrix}x_3\cos\phi - y_3\sin\phi\\x_3\sin\phi + y_3\cos\phi\end{bmatrix}.$$

Then

area of $X'Y'Z'$

$$= \frac{1}{2}\left|\,\begin{vmatrix}x_1\cos\phi - y_1\sin\phi & x_1\sin\phi + y_1\cos\phi & 1\\x_2\cos\phi - y_2\sin\phi & x_2\sin\phi + y_2\cos\phi & 1\\x_3\cos\phi - y_3\sin\phi & x_3\sin\phi + y_3\cos\phi & 1\end{vmatrix}\,\right|$$

(expand the determinant using the "3 × 3 trick ")

$$= \frac{1}{2}\ \big|(x_1\cos\phi - y_1\sin\phi)(x_2\sin\phi + y_2\cos\phi)$$
$$+ (x_1\sin\phi + y_1\cos\phi)(x_3\cos\phi - y_3\sin\phi)$$
$$+ (x_2\cos\phi - y_2\sin\phi)(x_3\sin\phi + y_3\cos\phi)$$
$$- (x_3\cos\phi - y_3\sin\phi)(x_2\sin\phi + y_2\cos\phi)$$
$$- (x_3\sin\phi + y_3\cos\phi)(x_1\cos\phi - y_1\sin\phi)$$
$$- (x_2\cos\phi - y_2\sin\phi)(x_1\sin\phi + y_1\cos\phi)\big|$$

(factor as follows)

$$= \frac{1}{2}\ \big|(x_1\cos\phi - y_1\sin\phi)\ [x_2\sin\phi + y_2\cos\phi - x_3\sin\phi - y_3\cos\phi]$$
$$+ (x_2\cos\phi - y_2\sin\phi)\ [x_3\sin\phi + y_3\cos\phi - x_1\sin\phi - y_1\cos\phi]$$
$$+ (x_3\cos\phi - y_3\sin\phi)\ [x_1\sin\phi + y_1\cos\phi - x_2\sin\phi - y_2\cos\phi]\big|$$

(expand the preceding products, collect like terms containing x's and y's, and use the identity $\sin^2\phi + \cos^2\phi = 1$ to obtain)

$$= \frac{1}{2}\big|x_1y_2 + y_1x_3 + x_2y_3 - x_3y_2 - y_3x_1 - x_2y_1\big|$$
$$= \frac{1}{2}\ |B|\ = \text{area of XYZ.}$$

Chapter 4 Supplementary Exercises

1. Let $\mathbf{L}:\mathbf{R}^2 \to \mathbf{R}^2$ be defined by $\mathbf{L}(x,y) = (x-1,y-x)$. Also let $\mathbf{X} = (x_1,y_1)$ and $\mathbf{Y} = (x_2,y_2)$ be any vectors in $\mathbf{R}^2$ and c be any real scalar. Then

$$\mathbf{L}(\mathbf{X} + \mathbf{Y}) = \mathbf{L}((x_1,y_1) + (x_2,y_2)) = \mathbf{L}(x_1+x_2,\ y_1+y_2)$$

$$= (x_1+x_2-1,\ y_1+y_2-x_1-x_2)$$

On the other hand,

$$\mathbf{L}(\mathbf{X}) + \mathbf{L}(\mathbf{Y}) = (x_1-1,y_1-x_1) + (x_2-1,y_2-x_2)$$

$$= (x_1+x_2-2,\ y_1+y_2-x_1-x_2).$$

Since, $\mathbf{L}(\mathbf{X} + \mathbf{Y}) \neq \mathbf{L}(\mathbf{X}) + \mathbf{L}(\mathbf{Y})$, $\mathbf{L}$ is not a linear transformation.

3. Since $S = \{t - 1,\ t + 1\}$ is a basis for $\mathbf{P}_1$ we determine coefficients c_1 and c_2 such that

$$c_1(t - 1) + c_2(t + 1) = 5t + 1.$$

Expanding, collecting like terms, and equating coefficients of like powers of t from both sides of the equation we obtain the linear system

$$\begin{bmatrix} 1 & 1 \\ -1 & 1 \end{bmatrix} \begin{bmatrix} c_1 \\ c_2 \end{bmatrix} = \begin{bmatrix} 5 \\ 1 \end{bmatrix}$$

whose solution is $c_1 = 2$ and $c_2 = 3$. Then

$$\mathbf{L}(5t + 1) = \mathbf{L}(2(t - 1) + 3(t + 1)) = 2\mathbf{L}(t - 1) + 3\mathbf{L}(t + 1)$$

$$= 2(t + 2) + 3(2t + 1) = 8t + 7.$$

5. Let $\mathbf{L}:\mathbf{R}^2 \to \mathbf{R}^3$ be defined by

$$\mathbf{L}(x,y) = (x + y,\ x - y,\ x + 2y).$$

(a) Following the technique in Example 11 in Section 4.2, we have

$$\mathbf{L}(x,y) = (x + y,\ x - y,\ x + 2y)$$

$$= x(1,\ 1,\ 1) + y(1,\ -1,\ 2).$$

Thus $S = \{\mathbf{X}_1,\ \mathbf{X}_2\} = \{(1,\ 1,\ 1),\ (1,\ -1,\ 2)\}$ spans range $\mathbf{L}$. Since $\mathbf{X}_2$ is not a scalar multiple of $\mathbf{X}_1$, S is also linearly independent. Thus S is a basis for range $\mathbf{L}$. (There are many other bases for range $\mathbf{L}$.)

(b) $\mathbf{L}$ is not onto since range $\mathbf{L}$ is not identically $\mathbf{R}^3$. This follows since dim(range $\mathbf{L}$) = 2, but $\dim(\mathbf{R}^3) = 3$.

7. Let $S = \{p_1(t), p_2(t), p_3(t)\} = \{t^2 - 1, t + 2, t - 1\}$ be a basis for P_2.

   (a) Following Example 5 of Section 4.3, we determine coefficients c_1, c_2, and c_3 such that

   $$p(t) = 2t^2 - 2t + 6 = c_1p_1(t) + c_2p_2(t) + c_3p_3(t).$$

   Expanding, collecting like terms, and equating coefficients of like powers of t from both sides of the equation yields the linear system

   $$\begin{aligned} c_1 \qquad\qquad\quad &= 2 \\ c_2 + c_3 &= -2 \\ -c_1 + 2c_2 - c_3 &= 6. \end{aligned}$$

   Form the augmented matrix and determine its reduced row echelon form. We obtain the solution as $c_1 = 2$, $c_2 = 2$, $c_3 = -4$. Then $[p(t)]_S = \begin{bmatrix} 2 \\ 2 \\ -4 \end{bmatrix}$.

   (b) We have $[p(t)]_S = \begin{bmatrix} 2 \\ -1 \\ 3 \end{bmatrix}$. Then $p(t) = 2p_1(t) - p_2(t) + 3p_3(t) = 2t^2 + 2t - 7$.

9. Let $\mathbf{L}:\mathbf{P}_1 \to \mathbf{P}_1$ be a linear transformation which is represented by the matrix $\mathbf{A} = \begin{bmatrix} 2 & -3 \\ 1 & 2 \end{bmatrix}$ with respect to the basis $S = \{p_1(t), p_2(t)\} = \{t - 2, t + 1\}$.

   (a) $[p_1(t)]_S = \begin{bmatrix} 1 \\ 0 \end{bmatrix}$, hence $[\mathbf{L}(p_1(t))]_S = \mathbf{A}[p_1(t)]_S = \begin{bmatrix} 2 \\ 1 \end{bmatrix}$.

   $[p_2(t)]_S = \begin{bmatrix} 0 \\ 1 \end{bmatrix}$, hence $[\mathbf{L}(p_2(t))]_S = \mathbf{A}[p_2(t)]_S = \begin{bmatrix} -3 \\ 2 \end{bmatrix}$.

   (b) Using the coordinates from part (a),

   $$\mathbf{L}(p_1(t)) = 2p_1(t) + 1p_2(t) = 3t - 3$$

   and

   $$\mathbf{L}(p_2(t)) = -3p_1(t) + 2p_2(t) = -t + 8.$$

(c) We first compute the coordinates of $p(t) = t + 2$ with respect to S. We have,

$$t + 2 = (-1/3)(t - 2) + (4/3)(t + 1),$$

thus $[p(t)]_S = \begin{bmatrix} -1/3 \\ 4/3 \end{bmatrix}$. Then,

$$\begin{aligned} \mathbf{L}(p(t)) &= \mathbf{L}((-1/3)(t - 2) + (4/3)(t + 1)) \\ &= (-1/3)\mathbf{L}(t - 2) + (4/3)\mathbf{L}(t + 1) \\ &= (-1/3)(3t - 3) + (4/3)(-t + 8) \\ &= -(7/3)t + 35/3. \end{aligned}$$

11. Let the natural basis $S = \{\mathbf{E}_1, \mathbf{E}_2, \mathbf{E}_3\}$ and $T = \{\mathbf{Y}_1, \mathbf{Y}_2, \mathbf{Y}_3\}$ $= \left\{\begin{bmatrix} 0 \\ 1 \\ 1 \end{bmatrix}, \begin{bmatrix} 1 \\ 0 \\ 1 \end{bmatrix}, \begin{bmatrix} 1 \\ 1 \\ 0 \end{bmatrix}\right\}$. We have $\mathbf{L}(\mathbf{E}_1) = \begin{bmatrix} 0 \\ 1 \\ 1 \end{bmatrix} = \mathbf{Y}_1$, $\mathbf{L}(\mathbf{E}_2) = \begin{bmatrix} 1 \\ 0 \\ 1 \end{bmatrix}$ $= \mathbf{Y}_2$, $\mathbf{L}(\mathbf{E}_3) = \begin{bmatrix} 1 \\ 1 \\ 0 \end{bmatrix} = \mathbf{Y}_3$. Hence $[\mathbf{L}(\mathbf{E}_1)]_T = \begin{bmatrix} 1 \\ 0 \\ 0 \end{bmatrix}$, $[\mathbf{L}(\mathbf{E}_2)]_T = \begin{bmatrix} 0 \\ 1 \\ 0 \end{bmatrix}$, $[\mathbf{L}(\mathbf{E}_3)]_T = \begin{bmatrix} 0 \\ 0 \\ 1 \end{bmatrix}$. The matrix of $\mathbf{L}$ with respect to S and T is $\mathbf{I}_3$.

13. $\mathbf{L}(\mathbf{X}) = a\mathbf{X} + b$. Then

$$\mathbf{L}(\mathbf{X} + \mathbf{Y}) = a(\mathbf{X} + \mathbf{Y}) + b = a\mathbf{X} + a\mathbf{Y} + b,$$

while

$$\mathbf{L}(\mathbf{X}) + \mathbf{L}(\mathbf{Y}) = a\mathbf{X} + b + a\mathbf{Y} + b = a\mathbf{X} + a\mathbf{Y} + 2b.$$

Thus $\mathbf{L}(\mathbf{X} + \mathbf{Y}) = \mathbf{L}(\mathbf{X}) + \mathbf{L}(\mathbf{Y})$ if and only if $b = 0$. Also,

$$\mathbf{L}(k\mathbf{X}) = a(k\mathbf{X} + 0) = k(a\mathbf{X} + 0) = k\mathbf{L}(\mathbf{X})$$

for any a. Thus $\mathbf{L}(\mathbf{X})$ is a linear transformation provided $b = 0$.

Supplementary Exercises 4

15. Let $\mathbf{A}$ and $\mathbf{B}$ be in V.

$$\mathbf{L}(\mathbf{A} + \mathbf{B}) = \mathbf{P}^{-1}(\mathbf{A} + \mathbf{B})\mathbf{P} = \mathbf{P}^{-1}\mathbf{AP} + \mathbf{P}^{-1}\mathbf{BP} = \mathbf{L}(\mathbf{A}) + \mathbf{L}(\mathbf{B})$$

and

$$\mathbf{L}(k\mathbf{A}) = \mathbf{P}^{-1}(k\mathbf{A})\mathbf{P} = k(\mathbf{P}^{-1}\mathbf{AP}) = k\mathbf{L}(\mathbf{A}).$$

Hence $\mathbf{L}$ is a linear transformation.

T.1. Since $[\mathbf{X}_j]_S = \mathbf{E}_j$, column j of $\mathbf{I}_n$, $\{[\mathbf{X}_1]_S, \ldots, [\mathbf{X}_n]_S\}$ is the standard basis for $\mathbf{R}^n$.

T.3. (a) $(\mathbf{L}_1 [\mathbf{L}_2)(\mathbf{X} + \mathbf{Y}) = \mathbf{L}_1(\mathbf{X} + \mathbf{Y}) + \mathbf{L}_2(\mathbf{X} + \mathbf{Y})$

$= \mathbf{L}_1(\mathbf{X}) + \mathbf{L}_1(\mathbf{Y}) + \mathbf{L}_2(\mathbf{X}) + \mathbf{L}_2(\mathbf{Y}) = (\mathbf{L}_1 [\mathbf{L}_2)(\mathbf{X}) + (\mathbf{L}_1 [\mathbf{L}_2)(\mathbf{Y})$

$(\mathbf{L}_1 [\mathbf{L}_2)(k\mathbf{X}) = \mathbf{L}_1(k\mathbf{X}) + \mathbf{L}_2(k\mathbf{X}) = k\mathbf{L}_1(\mathbf{X}) + k\mathbf{L}_2(\mathbf{X})$

$= k(\mathbf{L}_1 [\mathbf{L}_2)(\mathbf{X})$

(b) $(c] \mathbf{L})(\mathbf{X} + \mathbf{Y}) = c\mathbf{L}(\mathbf{X} + \mathbf{Y}) = c\mathbf{L}(\mathbf{X}) + c\mathbf{L}(\mathbf{Y})$

$= (c] \mathbf{L})(\mathbf{X}) + (c] \mathbf{L})(\mathbf{Y})$

$(c] \mathbf{L})(k\mathbf{X}) = c\mathbf{L}(k\mathbf{X}) = ck\mathbf{L}(\mathbf{X}) = kc\mathbf{L}(\mathbf{X}) = k(c] \mathbf{L})(\mathbf{X})$

(c) $(\mathbf{L}_1 [\mathbf{L}_2)(\mathbf{X}) = \mathbf{L}_1(\mathbf{X}) + \mathbf{L}_2(\mathbf{X})$

$= (\mathbf{X}_1 + \mathbf{X}_2, \mathbf{X}_2 + \mathbf{X}_3) + (\mathbf{X}_1 + \mathbf{X}_3, \mathbf{X}_2)$

$= (2\mathbf{X}_1 + \mathbf{X}_2 + \mathbf{X}_3, 2\mathbf{X}_2 + \mathbf{X}_3)$

$(-2] \mathbf{L})(\mathbf{X}) = -2\mathbf{L}(\mathbf{X}) = -2(\mathbf{X}_1 + \mathbf{X}_2, \mathbf{X}_2 + \mathbf{X}_3)$

$= (-2\mathbf{X}_1 - 2\mathbf{X}_2, -2\mathbf{X}_2 - 2\mathbf{X}_3)$

T.5. (a) $(\mathbf{L}_2 \circ \mathbf{L}_1)(\mathbf{X} + \mathbf{Y}) = \mathbf{L}_2(\mathbf{L}_1(\mathbf{X} + \mathbf{Y})) = \mathbf{L}_2(\mathbf{L}_1(\mathbf{X}) + \mathbf{L}_1(\mathbf{Y}))$

$= \mathbf{L}_2(\mathbf{L}_1(\mathbf{X})) + \mathbf{L}_2(\mathbf{L}_1(\mathbf{Y})) = (\mathbf{L}_2 \circ \mathbf{L}_1)(\mathbf{X}) + (\mathbf{L}_2 \circ \mathbf{L}_1)(\mathbf{Y})$

and

$(\mathbf{L}_2 \circ \mathbf{L}_1)(k\mathbf{X}) = \mathbf{L}_2(\mathbf{L}_1(k\mathbf{X})) = \mathbf{L}_2(k\mathbf{L}_1(\mathbf{X}))$

$= k\mathbf{L}_2(\mathbf{L}_1(\mathbf{X})) = k(\mathbf{L}_2 \circ \mathbf{L}_1)(\mathbf{X}).$

Thus $\mathbf{L}_2 \circ \mathbf{L}_1$ is a linear transformation.

(b) $(\mathbf{L}_1 \circ \mathbf{L}_2)((x_1, x_2)) = \mathbf{L}_1(\mathbf{L}_2(x_1, x_2)) = \mathbf{L}_1(x_2, x_1) = (x_2, -x_1)$

$(\mathbf{L}_2 \circ \mathbf{L}_1)((x_1, x_2)) = \mathbf{L}_2(\mathbf{L}_1(x_1, x_2)) = \mathbf{L}_2(x_1, -x_2) = (-x_2, x_1)$

T.7. For $\mathbf{X} = (a_1, a_2, \ldots, a_n)$, $\mathbf{L}(\mathbf{X}) = a_1\mathbf{X}_1 + a_2\mathbf{X}_2 + \cdots + a_n\mathbf{X}_n$.

(a) Let $\mathbf{Y} = (b_1, b_2, \ldots, b_n)$. Then

$$\begin{aligned}\mathbf{L}(\mathbf{X}+\mathbf{Y}) &= (a_1+b_1)\mathbf{X}_1 + (a_2+b_2)\mathbf{X}_2 + \cdots + (a_n+b_n)\mathbf{X}_n \\ &= (a_1\mathbf{X}_1 + a_2\mathbf{X}_2 + \cdots + a_n\mathbf{X}_n) \\ &\quad + (b_1\mathbf{X}_1 + b_2\mathbf{X}_2 + \cdots + b_n\mathbf{X}_n) \\ &= \mathbf{L}(\mathbf{X}) + \mathbf{L}(\mathbf{Y})\end{aligned}$$

$$\begin{aligned}\mathbf{L}(k\mathbf{X}) &= (ka_1)\mathbf{X}_1 + (ka_2)\mathbf{X}_2 + \cdots + (ka_n)\mathbf{X}_n \\ &= k(a_1\mathbf{X}_1 + a_2\mathbf{X}_2 + \cdots + a_n\mathbf{X}_n) \\ &= k\mathbf{L}(\mathbf{X})\end{aligned}$$

Thus $\mathbf{L}$ is a linear transformation.

(b) From Theorem 4.5, $\mathbf{L}$ is one-to-one provided ker $\mathbf{L} = \mathbf{0}$. Assume that $\mathbf{L}(\mathbf{X}) = \mathbf{0}$ where

$$\mathbf{X} = a_1\mathbf{X}_1 + a_2\mathbf{X}_2 + \cdots + a_n\mathbf{X}_n .$$

Then we have

$$a_1\mathbf{X}_1 + a_2\mathbf{X}_2 + \cdots + a_n\mathbf{X}_n = \mathbf{0}.$$

Since S is a basis, $\{\mathbf{X}_1, \mathbf{X}_2, \ldots, \mathbf{X}_n\}$ are linearly independent so $a_1 = a_2 = \cdots = a_n = 0$ and it follows that $\mathbf{X} = \mathbf{0}$. Hence ker $\mathbf{L} = \mathbf{0}$ and $\mathbf{L}$ is one-to-one.

(c) That $\mathbf{L}$ is onto follows from part b and Corollary 4.1(a).

Exercises 5.1

1. $\det(\lambda \mathbf{I}_3 - \mathbf{A}) = \begin{vmatrix} \lambda-1 & -2 & -1 \\ 0 & \lambda-1 & -2 \\ 1 & -3 & \lambda-2 \end{vmatrix}$

$= (\lambda-1)(\lambda-1)(\lambda-2) + 4 + 0 + (\lambda-1) - 6(\lambda-1) - 0$
$= \lambda^3 - 4\lambda^2 + 7$

3. $\det(\lambda \mathbf{I}_3 - \mathbf{A}) = \begin{vmatrix} \lambda-4 & 1 & -3 \\ 0 & \lambda-2 & -1 \\ 0 & 0 & \lambda-3 \end{vmatrix} = (\lambda-4)(\lambda-2)(\lambda-3)$

$= \lambda^3 - 9\lambda^2 + 26\lambda - 24$

5. $\det(\lambda \mathbf{I}_3 - \mathbf{A}) = \begin{vmatrix} \lambda-1 & 0 & 0 \\ 1 & \lambda-3 & 0 \\ -3 & -2 & \lambda+2 \end{vmatrix} = (\lambda-1)(\lambda-3)(\lambda+2)$

Thus the eigenvalues of A are $\lambda_1 = 1$, $\lambda_2 = 3$, $\lambda_3 = -2$. To find the corresponding eigenvectors, solve the linear systems $(\lambda_j \mathbf{I}_3 - \mathbf{A})\mathbf{X} = \mathbf{0}$. For $\lambda_1 = 1$, we have $(1\mathbf{I}_3 - \mathbf{A})\mathbf{X} = \mathbf{0}$:

$$\begin{bmatrix} 0 & 0 & 0 \\ 1 & -2 & 0 \\ -3 & -2 & 3 \end{bmatrix} \begin{bmatrix} x_1 \\ x_2 \\ x_3 \end{bmatrix} = \begin{bmatrix} 0 \\ 0 \\ 0 \end{bmatrix}.$$

The reduced row echelon form of the coefficient matrix is

$$\begin{bmatrix} 1 & 0 & -3/4 \\ 0 & 1 & -3/8 \\ 0 & 0 & 0 \end{bmatrix}.$$

Thus the solution is $x_1 = (3/4)r$, $x_2 = (3/8)r$, $x_3 = r$, where r is any real number and every vector of the form

$\begin{bmatrix} (3/4)r \\ (3/8)r \\ r \end{bmatrix}$, $r \neq 0$, is an eigenvector associated with eigenvalue

$\lambda_1 = 1$. Let $r = 8$, and we have eigenvector $\begin{bmatrix} 6 \\ 3 \\ 8 \end{bmatrix}$. For $\lambda_2 = 3$, we

have $(3\mathbf{I}_3 - \mathbf{A})\mathbf{X} = \mathbf{0}$:

Exercises 5.1

$$\begin{bmatrix} 2 & 0 & 0 \\ 1 & 0 & 0 \\ -3 & -2 & 5 \end{bmatrix}\begin{bmatrix} x_1 \\ x_2 \\ x_3 \end{bmatrix} = \begin{bmatrix} 0 \\ 0 \\ 0 \end{bmatrix}.$$

The reduced row echelon form of the coefficient matrix is

$$\begin{bmatrix} 1 & 0 & 0 \\ 0 & 1 & -5/2 \\ 0 & 0 & 0 \end{bmatrix}.$$

Thus the solution is $x_1 = 0$, $x_2 = (5/2)r$, $x_3 = r$, where r is any real number and every vector of the form $\begin{bmatrix} 0 \\ (5/2)r \\ r \end{bmatrix}$, $r \neq 0$, is an eigenvector associated with eigenvalue $\lambda_2 = 3$. Let $r = 2$, and we have eigenvector $\begin{bmatrix} 0 \\ 5 \\ 2 \end{bmatrix}$. For $\lambda_3 = -2$, we have $(-2\mathbf{I}_3 - \mathbf{A})\mathbf{X} = \mathbf{0}$:

$$\begin{bmatrix} -3 & 0 & 0 \\ 1 & -5 & 0 \\ -3 & -2 & 0 \end{bmatrix}\begin{bmatrix} x_1 \\ x_2 \\ x_3 \end{bmatrix} = \begin{bmatrix} 0 \\ 0 \\ 0 \end{bmatrix}.$$

The reduced row echelon form of the coefficient matrix is

$$\begin{bmatrix} 1 & 0 & 0 \\ 0 & 1 & 0 \\ 0 & 0 & 0 \end{bmatrix}.$$

Thus the solution is $x_1 = 0$, $x_2 = 0$, $x_3 = r$, where r is any real number and every vector of the form $\begin{bmatrix} 0 \\ 0 \\ r \end{bmatrix}$, $r \neq 0$, is an eigenvector associated with eigenvalue $\lambda_3 = -2$. Let $r = 1$, and we have eigenvector $\begin{bmatrix} 0 \\ 0 \\ 1 \end{bmatrix}$.

7. $\det(\lambda\mathbf{I}_2 - \mathbf{A}) = \begin{vmatrix} \lambda-1 & 1 \\ -2 & \lambda-4 \end{vmatrix} = (\lambda-1)(\lambda-4) + 2 = \lambda^2 - 5\lambda + 6$

$= (\lambda-2)(\lambda-3)$

Thus the eigenvalues of A are $\lambda_1 = 2$, $\lambda_2 = 3$,. To find the corresponding eigenvectors, solve the linear systems $(\lambda_j\mathbf{I}_2 - \mathbf{A})\mathbf{X} = \mathbf{0}$. For $\lambda_1 = 2$, we have $(2\mathbf{I}_2 - \mathbf{A})\mathbf{X} = \mathbf{0}$:

$$\begin{bmatrix} 1 & 1 \\ -2 & -2 \end{bmatrix}\begin{bmatrix} x_1 \\ x_2 \end{bmatrix} = \begin{bmatrix} 0 \\ 0 \end{bmatrix}.$$

The reduced row echelon form of the coefficient matrix is

$$\begin{bmatrix} 1 & 1 \\ 0 & 0 \end{bmatrix}.$$

Thus the solution is $x_1 = -r$, $x_2 = r$, where r is any real number and every vector of the form $\begin{bmatrix} -r \\ r \end{bmatrix}$, $r \neq 0$, is an eigenvector associated with eigenvalue $\lambda_1 = 2$. Let $r = -1$, and we have eigenvector $\begin{bmatrix} 1 \\ -1 \end{bmatrix}$. For $\lambda_2 = 3$, we have $(3\mathbf{I}_2 - \mathbf{A})\mathbf{X} = \mathbf{0}$:

$$\begin{bmatrix} 2 & 1 \\ -2 & -1 \end{bmatrix}\begin{bmatrix} x_1 \\ x_2 \end{bmatrix} = \begin{bmatrix} 0 \\ 0 \end{bmatrix}.$$

The reduced row echelon form of the coefficient matrix is

$$\begin{bmatrix} 1 & 1/2 \\ 0 & 0 \end{bmatrix}.$$

Thus the solution is $x_1 = (-1/2)r$, $x_2 = r$, where r is any real number and every vector of the form $\begin{bmatrix} (-1/2)r \\ r \end{bmatrix}$, $r \neq 0$, is an eigenvector associated with eigenvalue $\lambda_2 = 3$. Let $r = -2$, and we have eigenvector $\begin{bmatrix} 1 \\ -2 \end{bmatrix}$.

9. $\det(\lambda\mathbf{I}_3 - \mathbf{A}) = \begin{vmatrix} \lambda-2 & -2 & -3 \\ -1 & \lambda-2 & -1 \\ -2 & 2 & \lambda-1 \end{vmatrix} = (\lambda-2)(\lambda-2)(\lambda-1) - 4 + 6$

$- 6(\lambda-2) + 2(\lambda-2) - 2(\lambda-1) = \lambda^3 - 5\lambda^2 + 2\lambda + 8$
$= (\lambda+1)(\lambda-2)(\lambda-4)$

Thus the eigenvalues of A are $\lambda_1 = -1$, $\lambda_2 = 2$, $\lambda_3 = 4$. To find the corresponding eigenvectors, solve the linear systems $(\lambda_j\mathbf{I}_3 - \mathbf{A})\mathbf{X} = \mathbf{0}$. For $\lambda_1 = -1$, we have $(-1\mathbf{I}_3 - \mathbf{A})\mathbf{X} = \mathbf{0}$:

$$\begin{bmatrix} -3 & -2 & -3 \\ -1 & -3 & -1 \\ -2 & 2 & -2 \end{bmatrix}\begin{bmatrix} x_1 \\ x_2 \\ x_3 \end{bmatrix} = \begin{bmatrix} 0 \\ 0 \\ 0 \end{bmatrix}.$$

The reduced row echelon form of the coefficient matrix is

$$\begin{bmatrix} 1 & 0 & 1 \\ 0 & 1 & 0 \\ 0 & 0 & 0 \end{bmatrix}.$$

Thus the solution is $x_1 = -r$, $x_2 = 0$, $x_3 = r$, where r is any real number and every vector of the form $\begin{bmatrix} -r \\ 0 \\ r \end{bmatrix}$, $r \neq 0$, is an eigenvector associated with eigenvalue $\lambda_1 = -1$. Let $r = -1$, and we have eigenvector $\begin{bmatrix} 1 \\ 0 \\ -1 \end{bmatrix}$. For $\lambda_2 = 2$, we have

$(2\mathbf{I}_3 - \mathbf{A})\mathbf{X} = \mathbf{0}$:

$$\begin{bmatrix} 0 & -2 & -3 \\ -1 & 0 & -1 \\ -2 & 2 & 1 \end{bmatrix}\begin{bmatrix} x_1 \\ x_2 \\ x_3 \end{bmatrix} = \begin{bmatrix} 0 \\ 0 \\ 0 \end{bmatrix}.$$

The reduced row echelon form of the coefficient matrix is

$$\begin{bmatrix} 1 & 0 & 1 \\ 0 & 1 & 3/2 \\ 0 & 0 & 0 \end{bmatrix}.$$

Thus the solution is $x_1 = -r$, $x_2 = (-3/2)r$, $x_3 = r$, where r is any real number and every vector of the form $\begin{bmatrix} -r \\ (-3/2)r \\ r \end{bmatrix}$, $r \neq 0$, is an eigenvector associated with eigenvalue $\lambda_2 = 2$. Let $r = 2$, and we have eigenvector $\begin{bmatrix} -2 \\ -3 \\ 2 \end{bmatrix}$. For $\lambda_3 = 4$, we have

$(4\mathbf{I}_3 - \mathbf{A})\mathbf{X} = \mathbf{0}$:

$$\begin{bmatrix} 2 & -2 & -3 \\ -1 & 2 & -1 \\ -2 & 2 & 3 \end{bmatrix}\begin{bmatrix} x_1 \\ x_2 \\ x_3 \end{bmatrix} = \begin{bmatrix} 0 \\ 0 \\ 0 \end{bmatrix}.$$

The reduced row echelon form of the coefficient matrix is

$$\begin{bmatrix} 1 & 0 & -4 \\ 0 & 1 & -5/2 \\ 0 & 0 & 0 \end{bmatrix}.$$

Thus the solution is $x_1 = 4r$, $x_2 = (5/2)r$, $x_3 = r$, where r is any real number and every vector of the form $\begin{bmatrix} 4r \\ (5/2)r \\ r \end{bmatrix}$, $r \neq 0$, is an eigenvector associated with eigenvalue $\lambda_3 = 4$. Let $r = 2$, and we have eigenvector $\begin{bmatrix} 8 \\ 5 \\ 2 \end{bmatrix}$.

11. $\det(\lambda \mathbf{I}_4 - \mathbf{A}) = \begin{vmatrix} \lambda-1 & -2 & -3 & -4 \\ 0 & \lambda+1 & -3 & -2 \\ 0 & 0 & \lambda-3 & -3 \\ 0 & 0 & 0 & \lambda-2 \end{vmatrix}$

$= (\lambda-1)(\lambda+1)(\lambda-3)(\lambda-2)$

Thus the eigenvalues of A are $\lambda_1 = 1$, $\lambda_2 = -1$, $\lambda_3 = 3$, $\lambda_4 = 2$. To find the corresponding eigenvectors, solve the linear systems $(\lambda_j \mathbf{I}_4 - \mathbf{A})\mathbf{X} = \mathbf{0}$. For $\lambda_1 = 1$, we have $(1\mathbf{I}_4 - \mathbf{A})\mathbf{X} = \mathbf{0}$:

$$\begin{bmatrix} 0 & -2 & -3 & -4 \\ 0 & 2 & -3 & -2 \\ 0 & 0 & -2 & -3 \\ 0 & 0 & 0 & -1 \end{bmatrix} \begin{bmatrix} x_1 \\ x_2 \\ x_3 \\ x_4 \end{bmatrix} = \begin{bmatrix} 0 \\ 0 \\ 0 \\ 0 \end{bmatrix}.$$

The reduced row echelon form of the coefficient matrix is

$$\begin{bmatrix} 0 & 1 & 0 & 0 \\ 0 & 0 & 1 & 0 \\ 0 & 0 & 0 & 1 \\ 0 & 0 & 0 & 0 \end{bmatrix}.$$

Thus the solution is $x_1 = r$, $x_2 = 0$, $x_3 = 0$, $x_4 = 0$, where r is any real number and every vector of the form $\begin{bmatrix} r \\ 0 \\ 0 \\ 0 \end{bmatrix}$, $r \neq 0$, is an eigenvector associated with eigenvalue $\lambda_1 = 1$. Let

$r = 1$, and we have eigenvector $\begin{bmatrix} 1 \\ 0 \\ 0 \\ 0 \end{bmatrix}$. For $\lambda_2 = -1$, we have

$(-1\mathbf{I}_4 - \mathbf{A})\mathbf{X} = \mathbf{0}$:

$$\begin{bmatrix} -2 & -2 & -3 & -4 \\ 0 & 0 & -3 & -2 \\ 0 & 0 & -4 & -3 \\ 0 & 0 & 0 & -3 \end{bmatrix} \begin{bmatrix} x_1 \\ x_2 \\ x_3 \\ x_4 \end{bmatrix} = \begin{bmatrix} 0 \\ 0 \\ 0 \\ 0 \end{bmatrix}.$$

The reduced row echelon form of the coefficient matrix is

$$\begin{bmatrix} 1 & 1 & 0 & 0 \\ 0 & 0 & 1 & 0 \\ 0 & 0 & 0 & 1 \\ 0 & 0 & 0 & 0 \end{bmatrix}.$$

Thus the solution is $x_1 = -r$, $x_2 = r$, $x_3 = 0$, $x_4 = 0$, where r is any real number and every vector of the form

$\begin{bmatrix} -r \\ r \\ 0 \\ 0 \end{bmatrix}$, $r \neq 0$, is an eigenvector associated with eigenvalue

$\lambda_2 = -1$. Let $r = -1$, and we have eigenvector $\begin{bmatrix} 1 \\ -1 \\ 0 \\ 0 \end{bmatrix}$. For $\lambda_3 = 3$,

we have $(3\mathbf{I}_4 - \mathbf{A})\mathbf{X} = \mathbf{0}$:

$$\begin{bmatrix} 2 & -2 & -3 & -4 \\ 0 & 4 & -3 & -2 \\ 0 & 0 & 0 & -3 \\ 0 & 0 & 0 & 1 \end{bmatrix} \begin{bmatrix} x_1 \\ x_2 \\ x_3 \\ x_4 \end{bmatrix} = \begin{bmatrix} 0 \\ 0 \\ 0 \\ 0 \end{bmatrix}.$$

The reduced row echelon form of the coefficient matrix is

$$\begin{bmatrix} 1 & 0 & -9/4 & 0 \\ 0 & 1 & -3/4 & 0 \\ 0 & 0 & 0 & 1 \\ 0 & 0 & 0 & 0 \end{bmatrix}.$$

Thus the solution is $x_1 = (9/4)r$, $x_2 = (3/4)r$, $x_3 = r$, $x_4 = 0$, where r is any real number and every vector of the form $\begin{bmatrix} (9/4)r \\ (3/4)r \\ r \\ 0 \end{bmatrix}$, $r \neq 0$, is an eigenvector associated with

eigenvalue $\lambda_3 = 3$. Let $r = 4$, and we have eigenvector $\begin{bmatrix} 9 \\ 3 \\ 4 \\ 0 \end{bmatrix}$.

For $\lambda_4 = 2$, we have $(2\mathbf{I}_4 - \mathbf{A})\mathbf{X} = \mathbf{0}$:

$$\begin{bmatrix} 1 & -2 & -3 & -4 \\ 0 & 3 & -3 & -2 \\ 0 & 0 & -1 & -3 \\ 0 & 0 & 0 & 0 \end{bmatrix}\begin{bmatrix} x_1 \\ x_2 \\ x_3 \\ x_4 \end{bmatrix} = \begin{bmatrix} 0 \\ 0 \\ 0 \\ 0 \end{bmatrix}.$$

The reduced row echelon form of the coefficient matrix is

$$\begin{bmatrix} 1 & 0 & 0 & 29/3 \\ 0 & 1 & 0 & 7/3 \\ 0 & 0 & 1 & 3 \\ 0 & 0 & 0 & 0 \end{bmatrix}.$$

Thus the solution is $x_1 = (-29/3)r$, $x_2 = (-7/3)r$, $x_3 = -3r$, $x_4 = r$, where r is any real number and every vector of the

form $\begin{bmatrix} (-29/3)r \\ (-7/3)r \\ -3r \\ r \end{bmatrix}$, $r \neq 0$, is an eigenvector associated with

eigenvalue $\lambda_4 = 2$. Let $r = -3$, and we have eigenvector $\begin{bmatrix} 29 \\ 7 \\ 9 \\ -3 \end{bmatrix}$.

13. Let $\mathbf{A} = \begin{bmatrix} 1 & 0 \\ -2 & 1 \end{bmatrix}$. Then $\det(\lambda\mathbf{I}_2 - \mathbf{A}) = \begin{vmatrix} \lambda-1 & 0 \\ 2 & \lambda-1 \end{vmatrix} = (\lambda-1)(\lambda-1)$.

Thus the eigenvalues are $\lambda = 1$ and $\lambda = 1$. Since the eigenvalue has multiplicity 2, we must determine the number of linearly independent eigenvectors associated with $\lambda = 1$. They are obtained by solving the linear system $(1\mathbf{I}_2 - \mathbf{A})\mathbf{X} = \mathbf{0}$:

$$\begin{bmatrix} 0 & 0 \\ 2 & 0 \end{bmatrix}\begin{bmatrix} x_1 \\ x_2 \end{bmatrix} = \begin{bmatrix} 0 \\ 0 \end{bmatrix}.$$

The reduced row echelon form of the coefficient matrix is $\begin{bmatrix} 1 & 0 \\ 0 & 0 \end{bmatrix}$. Thus the solution is $x_1 = 0$, $x_2 = r$, where r is any real number and every vector of the form $\begin{bmatrix} 0 \\ r \end{bmatrix}$, $r \neq 0$, is an eigenvector associated with eigenvalue $\lambda = 1$. The dimension of the solution space of $(1\mathbf{I}_2 - \mathbf{A})\mathbf{X} = \mathbf{0}$ is 1 (a basis consists of the vector $\begin{bmatrix} 0 \\ 1 \end{bmatrix}$). Hence there do not exist two linearly independent eigenvectors associated with $\lambda = 1$. Thus **A** is not diagonalizable.

15. Let $\mathbf{A} = \begin{bmatrix} 1 & 2 & 3 \\ 0 & -1 & 2 \\ 0 & 0 & 2 \end{bmatrix}$. Then $\det(\lambda\mathbf{I}_3 - \mathbf{A}) = \begin{vmatrix} \lambda-1 & -2 & -3 \\ 0 & \lambda+1 & -2 \\ 0 & 0 & \lambda-2 \end{vmatrix}$
$= (\lambda-1)(\lambda+1)(\lambda-2)$. Thus the eigenvalues are $\lambda_1 = 1$, $\lambda_2 = -1$, and $\lambda_3 = 2$. Since the eigenvalues are real and distinct, **A** is diagonalizable. See Theorem 5.3.

17. Let $\mathbf{A} = \begin{bmatrix} 4 & 2 & 3 \\ 2 & 1 & 2 \\ -1 & -2 & 0 \end{bmatrix}$. Then $\det(\lambda\mathbf{I}_3 - \mathbf{A}) = \begin{vmatrix} \lambda-4 & -2 & -3 \\ -2 & \lambda-1 & -2 \\ 1 & 2 & \lambda \end{vmatrix}$
$= (\lambda-1)(\lambda-1)(\lambda-3)$. Thus the eigenvalues of **A** are $\lambda_1 = 1$, $\lambda_2 = 1$, $\lambda_3 = 3$. Determine the eigenvectors associated with the eigenvalue $\lambda = 1$ which has multiplicity 2. They are obtained by solving the linear system $(1\mathbf{I}_3 - \mathbf{A})\mathbf{X} = \mathbf{0}$:

$$\begin{bmatrix} -3 & -2 & -3 \\ -2 & 0 & -2 \\ 1 & 2 & 1 \end{bmatrix}\begin{bmatrix} x_1 \\ x_2 \\ x_3 \end{bmatrix} = \begin{bmatrix} 0 \\ 0 \\ 0 \end{bmatrix}.$$

The reduced row echelon form of the coefficient matrix is $\begin{bmatrix} 1 & 0 & 1 \\ 0 & 1 & 0 \\ 0 & 0 & 0 \end{bmatrix}$. Thus the solution is $x_1 = -r$, $x_2 = 0$, $x_3 = r$, where r is any real number and every vector of the form $\begin{bmatrix} -r \\ 0 \\ r \end{bmatrix}$, $r \neq 0$, is an eigenvector associated with eigenvalue $\lambda = 1$. Hence the dimension of the solution space of $(1\mathbf{I}_3 - \mathbf{A})\mathbf{X} = \mathbf{0}$ is 1 (a basis consists of the vector $\begin{bmatrix} -1 \\ 0 \\ 1 \end{bmatrix}$); there are not two linearly independent eigenvectors associated with the eigenvalue $\lambda = 1$, and $\mathbf{A}$ is not diagonalizable.

19. Let $\mathbf{A} = \begin{bmatrix} 1 & 2 & 3 \\ 0 & 1 & 0 \\ 2 & 1 & 2 \end{bmatrix}$. Then $\det(\lambda\mathbf{I}_3 - \mathbf{A}) = \begin{vmatrix} \lambda-1 & -2 & -3 \\ 0 & \lambda-1 & 0 \\ -2 & -1 & \lambda-2 \end{vmatrix}$ $= (\lambda-4)(\lambda+1)(\lambda-1)$. The eigenvalues are $\lambda_1 = 4$, $\lambda_2 = -1$, $\lambda_3 = 1$. Since the eigenvalues are distinct, Theorem 5.3 guarantees that $\mathbf{A}$ is diagonalizable. To find $\mathbf{P}$, we determine the corresponding eigenvectors. Solve $(4\mathbf{I}_3 - \mathbf{A})\mathbf{X} = \mathbf{0}$:

$$\begin{bmatrix} 3 & -2 & -3 \\ 0 & 3 & 0 \\ -2 & -1 & 2 \end{bmatrix} \begin{bmatrix} x_1 \\ x_2 \\ x_3 \end{bmatrix} = \begin{bmatrix} 0 \\ 0 \\ 0 \end{bmatrix}.$$

The reduced row echelon form of the coefficient matrix is $\begin{bmatrix} 1 & 0 & -1 \\ 0 & 1 & 0 \\ 0 & 0 & 0 \end{bmatrix}$. The solution is $x_1 = r$, $x_2 = 0$, $x_3 = r$, where r is any real number. Let $r = 1$, then we can take $\mathbf{X}_1 = \begin{bmatrix} 1 \\ 0 \\ 1 \end{bmatrix}$ as an eigenvector associated with $\lambda_1 = 4$. Solve $(-1\mathbf{I}_3 - \mathbf{A})\mathbf{X} = \mathbf{0}$:

$$\begin{bmatrix} -2 & -2 & -3 \\ 0 & -2 & 0 \\ -2 & -1 & -3 \end{bmatrix} \begin{bmatrix} x_1 \\ x_2 \\ x_3 \end{bmatrix} = \begin{bmatrix} 0 \\ 0 \\ 0 \end{bmatrix}.$$

The reduced row echelon form of the coefficient matrix is $\begin{bmatrix} 1 & 0 & 3/2 \\ 0 & 1 & 0 \\ 0 & 0 & 0 \end{bmatrix}$. The solution is $x_1 = (-3/2)r$, $x_2 = 0$, $x_3 = r$, where r is any real number. Let $r = 2$, then we can take

$\mathbf{X}_2 = \begin{bmatrix} -3 \\ 0 \\ 2 \end{bmatrix}$ as an eigenvector associated with $\lambda_2 = -1$. Solve $(1\mathbf{I}_3 - \mathbf{A})\mathbf{X} = \mathbf{0}$:

$$\begin{bmatrix} 0 & -2 & -3 \\ 0 & 0 & 0 \\ -2 & -1 & -1 \end{bmatrix} \begin{bmatrix} x_1 \\ x_2 \\ x_3 \end{bmatrix} = \begin{bmatrix} 0 \\ 0 \\ 0 \end{bmatrix}.$$

The reduced row echelon form of the coefficient matrix is $\begin{bmatrix} 1 & 0 & -1/4 \\ 0 & 1 & 3/2 \\ 0 & 0 & 0 \end{bmatrix}$. The solution is $x_1 = (1/4)r$, $x_2 = (-3/2)r$, $x_3 = r$, where r is any real number. Let r = 4, then we can take $\mathbf{X}_3 = \begin{bmatrix} 1 \\ -6 \\ 4 \end{bmatrix}$ as an eigenvector associated with $\lambda_3 = 1$. Let

$\mathbf{P} = [\mathbf{X}_1\ \mathbf{X}_2\ \mathbf{X}_3] = \begin{bmatrix} 1 & -3 & 1 \\ 0 & 0 & -6 \\ 1 & 2 & 4 \end{bmatrix}$. Then $\mathbf{P}$ is nonsingular and

$$\mathbf{P}^{-1}\mathbf{A}\mathbf{P} = \begin{bmatrix} 4 & 0 & 0 \\ 0 & -1 & 0 \\ 0 & 0 & 1 \end{bmatrix}.$$

21. Let $\mathbf{A} = \begin{bmatrix} 3 & -2 & 1 \\ 0 & 2 & 0 \\ 0 & 0 & 0 \end{bmatrix}$. Then $\det(\lambda\mathbf{I}_3 - \mathbf{A}) = \begin{vmatrix} \lambda-3 & 2 & -1 \\ 0 & \lambda-2 & 0 \\ 0 & 0 & \lambda \end{vmatrix}$
$= (\lambda-3)(\lambda-2)\lambda$. The eigenvalues are $\lambda_1 = 3$, $\lambda_2 = 2$, $\lambda_3 = 0$. Since the eigenvalues are distinct, Theorem 5.3 guarantees that $\mathbf{A}$ is diagonalizable. To find $\mathbf{P}$, we determine the corresponding eigenvectors. Solve $(3\mathbf{I}_3 - \mathbf{A})\mathbf{X} = \mathbf{0}$:

$$\begin{bmatrix} 0 & 2 & -1 \\ 0 & 1 & 0 \\ 0 & 0 & 3 \end{bmatrix} \begin{bmatrix} x_1 \\ x_2 \\ x_3 \end{bmatrix} = \begin{bmatrix} 0 \\ 0 \\ 0 \end{bmatrix}.$$

The reduced row echelon form of the coefficient matrix is

$\begin{bmatrix} 0 & 1 & 0 \\ 0 & 0 & 1 \\ 0 & 0 & 0 \end{bmatrix}$. The solution is $x_1 = r$, $x_2 = 0$, $x_3 = 0$, where r is any real number. Let r = 1, then we can take $\mathbf{X}_1 = \begin{bmatrix} 1 \\ 0 \\ 0 \end{bmatrix}$ as an eigenvector associated with $\lambda_1 = 3$. Solve $(2\mathbf{I}_3 - \mathbf{A})\mathbf{X} = \mathbf{0}$:

$$\begin{bmatrix} -1 & 2 & -1 \\ 0 & 0 & 0 \\ 0 & 0 & 2 \end{bmatrix}\begin{bmatrix} x_1 \\ x_2 \\ x_3 \end{bmatrix} = \begin{bmatrix} 0 \\ 0 \\ 0 \end{bmatrix}.$$

The reduced row echelon form of the coefficient matrix is $\begin{bmatrix} 1 & -2 & 0 \\ 0 & 0 & 1 \\ 0 & 0 & 0 \end{bmatrix}$. The solution is $x_1 = 2r$, $x_2 = r$, $x_3 = 0$, where r is any real number. Let r = 1, then we can take $\mathbf{X}_2 = \begin{bmatrix} 2 \\ 1 \\ 0 \end{bmatrix}$ as an eigenvector associated with $\lambda_2 = 2$. Solve $(0\mathbf{I}_3 - \mathbf{A})\mathbf{X} = \mathbf{0}$:

$$\begin{bmatrix} -3 & 2 & -1 \\ 0 & -2 & 0 \\ 0 & 0 & 0 \end{bmatrix}\begin{bmatrix} x_1 \\ x_2 \\ x_3 \end{bmatrix} = \begin{bmatrix} 0 \\ 0 \\ 0 \end{bmatrix}.$$

The reduced row echelon form of the coefficient matrix is $\begin{bmatrix} 1 & 0 & 1/3 \\ 0 & 1 & 0 \\ 0 & 0 & 0 \end{bmatrix}$. The solution is $x_1 = (-1/3)r$, $x_2 = 0$, $x_3 = r$, where r is any real number. Let r = -3, then we can take $\mathbf{X}_3 = \begin{bmatrix} 1 \\ 0 \\ -3 \end{bmatrix}$ as an eigenvector associated with $\lambda_3 = 0$. Let

$\mathbf{P} = [\mathbf{X}_1\ \mathbf{X}_2\ \mathbf{X}_3] = \begin{bmatrix} 1 & 2 & 1 \\ 0 & 1 & 0 \\ 0 & 0 & -3 \end{bmatrix}$. Then $\mathbf{P}$ is nonsingular and

$$P^{-1}AP = \begin{bmatrix} 3 & 0 & 0 \\ 0 & 2 & 0 \\ 0 & 0 & 0 \end{bmatrix}.$$

23. Let $\mathbf{A} = \begin{bmatrix} 2 & 2 & 3 & 4 \\ 0 & 2 & 3 & 2 \\ 0 & 0 & 1 & 1 \\ 0 & 0 & 0 & 1 \end{bmatrix}$. Since $\mathbf{A}$ is upper triangular, its eigenvalues are its diagonal entries: $\lambda_1 = 2$, $\lambda_2 = 2$, $\lambda_3 = 1$, $\lambda_4 = 1$. (see Exercise T.4.) To find the corresponding eigenvectors, solve $(2\mathbf{I}_4 - \mathbf{A})\mathbf{X} = \mathbf{0}$:

$$\begin{bmatrix} 0 & -2 & -3 & -4 \\ 0 & 0 & -3 & -2 \\ 0 & 0 & 1 & -1 \\ 0 & 0 & 0 & 1 \end{bmatrix}\begin{bmatrix} x_1 \\ x_2 \\ x_3 \\ x_4 \end{bmatrix} = \begin{bmatrix} 0 \\ 0 \\ 0 \\ 0 \end{bmatrix}.$$

The reduced row echelon form of the coefficient matrix is $\begin{bmatrix} 0 & 1 & 0 & 0 \\ 0 & 0 & 1 & 0 \\ 0 & 0 & 0 & 1 \\ 0 & 0 & 0 & 0 \end{bmatrix}$. Thus the solution is $x_1 = r$, $x_2 = 0$, $x_3 = 0$, $x_4 = 0$, where r is are any real number. The solution in vector form is $\begin{bmatrix} r \\ 0 \\ 0 \\ 0 \end{bmatrix} = r\begin{bmatrix} 1 \\ 0 \\ 0 \\ 0 \end{bmatrix}$. Thus $S = \{\mathbf{X}_1\} = \left\{\begin{bmatrix} 1 \\ 0 \\ 0 \\ 0 \end{bmatrix}\right\}$ is a basis for the eigenspace for eigenvalue $\lambda = 2$. Next we solve $(1\mathbf{I}_4 - \mathbf{A})\mathbf{X} = \mathbf{0}$:

$$\begin{bmatrix} -1 & -2 & -3 & -4 \\ 0 & -1 & -3 & -2 \\ 0 & 0 & 0 & -1 \\ 0 & 0 & 0 & 0 \end{bmatrix}\begin{bmatrix} x_1 \\ x_2 \\ x_3 \\ x_4 \end{bmatrix} = \begin{bmatrix} 0 \\ 0 \\ 0 \\ 0 \end{bmatrix}.$$

The row reduced echelon form of the coefficient matrix is

$\begin{bmatrix} 1 & 0 & -3 & 0 \\ 0 & 1 & 3 & 0 \\ 0 & 0 & 0 & 1 \\ 0 & 0 & 0 & 0 \end{bmatrix}$. Thus the solution is $x_1 = 3r$, $x_2 = -3r$, $x_3 = r$, $x_4 = 0$, where r is any real number. In vector form the solution is $\begin{bmatrix} 3r \\ -3r \\ r \\ 0 \end{bmatrix} = r\begin{bmatrix} 3 \\ -3 \\ 1 \\ 0 \end{bmatrix}$. Thus we have $\mathbf{X}_2 = \begin{bmatrix} 3 \\ -3 \\ 1 \\ 0 \end{bmatrix}$ as a basis for the eigenspace for eigenvalue $\lambda = 1$.

25. Let $\mathbf{A} = \begin{bmatrix} 3 & -5 \\ 1 & -3 \end{bmatrix}$. To find $\mathbf{P}$ so that $\mathbf{P}^{-1}\mathbf{AP} = \mathbf{D}$, a diagonal matrix, we find the eigenvalues and corresponding eigenvectors.We have,

$$\det(\lambda \mathbf{I}_2 - \mathbf{A}) = \begin{vmatrix} \lambda-3 & 5 \\ -1 & \lambda+3 \end{vmatrix} = (\lambda+2)(\lambda-2).$$

Thus the eigenvalues are $\lambda_1 = -2$ and $\lambda_2 = 2$. Solve $(-2\mathbf{I}_2 - \mathbf{A})\mathbf{X} = \mathbf{0}$:

$$\begin{bmatrix} -5 & 5 \\ -1 & 1 \end{bmatrix}\begin{bmatrix} x_1 \\ x_2 \end{bmatrix} = \begin{bmatrix} 0 \\ 0 \end{bmatrix}.$$

The reduced row echelon form of the coefficient matrix is $\begin{bmatrix} 1 & -1 \\ 0 & 0 \end{bmatrix}$. Thus an eigenvector is $\mathbf{X}_1 = \begin{bmatrix} 1 \\ 1 \end{bmatrix}$. Solve $(2\mathbf{I}_2 - \mathbf{A})\mathbf{X} = \mathbf{0}$:

$$\begin{bmatrix} -1 & 5 \\ -1 & 5 \end{bmatrix}\begin{bmatrix} x_1 \\ x_2 \end{bmatrix} = \begin{bmatrix} 0 \\ 0 \end{bmatrix}.$$

The reduced row echelon form of the coefficient matrix is $\begin{bmatrix} 1 & -5 \\ 0 & 0 \end{bmatrix}$. Thus an eigenvector is $\mathbf{X}_2 = \begin{bmatrix} 5 \\ 1 \end{bmatrix}$. Set $\mathbf{P} = [\mathbf{X}_1\ \mathbf{X}_2] = \begin{bmatrix} 1 & 5 \\ 1 & 1 \end{bmatrix}$. Then $\mathbf{P}^{-1} = \begin{bmatrix} -1/4 & 5/4 \\ 1/4 & -1/4 \end{bmatrix}$ and $\mathbf{P}^{-1}\mathbf{AP} = \mathbf{D} = \begin{bmatrix} -2 & 0 \\ 0 & 2 \end{bmatrix}$. It follows that $\mathbf{A}^9 = \mathbf{PD}^9\mathbf{P}^{-1} = \mathbf{P}\begin{bmatrix} -512 & 0 \\ 0 & 512 \end{bmatrix}\mathbf{P}^{-1} = 256\begin{bmatrix} 3 & -5 \\ 1 & -3 \end{bmatrix}$.

Exercises 5.1

T.1. (a) $\mathbf{A} = \mathbf{P}^{-1}\mathbf{AP}$ for $\mathbf{P} = \mathbf{I}_n$.

(b) If $\mathbf{B} = \mathbf{P}^{-1}\mathbf{AP}$, then $\mathbf{A} = \mathbf{PBP}^{-1}$ and so $\mathbf{A}$ is similar to $\mathbf{B}$.

(c) If $\mathbf{B} = \mathbf{P}^{-1}\mathbf{AP}$ and $\mathbf{C} = \mathbf{Q}^{-1}\mathbf{BQ}$ then

$$\mathbf{C} = \mathbf{Q}^{-1}\mathbf{P}^{-1}\mathbf{APQ} = (\mathbf{PQ})^{-1}\mathbf{A}(\mathbf{PQ}).$$

Thus $\mathbf{A}$ is similar to $\mathbf{C}$.

T.3. Suppose $\mathbf{B} = \mathbf{P}^{-1}\mathbf{AP}$ for some nonsingular matrix $\mathbf{P}$. The characteristic polynomial of $\mathbf{B}$ is

$$\begin{aligned} \det(\lambda\mathbf{I}_n - \mathbf{B}) &= \det(\lambda\mathbf{I}_n - \mathbf{P}^{-1}\mathbf{AP}) = \det(\lambda\mathbf{P}^{-1}\mathbf{P} - \mathbf{P}^{-1}\mathbf{AP}) \\ &= \det(\mathbf{P}^{-1}(\lambda\mathbf{I}_n - \mathbf{A})\mathbf{P}) \\ &= \det(\mathbf{P}^{-1}) \cdot \det(\lambda\mathbf{I}_n - \mathbf{A}) \cdot \det(\mathbf{P}) \\ &= \det(\lambda\mathbf{I}_n - \mathbf{A}) \end{aligned}$$

since $\det(\mathbf{P}^{-1}) = 1/\det(\mathbf{P})$. Thus $\mathbf{A}$ and $\mathbf{B}$ have the same characteristic polynomials. Hence it follows that $\mathbf{A}$ and $\mathbf{B}$ have the same eigenvalues.

T.5. We have,

$$\begin{aligned} \det(\lambda\mathbf{I}_n - \mathbf{A}^T) &= \det(\lambda\mathbf{I}_n{}^T - \mathbf{A}^T) = \det(\lambda\mathbf{I}_n - \mathbf{A})^T) \\ &= \det(\lambda\mathbf{I}_n - \mathbf{A}). \end{aligned}$$

Thus the characteristic polynomials of $\mathbf{A}$ and $\mathbf{A}^T$ are the same and it follows that their eigenvalues are the same. The associated eigenvectors need not be the same. (But the dimensions of the eigenspace associated with λ, for $\mathbf{A}$ and for $\mathbf{A}^T$, are equal.)

T.7. If $\mathbf{A}$ is nilpotent and $\mathbf{A}^k = \mathbf{0}$, and if λ is an eigenvalue for $\mathbf{A}$ with associated eigenvector $\mathbf{X}$, then $\mathbf{0} = \mathbf{A}^k\mathbf{X} = \lambda^k\mathbf{X}$ (see Exercise T.6). Since $\mathbf{X}$ is an eigenvector $\mathbf{X} \neq \mathbf{0}$, hence $\lambda^k = 0$, so $\lambda = 0$. Thus the only eigenvalue of a nilpotent matrix is zero.

T.9. Suppose that $\mathbf{A}$ is nonsingular and $\mathbf{AX} = \lambda\mathbf{X}$. From Exercise T.8, $\lambda \neq 0$. Then

$$\lambda^{-1}\mathbf{X} = \lambda^{-1}\mathbf{A}^{-1}\mathbf{AX} = \lambda^{-1}\mathbf{A}^{-1}(\lambda\mathbf{X}) = (\lambda^{-1}\lambda)\mathbf{A}^{-1}\mathbf{X} = \mathbf{A}^{-1}\mathbf{X}.$$

Thus $\lambda^{-1} = 1/\lambda$ is an eigenvalue of $\mathbf{A}^{-1}$ with associated eigenvector $\mathbf{X}$.

T.11. We have $\mathbf{BA} = \mathbf{A}^{-1}(\mathbf{AB})\mathbf{A}$, so **AB** and **BA** are similar. By Exercise T.3, **AB** and **BA** have the same characteristic polynomials and thus the same eigenvalues.

T.13. We have $\mathbf{B} = \mathbf{P}^{-1}\mathbf{AP}$ and $\mathbf{AX} = \lambda\mathbf{X}$. Hence,

$$\mathbf{B}(\mathbf{P}^{-1}\mathbf{X}) = (\mathbf{BP}^{-1})\mathbf{X} = \mathbf{P}^{-1}\mathbf{AX} \quad (\text{since } \mathbf{B} = \mathbf{P}^{-1}\mathbf{AP},\ \mathbf{BP}^{-1} = \mathbf{P}^{-1}\mathbf{APP}^{-1} = \mathbf{P}^{-1}\mathbf{A})$$

$$= \mathbf{P}^{-1}(\lambda\mathbf{X})$$

$$= \lambda(\mathbf{P}^{-1}\mathbf{X})$$

which shows that $\mathbf{P}^{-1}\mathbf{X}$ is an eigenvector of **B** associated with eigenvalue λ.

ML.1. Enter each matrix **A** into MATLAB and use command **poly(A)**.

(a)
```
A=[1 2;2 -1];
v=poly(A)

v =

    1.0000         0   -5.0000
```

The characteristic polynomial is $\lambda^2 - 5$.

(b)
```
A=[2 4 0;1 2 1;0 4 2];
v=poly(A)

v =

    1.0000   -6.0000    4.0000    8.0000
```

The characteristic polynomial is $\lambda^3 - 6\lambda^2 + 4\lambda + 8$.

(c)
```
A=[1 0 0 0;2 -2 0 0;0 0 2 -1;0 0 -1 2];
v=poly(A)

v =

    1    -3    -3    11    -6
```

The characteristic polynomial is $\lambda^4 - 3\lambda^3 - 3\lambda^2 + 11\lambda - 6$.

ML.2. The eigenvalues of matrix A will be computed using MATLAB command roots(poly(A)).

(a)
```
A=[1 -3;3 -5];
r=roots(poly(A))
```

```
r =

   -2
   -2
```

(b)
```
A=[3 -1 4;-1 0 1;4 1 2];
r=roots(poly(A))

r =

    6.5324
   -2.3715
    0.8392
```

(c)
```
A=[2 -2 0;1 -1 0;1 -1 0];
r=roots(poly(A))

r =

   0
   0
   1
```

(d)
```
A=[2 4;3 6];
r=roots(poly(A))

r =

   0
   8
```

ML.3. We solve the homogeneous system (lambdaI-A)X = 0 by finding the reduced row echelon form of the corresponding augmented matrix and then writing out the general solution.

(a)
```
A=[1 2;-1 4];
M=(3*eye(A)-A);
rref([M [0 0]'])

ans =

     1    -1     0
     0     0     0
```

The general solution is $x_2 = r$, $x_1 = x_2 = r$. Let r = 1 and we have that [1 1]′ is an eigenvector.

(b)
```
A=[4 0 0;1 3 0;2 1 -1];
M=(-1*eye(A)-A);
rref([M [0 0 0]'])
```

```
ans =

        1     0     0     0
        0     1     0     0
        0     0     0     0
```

The general solution is $x_3 = r$, $x_2 = 0$, $x_1 = 0$. Let $r = 1$ and we have that [0 0 1]' is an eigenvector.

(c)
```
A=[2 1 2;2 2 -2;3 1 1];
M=(2*eye(A)-A);
rref([M [0 0 0]'])

ans =

        1     0    -1     0
        0     1     2     0
        0     0     0     0
```

The general solution is $x_3 = r$, $x_2 = -2x_3 = -2r$, $x_1 = x_3 = r$. Let $r = 1$ and we have that [1 -2 1]' is an eigenvector.

ML.4. (a)
```
A=[0 2;-1 3];
r=roots(poly(A))
r =

     2
     1
```

The eigenvalues are distinct, so **A** is diagonalizable. We find the corresponding eigenvectors.

```
M=(2*eye(A)-A);
rref([M [0 0]'])

ans =

        1    -1     0
        0     0     0
```

The general solution is $x_2 = r$, $x_1 = x_2 = r$. Let $r = 1$ and we have that [1 1]' is an eigenvector.

```
M=(1*eye(A)-A);
rref([M [0 0]'])

ans =

        1    -2     0
        0     0     0
```

Exercises 5.1

The general solution is $x_2 = r$, $x_1 = 2x_2 = 2r$. Let $r = 1$ and we have that [2 1]' is an eigenvector.

```
P=[1 1;2 1]'

P =

     1     2
     1     1

inverse(P)*A*P

ans =

      2     0
      0     1
```

(b)
```
A=[1 -3;3 -5];
r=roots(poly(A))

r =

    -2
    -2

M=(-2*eye(A)-A);
rref([M [0 0]'])

ans =

      1    -1     0
      0     0     0
```

The general solution is $x_2 = r$, $x_1 = x_2 = r$. Let $r = 1$ and it follows that [1 1]' is an eigenvector, but there is only one linearly independent eigenvector. Hence **A** is not diagonalizable.

(c)
```
A=[0 0 4;5 3 6;6 0 5];
r=roots(poly(A))

r =

   8.0000
   3.0000
  -3.0000
```

The eigenvalues are distinct, thus A is diagonalizable. We find the corresponding eigenvectors.

```
M=(8*eye(A)-A);
rref([M [0 0 0]'])
```

```
ans =

    1.0000         0   -0.5000         0
         0    1.0000   -1.7000         0
         0         0         0         0
```

The general solution is $x_3 = r$, $x_2 = 1.7*x_3 = 1.7*r$, $x_1 = .5x_3 = .5r$. Let $r = 1$ and we have that [.5 1.7 1]' is an eigenvector.

```
M=(3*eye(A)-A);
rref([M [0 0 0]'])

ans =

     1     0     0     0
     0     0     1     0
     0     0     0     0
```

Thus [0 1 0]' is an eigenvector.

```
M=(-3*eye(A)-A);
rref([M [0 0 0]'])

ans =

    1.0000         0    1.3333         0
         0    1.0000   -0.1111         0
         0         0         0         0
```

The general solution is $x_3 = r$, $x_2 = (1/9)x_3 = (1/9)r$, $x_1 = (-4/3)x_3 = (-4/3)r$. Let $r = 1$ and we have that [-4/3 1/9 1]' is an eigenvector. Thus **P** is

```
P = [.5 1.7 1;0 1 0;-4/3 1/9 1]';

inverse(P)*A*P

ans =

     8     0     0
     0     3     0
     0     0    -3
```

ML.5. We find the eigenvalues and corresponding eigenvectors.

```
A=[-1 1 -1;-2 2 -1;-2 2 -1];
r=roots(poly(A))

r =

        0
  -1.0000
   1.0000
```

Exercises 5.1

The eigenvalues are distinct, hence **A** is diagonalizable.

```
M=(0*eye(A)-A);
rref([M [0 0 0]'])

ans =

     1    -1     0     0
     0     0     1     0
     0     0     0     0
```

The general solution is x3 = 0, x2 = r, x1 = x2 = r. Let r =1 and we have that [1 1 0]' is an eigenvector.

```
M=(-1*eye(A)-A);
rref([M [0 0 0]'])

ans =

     1     0    -1     0
     0     1    -1     0
     0     0     0     0
```

The general solution is x3 = r, x2 = x3 = r, x1 = x3 = r. Let r = 1 and we have that [1 1 1]' is an eigenvector.

```
M=(1*eye(A)-A);
rref([M [0 0 0]'])

ans =

     1     0     0     0
     0     1    -1     0
     0     0     0     0
```

The general solution is x3 = r, x2 = x3 = r, x1 = 0. Let r = 1 and we have that [0 1 1]' is an eigenvector.

```
P = [1 1 0;1 1 1;0 1 1]'

P =

     1     1     0
     1     1     1
     0     1     1

A30=P*diag([0 -1 1])^30*inverse(P)

A30 =

     1    -1     1
     0     0     1
     0     0     1
```

ML.6.
```
A=[-1 1.5 -1.5;-2 2.5 -1.5;-2 2.0 -1.0];
r=roots(poly(A))

r =

    1.0000
   -1.0000
    0.5000
```

The eigenvalues are distinct, hence **A** is diagonalizable.

```
M=(1*eye(A)-A);
rref([M [0 0 0]'])

ans =

     1     0     0     0
     0     1    -1     0
     0     0     0     0
```

The general solution is $x_3 = r$, $x_2 = r$, $x_1 = 0$. Let $r = 1$ and we have that [0 1 1]' is an eigenvector.

```
M=(-1*eye(A)-A);
rref([M [0 0 0]'])

ans =

     1     0    -1     0
     0     1    -1     0
     0     0     0     0
```

The general solution is $x_3 = r$, $x_2 = r$, $x_1 = r$. Let $r = 1$ and we have that [1 1 1]' is an eigenvector.

```
M=(.5*eye(A)-A);
rref([M [0 0 0]'])

ans =

     1    -1     0     0
     0     0     1     0
     0     0     0     0
```

The general solution is $x_3 = 0$, $x_2 = r$, $x_1 = r$. Let $r = 1$ and we have that [1 1 0]' is an eigenvector. Hence let

```
P = [0 1 1;1 1 1;1 1 0]'

P =

     0     1     1
     1     1     1
     1     1     0
```

then we have

```
A30=P*diag([1 -1 .5])^30*inverse(P)

A30 =

        1.0000   -1.0000    1.0000
             0    0.0000    1.0000
             0         0    1.0000
```

Since all the entries are not displayed as integers we set the format to long and redisplay the matrix to view its contents for more detail.

```
format long
A30

A30 =

        1.00000000000000  -0.99999999906868   0.99999999906868
                       0   0.00000000093132   0.99999999906868
                       0                  0   1.00000000000000
```

Note that this is not the same as the matrix A30 in Exercise ML.5.

ML.7.
```
A=[-1 1 -1;-2 2 -1;-2 2 -1];
A,A^3,A^5

A =

    -1     1    -1
    -2     2    -1
    -2     2    -1

ans =

    -1     1    -1
    -2     2    -1
    -2     2    -1

ans =

    -1     1    -1
    -2     2    -1
    -2     2    -1
```

Further computation shows that A raised to an odd power gives A, hence the sequence A, A^3, A^5, ... converges to A.

```
A^2,A^4,A^6
```

```
ans =

     1    -1     1
     0     0     1
     0     0     1

ans =

     1    -1     1
     0     0     1
     0     0     1


ans =

     1    -1     1
     0     0     1
     0     0     1
```

Further investigation shows that A raised to an even power gives the same matrix as displayed above. Hence the sequence A^2, A^4, A^6, ... converges to this matrix.

Exercises 5.2

1. We show that $\mathbf{PP}^T = \mathbf{I}_3$:

$$\mathbf{PP}^T = \begin{bmatrix} 2/3 & -2/3 & 1/3 \\ 2/3 & 1/3 & -2/3 \\ 1/3 & 2/3 & 2/3 \end{bmatrix}\begin{bmatrix} 2/3 & 2/3 & 1/3 \\ -2/3 & 1/3 & 2/3 \\ 1/3 & -2/3 & 2/3 \end{bmatrix} = \begin{bmatrix} 1 & 0 & 0 \\ 0 & 1 & 0 \\ 0 & 0 & 1 \end{bmatrix}.$$

3. (a) Let the columns of $\mathbf{A}$ be denoted $\mathbf{X}_1 = \begin{bmatrix} 1 \\ 0 \\ 0 \end{bmatrix}$,

$\mathbf{X}_2 = \begin{bmatrix} 0 \\ \cos\theta \\ -\sin\theta \end{bmatrix}$, $\mathbf{X}_3 = \begin{bmatrix} 0 \\ \sin\theta \\ \cos\theta \end{bmatrix}$. Then $\mathbf{X}_1 \cdot \mathbf{X}_2 = \mathbf{X}_1 \cdot \mathbf{X}_3 =$ $\mathbf{X}_2 \cdot \mathbf{X}_3 = 0$ and $\mathbf{X}_1 \cdot \mathbf{X}_1 = \mathbf{X}_2 \cdot \mathbf{X}_2 = \mathbf{X}_3 \cdot \mathbf{X}_3 = 1$. Thus $\{\mathbf{X}_1, \mathbf{X}_2, \mathbf{X}_3\}$ is an orthonormal set.

(b) Let the columns of $\mathbf{B}$ be denoted $\mathbf{Y}_1 = \begin{bmatrix} 1 \\ 0 \\ 0 \end{bmatrix}$, $\mathbf{Y}_2 = \begin{bmatrix} 0 \\ 1/\sqrt{2} \\ -1/\sqrt{2} \end{bmatrix}$,

$\mathbf{Y}_3 = \begin{bmatrix} 0 \\ -1/\sqrt{2} \\ -1/\sqrt{2} \end{bmatrix}$. Then $\mathbf{Y}_1 \cdot \mathbf{Y}_2 = \mathbf{Y}_1 \cdot \mathbf{Y}_3 = \mathbf{Y}_2 \cdot \mathbf{Y}_3 = 0$ and $\mathbf{Y}_1 \cdot \mathbf{Y}_1 = \mathbf{Y}_2 \cdot \mathbf{Y}_2 = \mathbf{Y}_3 \cdot \mathbf{Y}_3 = 1$. Thus $\{\mathbf{Y}_1, \mathbf{Y}_2, \mathbf{Y}_3\}$ is an orthonormal set.

5. Let $\mathbf{X} = \begin{bmatrix} x_1 \\ x_2 \end{bmatrix}$ and $\mathbf{Y} = \begin{bmatrix} y_1 \\ y_2 \end{bmatrix}$. Then $\mathbf{X} \cdot \mathbf{Y} = x_1y_1 + x_2y_2$ and

$$(\mathbf{AX}) \cdot (\mathbf{AY}) = \begin{bmatrix} \dfrac{x_1-x_2}{\sqrt{2}} \\ \dfrac{-x_1-x_2}{\sqrt{2}} \end{bmatrix} \cdot \begin{bmatrix} \dfrac{y_1-y_2}{\sqrt{2}} \\ \dfrac{-y_1-y_2}{\sqrt{2}} \end{bmatrix} = (1/2)(x_1-x_2)(y_1-y_2) +$$

$$(1/2)(x_1+x_2)(y_1+y_2) = x_1y_1 + x_2y_2.$$

Thus $(\mathbf{AX}) \cdot (\mathbf{AY}) = \mathbf{X} \cdot \mathbf{Y}$ for any $\mathbf{X}$ and $\mathbf{Y}$ in $\mathbf{R}^2$.

<<**Strategy:** In Exercises 7, 9, and 11, we follow the techniques used in Examples 4 and 5.>>

7. Let $\mathbf{A} = \begin{bmatrix} 2 & 2 \\ 2 & 2 \end{bmatrix}$. Then $\det(\lambda \mathbf{I}_2 - \mathbf{A}) = \lambda(\lambda - 4)$ and the eigenvalues of $\mathbf{A}$ are $\lambda_1 = 0$ and $\lambda_2 = 4$. To find an eigenvector $\mathbf{X}_1$ associated with $\lambda_1 = 0$, we solve $(0\mathbf{I}_2 - \mathbf{A})\mathbf{X} = \mathbf{0}$:

$$\begin{bmatrix} -2 & -2 \\ -2 & -2 \end{bmatrix}\begin{bmatrix} x_1 \\ x_2 \end{bmatrix} = \begin{bmatrix} 0 \\ 0 \end{bmatrix}.$$

The reduced row echelon form of the coefficient matrix is $\begin{bmatrix} 1 & 1 \\ 0 & 0 \end{bmatrix}$. Thus we can take $\mathbf{X}_1 = \begin{bmatrix} 1 \\ -1 \end{bmatrix}$. To find an eigenvector $\mathbf{X}_2$ associated with $\lambda_2 = 4$, we solve $(4\mathbf{I}_2 - \mathbf{A})\mathbf{X} = \mathbf{0}$:

$$\begin{bmatrix} 2 & -2 \\ -2 & 2 \end{bmatrix}\begin{bmatrix} x_1 \\ x_2 \end{bmatrix} = \begin{bmatrix} 0 \\ 0 \end{bmatrix}.$$

The reduced row echelon form of the coefficient matrix is $\begin{bmatrix} 1 & -1 \\ 0 & 0 \end{bmatrix}$. Thus we can take $\mathbf{X}_2 = \begin{bmatrix} 1 \\ 1 \end{bmatrix}$. We note that $\mathbf{X}_1 \cdot \mathbf{X}_2 = 0$; that is, $\mathbf{X}_1$ and $\mathbf{X}_2$ are orthogonal. To obtain an orthonormal basis of eigenvectors we normalize $\mathbf{X}_1$ and $\mathbf{X}_2$ as

$$\mathbf{Z}_1 = (1/\|\mathbf{X}_1\|)\ \mathbf{X}_1 = \begin{bmatrix} 1/\sqrt{2} \\ -1/\sqrt{2} \end{bmatrix}, \quad \mathbf{Z}_2 = (1/\|\mathbf{X}_2\|)\ \mathbf{X}_2 = \begin{bmatrix} 1/\sqrt{2} \\ 1/\sqrt{2} \end{bmatrix}.$$

Then set $\mathbf{P} = [\mathbf{Z}_1\ \mathbf{Z}_2] = \begin{bmatrix} 1/\sqrt{2} & 1/\sqrt{2} \\ -1/\sqrt{2} & 1/\sqrt{2} \end{bmatrix}$ and hence

$$\mathbf{P}^{-1}\mathbf{A}\mathbf{P} = \begin{bmatrix} 0 & 0 \\ 0 & 4 \end{bmatrix}.$$

9. Let $\mathbf{A} = \begin{bmatrix} 0 & 0 & 0 \\ 0 & 2 & 2 \\ 0 & 2 & 2 \end{bmatrix}$. Then $\det(\lambda \mathbf{I}_3 - \mathbf{A}) = \lambda^2(\lambda - 4)$ and the eigenvalues of $\mathbf{A}$ are $\lambda_1 = \lambda_2 = 0$, $\lambda_3 = 4$. To find an eigenvector $\mathbf{X}_1$ associated with $\lambda_1 = 0$, we solve $(0\mathbf{I}_3 - \mathbf{A})\mathbf{X} = \mathbf{0}$:

Exercises 5.2

$$\begin{bmatrix} 0 & 0 & 0 \\ 0 & -2 & -2 \\ 0 & -2 & -2 \end{bmatrix} \begin{bmatrix} x_1 \\ x_2 \\ x_3 \end{bmatrix} = \begin{bmatrix} 0 \\ 0 \\ 0 \end{bmatrix}.$$

The reduced row echelon form of the coefficient matrix is $\begin{bmatrix} 0 & 1 & 1 \\ 0 & 0 & 0 \\ 0 & 0 & 0 \end{bmatrix}$. The general solution is $x_1 = r$, $x_2 = -s$, $x_3 = s$, where r and s are any real numbers. In vector form the solution is $\begin{bmatrix} r \\ -s \\ s \end{bmatrix} = r\begin{bmatrix} 1 \\ 0 \\ 0 \end{bmatrix} + s\begin{bmatrix} 0 \\ -1 \\ 1 \end{bmatrix}$. Thus there are two linearly independent eigenvectors associated with $\lambda_1 = \lambda_2 = 0$; $\mathbf{X}_1 = \begin{bmatrix} 1 \\ 0 \\ 0 \end{bmatrix}$ and $\mathbf{X}_2 = \begin{bmatrix} 0 \\ -1 \\ 1 \end{bmatrix}$. To find an eigenvector $\mathbf{X}_3$ associated with $\lambda_3 = 4$, we solve $(4\mathbf{I}_3 - \mathbf{A})\mathbf{X} = \mathbf{0}$:

$$\begin{bmatrix} 4 & 0 & 0 \\ 0 & 2 & -2 \\ 0 & -2 & 2 \end{bmatrix} \begin{bmatrix} x_1 \\ x_2 \\ x_3 \end{bmatrix} = \begin{bmatrix} 0 \\ 0 \\ 0 \end{bmatrix}.$$

The reduced row echelon form of the coefficient matrix is $\begin{bmatrix} 1 & 0 & 0 \\ 0 & 1 & -1 \\ 0 & 0 & 0 \end{bmatrix}$. Thus we can take $\mathbf{X}_3 = \begin{bmatrix} 0 \\ 1 \\ 1 \end{bmatrix}$. Since **A** is symmetric $\mathbf{X}_1$ and $\mathbf{X}_2$ are orthogonal to $\mathbf{X}_3$ (see Theorem 5.5). For this matrix **A** it is easily checked that $\mathbf{X}_1 \cdot \mathbf{X}_2 = 0$. (Note: this need not be true when the eigenvalue has multiplicity greater than 1.) Hence, $\{\mathbf{X}_1, \mathbf{X}_2, \mathbf{X}_3\}$ is orthogonal. To obtain an orthonormal basis of eigenvectors we normalize $\mathbf{X}_1$, $\mathbf{X}_2$, $\mathbf{X}_3$ as

$$\mathbf{Z}_1 = (1/\|\mathbf{X}_1\|)\ \mathbf{X}_1 = \mathbf{X}_1,\quad \mathbf{Z}_2 = (1/\|\mathbf{X}_2\|)\ \mathbf{X}_2 = (1/\sqrt{2})\begin{bmatrix} 0 \\ -1 \\ 1 \end{bmatrix},$$

$$\mathbf{Z}_3 = (1/\|\mathbf{X}_3\|)\ \mathbf{X}_3 = (1/\sqrt{2})\begin{bmatrix} 0 \\ 1 \\ 1 \end{bmatrix}.$$

Then set $\mathbf{P} = [\mathbf{Z}_1\ \mathbf{Z}_2\ \mathbf{Z}_3]$ and hence $\mathbf{P}^{-1}\mathbf{AP} = \begin{bmatrix} 0 & 0 & 0 \\ 0 & 0 & 0 \\ 0 & 0 & 4 \end{bmatrix}$.

11. Let $\mathbf{A} = \begin{bmatrix} 0 & -1 & -1 \\ -1 & 0 & -1 \\ -1 & -1 & 0 \end{bmatrix}$. Then $\det(\lambda\mathbf{I}_3 - \mathbf{A}) = (\lambda+2)(\lambda-1)^2$ and the eigenvalues of $\mathbf{A}$ are $\lambda_1 = \lambda_2 = 1$, $\lambda_3 = -2$. To find an eigenvector $\mathbf{X}_1$ associated with $\lambda_1 = 1$, we solve $(1\mathbf{I}_3 - \mathbf{A})\mathbf{X} = \mathbf{0}$:

$$\begin{bmatrix} 1 & 1 & 1 \\ 1 & 1 & 1 \\ 1 & 1 & 1 \end{bmatrix} \begin{bmatrix} x_1 \\ x_2 \\ x_3 \end{bmatrix} = \begin{bmatrix} 0 \\ 0 \\ 0 \end{bmatrix}.$$

The reduced row echelon form of the coefficient matrix is $\begin{bmatrix} 1 & 1 & 1 \\ 0 & 0 & 0 \\ 0 & 0 & 0 \end{bmatrix}$. The solution is $x_1 = -r - s$, $x_2 = r$, $x_3 = s$, where r and s are any real numbers. In vector form the solution is $\begin{bmatrix} -r-s \\ r \\ s \end{bmatrix} = r\begin{bmatrix} -1 \\ 1 \\ 0 \end{bmatrix} + s\begin{bmatrix} -1 \\ 0 \\ 1 \end{bmatrix}$. Thus there are two linearly independent eigenvectors associated with $\lambda_1 = \lambda_2 = 1$; $\mathbf{X}_1 = \begin{bmatrix} -1 \\ 1 \\ 0 \end{bmatrix}$ and $\mathbf{X}_2 = \begin{bmatrix} -1 \\ 0 \\ 1 \end{bmatrix}$. To find an eigenvector $\mathbf{X}_3$ associated with $\lambda_3 = -2$, we solve $(-2\mathbf{I}_3 - \mathbf{A})\mathbf{X} = \mathbf{0}$:

$$\begin{bmatrix} -2 & 1 & 1 \\ 1 & -2 & 1 \\ 1 & 1 & -2 \end{bmatrix} \begin{bmatrix} x_1 \\ x_2 \\ x_3 \end{bmatrix} = \begin{bmatrix} 0 \\ 0 \\ 0 \end{bmatrix}.$$

The reduced row echelon form of the coefficient matrix is $\begin{bmatrix} 1 & 0 & -1 \\ 0 & 1 & -1 \\ 0 & 0 & 0 \end{bmatrix}$. Thus we can take $\mathbf{X}_3 = \begin{bmatrix} 1 \\ 1 \\ 1 \end{bmatrix}$. Since $\mathbf{A}$ is symmetric $\mathbf{X}_1$ and $\mathbf{X}_2$ are orthogonal with $\mathbf{X}_3$ (see Theorem 5.5). For this matrix $\mathbf{A}$, $\mathbf{X}_1 \cdot \mathbf{X}_2 = 1$, hence $\mathbf{X}_1$ and $\mathbf{X}_2$ are not orthogonal. We use the Gram-Schmidt process to obtain an orthogonal basis for the eigenspace associated with $\lambda_1 = 1$. Let

$$\mathbf{Y}_1 = \mathbf{X}_1 = \begin{bmatrix} -1 \\ 1 \\ 0 \end{bmatrix}$$

and

$$\mathbf{Y}_2 = \mathbf{X}_2 - \left(\frac{\mathbf{X}_2 \cdot \mathbf{Y}_1}{\mathbf{Y}_1 \cdot \mathbf{Y}_1}\right) \mathbf{Y}_1 = \begin{bmatrix} -1/2 \\ -1/2 \\ 1 \end{bmatrix}.$$

Let

$$\mathbf{Y}_1' = \mathbf{Y}_1 \text{ and } \mathbf{Y}_2' = 2\mathbf{Y}_2 = \begin{bmatrix} -1 \\ -1 \\ 2 \end{bmatrix}.$$

The set $\{\mathbf{Y}_1', \mathbf{Y}_2'\}$ is an orthogonal basis for the eigenspace associated with $\lambda_1 = 1$. Then we normalize $\{\mathbf{Y}_1', \mathbf{Y}_2', \mathbf{X}_3\}$ as,

$$\mathbf{Z}_1 = (1/\|\mathbf{Y}_1'\|)\ \mathbf{Y}_1' = (1/\sqrt{2}) \begin{bmatrix} -1 \\ 1 \\ 0 \end{bmatrix},$$

$$\mathbf{Z}_2 = (1/\|\mathbf{Y}_2'\|)\ \mathbf{Y}_2' = (1/\sqrt{6}) \begin{bmatrix} -1 \\ -1 \\ 2 \end{bmatrix},$$

$$\mathbf{Z}_3 = (1/\|\mathbf{X}_3\|)\ \mathbf{X}_3 = (1/\sqrt{3}) \begin{bmatrix} 1 \\ 1 \\ 1 \end{bmatrix}.$$

Then set $\mathbf{P} = [\mathbf{Z}_1\ \mathbf{Z}_2\ \mathbf{Z}_3]$ and hence $\mathbf{P}^{-1}\mathbf{AP} = \begin{bmatrix} 1 & 0 & 0 \\ 0 & 1 & 0 \\ 0 & 0 & -2 \end{bmatrix}.$

(The order of the columns of **P** and the order of the diagonal enties may vary.)

13. Let $\mathbf{A} = \begin{bmatrix} 2 & 1 \\ 1 & 2 \end{bmatrix}$. Then $\det(\lambda\mathbf{I}_2 - \mathbf{A}) = (\lambda-3)(\lambda-1)$ and the eigenvalues of **A** are $\lambda_1 = 3$ and $\lambda_2 = 1$. To find an eigenvector $\mathbf{X}_1$ associated with $\lambda_1 = 3$, we solve $(3\mathbf{I}_2 - \mathbf{A})\mathbf{X} = \mathbf{0}$:

$$\begin{bmatrix} 1 & -1 \\ -1 & 1 \end{bmatrix} \begin{bmatrix} x_1 \\ x_2 \end{bmatrix} = \begin{bmatrix} 0 \\ 0 \end{bmatrix}.$$

The reduced row echelon form of the coefficient matrix is

$\begin{bmatrix} 1 & -1 \\ 0 & 0 \end{bmatrix}$. Thus we can take $\mathbf{X}_1 = \begin{bmatrix} 1 \\ 1 \end{bmatrix}$. To find an eigenvector $\mathbf{X}_2$ associated with $\lambda_2 = 1$, we solve $(1\mathbf{I}_2 - \mathbf{A})\mathbf{X} = \mathbf{0}$:

$$\begin{bmatrix} -1 & -1 \\ -1 & -1 \end{bmatrix}\begin{bmatrix} x_1 \\ x_2 \end{bmatrix} = \begin{bmatrix} 0 \\ 0 \end{bmatrix}.$$

The reduced row echelon form of the coefficient matrix is $\begin{bmatrix} 1 & 1 \\ 0 & 0 \end{bmatrix}$. Thus we can take $\mathbf{X}_2 = \begin{bmatrix} 1 \\ -1 \end{bmatrix}$. Since $\mathbf{A}$ is symmetric and $\lambda_1 \neq \lambda_2$, $\mathbf{X}_1$ and $\mathbf{X}_2$ are orthogonal. (See Theorem 5.5.) To obtain an orthonormal basis of eigenvectors we normalize $\mathbf{X}_1$ and $\mathbf{X}_2$ as

$$\mathbf{Z}_1 = (1/\|\mathbf{X}_1\|)\ \mathbf{X}_1 = (1/\sqrt{2})\mathbf{X}_1$$

and

$$\mathbf{Z}_2 = (1/\|\mathbf{X}_2\|)\ \mathbf{X}_2 = (1/\sqrt{2})\mathbf{X}_2.$$

Then set $\mathbf{P} = [\mathbf{Z}_1\ \mathbf{Z}_2]$ and we have $\mathbf{P}^{-1}\mathbf{A}\mathbf{P} = \begin{bmatrix} 3 & 0 \\ 0 & 1 \end{bmatrix}$.

15. Let $\mathbf{A} = \begin{bmatrix} 1 & 1 & 0 \\ 1 & 1 & 0 \\ 0 & 0 & 1 \end{bmatrix}$. Then $\det(\lambda\mathbf{I}_3 - \mathbf{A}) = (\lambda-1)(\lambda-2)\lambda$ and the eigenvalues of $\mathbf{A}$ are $\lambda_1 = 1$, $\lambda_2 = 2$, and $\lambda_3 = 0$. To find an eigenvector $\mathbf{X}_1$ associated with $\lambda_1 = 1$, we solve $(1\mathbf{I}_3 - \mathbf{A})\mathbf{X} = \mathbf{0}$:

$$\begin{bmatrix} 0 & -1 & 0 \\ -1 & 0 & 0 \\ 0 & 0 & 0 \end{bmatrix}\begin{bmatrix} x_1 \\ x_2 \\ x_3 \end{bmatrix} = \begin{bmatrix} 0 \\ 0 \\ 0 \end{bmatrix}.$$

The reduced row echelon form of the coefficient matrix is $\begin{bmatrix} 1 & 0 & 0 \\ 0 & 1 & 0 \\ 0 & 0 & 0 \end{bmatrix}$. Thus we can take $\mathbf{X}_1 = \begin{bmatrix} 0 \\ 0 \\ 1 \end{bmatrix}$. To find an eigenvector $\mathbf{X}_2$ associated with $\lambda_2 = 2$, we solve $(2\mathbf{I}_3 - \mathbf{A})\mathbf{X} = \mathbf{0}$:

$$\begin{bmatrix} 1 & -1 & 0 \\ -1 & 1 & 0 \\ 0 & 0 & 1 \end{bmatrix}\begin{bmatrix} x_1 \\ x_2 \\ x_3 \end{bmatrix} = \begin{bmatrix} 0 \\ 0 \\ 0 \end{bmatrix}.$$

The reduced row echelon form of the coefficient matrix is

$\begin{bmatrix} 1 & -1 & 0 \\ 0 & 0 & 1 \\ 0 & 0 & 0 \end{bmatrix}$. Thus we can take $\mathbf{X}_2 = \begin{bmatrix} 1 \\ 1 \\ 0 \end{bmatrix}$. To find an eigenvector $\mathbf{X}_3$ associated with $\lambda_3 = 0$, we solve $(0\mathbf{I}_3 - \mathbf{A})\mathbf{X} = \mathbf{0}$:

$$\begin{bmatrix} -1 & -1 & 0 \\ -1 & -1 & 0 \\ 0 & 0 & -1 \end{bmatrix} \begin{bmatrix} x_1 \\ x_2 \\ x_3 \end{bmatrix} = \begin{bmatrix} 0 \\ 0 \\ 0 \end{bmatrix}.$$

The reduced row echelon form of the coefficient matrix is $\begin{bmatrix} 1 & 1 & 0 \\ 0 & 0 & 1 \\ 0 & 0 & 0 \end{bmatrix}$. Thus we can take $\mathbf{X}_3 = \begin{bmatrix} 1 \\ -1 \\ 0 \end{bmatrix}$. Since $\mathbf{A}$ is symmetric and the eigenvalues are distinct, Theorem 5.5 implies that $\{\mathbf{X}_1, \mathbf{X}_2, \mathbf{X}_3\}$ is orthogonal. To obtain an orthonormal basis of eigenvectors we normalize $\mathbf{X}_1$, $\mathbf{X}_2$, $\mathbf{X}_3$ as

$$\mathbf{Z}_1 = (1/\|\mathbf{X}_1\|)\ \mathbf{X}_1 = \mathbf{X}_1,\quad \mathbf{Z}_2 = (1/\|\mathbf{X}_2\|)\ \mathbf{X}_2 = (1/\sqrt{2}) \begin{bmatrix} 1 \\ 1 \\ 0 \end{bmatrix},$$

$$\mathbf{Z}_3 = (1/\|\mathbf{X}_3\|)\ \mathbf{X}_3 = (1/\sqrt{2}) \begin{bmatrix} 1 \\ -1 \\ 0 \end{bmatrix}.$$

Then set $\mathbf{P} = [\mathbf{Z}_1\ \mathbf{Z}_2\ \mathbf{Z}_3]$ and hence $\mathbf{P}^{-1}\mathbf{A}\mathbf{P} = \begin{bmatrix} 1 & 0 & 0 \\ 0 & 2 & 0 \\ 0 & 0 & 0 \end{bmatrix}$.

17. Let $\mathbf{A} = \begin{bmatrix} 1 & 0 & 0 \\ 0 & 1 & 1 \\ 0 & 1 & 1 \end{bmatrix}$. Then $\det(\lambda\mathbf{I}_3 - \mathbf{A}) = (\lambda-1)\lambda(\lambda-2)$ and the eigenvalues of $\mathbf{A}$ are $\lambda_1 = 1$, $\lambda_2 = 0$, and $\lambda_3 = 2$. To find an eigenvector $\mathbf{X}_1$ associated with $\lambda_1 = 1$, we solve $(1\mathbf{I}_3 - \mathbf{A})\mathbf{X} = \mathbf{0}$:

$$\begin{bmatrix} 0 & 0 & 0 \\ 0 & 0 & -1 \\ 0 & -1 & 0 \end{bmatrix} \begin{bmatrix} x_1 \\ x_2 \\ x_3 \end{bmatrix} = \begin{bmatrix} 0 \\ 0 \\ 0 \end{bmatrix}.$$

The reduced row echelon form of the coefficient matrix is

$\begin{bmatrix} 0 & 1 & 0 \\ 0 & 0 & 1 \\ 0 & 0 & 0 \end{bmatrix}$. Thus we can take $\mathbf{X}_1 = \begin{bmatrix} 1 \\ 0 \\ 0 \end{bmatrix}$. To find an eigenvector $\mathbf{X}_2$ associated with $\lambda_2 = 0$, we solve $(0\mathbf{I}_3 - \mathbf{A})\mathbf{X} = \mathbf{0}$:

$$\begin{bmatrix} -1 & 0 & 0 \\ 0 & -1 & -1 \\ 0 & -1 & -1 \end{bmatrix} \begin{bmatrix} x_1 \\ x_2 \\ x_3 \end{bmatrix} = \begin{bmatrix} 0 \\ 0 \\ 0 \end{bmatrix}.$$

The reduced row echelon form of the coefficient matrix is

$\begin{bmatrix} 1 & 0 & 0 \\ 0 & 1 & 1 \\ 0 & 0 & 0 \end{bmatrix}$. Thus we can take $\mathbf{X}_2 = \begin{bmatrix} 0 \\ 1 \\ -1 \end{bmatrix}$. To find an eigenvector $\mathbf{X}_3$ associated with $\lambda_3 = 2$, we solve $(2\mathbf{I}_3 - \mathbf{A})\mathbf{X} = \mathbf{0}$:

$$\begin{bmatrix} 1 & 0 & 0 \\ 0 & 1 & -1 \\ 0 & -1 & 1 \end{bmatrix} \begin{bmatrix} x_1 \\ x_2 \\ x_3 \end{bmatrix} = \begin{bmatrix} 0 \\ 0 \\ 0 \end{bmatrix}.$$

The reduced row echelon form of the coefficient matrix is

$\begin{bmatrix} 1 & 0 & 0 \\ 0 & 1 & -1 \\ 0 & 0 & 0 \end{bmatrix}$. Thus we can take $\mathbf{X}_3 = \begin{bmatrix} 0 \\ 1 \\ 1 \end{bmatrix}$. Since $\mathbf{A}$ is symmetric and the eigenvalues are distinct, Theorem 5.5 implies that $\{\mathbf{X}_1, \mathbf{X}_2, \mathbf{X}_3\}$ is orthogonal. To obtain an orthonormal basis of eigenvectors we normalize $\mathbf{X}_1$, $\mathbf{X}_2$, $\mathbf{X}_3$ as

$$\mathbf{Z}_1 = (1/\|\mathbf{X}_1\|)\ \mathbf{X}_1 = \mathbf{X}_1,\ \mathbf{Z}_2 = (1/\|\mathbf{X}_2\|)\ \mathbf{X}_2 = (1/\sqrt{2}) \begin{bmatrix} 0 \\ 1 \\ -1 \end{bmatrix},$$

$$\mathbf{Z}_3 = (1/\|\mathbf{X}_3\|)\ \mathbf{X}_3 = (1/\sqrt{2}) \begin{bmatrix} 0 \\ 1 \\ 1 \end{bmatrix}.$$

Then set $\mathbf{P} = [\mathbf{Z}_1\ \mathbf{Z}_2\ \mathbf{Z}_3]$ and hence $\mathbf{P}^{-1}\mathbf{AP} = \begin{bmatrix} 1 & 0 & 0 \\ 0 & 0 & 0 \\ 0 & 0 & 2 \end{bmatrix}$.

19. Let $\mathbf{A} = \begin{bmatrix} 2 & 1 & 1 \\ 1 & 2 & 1 \\ 1 & 1 & 2 \end{bmatrix}$. Then $\det(\lambda\mathbf{I}_3 - \mathbf{A}) = (\lambda-1)^2(\lambda-4)$ and the eigenvalues of $\mathbf{A}$ are $\lambda_1 = \lambda_2 = 1$, $\lambda_3 = 4$. To find an eigenvector $\mathbf{X}_1$ associated with $\lambda_1 = 1$, we solve $(1\mathbf{I}_3 - \mathbf{A})\mathbf{X} = \mathbf{0}$:

$$\begin{bmatrix} -1 & -1 & -1 \\ -1 & -1 & -1 \\ -1 & -1 & -1 \end{bmatrix}\begin{bmatrix} x_1 \\ x_2 \\ x_3 \end{bmatrix} = \begin{bmatrix} 0 \\ 0 \\ 0 \end{bmatrix}.$$

The reduced row echelon form of the coefficient matrix is $\begin{bmatrix} 1 & 1 & 1 \\ 0 & 0 & 0 \\ 0 & 0 & 0 \end{bmatrix}$. The solution is $x_1 = -r - s$, $x_2 = r$, $x_3 = s$, where r and s are any real numbers. In vector form the solution is $\begin{bmatrix} -r-s \\ r \\ s \end{bmatrix} = r\begin{bmatrix} -1 \\ 1 \\ 0 \end{bmatrix} + s\begin{bmatrix} -1 \\ 0 \\ 1 \end{bmatrix}$. Thus there are two linearly independent eigenvectors, $\mathbf{X}_1 = \begin{bmatrix} -1 \\ 1 \\ 0 \end{bmatrix}$ and $\mathbf{X}_2 = \begin{bmatrix} -1 \\ 0 \\ 1 \end{bmatrix}$.

To find an eigenvector $\mathbf{X}_3$ associated with $\lambda_3 = 4$, we solve $(4\mathbf{I}_3 - \mathbf{A})\mathbf{X} = \mathbf{0}$:

$$\begin{bmatrix} 2 & -1 & -1 \\ -1 & 2 & -1 \\ -1 & -1 & 2 \end{bmatrix}\begin{bmatrix} x_1 \\ x_2 \\ x_3 \end{bmatrix} = \begin{bmatrix} 0 \\ 0 \\ 0 \end{bmatrix}.$$

The reduced row echelon form of the coefficient matrix is $\begin{bmatrix} 1 & 0 & -1 \\ 0 & 1 & -1 \\ 0 & 0 & 0 \end{bmatrix}$. Thus we can take $\mathbf{X}_3 = \begin{bmatrix} 1 \\ 1 \\ 1 \end{bmatrix}$. Since $\mathbf{A}$ is symmetric $\mathbf{X}_1$ and $\mathbf{X}_2$ are orthogonal with $\mathbf{X}_3$ (see Theorem 5.5). For this matrix $\mathbf{A}$, vectors $\mathbf{X}_1$ and $\mathbf{X}_2$ are not orthogonal. We use the Gram-Schmidt process to obtain an orthogonal basis for the eigenspace associated with $\lambda_1 = 1$ as follows. Let

$$\mathbf{Y}_1 = \mathbf{X}_1 = \begin{bmatrix} -1 \\ 1 \\ 0 \end{bmatrix}$$

and

$$\mathbf{Y}_2 = \mathbf{X}_2 - \left(\frac{\mathbf{X}_2\cdot\mathbf{Y}_1}{\mathbf{Y}_1\cdot\mathbf{Y}_1}\right)\mathbf{Y}_1 = \begin{bmatrix}-1/2\\-1/2\\1\end{bmatrix}.$$

Let

$$\mathbf{Y}_1' = \mathbf{Y}_1 \text{ and } \mathbf{Y}_2' = 2\mathbf{Y}_2 = \begin{bmatrix}-1\\-1\\2\end{bmatrix}.$$

The set $\{\mathbf{Y}_1', \mathbf{Y}_2'\}$ is an orthogonal basis for the eigenspace associated with $\lambda_1 = 1$. Then we normalize $\{\mathbf{Y}_1', \mathbf{Y}_2', \mathbf{X}_3\}$ as,

$$\mathbf{Z}_1 = (1/\|\mathbf{Y}_1'\|)\ \mathbf{Y}_1' = (1/\sqrt{2})\begin{bmatrix}-1\\1\\0\end{bmatrix},$$

$$\mathbf{Z}_2 = (1/\|\mathbf{Y}_2'\|)\ \mathbf{Y}_2' = (1/\sqrt{6})\begin{bmatrix}-1\\-1\\2\end{bmatrix},$$

$$\mathbf{Z}_3 = (1/\|\mathbf{X}_3\|)\ \mathbf{X}_3 = (1/\sqrt{3})\begin{bmatrix}1\\1\\1\end{bmatrix}.$$

Then set $\mathbf{P} = [\mathbf{Z}_1\ \mathbf{Z}_2\ \mathbf{Z}_3]$ and hence $\mathbf{P}^{-1}\mathbf{AP} = \begin{bmatrix}1&0&0\\0&1&0\\0&0&4\end{bmatrix}$.

21. We show that $\mathbf{L}$ is an isometry by verifying that Equation (4) holds. Let $\mathbf{A} = \begin{bmatrix}1/\sqrt{2} & 1/\sqrt{2}\\ 1/\sqrt{2} & -1/\sqrt{2}\end{bmatrix} = 1/\sqrt{2}\begin{bmatrix}1&1\\1&-1\end{bmatrix}$. Then

$$\mathbf{L}(\mathbf{X})\cdot\mathbf{L}(\mathbf{Y}) = (\mathbf{AX})\cdot(\mathbf{AY}) = \mathbf{X}\cdot(\mathbf{A}^T\mathbf{AY})$$

We see that $\mathbf{A}^T\mathbf{A} = (1/\sqrt{2})^2\begin{bmatrix}1&1\\1&-1\end{bmatrix}^2 = 1/2\begin{bmatrix}2&0\\0&2\end{bmatrix} = \mathbf{I}_2$. Thus

$$\mathbf{L}(\mathbf{X})\cdot\mathbf{L}(\mathbf{Y}) = \mathbf{X}\cdot(\mathbf{I}_2\mathbf{Y}) = \mathbf{X}\cdot\mathbf{Y}$$

T.1. $(\mathbf{AX})\cdot\mathbf{Y} = (\mathbf{AX})^T\mathbf{Y} = \mathbf{X}^T\mathbf{A}^T\mathbf{Y} = \mathbf{X}\cdot(\mathbf{A}^T\mathbf{Y})$.

Exercises 5.2

T.3. If $\mathbf{A}^T\mathbf{A} = \mathbf{I}_n$, then

$$1 = \det(\mathbf{I}_n) = \det(\mathbf{A}^T\mathbf{A}) = \det(\mathbf{A}^T)\cdot\det(\mathbf{A}) = \det(\mathbf{A})^2.$$

Thus $\det(\mathbf{A}) = \pm 1$.

T.5. Let $\mathbf{A} = \begin{bmatrix} a & b \\ b & d \end{bmatrix}$ be a 2×2 symmetric matrix. Then its characteristic polynomial is $\lambda^2 - (a+d)\lambda + (ad-b^2)$. The roots of this polynomial are

$$\lambda = \frac{a+d \pm\sqrt{(a+d)^2 - 4(ad-b^2)}}{2} = \frac{a+d \pm \sqrt{(a-d)^2 + 4b^2}}{2}.$$

If $b = 0$, $\mathbf{A}$ is already diagonal. If $b \neq 0$, the discriminant $(a-d)^2 + 4b^2$ is positive and there are two distinct real eigenvalues. Thus $\mathbf{A}$ is diagonalizable. (See Theorem 5.3.) By Theorem 5.2, there is a diagonalizing matrix $\mathbf{P}$ whose columns are linearly independent eigenvectors of $\mathbf{A}$. We may assume further that those columns are unit vectors in $\mathbf{R}^2$. By Theorem 5.5, the two columns are orthogonal. Thus $\mathbf{P}$ is an orthogonal matrix.

T.7. We show that $(\mathbf{A}^{-1})^T = (\mathbf{A}^{-1})^{-1}$:

$$(\mathbf{A}^{-1})^T\mathbf{A}^{-1} = (\mathbf{A}^T)^{-1}\mathbf{A}^{-1} = (\mathbf{A}\mathbf{A}^T)^{-1} = \mathbf{I}_n^{-1} = \mathbf{I}_n.$$

T.9. If $\mathbf{A}^T\mathbf{A}\mathbf{Y} = \mathbf{Y}$ for any $\mathbf{Y}$ in $\mathbf{R}^n$, then $(\mathbf{A}^T\mathbf{A} - \mathbf{I}_n)\mathbf{Y} = \mathbf{0}$ for all $\mathbf{Y}$ in $\mathbf{R}^n$. Thus $\mathbf{A}^T\mathbf{A} - \mathbf{I}_n$ is the zero matrix , so $\mathbf{A}^T\mathbf{A} = \mathbf{I}_n$.

ML.1. (a)

```
A=[6 6;6 6];
[V,D]=eig(A)

V =

    0.7071    0.7071
   -0.7071    0.7071

D =

    0     0
    0    12
```

Let **P** = **V**, then

```
P=V;P'*A*P

ans =

     0          0
     0    12.0000
```

(b)
```
A=[1 2 2;2 1 2;2 2 1];
[V,D]=eig(A)

V =

    0.7743   -0.2590    0.5774
   -0.6115   -0.5411    0.5774
   -0.1629    0.8001    0.5774

D =

   -1.0000         0         0
         0   -1.0000         0
         0         0    5.0000
```

Let **P** = **V**, then

```
P=V;P'*A*P

ans =

   -1.0000    0.0000    0.0000
    0.0000   -1.0000   -0.0000
    0.0000   -0.0000    5.0000
```

(c)
```
A=[4 1 0;1 4 1;0 1 4];
[V,D]=eig(A)

V =

    0.5000   -0.7071   -0.5000
    0.7071   -0.0000    0.7071
    0.5000    0.7071   -0.5000

D =

    5.4142         0         0
         0    4.0000         0
         0         0    2.5858
```

Let **P** = **V**, then

```
P=V;P'*A*P
```

```
ans =

   5.4142   -0.0000   -0.0000
  -0.0000    4.0000    0.0000
  -0.0000    0.0000    2.5858
```

ML.2. (a)
```
A=[1 2;-1 4];
[V,D]=eig(A)

V =

  -0.8944   -0.7071
  -0.4472   -0.7071

D =

     2     0
     0     3

V'*V

ans =

   1.0000    0.9487
   0.9487    1.0000
```

Hence **V** is not orthogonal. However, since the eigenvalues are distinct **A** is diagonalizable, so **V** can be replaced by an orthogonal matrix.

(b)
```
A=[2 1 2;2 2 -2;3 1 1];
[V,D]=eig(A)

V =

  -0.5482    0.7071    0.4082
   0.6852   -0.0000   -0.8165
   0.4796    0.7071    0.4082

D =

  -1.0000         0         0
        0    4.0000         0
        0         0    2.0000

V'*V

ans =

   1.0000   -0.0485   -0.5874
  -0.0485    1.0000    0.5774
  -0.5874    0.5774    1.0000
```

Hence **V** is not orthogonal. However, since the eigenvalues are distinct **A** is diagonalizable, so **V** can be replaced by an orthogonal matrix.

(c)
```
A=[1 -3;3 -5];
[V,D]=eig(A)

V =

    0.7071    0.7071
    0.7071    0.7071

D =

    -2     0
     0    -2
```

Inspecting **V**, we see that there is only one linearly independent eigenvector, so **A** is not diagonalizable.

(d)
```
A=[1 0 0;0 1 1;0 1 1];
[V,D]=eig(A)

V =

    1.0000         0         0
         0    0.7071    0.7071
         0    0.7071   -0.7071

D =

    1.0000         0         0
         0    2.0000         0
         0         0    0.0000

V'*V

ans =
    1.0000         0         0
         0    1.0000         0
         0         0    1.0000
```

Hence **V** is orthogonal. We should have expected this since **A** is symmetric.

Chapter 5 Supplementary Exercises

1. Let $\mathbf{A} = \begin{bmatrix} -2 & 0 & 0 \\ 3 & 2 & 3 \\ 4 & -1 & 6 \end{bmatrix}$. The characteristic polynomial is given by

$|\lambda \mathbf{I}_3 - \mathbf{A}| = \begin{vmatrix} \lambda+2 & 0 & 0 \\ -3 & \lambda-2 & -3 \\ -4 & 1 & \lambda-6 \end{vmatrix} = (\lambda+2)(\lambda-3)(\lambda-5)$. Thus the eigenvalues are $\lambda_1 = -2$, $\lambda_2 = 3$, $\lambda_3 = 5$. To find an eigenvector $\mathbf{X}_1$ associated with $\lambda_1 = -2$, solve $(-2\mathbf{I}_3 - \mathbf{A})\mathbf{X} = \mathbf{0}$:

$$\begin{bmatrix} 0 & 0 & 0 \\ -3 & -4 & -3 \\ -4 & 1 & -8 \end{bmatrix}\begin{bmatrix} x_1 \\ x_2 \\ x_3 \end{bmatrix} = \begin{bmatrix} 0 \\ 0 \\ 0 \end{bmatrix}.$$

The reduced row echelon form of the coefficient matrix is $\begin{bmatrix} 1 & 0 & 35/19 \\ 0 & 1 & -12/19 \\ 0 & 0 & 0 \end{bmatrix}$. Thus the solution is $x_1 = (-35/19)r$, $x_2 = (12/19)r$, $x_3 = r$, where r is any real number. Let r = 19 and we have eigenvector $\mathbf{X}_1 = \begin{bmatrix} -35 \\ 12 \\ 19 \end{bmatrix}$. To find an eigenvector $\mathbf{X}_2$ associated with $\lambda_2 = 3$, solve $(3\mathbf{I}_3 - \mathbf{A})\mathbf{X} = \mathbf{0}$:

$$\begin{bmatrix} 5 & 0 & 0 \\ -3 & 1 & -3 \\ -4 & 1 & -3 \end{bmatrix}\begin{bmatrix} x_1 \\ x_2 \\ x_3 \end{bmatrix} = \begin{bmatrix} 0 \\ 0 \\ 0 \end{bmatrix}.$$

The reduced row echelon form of the coefficient matrix is $\begin{bmatrix} 1 & 0 & 0 \\ 0 & 1 & -3 \\ 0 & 0 & 0 \end{bmatrix}$. Thus the solution is $x_1 = 0$, $x_2 = 3r$, $x_3 = r$, where r is any real number. Let r = 1 and we have eigenvector $\mathbf{X}_2 = \begin{bmatrix} 0 \\ 3 \\ 1 \end{bmatrix}$. To find an eigenvector $\mathbf{X}_3$ associated with $\lambda_3 = 5$, solve $(5\mathbf{I}_3 - \mathbf{A})\mathbf{X} = \mathbf{0}$:

$$\begin{bmatrix} 7 & 0 & 0 \\ -3 & 3 & -3 \\ -4 & 1 & -1 \end{bmatrix}\begin{bmatrix} x_1 \\ x_2 \\ x_3 \end{bmatrix} = \begin{bmatrix} 0 \\ 0 \\ 0 \end{bmatrix}.$$

The reduced row echelon form of the coefficient matrix is

$\begin{bmatrix} 1 & 0 & 0 \\ 0 & 1 & -1 \\ 0 & 0 & 0 \end{bmatrix}$. Thus the solution is $x_1 = 0$, $x_2 = r$, $x_3 = r$, where r is any real number. Let r = 1 and we have eigenvector

$\mathbf{X}_3 = \begin{bmatrix} 0 \\ 1 \\ 1 \end{bmatrix}$.

3. Let $\mathbf{A} = \begin{bmatrix} 2 & 2 & 0 \\ 5 & -1 & 3 \\ 0 & 0 & 0 \end{bmatrix}$. We have $|\lambda\mathbf{I}_3 - \mathbf{A}| = \begin{vmatrix} \lambda-2 & -2 & 0 \\ -5 & \lambda+1 & -3 \\ 0 & 0 & \lambda \end{vmatrix}$

$= \lambda(\lambda+3)(\lambda-4)$, thus the eigenvalues are $\lambda_1 = 0$, $\lambda_2 = -3$, $\lambda_3 = 4$. Since the eigenvalues are real and distinct, **A** is diagonalizable. (See Theorem 5.3.)

5. Let $\mathbf{A} = \begin{bmatrix} 1 & 0 & 1 \\ 0 & 1 & 0 \\ -1 & 0 & 1 \end{bmatrix}$. We have $|\lambda\mathbf{I}_3 - \mathbf{A}| = \begin{vmatrix} \lambda-1 & 0 & -1 \\ 0 & \lambda-1 & 0 \\ 1 & 0 & \lambda-1 \end{vmatrix}$

$= (\lambda-1)(\lambda^2 - 2\lambda + 2)$, thus the eigenvalues are $\lambda_1 = 1$, $\lambda_2 = 1 + \sqrt{-1}$, $\lambda_3 = 1 - \sqrt{-1}$. Since all the roots of the characteristic polynomial are not real, we say **A** is not diagonalizable.

7. Let $\mathbf{A} = \begin{bmatrix} 0 & 0 & 1 \\ 0 & 2 & 0 \\ 0 & 0 & 2 \end{bmatrix}$. Since **A** is upper triangular its eigenvalues

are $\lambda_1 = 0$, $\lambda_2 = \lambda_3 = 2$. To find an eigenvector associated with $\lambda_1 = 0$, solve $(0\mathbf{I}_3 - \mathbf{A})\ \mathbf{X} = \mathbf{0}$:

$$\begin{bmatrix} 0 & 0 & -1 \\ 0 & -2 & 0 \\ 0 & 0 & -2 \end{bmatrix}\begin{bmatrix} x_1 \\ x_2 \\ x_3 \end{bmatrix} = \begin{bmatrix} 0 \\ 0 \\ 0 \end{bmatrix}.$$

The reduced row echelon form of the coefficient matrix is $\begin{bmatrix} 0 & 1 & 0 \\ 0 & 0 & 1 \\ 0 & 0 & 0 \end{bmatrix}$. The solution is $x_1 = r$, $x_2 = x_3 = 0$, where r is any real number. Let r = 1, then we have eigenvector $\mathbf{X}_1 = \begin{bmatrix} 1 \\ 0 \\ 0 \end{bmatrix}$, which is a basis for the eigenspace associated with $\lambda_1 = 0$. To find an eigenvector associated with $\lambda_2 = 2$, solve $(2\mathbf{I}_3 - \mathbf{A})\mathbf{X} = \mathbf{0}$:

$$\begin{bmatrix} 2 & 0 & -1 \\ 0 & 0 & 0 \\ 0 & 0 & 0 \end{bmatrix} \begin{bmatrix} x_1 \\ x_2 \\ x_3 \end{bmatrix} = \begin{bmatrix} 0 \\ 0 \\ 0 \end{bmatrix}.$$

The reduced row echelon form of the coefficient matrix is $\begin{bmatrix} 1 & 0 & -1/2 \\ 0 & 0 & 0 \\ 0 & 0 & 0 \end{bmatrix}$. The solution is $x_1 = (1/2)s$, $x_2 = r$, $x_3 = s$, where r and s are any real numbers. In vector form, the solution is $\begin{bmatrix} (1/2)s \\ r \\ s \end{bmatrix} = r\begin{bmatrix} 0 \\ 1 \\ 0 \end{bmatrix} + s\begin{bmatrix} 1/2 \\ 0 \\ 1 \end{bmatrix}$. Thus $\mathbf{X}_2 = \begin{bmatrix} 0 \\ 1 \\ 0 \end{bmatrix}$ and $\mathbf{X}_3 = \begin{bmatrix} 1/2 \\ 0 \\ 1 \end{bmatrix}$ are linearly independent eigenvectors associated with $\lambda_2 = 2$ and form a basis for the eigenspace. (Many other bases are possible.)

9. Let $\mathbf{A} = \begin{bmatrix} 1 & 1 & 1 \\ 1 & 1 & 1 \\ 1 & 1 & 1 \end{bmatrix}$. We have $|\lambda\mathbf{I}_3 - \mathbf{A}| = \lambda^2(\lambda-3)$, thus the eigenvalues are $\lambda_1 = \lambda_2 = 0$ and $\lambda_3 = 3$. To find eigenvectors associated with eigenvalue λ_1, solve $(0\mathbf{I}_3 - \mathbf{A})\mathbf{X} = \mathbf{0}$:

$$\begin{bmatrix} -1 & -1 & -1 \\ -1 & -1 & -1 \\ -1 & -1 & -1 \end{bmatrix} \begin{bmatrix} x_1 \\ x_2 \\ x_3 \end{bmatrix} = \begin{bmatrix} 0 \\ 0 \\ 0 \end{bmatrix}.$$

The reduced row echelon form of the coefficient matrix is

$\begin{bmatrix} 1 & 1 & 1 \\ 0 & 0 & 0 \\ 0 & 0 & 0 \end{bmatrix}$. The solution is $x_1 = -r-s$, $x_2 = r$, $x_3 = s$, where r and s are any real numbers. In vector form the solution is

$\begin{bmatrix} -r-s \\ r \\ s \end{bmatrix} = r\begin{bmatrix} -1 \\ 1 \\ 0 \end{bmatrix} + s\begin{bmatrix} -1 \\ 0 \\ 1 \end{bmatrix}$. Then $\mathbf{X}_1 = \begin{bmatrix} -1 \\ 1 \\ 0 \end{bmatrix}$ and $\mathbf{X}_2 = \begin{bmatrix} -1 \\ 0 \\ 1 \end{bmatrix}$ are linearly independent eigenvectors associated with eigenvalue $\lambda = 0$. To find an eigenvector associated with $\lambda_3 = 3$, solve $(3\mathbf{I}_3 - \mathbf{A})\mathbf{X} = \mathbf{0}$:

$$\begin{bmatrix} 2 & -1 & -1 \\ -1 & 2 & -1 \\ -1 & -1 & 2 \end{bmatrix}\begin{bmatrix} x_1 \\ x_2 \\ x_3 \end{bmatrix} = \begin{bmatrix} 0 \\ 0 \\ 0 \end{bmatrix}.$$

The reduced row echelon form of the coefficient matrix is

$\begin{bmatrix} 1 & 0 & -1 \\ 0 & 1 & -1 \\ 0 & 0 & 0 \end{bmatrix}$. The solution is $x_1 = x_2 = x_3 = r$, where r is any real number. Let $r = 1$, then $\mathbf{X}_3 = \begin{bmatrix} 1 \\ 1 \\ 1 \end{bmatrix}$ is an eigenvector associated with $\lambda_3 = 3$. Theorem 5.5 implies that $\mathbf{X}_3$ is orthogonal to both $\mathbf{X}_1$ and $\mathbf{X}_2$. However, $\mathbf{X}_1 \cdot \mathbf{X}_2 \neq 0$, thus we use the Gram-Schmidt process to determine an orthogonal pair of eigenvectors associated with $\lambda_1 = 0$. Let $\mathbf{Y}_1 = \mathbf{X}_1$, and set

$$\mathbf{Y}_2 = \mathbf{X}_2 - \left(\frac{\mathbf{X}_2 \cdot \mathbf{Y}_1}{\mathbf{Y}_1 \cdot \mathbf{Y}_1}\right)\mathbf{Y}_1 = \begin{bmatrix} -1/2 \\ -1/2 \\ 1 \end{bmatrix}.$$

To simplify the computations, let $\mathbf{Y}_1' = \mathbf{Y}_1$ and $\mathbf{Y}_2' = 2\mathbf{Y}_2 = \begin{bmatrix} -1 \\ -1 \\ 2 \end{bmatrix}$. We have that $\{\mathbf{Y}_1', \mathbf{Y}_2', \mathbf{X}_3\}$ is an orthogonal basis of eigenvectors. Next we normalize these as

$$\mathbf{Z}_1 = (1/\|\mathbf{Y}_1'\|)\ \mathbf{Y}_1' = \begin{bmatrix} -1/\sqrt{2} \\ 1/\sqrt{2} \\ 0 \end{bmatrix},\ \mathbf{Z}_2 = (1/\|\mathbf{Y}_2'\|)\ \mathbf{Y}_2'$$

$$= \begin{bmatrix} -1/\sqrt{6} \\ -1/\sqrt{6} \\ 2/\sqrt{6} \end{bmatrix}, \quad \mathbf{Z}_3 = (1/\|\mathbf{X}_3\|)\,\mathbf{X}_3 = \begin{bmatrix} 1/\sqrt{3} \\ 1/\sqrt{3} \\ 1/\sqrt{3} \end{bmatrix}.$$

Set $\mathbf{P} = [\mathbf{Z}_1\ \mathbf{Z}_2\ \mathbf{Z}_3]$, then $\mathbf{P}^{-1}\mathbf{A}\mathbf{P} = \mathbf{D} = \begin{bmatrix} 0 & 0 & 0 \\ 0 & 0 & 0 \\ 0 & 0 & 3 \end{bmatrix}$.

11. If λ is an eigenvalue of $\mathbf{A}$ with corresponding eigenvector $\mathbf{X}$, then $\mathbf{AX} = \lambda\mathbf{X}$. Hence $\mathbf{A}^2\mathbf{X} = \mathbf{A}(\mathbf{AX}) = \mathbf{A}(\lambda\mathbf{X}) = \lambda(\mathbf{AX}) = \lambda^2\mathbf{X}$. If $\mathbf{A}^2 = \mathbf{A}$ then $\lambda^2 = \lambda$. Thus $\lambda = 0$ or $\lambda = 1$.

T.1. Let $\mathbf{P}$ be a nonsingular matix such that $\mathbf{P}^{-1}\mathbf{A}\mathbf{P} = \mathbf{D}$. Then

$$\begin{aligned} \mathrm{tr}(\mathbf{D}) &= \mathrm{tr}(\mathbf{P}^{-1}\mathbf{A}\mathbf{P}) = \mathrm{tr}(\mathbf{P}^{-1}(\mathbf{A}\mathbf{P})) = \mathrm{tr}((\mathbf{A}\mathbf{P})\mathbf{P}^{-1}) \\ &= \mathrm{tr}(\mathbf{A}\mathbf{P}\mathbf{P}^{-1}) = \mathrm{tr}(\mathbf{A}\mathbf{I}_n) = \mathrm{tr}(\mathbf{A}). \end{aligned}$$

T.3. Let $\mathbf{P}$ be such that $\mathbf{P}^{-1}\mathbf{A}\mathbf{P} = \mathbf{B}$.

(a) $\mathbf{B}^T = (\mathbf{P}^{-1}\mathbf{A}\mathbf{P})^T = \mathbf{P}^T\mathbf{A}^T(\mathbf{P}^{-1})^T = \mathbf{P}^T\mathbf{A}^T(\mathbf{P}^T)^{-1}$, hence $\mathbf{A}^T$ and $\mathbf{B}^T$ are similar.

(b) $\mathrm{rank}(\mathbf{B}) = \mathrm{rank}(\mathbf{P}^{-1}\mathbf{A}\mathbf{P}) = \mathrm{rank}(\mathbf{P}^{-1}\mathbf{A}) = \mathrm{rank}(\mathbf{A})$
(See Exercises T.7c and T.7d in the Supplementary Exercises to Chapter 3.)

(c) $\det(\mathbf{B}) = \det(\mathbf{P}^{-1}\mathbf{A}\mathbf{P}) = \det(\mathbf{P}^{-1})\det(\mathbf{A})\det(\mathbf{P})$
$= (1/\det(\mathbf{P}))\det(\mathbf{A})\det(\mathbf{P}) = \det(\mathbf{A})$
Thus $\det(\mathbf{B}) \neq 0$ if and only if $\det(\mathbf{A}) \neq 0$.

(d) Since $\mathbf{A}$ and $\mathbf{B}$ are nonsingular and $\mathbf{B} = \mathbf{P}^{-1}\mathbf{A}\mathbf{P}$,
$\mathbf{B}^{-1} = (\mathbf{P}^{-1}\mathbf{A}\mathbf{P})^{-1} = \mathbf{P}^{-1}\mathbf{A}^{-1}\mathbf{P}$.
That is, $\mathbf{A}^{-1}$ and $\mathbf{B}^{-1}$ are similar.

(e) $\mathrm{tr}(\mathbf{B}) = \mathrm{tr}(\mathbf{P}^{-1}\mathbf{A}\mathbf{P}) = \mathrm{tr}((\mathbf{P}^{-1}\mathbf{A})\mathbf{P}) = \mathrm{tr}(\mathbf{P}(\mathbf{P}^{-1}\mathbf{A})) = \mathrm{tr}(\mathbf{A})$
(See Exercise T.1 in the Supplementary Exercises to Chapter 1.)

T.5. $(c\mathbf{A}^T) = (c\mathbf{A})^{-1}$ if and only if $c\mathbf{A}^T = (1/c)\mathbf{A}^{-1} = (1/c)\mathbf{A}^T$. That is, $c = (1/c)$. Hence $c = \pm 1$.

Exercises 6.1

1. Let x be the number of tons of regular steel to be made and let y be the number of tons of special steel to made. Since each ton of regular steel requires 2 hours in the open-hearth furnace and each ton of special steel also requires 2 hours in the open-hearth furnace, the total amount of time all the steel produced is in the open-hearth furnace is

$$2x + 2y.$$

Similarly, each ton of regular steel requires 5 hours in the soaking pit and each ton of special steel requires 3 hours in the soaking pit, thus the total amount of time all the steel produced is in the soaking pit is

$$5x + 3y.$$

The open-hearth furnace is available 8 hours per day and the soaking pit is available 15 hours per day, hence we must have

$$\begin{aligned} 2x + 2y &\le 8 \\ 5x + 3y &\le 15. \end{aligned}$$

Since x and y cannot be negative, it follows that

$$x \ge 0 \quad \text{and} \quad y \ge 0.$$

The profit on a ton of regular steel is \$120 and it is \$100 on a ton of special steel, thus the total profit (in dollars) is

$$z = 120x + 100y.$$

The problem in mathematical form is: Find values of x and y that will maximize

$$z = 120x + 100\,y$$

subject to the following restrictions that must be satisfied by x and y:

$$\begin{aligned} 2x + 2y &\le 8 \\ 5x + 3y &\le 15 \\ x \ge 0 \quad &\text{and} \quad y \ge 0. \end{aligned}$$

3. Referring to Exercise 2, the additional rule would impose the further restriction

$$y \le (1/2)x.$$

5. Let x be the number of minutes of advertising and let y be the number of minutes of comedy. The total time of the program is

$$x + y,$$

which cannot exceed 30 minutes hence,

$$x + y \le 30$$

and of course

$$x \ge 0 \quad \text{and} \quad y \ge 0.$$

The advertiser requires at least 2 minutes of advertising time, while the station insists on no more than 4 minutes of advertising time hence,

$$x \ge 2 \quad \text{and} \quad x \le 4.$$

Similarly, the comedian insists on at least 24 minutes for the comedy portion of the show so,

$$y \ge 24.$$

Since each minute of advertising attracts 40,000 viewers and each minute of comedy attracts 45,000 viewers, the number of viewer-minutes is

$$z = 40{,}000x + 45{,}000y.$$

The problem in mathematical form is: Find values of x and y that will maximize

$$z = 40{,}000x + 45{,}000y$$

subject to the following restrictions that must be satisfied by x and y:

$$\begin{aligned} x + y &\le 30 \\ y &\ge 24 \\ x &\ge 2 \\ x &\le 4 \end{aligned}$$
$$x \ge 0 \quad \text{and} \quad y \ge 0.$$

7. Let x be the number of units of A and let y be the number of units of B. Of course, x and y cannot be negative, thus

$$x \ge 0 \quad \text{and} \quad y \ge 0.$$

Each unit of A has 1 gram of fat and each unit of B has 2 grams of fat, thus the total amount of fat provided is

$$x + 2y.$$

Since there is to be no more than 10 grams of fat, we have

$$x + 2y \le 10.$$

Similarly, each unit of A has 1 gram of carbohydrates and the same for a unit of B, hence the total amount of carbohydrate is

$$x + y.$$

There is to be no more than 7 grams of carbohydrate, thus we have

$$x + y \le 7.$$

Each unit of A has 4 grams of protein and each unit of B has 6 grams of protein, thus the total amount of protein provided is

$$z = 4x + 6y.$$

The problem in mathematical form is: Find values of x and y that will maximize

$$z = 4x + 6y$$

subject to the following restrictions that must be satisfied by x and y:

$$x + 2y \le 10$$
$$x + y \le 7$$
$$x \ge 0 \quad \text{and} \quad y \ge 0.$$

9. Let x be the number of units of type A grain and let y be the number of units of type B grain. Of course, x and y cannot be negative, so

$$x \ge 0 \quad \text{and} \quad y \ge 0.$$

Each unit of type A contains 2 grams of fat and each unit of type B contains 3 grams of fat, thus the total number of grams of fat is

$$2x + 3y.$$

Since at least 18 grams of fat is required, we have

$$2x + 3y \ge 18.$$

Similarly, each unit of type A contains 1 gram of protein and each unit of type B contains 3 grams of protein, hence the total number of grams of protein is

$$x + 3y.$$

Since there is to be at least 12 grams of protein, we have

$$x + 3y \ge 12.$$

Each unit of type A contains 80 calories and each unit of type B contains 60 calories, thus the total number of calories is

$$80x + 60\ y.$$

Since there is to be at least 480 calories, we have

$$80x + 60y \ge 480.$$

Each unit of type A costs 10 cents and each unit of type B costs 12 cents, then the total cost (in cents) is

$$z = 10x + 12y.$$

The problem in mathematical form is: Find values of x and y that will minimize

$$z = 10x + 12y$$

subject to the following restrictions that must be satisfied by x and y:

$$2x + 3y \geq 18$$
$$x + 3y \geq 12$$
$$80x + 60y \geq 480$$
$$x \geq 0 \quad \text{and} \quad y \geq 0.$$

11.

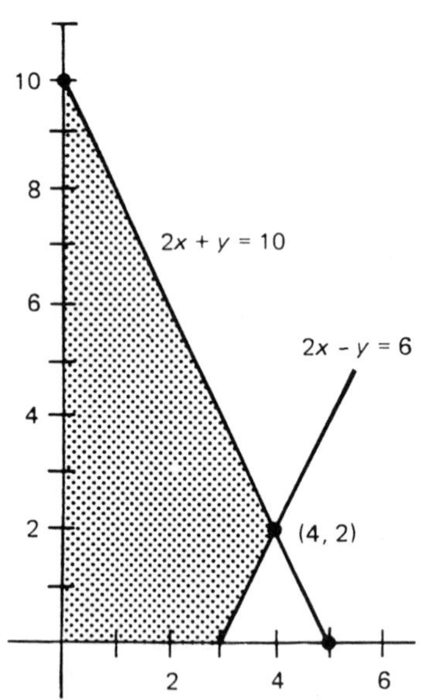

13.

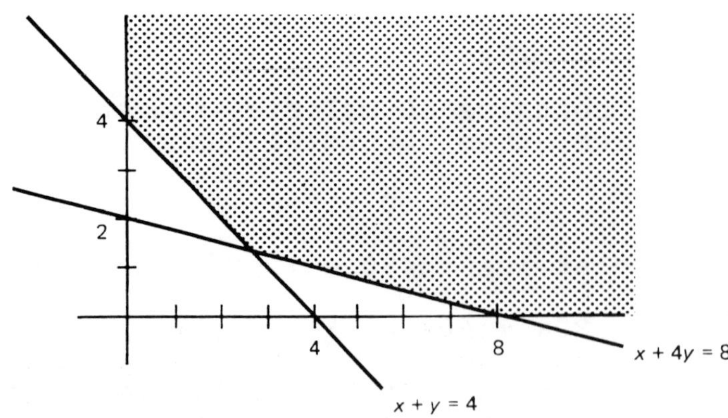

15. We follow the four step procedure given after Theorem 6.1. The feasible region S appears in the accompanying sketch.

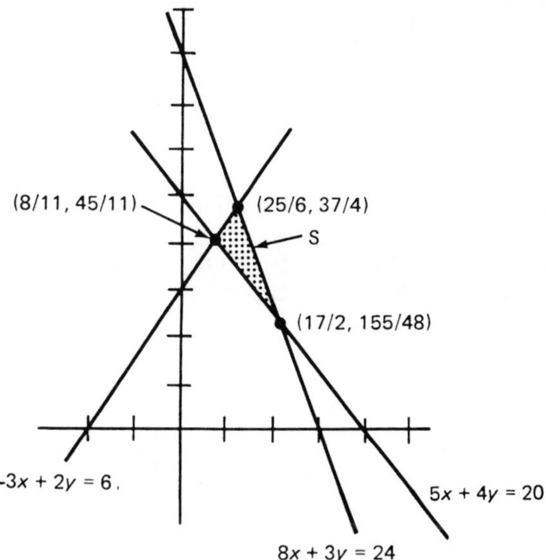

We see that S is bounded and it has extreme points (8/11,45/11), (25/6,37/4), and (17/2,155/48). The values of the objective function are shown in the table:

point	z = 3x - y
(8/11,45/11)	-21/11
(25/6,37/4)	39/12
(17/2,155/48)	1069/48

Thus the minimum occurs at extreme point (8/11,45/11). Hence the optimal solution is

$$x = 8/11, \quad y = 45/11.$$

17. We follow the four step procedure given after Theorem 6.1. The problem is to maximize

$$z = 0.08x + 0.10y$$

subject to the following restrictions that must be satisfied by x and y:

$$\begin{aligned} x + y &\le 6000 \\ y &\le 4000 \\ x &\ge 1500 \\ x \ge 0 \quad \text{and} \quad y &\ge 0. \end{aligned}$$

The feasible region S appears in the accompanying sketch. We see that S is bounded and it has extreme points (1500,0), (1500,4000),(2000,4000), and (6000,0). The values of the objective function are shown in the table:

Exercises 6.1

point	$z = 0.08x + 0.10y$
(1500,0)	120
(1500,4000)	520
(2000,4000)	560
(6000,0)	480

Thus the maximum occurs at extreme point (2000,4000). Hence the optimal solution is

$$x = 2000, \quad y = 4000.$$

The trust fund should invest $2000 in bond A and $4000 in bond B to give a maximum return on the investment of $560.

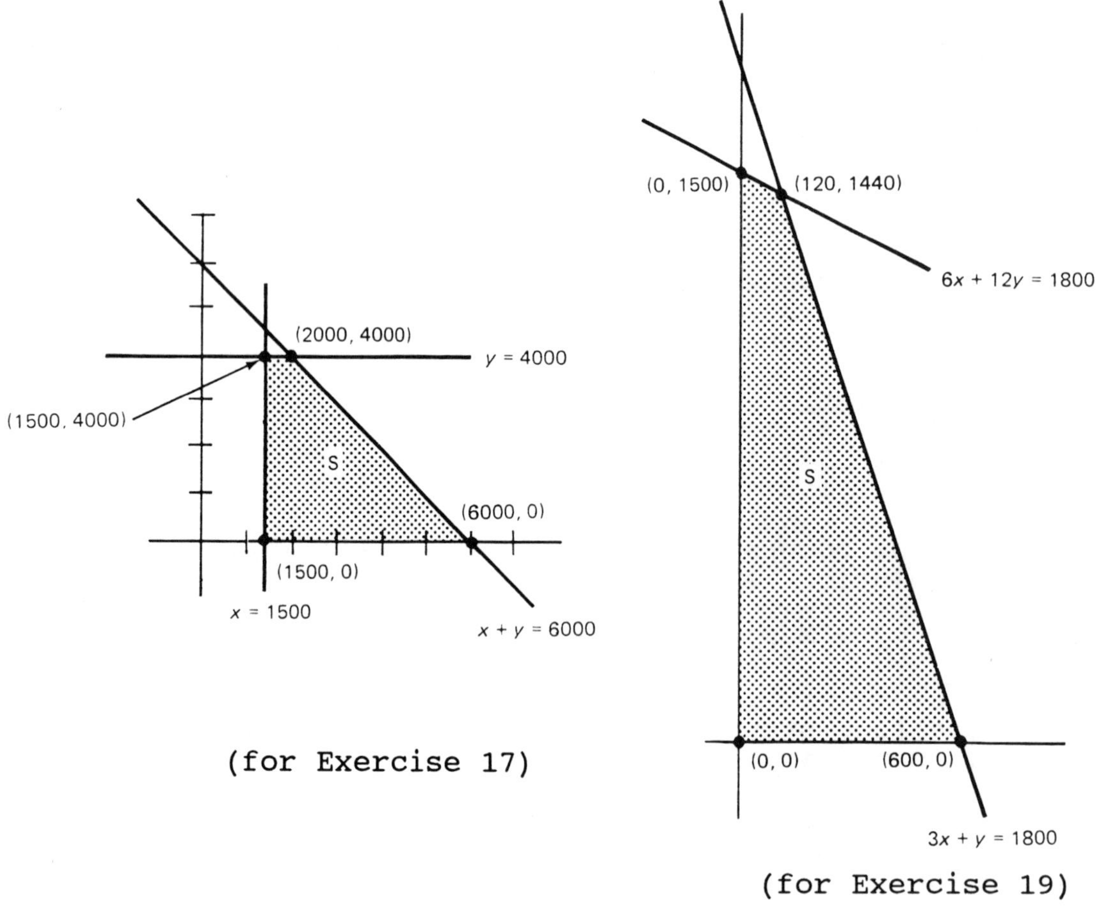

(for Exercise 17)

(for Exercise 19)

19. We follow the four step procedure given after Theorem 6.1. The problem is to maximize the revenue (in cents)

$$z = 30x + 60y$$

subject to the following restrictions that must be satisfied by x and y:

$$
\begin{aligned}
6x + 12y &\le 18000 \\
3x + y &\le 1800 \\
x \ge 0 \quad \text{and} \quad y &\ge 0.
\end{aligned}
$$

The feasible region S appears in the accompanying sketch. We see that S is bounded and it has extreme points (0,0), (0,1500),(120,1440), and (600,0). The values of the objective function are shown in the table:

point	$z = 30x + 60y$
(0,0)	0
(0,1500)	90000
(120,1440)	90000
(600,0)	18000

Thus a maximum occurs at extreme points (0,1500) and (120,1440). Hence there is an optimal solution at

$$x = 0, \quad y = 1500$$

and another at

$$x = 120, \; y = 1440.$$

In each case the maximum revenue is $900.

21. We follow the four step procedure given after Theorem 6.1. The problem is to minimize the cost (in cents)

$$z = 60x + 50y$$

subject to the following restrictions that must be satisfied by x and y:

$$
\begin{aligned}
3x + 5y &\le 15 \\
4x + 4y &\ge 16 \\
x \ge 0 \quad \text{and} \quad y &\ge 0.
\end{aligned}
$$

The feasible region S appears in the accompanying sketch.

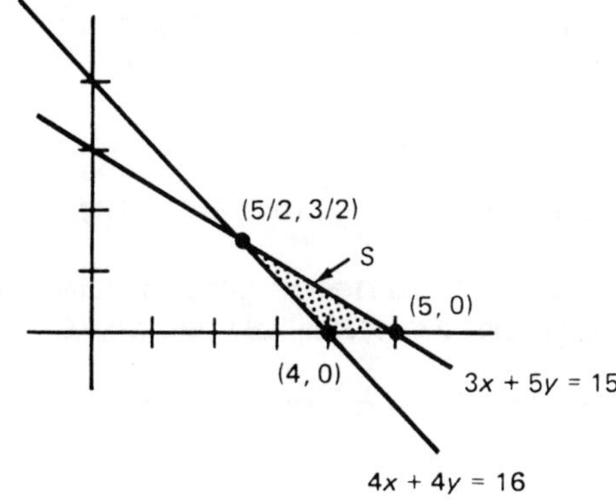

We see that S is bounded and it has extreme points (4,0), (5/2,3/2), and (5,0). The values of the objective function are shown in the table:

point	$z = 60x + 50y$
(4,0)	240
(5/2,3/2)	225
(5,0)	300

Thus the minimum occurs at extreme point (5/2,3/2). Hence the optimal solution is

$$x = 5/2, \quad y = 3/2.$$

The generators should burn 5/2 gallons of fuel L and 3/2 gallons of fuel H to achieve a minimum cost of $2.25 an hour.

23. We follow the four step procedure given after Theorem 6.1. The problem is to minimize

$$z = 20{,}000x + 25{,}000y$$

subject to the following restrictions that must be satisfied by x and y:

$$40x + 60y \geq 300$$
$$2x + 3y \leq 12$$
$$x \geq 0 \quad \text{and} \quad y \geq 0.$$

From the accompanying sketch we see that there is no feasible set of solutions. Hence there is no pair (x,y) which minimizes the objective function subject to the constraints.

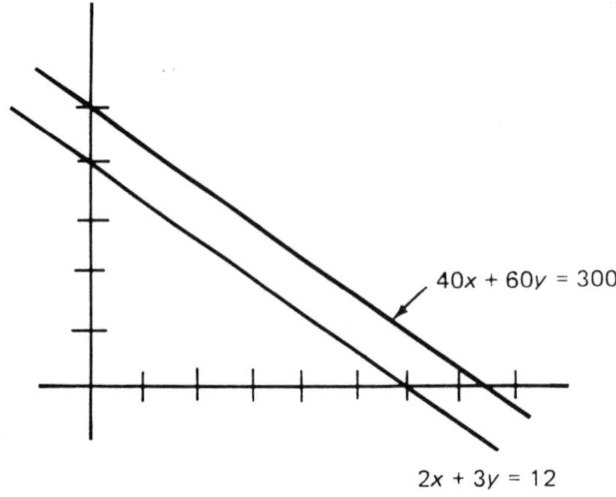

25. (a) This is a standard linear programming problem.

(b) This is not a standard linear programming problem because we are asked to minimize the objective function.

(c) This is not a standard linear programming problem because x_3 may be negative.

(d) This is not a standard linear programming problem because of the equality constraint $3x_1 - 2x_2 + 2x_3 = 4$.

27. The only change needed is to convert the constraint

$$-3x_1 + 2x_2 - 3x_3 \ge -4$$

to a $\le$ constraint. We replace the preceding constraint with

$$3x_1 - 2x_2 + 3x_3 \le 4.$$

29. We introduce slack variable x_4 to convert constraint

$$3x_1 + x_2 - 4x_3 \le 3$$

to the form

$$3x_1 + x_2 - 4x_3 + x_4 = 3.$$

Next introduce slack variable x_5 to convert constraint

$$x_1 - 2x_2 + 6x_3 \le 21$$

to the form

$$x_1 - 2x_2 + 6x_3 + x_5 = 21.$$

Finally introduce slack variable x_6 to convert constraint

$$x_1 - x_2 - x_3 \le 9$$

to the form

$$x_1 - x_2 - x_3 + x_6 = 9.$$

Then the new problem is

Maximize $z = 2x_1 + 3x_2 + 7x_3$
subject to

$$\begin{aligned} 3x_1 + x_2 - 4x_3 + x_4 \qquad\qquad &= 3 \\ x_1 - 2x_2 + 6x_3 \qquad + x_5 \qquad &= 21 \\ x_1 - x_2 - x_3 \qquad\qquad + x_6 &= 9 \end{aligned}$$

$$x_1 \ge 0,\ x_2 \ge 0,\ x_3 \ge 0,\ x_4 \ge 0,\ x_5 \ge 0,\ x_6 \ge 0.$$

Exercises 6.2

1. We first rewrite this standard linear programming problem using slack variables. We have:

Maximize $z = 3x + 7y$
subject to

$$\begin{aligned} 3x - 2y + u \qquad\qquad &= 7 \\ 2x + 5y \qquad + v \qquad &= 6 \\ 2x + 3y \qquad\qquad + w &= 8 \end{aligned}$$

$$x \geq 0,\ y \geq 0,\ u \geq 0,\ v \geq 0,\ w \geq 0.$$

To form the initial tableau we head the columns with the variable names and enter the coefficients of the constraints into the rows with the right-hand sides of the constraints in the rightmost column of the tableau. Next we rewrite the objective function into the form

$$-3x - 7y + z = 0$$

and place its coefficients into the bottom row (the objective row) of the tableau. Finally along the left side of the tableau we list the basic variable of the corresponding equation.

	x	y	u	v	w	z		
u	3	-2	1	0	0	0	7	44
v	2	5	0	1	0	0	6	
w	2	3	0	0	1	0	8	
	-3	-7	0	0	0	1	0	

3. We first rewrite this standard linear programming problem using slack variables. We have:

Maximize $z = 2x_1 + 2x_2 + 3x_3 + x_4$
subject to

$$\begin{aligned} 3x_1 - 2x_2 + x_3 + x_4 + x_5 \qquad\qquad &= 6 \\ x_1 + x_2 + x_3 + x_4 \qquad + x_6 \qquad &= 8 \\ 2x_1 - 3x_2 - x_3 + 2x_4 \qquad\qquad + x_7 &= 10 \end{aligned}$$

$$x_1 \geq 0,\ x_2 \geq 0,\ x_3 \geq 0,\ x_4 \geq 0,\ x_5 \geq 0,\ x_6 \geq 0,\ x_7 \geq 0.$$

To form the initial tableau we head the columns with the variable names and enter the coefficients of the constraints into the rows with the right-hand sides of the constraints in the rightmost column of the tableau. Next we rewrite the objective function in the form

$$-2x_1 - 2x_2 - 3x_3 - x_4 + z = 0$$

and place its coefficients into the bottom row (the objective row) of the tableau. Finally along the left side of the tableau we list the basic variable of the corresponding equation.

	x_1	x_2	x_3	x_4	x_5	x_6	x_7	z	
x_5	3	-2	1	1	1	0	0	0	6
x_6	1	1	1	1	0	1	0	0	8
x_7	2	-3	-1	2	0	0	1	0	10
	-2	-2	-3	-1	0	0	0	1	0

5. We follow the method used in Example 2. First reformulate the standard linear programming problem with slack variables:

Maximize $z = 2x + 3y$
subject to

$$\begin{aligned} 3x + 5y + u \quad &= 6 \\ 2x + 3y \quad + v &= 7 \end{aligned}$$
$$x \geq 0,\ y \geq 0,\ u \geq 0,\ v \geq 0.$$

Form the initial tableau:

	x	y	u	v	z	
u	3	5	1	0	0	6
v	2	3	0	1	0	7
	-2	-3	0	0	1	0

Next we determine the pivotal column and pivotal row. The most negative entry in the objective row is -3, hence the "y"-column is the pivotal column. Thus the θ-ratios are {6/5, 7/3} and 6/5 is the smallest. Hence the pivotal row is the "u"-row. It follows that the pivot has value 5. This information is marked in the next tableau.

	x	↓ y	u	v	z	
← u	3	5	1	0	0	6
v	2	3	0	1	0	7
	-2	-3	0	0	1	0

Next perform pivotal elimination. Take $(1/5)\mathbf{r}_1$, $-3\mathbf{r}_1+\mathbf{r}_2$, $3\mathbf{r}_1+\mathbf{r}_3$, and relabel the departing variable with the entering variable name. The resulting tableau is:

	x	y	u	v	z	
y	3/5	1	1/5	0	0	6/5
v	1/5	0	-3/5	1	0	17/5
	-1/5	0	3/5	0	1	18/5

Since there is a negative entry in the objective row, we do not have the optimal solution. We repeat the process.

Determine the pivotal column and pivotal row. The most negative entry in the objective row is -1/5, hence the "x"-column is the pivotal column. Thus the θ-ratios are {2, 17} and 2 is the smallest. Thus the pivotal row is the "y"-row. It follows that the pivot has value 3/5. This information is marked in the next tableau.

	↓ x	y	u	v	z	
← y	3/5	1	1/5	0	0	6/5
v	1/5	0	-3/5	1	0	17/5
	-1/5	0	3/5	0	1	18/5

Next perform pivotal elimination. Take $(5/3)\mathbf{r}_1$, $(-1/5)\mathbf{r}_1+\mathbf{r}_2$, $(1/5)\mathbf{r}_1+\mathbf{r}_3$, and relabel the departing variable with the entering variable name. The resulting tableau is:

	x	y	u	v	z	
x	1	5/3	1/3	0	0	2
v	0	-1/3	-2/3	1	0	3
	0	1/3	2/3	0	1	4

Since all the entries in the objective row are nonnegative, we have the optimal solution,

$$x = 2, \ y = 0.$$

The slack variables are

$$u = 0, \ v = 3,$$

and the optimal value of z is 4.

7. We follow the method used in Example 2. First reformulate the standard linear programming problem with slack variables:

Maximize $z = 2x + 5y$
subject to

$$\begin{aligned} 2x - 3y + u \quad\quad &= 4 \\ x - 2y \quad\quad + v &= 6 \\ x \geq 0,\ y \geq 0,\ u \geq 0,\ v &\geq 0. \end{aligned}$$

Form the initial tableau:

	x	y	u	v	z	
u	2	-3	1	0	0	4
v	1	-2	0	1	0	6
	-2	-5	0	0	1	0

Next we determine the pivotal column and pivotal row. The most negative entry in the objective row is -5, hence the "y"-column is the pivotal column. Since none of the entries in the pivotal column above the objective row is nonnegative, the problem has no finite optimum.

9. We follow the method used in Example 2. First reformulate the standard linear programming problem with slack variables:

Maximize $z = 2x_1 - 4x_2 + 5x_3$
subject to

$$\begin{aligned} 3x_1 + 2x_2 + x_3 + x_4 \qquad &= 6 \\ 3x_1 - 6x_2 + 7x_3 \qquad + x_5 &= 9 \end{aligned}$$
$$x_1 \geq 0,\ x_2 \geq 0,\ x_3 \geq 0,\ x_4 \geq 0,\ x_5 \geq 0.$$

Form the initial tableau:

	x_1	x_2	x_3	x_4	x_5	z	
x_4	3	2	1	1	0	0	6
x_5	3	-6	7	0	1	0	9
	-2	4	-5	0	0	1	0

Next we determine the pivotal column and pivotal row. The most negative entry in the objective row is -5, hence the "x_3"-column is the pivotal column. Thus the θ-ratios are {6, 9/7} and 9/7 is the smallest. Hence the pivotal row is the "x_5"-row. It follows that the pivot has value 7. This information is marked in the next tableau.

	x_1	x_2	↓ x_3	x_4	x_5	z	
x_4	3	2	1	1	0	0	6
← x_5	3	-6	7	0	1	0	9
	-2	4	-5	0	0	1	0

Next perform pivotal elimination. Take $(1/7)\mathbf{r}_2$, $-\mathbf{r}_2+\mathbf{r}_1$, $5\mathbf{r}_2+\mathbf{r}_3$, and relabel the departing variable with the entering variable name. The resulting tableau is:

	x_1	x_2	x_3	x_4	x_5	z	
x_4	18/7	20/7	0	1	-1/7	0	33/7
x_3	3/7	-6/7	1	0	1/7	0	9/7
	1/7	-2/7	0	0	5/7	1	45/7

There is a negative entry in the objective row, thus we do not have an optimal solution. Determine the next pivotal column and pivotal row. The most negative entry in the objective row is -2/7 hence the "x_2"-column is the pivotal column. The θ-ratios are {33/20} and 33/20 is the smallest. (Only the positive entries of the pivotal column are used in forming the θ-ratios.) Thus the pivot has value 20/7. This information is marked in the next tableau.

	x_1	↓ x_2	x_3	x_4	x_5	z	
← x_4	18/7	20/7	0	1	-1/7	0	33/7
x_3	3/7	-6/7	1	0	1/7	0	9/7
	1/7	-2/7	0	0	5/7	1	45/7

Next perform pivotal elimination. Take $(7/20)\mathbf{r}_1$, $(6/7)\mathbf{r}_1+\mathbf{r}_2$, $(2/7)\mathbf{r}_1+\mathbf{r}_3$, and relabel the departing variable with the entering variable name. The resulting tableau is:

	x_1	x_2	x_3	x_4	x_5	z	
x_2	9/10	1	0	7/20	-1/20	0	33/20
x_3	6/5	0	1	3/10	1/10	0	27/10
	2/5	0	0	1/10	7/10	1	69/10

Since all the entries of the objective row are nonnegative, we have an optimal solution,

$$x_1 = 0, \ x_2 = 33/20, \ x_3 = 27/10.$$

The slack variables are

$$x_4 = 0, \ x_5 = 0,$$

and the optimal value of z is 69/10.

11. We follow the method used in Example 2. First reformulate the standard linear programming problem with slack variables:

Maximize $z = x_1 + 2x_2 - x_3 + 5x_4$
subject to

$$2x_1 + 3x_2 + x_3 - x_4 + x_5 = 8$$
$$3x_1 + x_2 - 4x_3 + 5x_4 + x_6 = 9$$
$$x_1 \geq 0,\ x_2 \geq 0,\ x_3 \geq 0,\ x_4 \geq 0,\ x_5 \geq 0,\ x_6 \geq 0.$$

Form the initial tableau:

	x_1	x_2	x_3	x_4	x_5	x_6	z	
x_5	2	3	1	-1	1	0	0	8
x_6	3	1	-4	5	0	1	0	9
	-1	-2	1	-5	0	0	1	0

Next we determine the pivotal column and pivotal row. The most negative entry in the objective row is -5 hence the "x_4"-column is the pivotal column. Thus the θ-ratios are {9/5} and 9/5 is the smallest. (Only the positive entries of the pivotal column are used in forming the θ-ratios.) Hence the pivotal row is the "x_6"-row. It follows that the pivot has value 5. This information is marked in the next tableau.

	x_1	x_2	x_3	↓ x_4	x_5	x_6	z	
x_5	2	3	1	-1	1	0	0	8
← x_6	3	1	-4	5	0	1	0	9
	-1	-2	1	-5	0	0	1	0

Next perform pivotal elimination. Take $(1/5)\mathbf{r}_2$, $\mathbf{r}_2+\mathbf{r}_1$, $5\mathbf{r}_2+\mathbf{r}_3$, and relabel the departing variable with the entering variable name. The resulting tableau is:

	x_1	x_2	x_3	x_4	x_5	x_6	z	
x_5	13/5	16/5	1/5	0	1	1/5	0	49/5
x_4	3/5	1/5	-4/5	1	0	1/5	0	9/5
	2	-1	-3	0	0	1	1	9

Since an entry in the objective row is negative, we do not have an optimal solution. Determine the pivotal column and pivotal row. The most negative entry in the objective row is -3 hence the "x_3"-column is the pivotal column. Thus the θ-ratios are {49} and 49 is the smallest. (Only the positive entries of the pivotal column are used in forming the θ-ratios.) Hence the pivotal row is the "x_5"-row. It follows that the pivot has value 1/5. This information is marked in the next tableau.

	x_1	x_2	x_3 ↓	x_4	x_5	x_6	z	
← x_5	13/5	16/5	1/5	0	1	1/5	0	49/5
x_4	3/5	1/5	-4/5	1	0	1/5	0	9/5
	2	-1	-3	0	0	1	1	9

Next perform pivotal elimination. Take $5\mathbf{r}_1$, $(4/5)\mathbf{r}_1+\mathbf{r}_2$, $3\mathbf{r}_1+\mathbf{r}_3$, and relabel the departing variable with the entering variable name. The resulting tableau is:

	x_1	x_2	x_3	x_4	x_5	x_6	z	
x_3	13	16	1	0	5	1	0	49
x_4	11	13	0	1	4	1	0	41
	41	47	0	0	15	4	1	156

Since all the entries of the objective row are nonnegative, we have an optimal solution,

$$x_1 = 0,\ x_2 = 0,\ x_3 = 49,\ x_4 = 41.$$

The slack variables are

$$x_5 = 0,\ x_6 = 0$$

and the optimal value of z is 156.

13. The standard linear programming problem from Exercise 4 of Section 6.1 is:

Maximize $z = 30x + 60y$
subject to

$$\begin{aligned} 6x + 12y &\le 18000 \\ 3x + y &\le 1800 \\ x \ge 0 \text{ and } y &\ge 0. \end{aligned}$$

We first reformulate the problem introducing slack variables. In this form we have:

Maximize $z = 30x + 60y$
subject to

$$\begin{aligned} 6x + 12y + u \quad &= 18000 \\ 3x + y \quad + v &= 1800 \\ x \ge 0,\ y \ge 0,\ u \ge 0,\ v &\ge 0. \end{aligned}$$

Form the initial tableau:

	x	y	u	v	z		
u	6	12	1	0	0	18000	4
v	3	1	0	1	0	1800	
	-30	-60	0	0	1	0	

Next we determine the pivotal column and pivotal row. The most negative entry in the objective row is -60 hence the "y"-column is the pivotal column. Thus the θ-ratios are {18000/12, 1800} and 18000/12 is the smallest. Hence the pivotal row is the "u"-row. It follows that the pivot has value 12. This information is marked in the next tableau.

	x	↓ y	u	v	z		
← u	6	12	1	0	0	18000	4
v	3	1	0	1	0	1800	
	-30	-60	0	0	1	0	

Next perform pivotal elimination. Take $(1/12)\mathbf{r}_1$, $-\mathbf{r}_1+\mathbf{r}_2$, $60\mathbf{r}_1+\mathbf{r}_3$, and relabel the departing variable with the entering variable name. The resulting tableau is:

	x	y	u	v	z		
y	1/2	1	1/12	0	0	1500	4
v	-3	0	-1/12	1	0	300	
	0	0	5	0	1	90000	

Since all the entries of the objective row are nonnegative, we have an optimal solution,

$$x = 0, \ y = 1500.$$

The slack variables are

$$u = 0, \ v = 300$$

and the optimal value of z is 90,000. Thus this optimal solution implies that no containers from the the Smith Corporation and 1500 containers from the Johnson Corporation per truckload will give a maximum profit of $900. This problem has another optimal solution. See Exercise 19, Section 6.1.

15. Let x_1 be the number of tons of coal, x_2 be the number of tons of oil, and x_3 be the number of tons of gas. The problem is:

Exercises 6.2

Maximize $z = 600x_1 + 550x_2 + 500x_3$
subject to

$$\begin{aligned} 20x_1 + 18x_2 + 15x_3 &\le 60 \\ 15x_1 + 12x_2 + 10x_3 &\le 75 \\ 200x_1 + 220x_2 + 250x_3 &\le 2000 \\ x_1 \ge 0,\ x_2 \ge 0,\ x_3 &\ge 0. \end{aligned}$$

Reformulating the problem using slack variables we have:

Maximize $z = 600x_1 + 550x_2 + 500x_3$
subject to

$$\begin{aligned} 20x_1 + 18x_2 + 15x_3 + x_4 &= 60 \\ 15x_1 + 12x_2 + 10x_3 + x_5 &= 75 \\ 200x_1 + 220x_2 + 250x_3 + x_6 &= 2000 \\ x_1 \ge 0,\ x_2 \ge 0,\ x_3 \ge 0,\ x_4 \ge 0,\ x_5 \ge 0,\ x_6 &\ge 0. \end{aligned}$$

Form the initial tableau:

	x_1	x_2	x_3	x_4	x_5	x_6	z	
x_4	20	18	15	1	0	0	0	60
x_5	15	12	10	0	1	0	0	75
x_6	200	220	250	0	0	1	0	2000
	-600	-550	-500	0	0	0	1	0

Next we determine the pivotal column and pivotal row. The most negative entry in the objective row is -600, hence the "x_1"-column is the pivotal column. Thus the θ-ratios are {60/20,75/15, 2000/200} and 60/20 is the smallest. Hence the pivotal row is the "x_4"-row. It follows that the pivot has value 20. This information is marked in the next tableau.

	x_1 ↓	x_2	x_3	x_4	x_5	x_6	z	
← x_4	20	18	15	1	0	0	0	60
x_5	15	12	10	0	1	0	0	75
x_6	200	220	250	0	0	1	0	2000
	-600	-550	-500	0	0	0	1	0

Next perform pivotal elimination. Take $(1/20)\mathbf{r}_1$, $-15\mathbf{r}_1+\mathbf{r}_2$, $-200\mathbf{r}_1+\mathbf{r}_3$, $600\mathbf{r}_1+\mathbf{r}_4$ and relabel the departing variable with the entering variable name. The resulting tableau is:

	x_1	x_2	x_3	x_4	x_5	x_6	z	
x_1	1	9/10	3/4	1/20	0	0	0	3
x_5	0	-3/2	-5/4	-3/4	1	0	0	30
x_6	0	40	100	-10	0	1	0	1400
	0	-10	-50	30	0	0	1	1800

Since an entry in the objective row is negative, we do not have an optimal solution. Determine the pivotal column and pivotal row. The most negative entry in the objective row is -50 hence the "x_3"-column is the pivotal column. Thus the θ-ratios are {4, 14} and 4 is the smallest. Hence the pivotal row is the "x_1"-row. It follows that the pivot has value 3/4. This information is marked in the next tableau.

	x_1	x_2	x_3 ↓	x_4	x_5	x_6	z	
← x_1	1	9/10	3/4	1/20	0	0	0	3
x_5	0	-3/2	-5/4	-3/4	1	0	0	30
x_6	0	40	100	-10	0	1	0	1400
	0	-10	-50	30	0	0	1	1800

Next perform pivotal elimination. Take $(4/3)\mathbf{r}_1$, $5/4\mathbf{r}_1+\mathbf{r}_2$, $-100\mathbf{r}_1+\mathbf{r}_3$, $50\mathbf{r}_1+\mathbf{r}_4$ and relabel the departing variable with the entering variable name. The resulting tableau is:

	x_1	x_2	x_3	x_4	x_5	x_6	z	
x_3	4/3	6/5	1	1/15	0	0	0	4
x_5	5/3	0	0	-2/3	1	0	0	35
x_6	-400/3	-80	0	-50/3	0	1	0	1000
	200/3	50	0	65/2	0	0	1	2000

Since all the entries of the objective row are nonnegative, we have an optimal solution,

$$x_1 = 0, \ x_2 = 0, \ x_3 = 4.$$

The slack variables are

$$x_4 = 0, \ x_5 = 35, \ x_6 = 1000$$

and the optimal value of z is 2000. Thus 4 tons of gas and no oil and no coal should be used to have a maximum of 2000 kilowatts generated.

T.1. Let $\mathbf{X}$ and $\mathbf{Y}$ be any two feasible solutions of the standard linear programming problem. Let r be a scalar such that $0 \le r \le 1$. We show that $r\mathbf{X} + (1-r)\mathbf{Y}$ is also a feasible solution. First, since $r \ge 0$ and $(1-r) \ge 0$, and $\mathbf{AX} \le \mathbf{B}$, $\mathbf{AY} \le \mathbf{B}$,

$$\mathbf{A}[r\mathbf{X} + (1-r)\mathbf{Y}] = r\mathbf{AX} + (1-r)\mathbf{AY} \le r\mathbf{B} + (1-r)\mathbf{B} = \mathbf{B}.$$

Also since $\mathbf{X} \ge \mathbf{0}$, $\mathbf{Y} \ge \mathbf{0}$, $r\mathbf{X} + (1-r)\mathbf{Y} \ge r\mathbf{0} + (1-r)\mathbf{0} = \mathbf{0}$. Thus $r\mathbf{X} + (1-r)\mathbf{Y}$ is a feasible solution.

Exercises 6.2

ML.5. As a check, the solution is x4 = 2/3, x3 = 0, x2 = 1/3, x1 = 1, all other variables zero and the optimal value of z is 11. The final tableau is as follows:

0	0	1.6667	1	0.0000	-0.1667	0	0.6667
0	1	3.3333	0	1.0000	-0.3333	0	0.3333
1	0	-3.0000	0	-1.0000	0.5000	0	1.0000
0	0	1.0000	0	1.0000	1.0000	1	11.0000

ML.6. As a check, the solution is x3 = 4/3, x2 = 4, x7 = 22/3, all other variables zero, and the optimal value of z is 28/3. The final tableau is as follows:

0.4444	0	1	-0.1111	0.3333	-0.1111	0	0	1.3333
0.6667	1	0	1.3333	0	0.3333	0	0	4.0000
0.7778	0	0	-2.4444	-0.6667	-0.4444	1	0	7.3333
0.7778	0	0	1.5556	0.3333	0.5556	0	1	9.3333

Exercises 6.3

1. Let $\mathbf{C} = [3\ \ 2]$, $\mathbf{A} = \begin{bmatrix} 4 & 3 \\ 5 & -2 \\ 6 & 8 \end{bmatrix}$, $\mathbf{B} = \begin{bmatrix} 7 \\ 6 \\ 9 \end{bmatrix}$, $\mathbf{X} = \begin{bmatrix} x_1 \\ x_2 \end{bmatrix}$. Define

$\mathbf{Y} = \begin{bmatrix} y_1 \\ y_2 \\ y_3 \end{bmatrix}$. Then the primal problem is

$$\begin{aligned} &\text{Maximize} && z = \mathbf{CX} \\ &\text{subject to} && \\ &&& \mathbf{AX} \le \mathbf{B} \\ &&& \mathbf{X} \ge \mathbf{0}. \end{aligned}$$

The dual problem is

$$\begin{aligned} &\text{Minimize} && z' = \mathbf{B}^T\mathbf{Y} \\ &\text{subject to} && \\ &&& \mathbf{A}^T\mathbf{Y} \ge \mathbf{C}^T \\ &&& \mathbf{Y} \ge \mathbf{0}. \end{aligned}$$

That is,

$$\begin{aligned} &\text{Minimize} && z' = 7y_1 + 6y_2 + 9y_3 \\ &\text{subject to} && \\ &&& 4y_1 + 5y_2 + 6y_3 \ge 3 \\ &&& 3y_1 - 2y_2 + 8y_3 \ge 2 \\ &&& y_1 \ge 0,\ y_2 \ge 0,\ y_3 \ge 0. \end{aligned}$$

3. Let $\mathbf{C} = [3\ \ 5]$, $\mathbf{A} = \begin{bmatrix} 2 & 3 \\ 8 & -9 \\ 10 & 15 \end{bmatrix}$, $\mathbf{B} = \begin{bmatrix} 7 \\ 12 \\ 18 \end{bmatrix}$, $\mathbf{X} = \begin{bmatrix} x_1 \\ x_2 \end{bmatrix}$. Define

$\mathbf{Y} = \begin{bmatrix} y_1 \\ y_2 \\ y_3 \end{bmatrix}$. Then the primal problem is

$$\begin{aligned} &\text{Minimize} && z = \mathbf{CX} \\ &\text{subject to} && \\ &&& \mathbf{AX} \ge \mathbf{B} \\ &&& \mathbf{X} \ge \mathbf{0}. \end{aligned}$$

The dual problem is

$$\begin{aligned} &\text{Maximize} && z' = \mathbf{B}^T\mathbf{Y} \\ &\text{subject to} && \\ &&& \mathbf{A}^T\mathbf{Y} \le \mathbf{C}^T \\ &&& \mathbf{Y} \ge \mathbf{0}. \end{aligned}$$

That is,

$$\begin{aligned} &\text{Maximize} && z' = 7y_1 + 12y_2 + 18y_3 \\ &\text{subject to} && \\ &&& 2y_1 + 8y_2 + 10y_3 \le 3 \\ &&& 3y_1 - 9y_2 + 15y_3 \le 5 \\ &&& y_1 \ge 0,\ y_2 \ge 0,\ y_3 \ge 0. \end{aligned}$$

Exercises 6.3

5. The primal problem is

$$\begin{aligned}&\text{Maximize} \quad z = 3x_1 + 6x_2 + 9x_3\\&\text{subject to}\\&\quad 3x_1 + 2x_2 - 3x_3 \le 12\\&\quad 5x_1 + 4x_2 + 7x_3 \le 18\\&\quad x_1 \ge 0,\ x_2 \ge 0,\ x_3 \ge 0.\end{aligned}$$

In matrix form we have

$$\text{Maximize} \quad z = [3 \ \ 6 \ \ 9]\begin{bmatrix}x_1\\x_2\\x_3\end{bmatrix}$$

$$\text{subject to} \quad \begin{bmatrix}3 & 2 & -3\\5 & 4 & 7\end{bmatrix}\begin{bmatrix}x_1\\x_2\\x_3\end{bmatrix} \le \begin{bmatrix}12\\18\end{bmatrix} \quad \text{and} \quad \begin{bmatrix}x_1\\x_2\\x_3\end{bmatrix} \ge \mathbf{0}.$$

Then the dual problem is

$$\text{Minimize} \quad z' = [12 \ \ 18]\begin{bmatrix}y_1\\y_2\end{bmatrix}$$

$$\text{subject to} \quad \begin{bmatrix}3 & 5\\2 & 4\\-3 & 7\end{bmatrix}\begin{bmatrix}y_1\\y_2\end{bmatrix} \ge \begin{bmatrix}3\\6\\9\end{bmatrix} \quad \text{and} \quad \begin{bmatrix}y_1\\y_2\end{bmatrix} \ge \mathbf{0}.$$

Then the dual of the dual problem is

$$\text{Maximize} \quad z'' = [3 \ \ 6 \ \ 9\ ]\begin{bmatrix}w_1\\w_2\\w_3\end{bmatrix}$$

$$\text{subject to} \quad \begin{bmatrix}3 & 2 & -3\\5 & 4 & 7\end{bmatrix}\begin{bmatrix}w_1\\w_2\\w_3\end{bmatrix} \le \begin{bmatrix}12\\18\end{bmatrix} \quad \text{and} \quad \begin{bmatrix}w_1\\w_2\\w_3\end{bmatrix} \ge \mathbf{0}.$$

Let $z'' = z$ and $w_j = x_j$, $j = 1,2,3$ and we have the original problem.

7. From Exercise 6 of Section 6.2 we have the standard linear programming problem

$$\begin{aligned}&\text{Maximize} \quad z = 2x + 5y\\&\text{subject to}\\&\quad 3x + 7y \le 6\\&\quad 2x + 6y \le 7\\&\quad 3x + 2y \le 5\\&\quad x \ge 0,\ y \ge 0.\end{aligned}$$

This problem has an optimal solution and the final tableau is

	x	y	u	v	w	z	
y	3/7	1	1/7	0	0	0	6/7
v	-4/7	0	-6/7	1	0	0	13/7
w	15/7	0	-2/7	0	1	0	23/7
	1/7	0	5/7	0	0	1	30/7

Theorem 6.4 implies that the dual problem also has an optimal solution with the same optimal value. This final tableau contains the optimal solution to the dual problem in the objective row under the columns of the slack variables. Thus the optimal solution of the dual problem is

$$y_1 = 5/7,\ y_2 = 0,\ y_3 = 0.$$

The optimal value of the dual problem is

$$z' = 30/7.$$

9. From Exercise 10 of Section 6.2 we have the standard linear programming problem

$$\begin{aligned} \text{Maximize} \quad & z = 2x_1 + 4x_2 - 3x_3 \\ \text{subject to} \quad & \\ & 5x_1 + 2x_2 + x_3 \le 5 \\ & 3x_1 - 2x_2 + 3x_3 \le 10 \\ & 4x_1 + 5x_2 - x_3 \le 20 \\ & x_1 \ge 0,\ x_2 \ge 0,\ x_3 \ge 0. \end{aligned}$$

his problem has an optimal solution and the final tableau is

	x_1	x_2	x_3	x_4	x_5	x_6	z	
x_2	5/2	1	1/2	1/2	0	0	0	5/2
x_5	8	0	4	1	1	0	0	15
x_6	-17/2	0	-7/2	-5/2	0	1	0	15/2
	8	0	5	2	0	0	1	10

Theorem 6.4 implies that the dual problem also has an optimal solution with the same optimal value. This final tableau contains the optimal solution to the dual problem in the objective row under the columns of the slack variables. Thus the optimal solution of the dual problem is

$$y_1 = 2,\ y_2 = 0,\ y_3 = 0.$$

The optimal value of the dual problem is

$$z' = 10.$$

Chapter 6 Supplementary Exercises

1. The feasible region S appears in the accompanying figure.

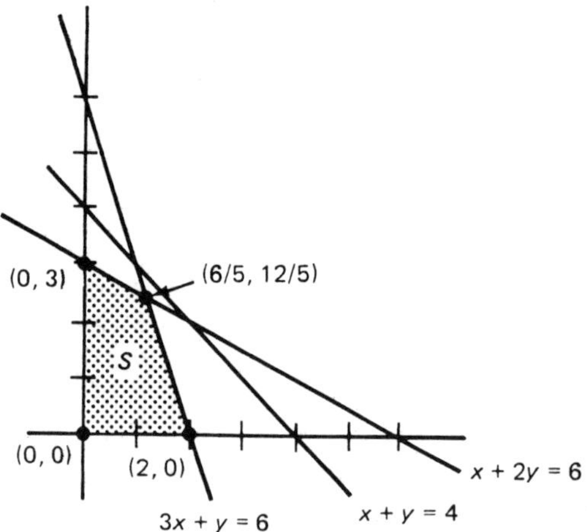

There are 4 extreme points, (0,0), (0,3), (6/5,12/5), and (2,0). The region S is bounded, thus by Theorem 6.1, the objective function z = 2x + 3y has a maximum at one of these points. The values of the objective function at the exteme points are shown in the following table.

point	z = 2x + 3y
(0,0)	0
(0,3)	9
(6/5,12/5)	48/5
(2,0)	4

There is a maximum at extreme point (6/5,12/5). The maximum is z = 48/5.

3. We first reformulate this standard linear programming problem using slack variables. We have

Maximize $z = 50x + 100y$
subject to

$$\begin{aligned} x + 2y + u \qquad\qquad &= 16 \\ 3x + 2y \qquad + v \qquad &= 24 \\ 2x + 2y \qquad\qquad + w &= 18 \end{aligned}$$
$$x \geq 0,\ y \geq 0,\ u \geq 0,\ v \geq 0,\ w \geq 0.$$

Form the initial tableau.

	x	y	u	v	w	z	
u	1	2	1	0	0	0	16
v	3	2	0	1	0	0	24
w	2	2	0	0	1	0	18
	-50	-100	0	0	0	1	0

Next we determine the pivotal column and pivotal row. The most negative entry in the objective row is -100, hence the "y"-column is the pivotal column. Thus the θ-ratios are {16/2, 24/2, 18/2} and 16/2 is the smallest. Thus the pivotal row is the "u"-row. It follows that the pivot has the value 2. This information is marked in the next tableau.

	x	y ↓	u	v	w	z	
← u	1	2	1	0	0	0	16
v	3	2	0	1	0	0	24
w	2	2	0	0	1	0	18
	-50	-100	0	0	0	1	0

Next we perform pivotal elimination. Take $(1/2)\mathbf{r}_1$, $-2\mathbf{r}_1+\mathbf{r}_2$, $-2\mathbf{r}_1+\mathbf{r}_3$, $100\mathbf{r}_1+\mathbf{r}_4$, and relabel the departing variable with the entering variable name. The resulting tableau is:

	x	y	u	v	w	z	
y	1/2	1	1/2	0	0	0	8
v	2	0	-1	1	0	0	8
w	1	0	-1	0	1	0	2
	0	0	50	0	0	1	800

Since all the entries in the objective row are nonnegative, we have an optimal solution

$$x = 0, \ y = 8.$$

The slack variables are

$$u = 0, \ v = 8, \ w = 2,$$

and the optimal value of z is 800.

5. Referring to Exercise 4, we reformulate the dual using slack variables and obtain

$$\begin{aligned}
&\text{Maximize} \quad z' = 6y_1 + 10y_2 \\
&\text{subject to} \\
&\qquad 2y_1 + 5y_2 + y_3 \phantom{{}+y_4} = 6 \\
&\qquad 3y_1 + 2y_2 \phantom{{}+y_3} + y_4 = 5 \\
&y_1 \geq 0,\ y_2 \geq 0,\ y_3 \geq 0,\ y_4 \geq 0.
\end{aligned}$$

Form the initial tableau.

	y_1	y_2	y_3	y_4	z'	
y_3	2	5	1	0	0	6
y_4	3	2	0	1	0	5
	-6	-10	0	0	1	0

Next we determine the pivotal column and pivotal row. The most negative entry in the objective row is -10, hence the "y_2"-column is the pivotal column. Thus the θ-ratios are {6/5, 5/2} and 6/5 is the smallest. Thus the pivotal row is the "y_3"-row. It follows that the pivot has the value 5. This information is marked in the next tableau.

	y_1	↓ y_2	y_3	y_4	z'	
← y_3	2	5	1	0	0	6
y_4	3	2	0	1	0	5
	-6	-10	0	0	1	0

Next we perform pivotal elimination. Take $(1/5)\mathbf{r}_1$, $-2\mathbf{r}_1+\mathbf{r}_2$, $10\mathbf{r}_1+\mathbf{r}_3$, and relabel the departing variable with the entering variable name. The resulting tableau is:

	y_1	y_2	y_3	y_4	z'	
y_2	2/5	1	1/5	0	0	6/5
y_4	11/5	0	-2/5	1	0	13/5
	-2	0	2	0	1	12

Since there is a negative entry in the objective row, we do not have an optimal solution. We repeat the process.

Determine the pivotal column and pivotal row. The most negative entry in the objective row is -2, hence the "y_1"-column is the pivotal column. Thus the θ-ratios are {3, 13/11} and 13/11 is the smallest. Thus the pivotal row is the "y_4"-row. It follows that the pivot has the value 11/5. This information is marked in the next tableau.

	↓ y_1	y_2	y_3	y_4	z'	
y_2	2/5	1	1/5	0	0	6/5
← y_4	11/5	0	-2/5	1	0	13/5
	-2	0	2	0	1	12

Next we perform pivotal elimination. Take $(5/11)\mathbf{r}_2$, $(-2/5)\mathbf{r}_2+\mathbf{r}_1$, $2\mathbf{r}_2+\mathbf{r}_3$, and relabel the departing variable with the entering variable name. The resulting tableau is:

	y_1	y_2	y_3	y_4	z'	
y_2	0	1	3/11	-2/11	0	8/11
y_1	1	0	-2/11	5/11	0	13/11
	0	0	18/11	10/11	1	158/11

Since all the entries in the objective row are nonnegative, we have an optimal solution

$$y_1 = 13/11, \; y_2 = 8/11.$$

The slack variables are

$$y_3 = 0, \; y_4 = 0,$$

and the optimal value of z is 158/11. An optimal solution of the original problem appears in the objective row under the columns of the slack variables y_3 and y_4. The original problem has optimal solution

$$x_1 = 18/11, \; x_2 = 10/11.$$

Exercises 7.1

1. In each of the following we substitute the coordinates of points $\mathbf{P}_1$ and $\mathbf{P}_2$ into Equation (5) and expand the determinant to find the equation of the line.

(a) $$\begin{vmatrix} x & y & 1 \\ -2 & -3 & 1 \\ 3 & 4 & 1 \end{vmatrix} = -7x + 5y + 1 = 0$$

(b) $$\begin{vmatrix} x & y & 1 \\ 2 & -5 & 1 \\ -3 & 4 & 1 \end{vmatrix} = -9x - 5y - 7 = 0$$

(c) $$\begin{vmatrix} x & y & 1 \\ 0 & 0 & 1 \\ -3 & 5 & 1 \end{vmatrix} = -5x - 3y = 0$$

(d) $$\begin{vmatrix} x & y & 1 \\ -3 & -5 & 1 \\ 0 & 2 & 1 \end{vmatrix} = -7x + 3y - 6 = 0$$

3. Convert the parametric form to the symmetric form by solving each equation for t and setting the results equal to one another. We obtain

$$\frac{x-3}{2} = \frac{y+2}{3} = \frac{z-4}{-3} \; .$$

A point is on the line provided the coordinates of the point satisfy the preceding string of equalities.

(a) For (1,1,1), we have

$$\frac{1-3}{2} = -1, \; \frac{1+2}{3} = 1, \; \frac{1-4}{-3} = 1.$$

Thus (1,1,1) is not on the line.

(b) For (1,-1,0), we have

$$\frac{1-3}{2} = -1, \; \frac{-1+2}{3} = 1/3, \; \frac{0-4}{-3} = 4/3.$$

Thus (1,-1,0) is not on the line.

(c) For (1,0,-2), we have

$$\frac{1-3}{2} = -1,\ \frac{0+2}{3} = 2/3,\ \frac{-2-4}{-3} = 2.$$

Thus (1,0,-2) is not on the line.

(d) For (4,-1/2,5/2), we have

$$\frac{4-3}{2} = 1/2,\ \frac{(-1/2)+2}{3} = 1/2,\ \frac{(5/2)-4}{-3} = 1/2.$$

Thus (4,-1/2,5/2) is on the line.

5. From Example 2 we have that the parametric equation of a line through $\mathbf{P}_0(x_0,y_0,z_0)$ which is parallel to vector $\mathbf{U} = (u,v,w)$ is given by

$$\begin{aligned} x &= x_0 + tv \\ y &= y_0 + tu \\ z &= z_0 + tw. \end{aligned} \qquad (-\infty < t < \infty)$$

(a) Let $\mathbf{P}_0 = (3,4,-2)$ and let $\mathbf{U} = (4,-5,2)$. Then the parametric equation is

$$\begin{aligned} x &= 3 + 4t \\ y &= 4 - 5t \\ z &= -2 + 2t. \end{aligned} \qquad (-\infty < t < \infty)$$

(b) Let $\mathbf{P}_0 = (3,2,4)$ and let $\mathbf{U} = (-2,5,1)$. Then the parametric equation is

$$\begin{aligned} x &= 3 - 2t \\ y &= 2 + 5t \\ z &= 4 + t. \end{aligned} \qquad (-\infty < t < \infty)$$

(c) Let $\mathbf{P}_0 = (0,0,0)$ and let $\mathbf{U} = (2,2,2)$. Then the parametric equation is

$$\begin{aligned} x &= 2t \\ y &= 2t \\ z &= 2t. \end{aligned} \qquad (-\infty < t < \infty)$$

Or equivalently $x = t$, $y = t$, $z = t$.

(d) Let $\mathbf{P}_0 = (-2,-3,1)$ and let $\mathbf{U} = (2,3,4)$. Then the parametric equation is

$$\begin{aligned} x &= -2 + 2t \\ y &= -3 + 3t \\ z &= 1 + 4t. \end{aligned} \qquad (-\infty < t < \infty)$$

7. Follow the method of Example 4.

(a) $\frac{x-2}{2} = \frac{y+3}{5} = \frac{z-1}{4}$ (b) $\frac{x+3}{8} = \frac{y+2}{7} = \frac{z+2}{6}$

(c) $\frac{x+2}{4} = \frac{y-3}{-6} = \frac{z-4}{1}$ (d) $\frac{x}{4} = \frac{y}{5} = \frac{z}{2}$

9. From Example 5 we have that the equation of a plane passing through point $\mathbf{P}_0(x_0,y_0,z_0)$ and perpendicular to vector $\mathbf{N} = (a,b,c)$ is

$$a(x-x_0) + b(y-y_0) + c(z-z_0) = 0.$$

(a) Let $\mathbf{P}_0 = (0,2,-3)$ and vector $\mathbf{N} = (3,-2,4)$. The equation of the plane is

$$3(x-0) - 2(y-2) + 4(z-(-3)) = 3x - 2y + 4z + 16 = 0.$$

(b) Let $\mathbf{P}_0 = (-1,3,2)$ and vector $\mathbf{N} = (0,1,-3)$. The equation of the plane is

$$0(x-(-1)) + 1(y-3) - 3(z-2) = y - 3z + 3 = 0.$$

(c) Let $\mathbf{P}_0 = (-2,3,4)$ and vector $\mathbf{N} = (0,0,-4)$. The equation of the plane is

$$0(x-(-2)) + 0(y-3) - 4(z-4) = -4z + 16 = 0.$$

Or equivalently, $-z + 4 = 0$.

(d) Let $\mathbf{P}_0 = (5,2,3)$ and vector $\mathbf{N} = (-1,-2,4)$. The equation of the plane is

$$-1(x-5) - 2(y-2) + 4(z-3) = -x - 2y + 4z - 3 = 0.$$

11. Following Example 11, we solve the pair of equations simultaneously.

(a) Form the augmented matrix of the linear system and row reduce it.

$$\left[\begin{array}{rrr|r} 2 & 3 & -4 & -5 \\ -3 & 2 & 5 & -6 \end{array}\right] \begin{array}{l} (1/2)\mathbf{r}_1 \\ 3\mathbf{r}_1+\mathbf{r}_2 \end{array} \rightarrow$$

$$\left[\begin{array}{rrr|r} 1 & 3/2 & -2 & -5/2 \\ 0 & 13/2 & -1 & -27/2 \end{array}\right] \begin{array}{l} (2/13)\mathbf{r}_2 \\ (-3/2)\mathbf{r}_2+\mathbf{r}_1 \end{array} \rightarrow$$

$$\left[\begin{array}{ccc|c} 1 & 0 & -23/13 & 8/13 \\ 0 & 1 & -2/13 & -27/13 \end{array}\right]$$

The solution of the linear system is

$x = (8/13) + (23/13)r, \; y = (-27/13) + (2/13)r, \; z = r,$

where r is any real number. Let r = 13t, where t is any real number. Then one way to write the parametric equation of the line of intersection is

$$\begin{aligned} x &= 8/13 + 23t \\ y &= -27/13 + 2t \qquad (-\infty < t < \infty) \\ z &= 13t. \end{aligned}$$

(b) Form the augmented matrix of the linear system and row reduce it.

$$\left[\begin{array}{ccc|c} 3 & -2 & -5 & -4 \\ 2 & 3 & 4 & -8 \end{array}\right] \begin{array}{l} (1/3)\mathbf{r}_1 \\ -2\mathbf{r}_1+\mathbf{r}_2 \end{array} \rightarrow$$

$$\left[\begin{array}{ccc|c} 1 & -2/3 & -5/3 & -4/3 \\ 0 & 13/3 & 22/3 & -16/3 \end{array}\right] \begin{array}{l} (3/13)\mathbf{r}_2 \\ (2/3)\mathbf{r}_2+\mathbf{r}_1 \end{array} \rightarrow$$

$$\left[\begin{array}{ccc|c} 1 & 0 & -7/13 & -28/13 \\ 0 & 1 & 22/13 & -16/13 \end{array}\right]$$

The solution of the linear system is

$x = (-28/13) + (7/13)r, \; y = (-16/13) - (22/13)r, \; z = r,$

where r is any real number. Let r = 13t, where t is any real number. Then one way to write the parametric equation of the line of intersection is

$$\begin{aligned} x &= -28/13 + 7t \\ y &= -16/13 - 22t \qquad (-\infty < t < \infty) \\ z &= 13t. \end{aligned}$$

(c) Form the augmented matrix of the linear system and row reduce it.

$$\left[\begin{array}{ccc|c} -1 & 2 & 1 & 0 \\ 2 & -1 & 2 & -8 \end{array}\right] \begin{array}{l} -1\mathbf{r}_1 \\ -2\mathbf{r}_1+\mathbf{r}_2 \end{array} \rightarrow \left[\begin{array}{ccc|c} 1 & -2 & -1 & 0 \\ 0 & 3 & 4 & -8 \end{array}\right] \begin{array}{l} (1/3)\mathbf{r}_2 \\ 2\mathbf{r}_2+\mathbf{r}_1 \end{array} \rightarrow$$

$$\left[\begin{array}{ccc|c} 1 & 0 & 5/3 & -16/3 \\ 0 & 1 & 4/3 & -8/3 \end{array}\right]$$

The solution of the linear system is

$x = (-16/3) - (5/3)r, \; y = (-8/3) - (4/3)r, \; z = r,$

where r is any real number. Let $r = -3t$, where t is any real number. Then one way to write the parametric equation of the line of intersection is

$$\begin{aligned} x &= -16/3 + 5t \\ y &= -8/3 + 4t \qquad (-\infty < t < \infty) \\ z &= -3t. \end{aligned}$$

13. Determine the equation of the line through $\mathbf{P}_1 = (2,3,-2)$ and $\mathbf{P}_2 = (4,-2,-3)$. Then $\mathbf{P}_3 = (0,8,-1)$ is on the line through $\mathbf{P}_1$ and $\mathbf{P}_2$ if it satisfies an equation of the line.

A vector parallel to the line through $\mathbf{P}_1$ and $\mathbf{P}_2$ is

$$\overrightarrow{\mathbf{P}_1\mathbf{P}_2} = (2,-5,-1).$$

Then a line through $\mathbf{P}_1$ parallel to vector $(2,-5,-1)$ is

$$\begin{aligned} x &= 2 + 2t \\ y &= 3 - 5t \qquad (-\infty < t < \infty) \\ z &= -2 - t. \end{aligned}$$

The symmetric form for the equation of this line is

$$\frac{x-2}{2} = \frac{y-3}{-5} = \frac{z+2}{-1}.$$

Substituting the coordinates of $\mathbf{P}_3$ into the equation we have,

$$\frac{0-2}{2} = -1, \quad \frac{8-3}{-5} = -1, \quad \frac{-1+2}{-1} = -1.$$

Thus, $\mathbf{P}_3$ is on the line through $\mathbf{P}_1$ and $\mathbf{P}_2$. All three points are on the same line.

15. Equate the expressions for x, y, and z respectively to form a system of equations in s and t. We obtain

$$\begin{aligned} 2 - 3s &= 5 + 2t \\ 3 + 2s &= 1 - 3t \\ 4 + 2s &= 2 + t \end{aligned} \quad \text{which is equivalent to} \quad \begin{aligned} -3s - 2t &= 3 \\ 2s + 3t &= -2 \\ 2s - t &= -2. \end{aligned}$$

Form the associated augmented matrix and row reduce it:

$$\left[\begin{array}{cc|c} -3 & -2 & 3 \\ 2 & 3 & -2 \\ 2 & -1 & -2 \end{array}\right] \text{ is equivalent to } \left[\begin{array}{cc|c} 1 & 0 & -1 \\ 0 & 1 & 0 \\ 0 & 0 & 0 \end{array}\right].$$

The solution is s = -1 and t = 0. Substituting s = -1 into the equation of the first line, we have that the point of intersection has coordinates

$$x = 5,\ y = 1,\ z = 2.$$

17. We show that the lines intersect in more than one point, hence they must be the same line. Rewrite the equation of the first line using the parameter s in place of t:

$$\begin{aligned} x &= 2 + 3s \\ y &= 3 - 2s \\ z &= -1 + 4s \end{aligned} \quad \text{and} \quad \begin{aligned} x &= -1 - 9t \\ y &= 5 + 6t \\ z &= -5 - 12t \end{aligned}.$$

Equating the expressions for x, y, and z respectively we have a system of equations which has augmented matrix

$$\left[\begin{array}{cc|c} 3 & 9 & -3 \\ -2 & -6 & 2 \\ 4 & 12 & -4 \end{array}\right] \text{ which row reduces to } \left[\begin{array}{cc|c} 1 & 3 & -1 \\ 0 & 0 & 0 \\ 0 & 0 & 0 \end{array}\right].$$

Thus there are infinitely many solutions, hence infinitely many intersections of the lines. It follows that the two lines are identical.

19. Let $\mathbf{P}_1 = (-2,3,4)$, $\mathbf{P}_2 = (4,-2,5)$, $\mathbf{P}_3 = (0,2,4)$. Also let L_2 denote the line through $\mathbf{P}_2$ and $\mathbf{P}_3$. Let a plane through $\mathbf{P}_1$ perpendicular to L_2 be denoted by π. Then a vector parallel to L_2 is $\mathbf{U} = (-4,4,-1)$. It follows that $\mathbf{U}$ is a normal to π. Thus for $\mathbf{P} = (x,y,z)$, any point in π, the equation of the plane is given by

$$\begin{aligned} \mathbf{U}\cdot\overrightarrow{\mathbf{P}_1\mathbf{P}} &= (-4,4,-1)\cdot(x+2,y-3,z-4) \\ &= -4x + 4y - z - 16 = 0. \end{aligned}$$

Or equivalently, $4x - 4y + z + 16 = 0$.

21. Let the lines be represented as follows:

$$L_1: \begin{aligned} x &= 3 + 2s \\ y &= 4 - 3s \\ z &= 5 + 4s \end{aligned} \quad \text{and} \quad L_2: \begin{aligned} x &= 1 - 2t \\ y &= 7 + 4t \\ z &= 1 - 3t. \end{aligned}$$

The vector $\mathbf{U} = (2,-3,4)$ is parallel to L_1 and vector $\mathbf{V} = (-2,4,-3)$ is parallel to L_2. The normal direction to a plane containing L_1 and L_2 is perpendicular to both $\mathbf{U}$ and $\mathbf{V}$. Thus $\mathbf{N} = \mathbf{U}\times\mathbf{V}$ is a normal to the plane. We have

$$\mathbf{N} = \mathbf{U}\times\mathbf{V} = \begin{vmatrix} \mathbf{i} & \mathbf{j} & \mathbf{k} \\ 2 & -3 & 4 \\ -2 & 4 & -3 \end{vmatrix} = -7\mathbf{i} - 2\mathbf{j} + 2\mathbf{k} = (-7,-2,2).$$

We need only determine a point $\mathbf{P}_0$ on one of the lines and then use Equation (8) to find the equation of the plane containing L_1 and L_2. Set s =0, then $\mathbf{P}_0 = (3,4,5)$. We have that the equation of the plane is given by

$$\begin{aligned}\mathbf{N}\cdot\overrightarrow{\mathbf{P}_0\mathbf{P}} &= (-7,-2,2)\cdot(x-3,y-4,z-5)\\ &= -7x - 2y + 2z + 19 = 0.\end{aligned}$$

Or Equivalently, $7x + 2y - 2z - 19 = 0$.

23. Let $\mathbf{P}_0 = (-2,5,-3)$ and let π denote the plane

$$2x - 3y + 4z + 7 = 0.$$

The vector $\mathbf{N} = (2,-3,4)$ is normal to π and hence any line perpendicular to π must be parallel to $\mathbf{N}$. Thus the line through $\mathbf{P}_0$ perpendicular to π in parametric form is

$$\begin{aligned} x &= -2 + 2t \\ y &= 5 - 3t \qquad (-\infty < t < \infty) \\ z &= -3 + 4t. \end{aligned}$$

T.1. Since by hypothesis a, b, and c are not all zero, take $a \neq 0$. Let $\mathbf{P}_0 = (-d/a,0,0)$. Then from (8) and (9), the equation of the plane through $\mathbf{P}_0$ with normal vector $\mathbf{N} = (a,b,c)$ is

$$a(x+(d/a)) + b(y-0) + c(z-0) = 0,$$

or

$$ax + by + cz + d = 0.$$

T.3. One possible solution is

$$L_1: \begin{aligned} x &= s \\ y &= 0 \\ z &= 0 \end{aligned} \qquad \text{and} \qquad L_2: \begin{aligned} x &= 0 \\ y &= 1 \\ z &= t \end{aligned}.$$

(the x-axis) — (a line in the yz-plane one unit to the right of the z axis and parallel to it)

T.5. The whole space $\mathbf{R}^3$, the zero subspace $\{\mathbf{0}\}$, all lines through the origin, and all planes through the origin.

Exercises 7.1

T.7. Expand the determinant about the first row:

$$\begin{vmatrix} x & y & z & 1 \\ a_1 & b_1 & c_1 & 1 \\ a_2 & b_2 & c_2 & 1 \\ a_3 & b_3 & c_3 & 1 \end{vmatrix} = xA_{11} + yA_{12} + zA_{13} + 1\cdot A_{14} \qquad (*)$$

where A_{1j} is the cofactor of the 1,jth element, and (being based on the second, third and fourth rows of the determinant) is a constant. (See Equation (1) in Theorem 2.9.) Thus (*) is an equation of the form

$$ax + by + cz + d = 0$$

and so is the equation of some plane. The noncolinearity of the three points insures that the three cofactors A_{11}, A_{12}, A_{13} are not all zero.

Next let $(x,y,z) = (a_i,b_i,c_i)$. The determinant has two equal rows, and so has the value zero. Thus the point P_i lies on the plane whose equation is (*). Thus (*) is an equation for the plane through P_1, P_2, P_3 .

Exercises 7.2

1. We use Equations (1) and (2) along with Examples 1 and 2.

(a) $-3x^2 + 5xy - 2y^2 = \begin{bmatrix} x & y \end{bmatrix} \begin{bmatrix} -3 & 5/2 \\ 5/2 & -2 \end{bmatrix} \begin{bmatrix} x \\ y \end{bmatrix}$

(b) $2x_1{}^2 + 3x_1x_2 - 5x_1x_3 + 7x_2x_3 = \begin{bmatrix} x_1 & x_2 & x_3 \end{bmatrix} \begin{bmatrix} 2 & 3/2 & -5/2 \\ 3/2 & 0 & 7/2 \\ -5/2 & 7/2 & 0 \end{bmatrix} \begin{bmatrix} x_1 \\ x_2 \\ x_3 \end{bmatrix}$

(c) $3x_1{}^2 + x_2{}^2 - 2x_3{}^2 + x_1x_2 - x_1x_3 - 4x_2x_3$

$= \begin{bmatrix} x_1 & x_2 & x_3 \end{bmatrix} \begin{bmatrix} 3 & 1/2 & -1/2 \\ 1/2 & 1 & -2 \\ -1/2 & -2 & -2 \end{bmatrix} \begin{bmatrix} x_1 \\ x_2 \\ x_3 \end{bmatrix}$

3. <<**Strategy:** Since each of the matrices **A** is symmetric we use Theorem 5.7. Hence the diagonal matrix **D** we seek has the eigenvalues of **A** as diagonal entries. Thus we compute the eigenvalues of **A** and form matrix **D**.>>

(a) Let $\mathbf{A} = \begin{bmatrix} -1 & 0 & 0 \\ 0 & 1 & 1 \\ 0 & 1 & 1 \end{bmatrix}$. The characteristic polynomial of **A** is

$$\det(\lambda\mathbf{I}_3 - \mathbf{A}) = \begin{vmatrix} \lambda+1 & 0 & 0 \\ 0 & \lambda-1 & -1 \\ 0 & -1 & \lambda-1 \end{vmatrix} = (\lambda + 1)((\lambda - 1)^2 - 1)$$

$$= \lambda(\lambda - 2)(\lambda + 1)$$

Hence the eigenvalues of **A** are $\lambda_1 = 0$, $\lambda_2 = 2$, and $\lambda_3 = -1$. Thus a diagonal matrix **D** congruent to **A** is

$$\mathbf{D} = \begin{bmatrix} 0 & 0 & 0 \\ 0 & 2 & 0 \\ 0 & 0 & -1 \end{bmatrix}$$

There is more than one diagonal matrix **D** congruent to **A**. Others are found by reordering the eigenvalues of **A** on the diagonal.

(b) Let $\mathbf{A} = \begin{bmatrix} 1 & 1 & 1 \\ 1 & 1 & 1 \\ 1 & 1 & 1 \end{bmatrix}$. The characteristic polynomial of **A** is

$$\det(\lambda\mathbf{I}_3 - \mathbf{A}) = \begin{vmatrix} \lambda-1 & -1 & -1 \\ -1 & \lambda-1 & -1 \\ -1 & -1 & \lambda-1 \end{vmatrix} = (\lambda-1)^3 - 1 - 1 - 3(\lambda-1)$$

$$= \lambda^3 - 3\lambda^2 = \lambda^2(\lambda - 3)$$

Hence the eigenvalues of **A** are $\lambda_1 = 0$, $\lambda_2 = 0$, and $\lambda_3 = 3$. Thus a diagonal matrix **D** congruent to **A** is

$$\mathbf{D} = \begin{bmatrix} 0 & 0 & 0 \\ 0 & 0 & 0 \\ 0 & 0 & 3 \end{bmatrix}$$

There is more than one diagonal matrix **D** congruent to **A**. Others are found by reordering the eigenvalues of **A** on the diagonal.

(c) Let $\mathbf{A} = \begin{bmatrix} 0 & 2 & 2 \\ 2 & 0 & 2 \\ 2 & 2 & 0 \end{bmatrix}$. The characteristic polynomial of **A** is

$$\det(\lambda \mathbf{I}_3 - \mathbf{A}) = \begin{vmatrix} \lambda & -2 & -2 \\ -2 & \lambda & -2 \\ -2 & -2 & \lambda \end{vmatrix} = \lambda^3 - 8 - 8 - 12\lambda$$

$$= \lambda^3 - 12\lambda - 16 = (\lambda - 4)(\lambda + 2)^2$$

Hence the eigenvalues of **A** are $\lambda_1 = 4$, $\lambda_2 = -2$, and $\lambda_3 = -2$. Thus a diagonal matrix **D** congruent to **A** is

$$\mathbf{D} = \begin{bmatrix} 4 & 0 & 0 \\ 0 & -2 & 0 \\ 0 & 0 & -2 \end{bmatrix}$$

There is more than one diagonal matrix **D** congruent to **A**. Others are found by reordering the eigenvalues of **A** on the diagonal.

<<**Strategy:** In Exercises 5-10 we form the matrix **A** of the quadratic form and find its eigenvalues $\lambda_1, \lambda_2, \ldots, \lambda_n$. Then Theorem 7.1 guarantees that the quadratic form

$$h(\mathbf{Y}) = \lambda_1 y_1^2 + \lambda_2 y_2^2 + \cdots + \lambda_n y_n^2$$

is equivalent to the quadratic form $g(\mathbf{X}) = \mathbf{X}^T\mathbf{A}\mathbf{X}$.>>

5. Let $g(\mathbf{X}) = 2x^2 - 4xy - y^2$. Then the matrix of the quadratic form is $\mathbf{A} = \begin{bmatrix} 2 & -2 \\ -2 & -1 \end{bmatrix}$. The characteristic polynomial of **A** is

$$\det(\lambda \mathbf{I}_2 - \mathbf{A}) = \begin{vmatrix} \lambda-2 & 2 \\ 2 & \lambda+1 \end{vmatrix} = (\lambda-2)(\lambda+1) - 4$$

$$= \lambda^2 - \lambda - 6 = (\lambda - 3)(\lambda + 2)$$

Thus the eigenvalues of **A** are $\lambda_1 = 3$, $\lambda_2 = -2$. Hence $g(\mathbf{X})$ is equivalent to $h(\mathbf{Y}) = 3x'^2 - 2y'^2$ where $\mathbf{Y} = \begin{bmatrix} x' & y' \end{bmatrix}^T$. (If the eigenvalues had been labeled $\lambda_1 = -2$, $\lambda_2 = 3$, then $h(\mathbf{Y}) = -2x'^2 + 3y'^2$. Hence $h(\mathbf{Y})$ is not unique.)

7. Let $g(\mathbf{X}) = 2x_1x_3$. Then the matrix of the quadratic form is $\mathbf{A} = \begin{bmatrix} 0 & 0 & 1 \\ 0 & 0 & 0 \\ 1 & 0 & 0 \end{bmatrix}$. The characteristic polynomial of **A** is

$$\det(\lambda\mathbf{I}_3 - \mathbf{A}) = \begin{bmatrix} \lambda & 0 & -1 \\ 0 & \lambda & 0 \\ -1 & 0 & \lambda \end{bmatrix} = \lambda^3 - \lambda$$

$$= \lambda(\lambda^2 - 1) = \lambda(\lambda - 1)(\lambda + 1)$$

Thus the eigenvalues of **A** are $\lambda_1 = 0$, $\lambda_2 = 1$, $\lambda_3 = -1$. Hence $g(\mathbf{X})$ is equivalent to $h(\mathbf{Y}) = y_2^2 - y_3^2$ where $\mathbf{Y} = \begin{bmatrix} y_1 & y_2 & y_3 \end{bmatrix}^T$. (Using the eigenvalues in a different order produces another equivalent quadratic form of the desired type.)

9. Let $g(\mathbf{X}) = -2x_1^2 - 4x_2^2 + 4x_3^2 - 6x_2x_3$. Then the matrix of the quadratic form is $\mathbf{A} = \begin{bmatrix} -2 & 0 & 0 \\ 0 & -4 & -3 \\ 0 & -3 & 4 \end{bmatrix}$. The characteristic polynomial of **A** is

$$\det(\lambda\mathbf{I}_3 - \mathbf{A}) = \begin{bmatrix} \lambda+2 & 0 & 0 \\ 0 & \lambda+4 & 3 \\ 0 & 3 & \lambda-4 \end{bmatrix} = (\lambda+2)(\lambda+4)(\lambda-4) - 9(\lambda+2)$$

$$= (\lambda + 2)(\lambda^2 - 25) = (\lambda + 2)(\lambda - 5)(\lambda + 5)$$

Thus the eigenvalues of **A** are $\lambda_1 = -2$, $\lambda_2 = 5$, $\lambda_3 = -5$. Hence $g(\mathbf{X})$ is equivalent to $h(\mathbf{Y}) = -2y_1^2 + 5y_2^2 - 5y_3^2$ where $\mathbf{Y} = \begin{bmatrix} y_1 & y_2 & y_3 \end{bmatrix}^T$ (Using the eigenvalues in a different order produces another equivalent quadratic form of the desired type.)

<<**Strategy:** In Exercises 11-16 we form the matrix **A** of the quadratic form and find its eigenvalues $\lambda_1, \lambda_2, \ldots, \lambda_n$. We determine r, the number of nonzero eigenvalues, p, the number of positive eigenvalues and r - p the number of negative eigenvalues. Then Theorem 7.2 guarantees that the quadratic form

$$h(\mathbf{Y}) = y_1^2 + y_2^2 + \cdots + y_p^2 - y_{p+1}^2 - \cdots - y_r^2$$

is equivalent to the quadratic form $g(\mathbf{X}) = \mathbf{X}^T\mathbf{A}\mathbf{X}$.>>

Exercises 7.2

11. Let $g(\mathbf{X}) = 2x^2 + 4xy + 2y^2$. Then the matrix of the quadratic form is $\mathbf{A} = \begin{bmatrix} 2 & 2 \\ 2 & 2 \end{bmatrix}$. The characteristic polynomial of **A** is

$$\det(\lambda\mathbf{I}_2 - \mathbf{A}) = \begin{vmatrix} \lambda-2 & -2 \\ -2 & \lambda-2 \end{vmatrix} = (\lambda-2)^2 - 4$$
$$= \lambda^2 - 4\lambda = \lambda(\lambda - 4)$$

Thus the eigenvalues of **A** are $\lambda_1 = 0$, $\lambda_2 = 4$. In this case $n = 2$, $r = 1$, $p = 1$, and $r - p = 0$. Hence $g(\mathbf{X})$ is equivalent to $h(\mathbf{Y}) = y_1^2$ where $\mathbf{Y} = [y_1 \ \ y_2]^T$.

13. Let $g(\mathbf{X}) = 2x_1^2 + 4x_2^2 + 4x_3^2 + 10x_2x_3$. Then the matrix of the quadratic form is $\mathbf{A} = \begin{bmatrix} 2 & 0 & 0 \\ 0 & 4 & 5 \\ 0 & 5 & 4 \end{bmatrix}$. The characteristic polynomial of **A** is

$$\det(\lambda\mathbf{I}_3 - \mathbf{A}) = \begin{bmatrix} \lambda-2 & 0 & 0 \\ 0 & \lambda-4 & -5 \\ 0 & -5 & \lambda-4 \end{bmatrix} = (\lambda-2)(\lambda-4)^2 - 25(\lambda-2)$$
$$= (\lambda - 2)(\lambda^2 - 8\lambda - 9) = (\lambda - 2)(\lambda - 9)(\lambda + 1)$$

Thus the eigenvalues are $\lambda_1 = 2$, $\lambda_2 = 9$, $\lambda_3 = -1$. In this case $n = 3$, $r = 3$, $p = 2$, and $r - p = 1$. Hence $g(\mathbf{X})$ is equivalent to $h(\mathbf{Y}) = y_1^2 + y_2^2 - y_3^2$ where $\mathbf{Y} = [y_1 \ \ y_2 \ \ y_3]^T$.

15. Let $g(\mathbf{X}) = -3x_1^2 + 2x_2^2 + 2x_3^2 + 4x_2x_3$. Then the matrix of the quadratic form is $\mathbf{A} = \begin{bmatrix} -3 & 0 & 0 \\ 0 & 2 & 2 \\ 0 & 2 & 2 \end{bmatrix}$. The characteristic polynomial of **A** is

$$\det(\lambda\mathbf{I}_3 - \mathbf{A}) = \begin{bmatrix} \lambda+3 & 0 & 0 \\ 0 & \lambda-2 & 2 \\ 0 & 2 & \lambda-2 \end{bmatrix} = (\lambda+3)(\lambda-2)^2 - 4(\lambda+3)$$
$$= (\lambda + 3)(\lambda^2 - 4\lambda) = (\lambda + 3)\lambda(\lambda - 4)$$

Thus the eigenvalues of **A** are $\lambda_1 = -3$, $\lambda_2 = 0$, $\lambda_3 = 4$. In this case $n = 3$, $r = 2$, $p = 1$, and $r - p = 1$. Hence $g(\mathbf{X})$ is equivalent to $h(\mathbf{Y}) = y_1^2 - y_2^2$ where $\mathbf{Y} = [y_1 \ \ y_2 \ \ y_3]^T$.

17. Let $g(\mathbf{X}) = 4x_2^2 + 4x_3^2 - 10x_2x_3$. Then the matrix of the quadratic form is $\mathbf{A} = \begin{bmatrix} 0 & 0 & 0 \\ 0 & 4 & -5 \\ 0 & -5 & 4 \end{bmatrix}$. The characteristic polynomial of $\mathbf{A}$ is

$$\det(\lambda\mathbf{I}_3 - \mathbf{A}) = \begin{bmatrix} \lambda & 0 & 0 \\ 0 & \lambda-4 & 5 \\ 0 & 5 & \lambda-4 \end{bmatrix} = \lambda(\lambda-4)^2 - 25\lambda$$

$$= \lambda(\lambda^2 - 8\lambda - 9) = \lambda(\lambda - 9)(\lambda + 1)$$

Thus the eigenvalues of $\mathbf{A}$ are $\lambda_1 = 0$, $\lambda_2 = 9$, $\lambda_3 = -1$. In this case $n = 3$, $r = 2$, $p = 1$, and $r - p = 1$. Hence $g(\mathbf{X})$ is equivalent to $h(\mathbf{Y}) = y_1^2 - y_2^2$ where $\mathbf{Y} = \begin{bmatrix} y_1 & y_2 & y_3 \end{bmatrix}^T$. The rank of g is 2 and the signature of g is 0.

19. For a 2×2 matrix $\mathbf{A}$ of a quadratic form $g(\mathbf{X}) = \mathbf{X}^T\mathbf{A}\mathbf{X}$ we have the following possibilities for its eigenvalues with the corresponding quadratic form of the type described in Theorem 7.2.

eigenvalues	quadratic form
both positive	$y_1^2 + y_2^2$
both negative	$-y_1^2 - y_2^2$
one positive, one negative	$y_1^2 - y_2^2$
one positive, one zero	y_1^2
one negative, one zero	$-y_1^2$

The conics for the equations $\mathbf{X}^T\mathbf{A}\mathbf{X} = 1$ are given next.

$y_1^2 + y_2^2 = 1$ is a circle

$-y_1^2 - y_2^2 = 1$ is empty; it represents no conic

$y_1^2 - y_2^2 = 1$ is a hyperbola

$y_1^2 = 1$ is a pair of lines; $y_1 = 1$, $y_1 = -1$

$-y_1^2 = 1$ is empty; it represents no conic

21. In the discussion preceding Example 7 it is stated that quadratic forms g and h are equivalent if and only if they have equal ranks and the same signature. We determine the rank and signature for each of the quadratic forms listed.

Let $g_1(\mathbf{X}) = x_1^2 + x_2^2 + x_3^2 + 2x_1x_2$. Then the matrix of the quadratic form is $\mathbf{A} = \begin{bmatrix} 1 & 1 & 0 \\ 1 & 1 & 0 \\ 0 & 0 & 1 \end{bmatrix}$. The characteristic polynomial of $\mathbf{A}$ is

$$\det(\lambda \mathbf{I}_3 - \mathbf{A}) = \begin{bmatrix} \lambda-1 & -1 & 0 \\ -1 & \lambda-1 & 0 \\ 0 & 0 & \lambda-1 \end{bmatrix} = (\lambda-1)^3 - (\lambda-1)$$
$$= (\lambda - 1)(\lambda^2 - 2\lambda) = \lambda(\lambda - 1)(\lambda - 2)$$

Thus the eigenvalues of $\mathbf{A}$ are $\lambda_1 = 0$, $\lambda_2 = 1$, $\lambda_3 = 2$. In this case $n = 3$, $r = 2$, $p = 2$, and $r - p = 0$. The rank of g is 2 and the signature of g_1 is 2.

Let $g_2(\mathbf{X}) = 2x_2^2 + 2x_3^2 + 2x_2x_3$. Then the matrix of the quadratic form is $\mathbf{A} = \begin{bmatrix} 0 & 0 & 0 \\ 0 & 2 & 1 \\ 0 & 1 & 2 \end{bmatrix}$. The characteristic polynomial of $\mathbf{A}$ is

$$\det(\lambda \mathbf{I}_3 - \mathbf{A}) = \begin{bmatrix} \lambda & 0 & 0 \\ 0 & \lambda-2 & 1 \\ 0 & 1 & \lambda-2 \end{bmatrix} = \lambda(\lambda-2)^2 - \lambda$$
$$= \lambda(\lambda^2 - 4\lambda + 3) = \lambda(\lambda - 1)(\lambda - 3)$$

Thus the eigenvalues of $\mathbf{A}$ are $\lambda_1 = 0$, $\lambda_2 = 1$, $\lambda_3 = 3$. In this case $n = 3$, $r = 2$, $p = 2$, and $r - p = 0$. The rank of g is 2 and the signature of g_2 is 2. Thus $g_1(\mathbf{X})$ and $g_2(\mathbf{X})$ are equivalent.

Let $g_3(\mathbf{X}) = 3x_2^2 - 3x_3^2 + 8x_2x_3$. Then the matrix of the quadratic form is $\mathbf{A} = \begin{bmatrix} 0 & 0 & 0 \\ 0 & 3 & 4 \\ 0 & 4 & -3 \end{bmatrix}$. The characteristic polynomial of $\mathbf{A}$ is

$$\det(\lambda \mathbf{I}_3 - \mathbf{A}) = \begin{bmatrix} \lambda & 0 & 0 \\ 0 & \lambda-3 & -4 \\ 0 & -4 & \lambda+3 \end{bmatrix} = \lambda(\lambda-3)(\lambda+3) - 16\lambda$$
$$= \lambda(\lambda^2 - 25) = \lambda(\lambda - 5)(\lambda + 5) = 0$$

Thus the eigenvalues of $\mathbf{A}$ are $\lambda_1 = 0$, $\lambda_2 = 5$, $\lambda_3 = -5$. In this case $n = 3$, $r = 2$, $p = 1$, and $r - p = 1$. The rank of g is 2 and the signature of g_2 is 0. Thus $g_3(\mathbf{X})$ is not equivalent to $g_1(\mathbf{X})$ or $g_2(\mathbf{X})$.

Let $g_4(\mathbf{X}) = 3x_2{}^2 + 3x_3{}^2 - 4x_2x_3$. Then the matrix of the quadratic form is $\mathbf{A} = \begin{bmatrix} 0 & 0 & 0 \\ 0 & 3 & -2 \\ 0 & -2 & 3 \end{bmatrix}$. The characteristic polynomial of $\mathbf{A}$ is

$$\det(\lambda\mathbf{I}_3 - \mathbf{A}) = \begin{bmatrix} \lambda & 0 & 0 \\ 0 & \lambda-3 & 2 \\ 0 & 2 & \lambda-3 \end{bmatrix} = \lambda(\lambda-3)^2 - 4\lambda$$

$$= \lambda(\lambda^2 - 6\lambda + 5) = \lambda(\lambda - 5)(\lambda - 1)$$

Thus the eigenvalues of $\mathbf{A}$ are $\lambda_1 = 0$, $\lambda_2 = 5$, $\lambda_3 = 1$. In this case $n = 3$, $r = 2$, $p = 2$, and $r - p = 0$. The rank of g is 2 and the signature of g_2 is 2. Thus $g_4(\mathbf{X})$ is equivalent to both $g_1(\mathbf{X})$ and $g_2(\mathbf{X})$.

23. Use Theorem 7.3 by computing the eigenvalues of each of the matrices.

(a) Let $\mathbf{A} = \begin{bmatrix} 2 & -1 \\ -1 & 2 \end{bmatrix}$. The characteristic polynomial of $\mathbf{A}$ is

$$\det(\lambda\mathbf{I}_2 - \mathbf{A}) = \begin{vmatrix} \lambda-2 & 1 \\ 1 & \lambda-2 \end{vmatrix} = (\lambda-2)^2 - 1 = \lambda^2 - 4\lambda + 3$$

$$= (\lambda - 1)(\lambda - 3)$$

Thus the eigenvalues of $\mathbf{A}$ are $\lambda_1 = 1$ and $\lambda_2 = 3$. Since all the eigenvalues are positive, $\mathbf{A}$ is positive definite.

(b) Let $\mathbf{A} = \begin{bmatrix} 2 & 1 \\ 1 & 2 \end{bmatrix}$. The characteristic polynomial of $\mathbf{A}$ is

$$\det(\lambda\mathbf{I}_2 - \mathbf{A}) = \begin{vmatrix} \lambda-2 & -1 \\ -1 & \lambda-2 \end{vmatrix} = (\lambda-2)^2 - 1 = \lambda^2 - 4\lambda + 3$$

$$= (\lambda - 1)(\lambda - 3) = 0$$

Thus the eigenvalues of $\mathbf{A}$ are $\lambda_1 = 1$ and $\lambda_2 = 3$. Since all the eigenvalues are positive, $\mathbf{A}$ is positive definite.

(c) Let $\mathbf{A} = \begin{bmatrix} 3 & 1 & 0 \\ 1 & 3 & 0 \\ 0 & 0 & 3 \end{bmatrix}$. The characteristic polynomial of $\mathbf{A}$ is

$$\det(\lambda\mathbf{I}_3 - \mathbf{A}) = \begin{vmatrix} \lambda-3 & -1 & 0 \\ -1 & \lambda-3 & 0 \\ 0 & 0 & \lambda-3 \end{vmatrix} = (\lambda-3)^3 - 1(\lambda-3)$$

$$= (\lambda - 3)(\lambda^2 - 6\lambda + 8)$$
$$= (\lambda - 3)(\lambda - 4)(\lambda - 2)$$

Thus the eigenvalues of **A** are $\lambda_1 = 3$, $\lambda_2 = 4$, and $\lambda_2 = 2$. Since all the eigenvalues are positive, **A** is positive definite.

(d) Let $\mathbf{A} = \begin{bmatrix} 1 & 0 & 0 \\ 0 & 2 & 0 \\ 0 & 0 & -3 \end{bmatrix}$. Since **A** is diagonal, its eigenvalues are its diagonal entries. It follows that **A** is not positive definite.

(e) Let $\mathbf{A} = \begin{bmatrix} 2 & 2 \\ 2 & 2 \end{bmatrix}$. Matrix **A** is singular, so one of its eigenvalues is zero. Thus **A** is not positive definite.

T.1. Let **A** be symmetric.
Prove: $\mathbf{P}^T\mathbf{AP}$ is symmetric.
<u>Proof:</u> A matrix is symmetric if it equals its transpose.

$$(\mathbf{P}^T\mathbf{AP})^T = \mathbf{P}^T\mathbf{A}^T(\mathbf{P}^T)^T = \mathbf{P}^T\mathbf{A}^T\mathbf{P} \quad \text{\{via properties of transposes\}}$$
$$= \mathbf{P}^T\mathbf{AP} \quad \text{\{since } \mathbf{A} \text{ is symmetric\}}$$

T.3. If **A** is symmetric the by Theorem 5.7 in Section 5.2 there exists an orthogonal matrix **P** such that $\mathbf{P}^{-1}\mathbf{AP} = \mathbf{D}$, a diagonal matrix. Since $\mathbf{P}^{-1} = \mathbf{P}^T$ we have $\mathbf{P}^T\mathbf{AP} = \mathbf{D}$. Thus **A** is congruent to a diagonal matrix.

T.5. Let **A** be the matrix of the quadratic form $g(\mathbf{X}) = \mathbf{X}^T\mathbf{AX}$. Then by Theorem 7.2 $g(\mathbf{X})$ is equivalent to

$$h(\mathbf{Y}) = y_1{}^2 + y_2{}^2 + \cdots + y_p{}^2 - y_{p+1}{}^2 - \cdots - y_r{}^2$$

Assume **A** is positive definite. Since g and h are equivalent $h(\mathbf{Y}) > 0$ for each $\mathbf{Y} \neq \mathbf{0}$. However, this can happen if and only if all the summands in $h(\mathbf{Y})$ are positive, that is, **A** is congruent to $\mathbf{I}_n$. Thus it follows that there exists a nonsingular matrix **P** such that

$$\mathbf{A} = \mathbf{P}^T\mathbf{I}_n\mathbf{P} = \mathbf{P}^T\mathbf{P}$$

The preceding steps are reversible to show that if $\mathbf{A} = \mathbf{P}^T\mathbf{P}$ for some nonsingular matrix **P**, then **A** is positive definite.

ML.1. (a)
```
A=[-1 0 0;0 1 1 ;0 1 1];
eig(A)

ans =

  -1.0000
   2.0000
   0.0000
```

If we set the format to long e, then the eigenvalues are displayed as

```
-1.000000000000000e+000
 2.000000000000000e+000
 2.220446049250313e-016
```

Since the last value is extremely small, we will consider it zero. The **eig** command approximates the eigenvalues, hence errors due to using machine arithmetic can occur. Thus it follows that rank(A) = 2 and the signature of the quadratic form is 0.

(b)
```
A=ones(3);
eig(A)

ans =

   0.0000
  -0.0000
   3.0000
```

If we set the format to long e, then the eigenvalues are displayed as

```
 2.343881062810587e-017
-7.011704839834072e-016
 2.999999999999999e+000
```

We will consider the first two eigenvalues zero. Hence rank(A) = 1 and the signature of the quadratic form is 1.

(c)
```
A=[2 1 0 -2;1 -1 1 3;0 1 2 -1;-2 3 -1 0];
eig(A)

ans =

   2.2896
   1.6599
   3.5596
  -4.5091
```

It follows that rank(A) = 4 and the signature of the quadratic form is 2.

(d)
```
A=[2 -1 0 0 ;-1 2 -1 0;0 -1 2 -1;0 0 -1 2];
eig(A)

ans =

   1.3820
   0.3820
   2.6180
   3.6180
```

It follows that rank(A) = 4 and the signature of the quadratic form is 4.

ML.2. By Theorem 7.3, only the matrix in part d is positive definite.

EXERCISES 7.3

<<**Strategy:** In Exercises 1-10, compare the given equations with those that appear in Figure 7.7. In some cases we may need to rearrange terms to properly identify the equation from the forms given in Figure 7.7.>>

1. Rewrite the equation $x^2 + 9y^2 - 9 = 0$ as

$$\frac{x^2}{9} + \frac{9y^2}{9} = \frac{9}{9} \implies \frac{x^2}{9} + \frac{y^2}{1} = 1$$

This is the equation of an ellipse in standard position with $a = 3$ and $b = 1$. The x-intercepts are (-3,0) and (3,0) and the y-intercepts are (0,-1) and (0,1).

3. Rewrite the equation $25y^2 - 4x^2 = 100$ as

$$\frac{25y^2}{100} - \frac{4x^2}{100} = \frac{100}{100} \implies \frac{y^2}{4} - \frac{x^2}{25} = 1$$

This is the equation of a hyperbola in standard position with $a = 2$ and $b = 5$. The y-intercepts are (0,-2) and (0,2).

5. Rewrite equation $3x^2 - y^2 = 0$ as

$$y^2 = 3x^2 \implies y = \sqrt{3}\,x \quad \text{and} \quad y = -\sqrt{3}\,x$$

This represents the graph of a pair of intersecting lines which is a degenerate conic section.

7. Rewrite equation $4x^2 + 4y^2 - 9 = 0$ as

$$\frac{4x^2}{9} + \frac{4y^2}{9} = \frac{9}{9} \implies \frac{x^2}{(3/2)^2} + \frac{y^2}{(3/2)^2} = 1$$

This is the equation of a circle in standard position with $a = 3/2$.

9. Upon inspection, equation $4x^2 + y^2 = 0$ is satisfied if and only if $x = y = 0$. Hence this is a degenerate conic section which represents the single point (0,0).

<<**Strategy:** In each of the Exercises 11-18 the equations do not contain cross-product terms. However, the equations do contain x^2 and x terms and/or y^2 and y terms. Thus we complete the square(s) and rewrite the equations as one of the forms in Figure 7.7.>>

11. In equation $x^2 + 2y^2 - 4x - 4y + 4 = 0$ we note that x^2 and x and y^2 and y terms appear. Hence we complete the square in both x and y. Rewrite the equation as

$$x^2 - 4x + 4 + 2y^2 - 4y = 0$$

$$(x - 2)^2 + 2(y^2 - 2y \quad) = 0$$

$$(x - 2)^2 + 2(y^2 - 2y + 1) = 2$$

$$(x - 2)^2 + 2(y - 1)^2 = 2$$

Let $x' = x - 2$ and $y' = y - 1$ then the preceding equation is written as

$$x'^2 + 2y'^2 = 2$$

Next we transform this equation to

$$\frac{x'^2}{2} + \frac{y'^2}{1} = 1$$

If we translate the xy-coordinate system to the x'y'-coordinate system, whose origin is at (2,1), then the graph is an ellipse in standard position with respect to the x'y'-coordinate system.

13. In equation $x^2 + y^2 - 8x - 6y = 0$ we note that x^2 and x and y^2 and y terms appear. Hence we complete the square in both x and y. Rewrite the equation as

$$x^2 - 8x + y^2 - 6y = 0$$

$$x^2 - 8x + 16 + y^2 - 6y + 9 = 16 + 9$$

$$(x - 4)^2 + (y - 3)^2 = 25$$

Let $x' = x - 4$ and $y' = y - 3$ then the preceding equation can be written as

$$x'^2 + y'^2 = 25$$

Next we transform this equation to

$$\frac{x'^2}{25} + \frac{y'^2}{25} = 1 \implies \frac{x'^2}{5^2} + \frac{y'^2}{5^2} = 1$$

If we translate the xy-coordinate system to the x'y'-coordinate system, whose origin is at (4,3), then the graph is a circle in standard position with respect to the x'y'-coordinate system.

15. In equation $y^2 - 4y = 0$ we note that only y^2 and y terms appear. Hence we complete the square only in y. Rewrite the equation as

$$y^2 - 4y + 4 = 4 \implies (y - 2)^2 = 4$$

Let $y' = y - 2$, then we have

$$y'^2 = 4 \implies y' = 2 \text{ and } y' = -2$$

If we translate the xy-coordinate system to the x'y'-coordinate system, whose origin is at (0,2), then the graph is a pair of parallel lines.

17. In equation $x^2 + y^2 - 2x - 6y + 10 = 0$ we note that x^2 and x and y^2 and y terms appear. Hence we complete the square in both x and y. Rewrite the equation as

$$\underbrace{x^2 - 2x} + \underbrace{y^2 - 6y} = -10$$

$$\underbrace{x^2 - 2x + 1} + \underbrace{y^2 - 6y + 9} = -10 + 1 + 9$$

$$(x - 1)^2 + (y - 3)^2 = 0$$

Let $x' = x - 1$ and $y' = y - 3$ then the preceding equation can be written as

$$x'^2 + y'^2 = 0$$

If we translate the xy-coordinate system to the x'y'-coordinate system, whose origin is at (1,3), then the graph is a single point which is the origin of the x'y'-coordinate system.

<<**Strategy:** In Exercises 19-24 each of the equations contains only an xy term. Thus a rotation will transform the graph to standard position in a new coordinate system. We determine the matrix **A** from Equation (5), find its eigenvalues, and construct the orthogonal matrix **P** that gives the correct rotation.>>

19. For equation $x^2 + xy + y^2 = 6$, we have $\mathbf{A} = \begin{bmatrix} 1 & 1/2 \\ 1/2 & 1 \end{bmatrix}$. The characteristic equation is

$$\begin{aligned} \det(\lambda \mathbf{I}_2 - \mathbf{A}) &= \begin{vmatrix} \lambda-1 & -1/2 \\ -1/2 & \lambda-1 \end{vmatrix} = \lambda^2 - 2\lambda + 3/4 \\ &= (\lambda - 1/2)(\lambda - 3/2) = 0 \end{aligned}$$

Thus the eigenvalues of **A** are $\lambda_1 = 1/2$ and $\lambda_2 = 3/2$. Next we determine an eigenvector $\mathbf{X}_1$ associated with eigenvalue λ_1:

Solving $(\lambda_1\mathbf{I}_2 - \mathbf{A})\mathbf{X} = \begin{bmatrix} -1/2 & -1/2 \\ -1/2 & -1/2 \end{bmatrix}\begin{bmatrix} x_1 \\ x_2 \end{bmatrix} = \begin{bmatrix} 0 \\ 0 \end{bmatrix}$

we find $\mathbf{X} = \begin{bmatrix} r \\ -r \end{bmatrix}$. For r = 1, we have eigenvector $\mathbf{X}_1 = \begin{bmatrix} 1 \\ -1 \end{bmatrix}$.

We determine an eigenvector $\mathbf{X}_2$ associated with eigenvalue λ_2:

Solving $(\lambda_2\mathbf{I}_2 - \mathbf{A})\mathbf{X} = \begin{bmatrix} 1/2 & -1/2 \\ -1/2 & 1/2 \end{bmatrix}\begin{bmatrix} x_1 \\ x_2 \end{bmatrix} = \begin{bmatrix} 0 \\ 0 \end{bmatrix}$

we find $\mathbf{X} = \begin{bmatrix} r \\ r \end{bmatrix}$. For r = 1, we have eigenvector $\mathbf{X}_2 = \begin{bmatrix} 1 \\ 1 \end{bmatrix}$.

Normalizing the eigenvectors and forming the matrix **P** we have

$$\mathbf{P} = \begin{bmatrix} 1/\sqrt{2} & 1/\sqrt{2} \\ -1/\sqrt{2} & 1/\sqrt{2} \end{bmatrix} \qquad \text{(Note: det}(\mathbf{P}) = 1.)$$

Let $\mathbf{X} = \mathbf{PY}$ where $\mathbf{Y} = \begin{bmatrix} x' \\ y' \end{bmatrix}$ then we can rewrite the original equation as in (6) to obtain

$$(1/2)x'^2 + (3/2)y'^2 = 6 \implies \frac{x'^2}{12} + \frac{y'^2}{4} = 1$$

Thus this conic section is an ellipse. The preceding is only a possible answer. The roles of x′ and y′ would be reversed if the eigenvalues of **A** were used in a different order.

21. For equation $9x^2 + y^2 + 6xy = 4$, we have $\mathbf{A} = \begin{bmatrix} 9 & 3 \\ 3 & 1 \end{bmatrix}$. The characteristic equation is

$$\det(\lambda\mathbf{I}_2 - \mathbf{A}) = \begin{vmatrix} \lambda-9 & -3 \\ -3 & \lambda-1 \end{vmatrix} = \lambda^2 - 10\lambda$$

$$= \lambda(\lambda - 10) = 0$$

Thus the eigenvalues of **A** are $\lambda_1 = 0$ and $\lambda_2 = 10$. Next we determine an eigenvector $\mathbf{X}_1$ associated with eigenvalue λ_1:

Solving $(\lambda_1\mathbf{I}_2 - \mathbf{A})\mathbf{X} = \begin{bmatrix} -9 & -3 \\ -3 & -1 \end{bmatrix}\begin{bmatrix} x_1 \\ x_2 \end{bmatrix} = \begin{bmatrix} 0 \\ 0 \end{bmatrix}$

we find $\mathbf{X} = \begin{bmatrix} (-1/3)r \\ r \end{bmatrix}$. For r = 3, we have eigenvector

$$\mathbf{X}_1 = \begin{bmatrix} -1 \\ 3 \end{bmatrix}.$$

We determine an eigenvector $\mathbf{X}_2$ associated with eigenvalue λ_2:

Solving $(\lambda_2\mathbf{I}_2 - \mathbf{A})\mathbf{X} = \begin{bmatrix} 1 & -3 \\ -3 & 9 \end{bmatrix}\begin{bmatrix} x_1 \\ x_2 \end{bmatrix} = \begin{bmatrix} 0 \\ 0 \end{bmatrix}$

we find $\mathbf{X} = \begin{bmatrix} 3r \\ r \end{bmatrix}$. For r = 1, we have eigenvector $\mathbf{X}_2 = \begin{bmatrix} 3 \\ 1 \end{bmatrix}$.

Normalizing the eigenvectors and forming the matrix **P** we have

$\mathbf{P} = \begin{bmatrix} -1/\sqrt{10} & 3/\sqrt{10} \\ 3/\sqrt{10} & 1/\sqrt{10} \end{bmatrix}$. Note that $\det(\mathbf{P}) = -1$. For a counterclockwise rotation we require $\det(\mathbf{P}) = 1$. Since any nonzero multiple of an eigenvector is still an eigenvector for the same eigenvalue we replace $\mathbf{X}_1$ by $-\mathbf{X}_1$ (an alternate procedure is to interchange columns of **P**) which results in redefining **P** as

$$\mathbf{P} = \begin{bmatrix} 1/\sqrt{10} & 3/\sqrt{10} \\ -3/\sqrt{10} & 1/\sqrt{10} \end{bmatrix}$$

Let $\mathbf{X} = \mathbf{PY}$ where $\mathbf{Y} = \begin{bmatrix} x' \\ y' \end{bmatrix}$ then we can rewrite the original equation as in (6) to obtain

$$0x'^2 + 10y'^2 = 4 \implies y'^2 = 4/10 \implies y' = \pm 2/5$$

Thus this conic section consists of two parallel lines. The preceding is only a possible answer. The roles of x' and y' would be reversed if the eigenvalues of **A** were used in a different order.

23. For equation $4x^2 + 4y^2 - 10xy = 0$, we have $\mathbf{A} = \begin{bmatrix} 4 & -5 \\ -5 & 4 \end{bmatrix}$. The characteristic equation is

$$\det(\lambda\mathbf{I}_2 - \mathbf{A}) = \begin{vmatrix} \lambda-4 & 5 \\ 5 & \lambda-4 \end{vmatrix} = \lambda^2 - 8\lambda - 9$$
$$= (\lambda - 9)(\lambda + 1) = 0$$

Thus the eigenvalues of **A** are $\lambda_1 = 9$ and $\lambda_2 = -1$. Next we determine an eigenvector $\mathbf{X}_1$ associated with eigenvalue λ_1:

Solving $(\lambda_1 \mathbf{I}_2 - \mathbf{A})\mathbf{X} = \begin{bmatrix} 5 & 5 \\ 5 & 5 \end{bmatrix}\begin{bmatrix} x_1 \\ x_2 \end{bmatrix} = \begin{bmatrix} 0 \\ 0 \end{bmatrix}$

we find $\mathbf{X} = \begin{bmatrix} -r \\ r \end{bmatrix}$. For r = 1, we have eigenvector

$$\mathbf{X}_1 = \begin{bmatrix} -1 \\ 1 \end{bmatrix}$$

We determine an eigenvector $\mathbf{X}_2$ associated with eigenvalue λ_2:

Solving $(\lambda_2 \mathbf{I}_2 - \mathbf{A})\mathbf{X} = \begin{bmatrix} -5 & 5 \\ 5 & -5 \end{bmatrix}\begin{bmatrix} x_1 \\ x_2 \end{bmatrix} = \begin{bmatrix} 0 \\ 0 \end{bmatrix}$

we find $\mathbf{X} = \begin{bmatrix} r \\ r \end{bmatrix}$. For r = 1, we have eigenvector $\mathbf{X}_2 = \begin{bmatrix} 1 \\ 1 \end{bmatrix}$.

Normalizing the eigenvectors and forming the matrix **P** we have

$\mathbf{P} = \begin{bmatrix} -1/\sqrt{2} & 1/\sqrt{2} \\ 1/\sqrt{2} & 1/\sqrt{2} \end{bmatrix}$. Note that $\det(\mathbf{P}) = -1$. For a counterclockwise rotation we require $\det(\mathbf{P}) = 1$. Since any nonzero multiple of an eigenvector is still an eigenvector for the same eigenvalue we replace $\mathbf{X}_1$ by $-\mathbf{X}_1$ which results in redefining **P** as

$$\mathbf{P} = \begin{bmatrix} 1/\sqrt{2} & 1/\sqrt{2} \\ -1/\sqrt{2} & 1/\sqrt{2} \end{bmatrix}$$

Let $\mathbf{X} = \mathbf{PY}$ where $\mathbf{Y} = \begin{bmatrix} x' \\ y' \end{bmatrix}$ then we can rewrite the original equation as in (6) to obtain

$$9x'^2 - 1y'^2 = 0 \implies y'^2 = 9x'^2 \implies y' = \pm 3x'$$

Thus this conic section is a pair of intersecting lines. The preceding is only a possible answer. The roles of x′ and y′ would be reversed if the eigenvalues of **A** were used in a different order.

<<**Strategy:** In Exercises 25-30, we see that both cross product terms and x or y terms appear. Thus we must combine the rotation and completing the square techniques used earlier.>>

25. For equation $9x^2 + y^2 + 6xy - 10\sqrt{10}\,x + 10\sqrt{10}\,y + 90 = 0$, we have from Equation (5)

Exercises 7.3

$$\mathbf{A} = \begin{bmatrix} 9 & 3 \\ 3 & 1 \end{bmatrix}, \quad \mathbf{B} = \begin{bmatrix} -10\sqrt{10} & 10\sqrt{10} \end{bmatrix}, \quad f = 90$$

The characteristic equation for matrix **A** is

$$\det(\lambda \mathbf{I}_2 - \mathbf{A}) = \begin{vmatrix} \lambda-9 & -3 \\ -3 & \lambda-1 \end{vmatrix} = \lambda^2 - 10\lambda = \lambda(\lambda - 10) = 0$$

Thus the eigenvalues are $\lambda_1 = 0$ and $\lambda = 10$. Next we determine an eigenvector $\mathbf{X}_1$ associated with eigenvalue λ_1:

Solving $(\lambda_1 \mathbf{I}_2 - \mathbf{A})\mathbf{X} = \begin{bmatrix} -9 & -3 \\ -3 & -1 \end{bmatrix}\begin{bmatrix} x_1 \\ x_2 \end{bmatrix} = \begin{bmatrix} 0 \\ 0 \end{bmatrix}$

we find $\mathbf{X} = \begin{bmatrix} (-1/3)r \\ r \end{bmatrix}$. For $r = 3$, we have eigenvector

$\mathbf{X}_1 = \begin{bmatrix} -1 \\ 3 \end{bmatrix}$.

We determine an eigenvector $\mathbf{X}_2$ associated with eigenvalue λ_2:

Solving $(\lambda_2 \mathbf{I}_2 - \mathbf{A})\mathbf{X} = \begin{bmatrix} 1 & -3 \\ -3 & 9 \end{bmatrix}\begin{bmatrix} x_1 \\ x_2 \end{bmatrix} = \begin{bmatrix} 0 \\ 0 \end{bmatrix}$

we find $\mathbf{X} = \begin{bmatrix} 3r \\ r \end{bmatrix}$. For $r = 1$, we have eigenvector $\mathbf{X}_2 = \begin{bmatrix} 3 \\ 1 \end{bmatrix}$.

Normalizing the eigenvectors and forming the matrix **P** we have

$\mathbf{P} = \begin{bmatrix} -1/\sqrt{10} & 3/\sqrt{10} \\ 3/\sqrt{10} & 1/\sqrt{10} \end{bmatrix}$. Note that $\det(\mathbf{P}) = -1$. For a counterclockwise rotation we require $\det(\mathbf{P}) = 1$. Since any nonzero multiple of an eigenvector is still an eigenvector for the same eigenvalue we replace $\mathbf{X}_1$ by $-\mathbf{X}_1$ which results in redefining **P** as

$$\mathbf{P} = \begin{bmatrix} 1/\sqrt{10} & 3/\sqrt{10} \\ -3/\sqrt{10} & 1/\sqrt{10} \end{bmatrix}$$

Let $\mathbf{X} = \mathbf{PY}$ where $\mathbf{Y} = \begin{bmatrix} x' \\ y' \end{bmatrix}$; then we can rewrite the original equation as in (6) to obtain

$$\begin{bmatrix} x' & y' \end{bmatrix}\begin{bmatrix} 0 & 0 \\ 0 & 10 \end{bmatrix}\begin{bmatrix} x' \\ y' \end{bmatrix}$$

$$+ \begin{bmatrix} -10\sqrt{10} & 10\sqrt{10} \end{bmatrix} \begin{bmatrix} 1/\sqrt{10} & 3/\sqrt{10} \\ -3/\sqrt{10} & 1/\sqrt{10} \end{bmatrix} \begin{bmatrix} x' \\ y' \end{bmatrix} + 90 = 0$$

Performing the matrix multiplies and simplifying gives

$$10y'^2 - 20y' + 40x' + 90 = 0$$

Dividing by 10 and completing the square in y' gives

$$(y' - 1)^2 + 4(x' + 2) = 0$$

Let $x'' = x' + 2$ and $y'' = y' - 1$; then we can write the equation as

$$y''^2 = -4x''$$

Thus this conic section is a parabola. The preceding is only a possible answer. The roles of x'' and y'' would be reversed if the eigenvalues of **A** were used in a different order.

27. For equation $5x^2 + 12xy - 12\sqrt{13}\,x = 36$ we have from Equation (5)

$$\mathbf{A} = \begin{bmatrix} 5 & 6 \\ 6 & 0 \end{bmatrix}, \quad \mathbf{B} = \begin{bmatrix} -12\sqrt{13} & 0 \end{bmatrix}, \quad f = -36$$

The characteristic equation for matrix **A** is

$$\det(\lambda\mathbf{I}_2 - \mathbf{A}) = \begin{vmatrix} \lambda-5 & -6 \\ -6 & \lambda \end{vmatrix} = \lambda^2 - 5\lambda - 36 = (\lambda - 9)(\lambda + 4) = 0$$

Thus the eigenvalues are $\lambda_1 = 9$ and $\lambda = -4$. Next we determine an eigenvector $\mathbf{X}_1$ associated with eigenvalue λ_1:

Solving $(\lambda_1\mathbf{I}_2 - \mathbf{A})\mathbf{X} = \begin{bmatrix} 4 & -6 \\ -6 & 9 \end{bmatrix} \begin{bmatrix} x_1 \\ x_2 \end{bmatrix} = \begin{bmatrix} 0 \\ 0 \end{bmatrix}$

we find $\mathbf{X} = \begin{bmatrix} (3/2)r \\ r \end{bmatrix}$. For $r = 2$, we have eigenvector

$\mathbf{X}_1 = \begin{bmatrix} 3 \\ 2 \end{bmatrix}$.

We determine an eigenvector $\mathbf{X}_2$ associated with eigenvalue λ_2:

Solving $(\lambda_2\mathbf{I}_2 - \mathbf{A})\mathbf{X} = \begin{bmatrix} -9 & -6 \\ -6 & -4 \end{bmatrix} \begin{bmatrix} x_1 \\ x_2 \end{bmatrix} = \begin{bmatrix} 0 \\ 0 \end{bmatrix}$

we find $\mathbf{X} = \begin{bmatrix} (-2/3)r \\ r \end{bmatrix}$. For $r = 3$, we have eigenvector

$$\mathbf{X}_2 = \begin{bmatrix} -2 \\ 3 \end{bmatrix}.$$

Normalizing the eigenvectors and forming the matrix **P** we have

$$\mathbf{P} = \begin{bmatrix} 3/\sqrt{13} & -2/\sqrt{13} \\ 2/\sqrt{13} & 3/\sqrt{13} \end{bmatrix}$$

Let $\mathbf{X} = \mathbf{PY}$ where $\mathbf{Y} = \begin{bmatrix} x' \\ y' \end{bmatrix}$; then we can rewrite the original equation as in (6) to obtain

$$[x' \quad y'] \begin{bmatrix} 9 & 0 \\ 0 & -4 \end{bmatrix} \begin{bmatrix} x' \\ y' \end{bmatrix}$$

$$+ \begin{bmatrix} -12\sqrt{13} & 0 \end{bmatrix} \begin{bmatrix} 3/\sqrt{13} & -2/\sqrt{13} \\ 2/\sqrt{13} & 3/\sqrt{13} \end{bmatrix} \begin{bmatrix} x' \\ y' \end{bmatrix} - 36 = 0$$

Performing the matrix multiplies and simplifying gives

$$9x'^2 - 4y'^2 - 36x' + 24y' - 36 = 0$$

Completing the square in x' and y' gives

$$9(x' - 2)^2 - 4(y' - 3)^2 = 36$$

Let $x'' = x' - 2$ and $y'' = y' - 3$; then we can write the equation as

$$9x''^2 - 4y''^2 = 36$$

or equivalently

$$\frac{x''^2}{4} - \frac{y''^2}{9} = 1$$

Thus this conic section is a hyperbola. The preceding is only a possible answer. The roles of x'' and y'' would be reversed if the eigenvalues of **A** were used in a different order.

29. For equation $x^2 - y^2 + 2\sqrt{3}\,xy + 6x = 0$ we have from Equation (5)

$$\mathbf{A} = \begin{bmatrix} 1 & \sqrt{3} \\ \sqrt{3} & -1 \end{bmatrix}, \quad \mathbf{B} = \begin{bmatrix} 6 & 0 \end{bmatrix}, \quad f = 0$$

The characteristic equation for matrix **A** is

$$\det(\lambda \mathbf{I}_2 - \mathbf{A}) = \begin{vmatrix} \lambda-1 & -\sqrt{3} \\ -\sqrt{3} & \lambda+1 \end{vmatrix} = \lambda^2 - 4 = (\lambda - 2)(\lambda + 2) = 0$$

Thus the eigenvalues are $\lambda_1 = 2$ and $\lambda = -2$. Next we determine an eigenvector $\mathbf{X}_1$ associated with eigenvalue λ_1:

$$\text{Solving } (\lambda_1\mathbf{I}_2 - \mathbf{A})\mathbf{X} = \begin{bmatrix} 1 & -\sqrt{3} \\ -\sqrt{3} & 3 \end{bmatrix}\begin{bmatrix} x_1 \\ x_2 \end{bmatrix} = \begin{bmatrix} 0 \\ 0 \end{bmatrix}$$

we find $\mathbf{X} = \begin{bmatrix} \sqrt{3}\,r \\ r \end{bmatrix}$. For r = 1, we have eigenvector

$$\mathbf{X}_1 = \begin{bmatrix} \sqrt{3} \\ 1 \end{bmatrix}.$$

We determine an eigenvector $\mathbf{X}_2$ associated with eigenvalue λ_2:

$$\text{Solving } (\lambda_2\mathbf{I}_2 - \mathbf{A})\mathbf{X} = \begin{bmatrix} -3 & -\sqrt{3} \\ -\sqrt{3} & -1 \end{bmatrix}\begin{bmatrix} x_1 \\ x_2 \end{bmatrix} = \begin{bmatrix} 0 \\ 0 \end{bmatrix}$$

we find $\mathbf{X} = \begin{bmatrix} (-\sqrt{3}/3)r \\ r \end{bmatrix}$. For r = 1, we have eigenvector

$$\mathbf{X}_2 = \begin{bmatrix} -\sqrt{3}/3 \\ 1 \end{bmatrix}.$$

Normalizing the eigenvectors and forming the matrix **P** we have

$$\mathbf{P} = \begin{bmatrix} \sqrt{3}/2 & -1/2 \\ 1/2 & \sqrt{3}/2 \end{bmatrix}$$

Let $\mathbf{X} = \mathbf{PY}$ where $\mathbf{Y} = \begin{bmatrix} x' \\ y' \end{bmatrix}$; then we can rewrite the original equation as in (6) to obtain

$$[x' \quad y']\begin{bmatrix} 2 & 0 \\ 0 & -2 \end{bmatrix}\begin{bmatrix} x' \\ y' \end{bmatrix} + [\,6 \quad 0\,]\begin{bmatrix} \sqrt{3}/2 & -1/2 \\ 1/2 & \sqrt{3}/2 \end{bmatrix}\begin{bmatrix} x' \\ y' \end{bmatrix} = 0$$

Performing the matrix multiplies and simplifying gives

$$2x'^2 - 2y'^2 + 3\sqrt{3}\,x' - 3y' = 0$$

Completing the square in x′ and y′ gives

$$2(x'^2 + (3/2)\sqrt{3}\,x' + 27/16) - 2(y' - (3/2)y' + 9/16) = 9/4$$

$$2(x' + \frac{3\sqrt{3}}{4})^2 - 2(y' - 3/4)^2 = 9/4$$

Let $x'' = x' - \frac{3\sqrt{3}}{4}$ and $y'' = y' + 3/4$; then we can write the equation as

$$2x''^2 - 2y''^2 = 9/4$$

or equivalently

$$\frac{x''^2}{(9/8)} - \frac{y''^2}{(9/8)} = 1$$

Thus this conic section is a hyperbola. The preceding is only a possible answer. The roles of x" and y" would be reversed if the eigenvalues of **A** were used in a different order.

Exercises 7.4

<<**Strategy:** In Exercises 1-14 we determine the matrix **A** of the quadratic surface, find its eigenvalues, and use the classifications from Table 7.2 .>>

1. For the quadric surface

$$x^2 + y^2 + 2z^2 - 2xy - 4xz - 4yz + 4x = 8$$

$\mathbf{A} = \begin{bmatrix} 1 & -1 & -2 \\ -1 & 1 & -2 \\ -2 & -2 & 2 \end{bmatrix}$. The characteristic equation is

$$\det(\lambda \mathbf{I}_3 - \mathbf{A}) = \begin{vmatrix} \lambda-1 & 1 & 2 \\ 1 & \lambda-1 & 2 \\ 2 & 2 & \lambda-2 \end{vmatrix} = \lambda^3 - 4\lambda^2 - 4\lambda + 16$$

$$= \lambda^2(\lambda - 4) - 4(\lambda - 4) = (\lambda - 4)(\lambda^2 - 4)$$
$$= (\lambda - 4)(\lambda - 2)(\lambda + 2) = 0$$

and hence the eigenvalues are $\lambda = 4, 2, -2$. Thus the inertia of **A** is (2,1,0) and it follows from Table 7.2 that this quadric surface is a hyperboloid of one sheet.

3. For the quadric surface $z = 4xy$ or equivalently

$-4xy + z = 0$, $\mathbf{A} = \begin{bmatrix} 0 & -2 & 0 \\ -2 & 0 & 0 \\ 0 & 0 & 0 \end{bmatrix}$. The characteristic equation is

$$\det(\lambda \mathbf{I}_3 - \mathbf{A}) = \begin{vmatrix} \lambda & 2 & 0 \\ 2 & \lambda & 0 \\ 0 & 0 & \lambda \end{vmatrix} = \lambda^3 - 4\lambda = \lambda(\lambda^2 - 4) = 0$$

and hence the eigenvalues are $\lambda = 2, -2, 0$. Thus the inertia of **A** is (1,1,1) and it follows from Table 7.2 that this quadric surface is a hyperbolic paraboloid.

5. For the quadric surface $x^2 - y = 0$, $\mathbf{A} = \begin{bmatrix} 1 & 0 & 0 \\ 0 & 0 & 0 \\ 0 & 0 & 0 \end{bmatrix}$. The

characteristic equation is

$$\det(\lambda \mathbf{I}_3 - \mathbf{A}) = \begin{vmatrix} \lambda-1 & 0 & 0 \\ 0 & \lambda & 0 \\ 0 & 0 & \lambda \end{vmatrix} = \lambda^2(\lambda - 1) = 0$$

and hence the eigenvalues are $\lambda = 1, 0, 0$. Thus the inertia of **A** is (1,0,2) and it follows from Table 7.2 that this quadric surface is a parabolic cylinder.

7. For the quadric surface $5y^2 + 20y + z - 23 = 0$,

$\mathbf{A} = \begin{bmatrix} 0 & 0 & 0 \\ 0 & 5 & 0 \\ 0 & 0 & 0 \end{bmatrix}$. The characteristic equation is

$$\det(\lambda\mathbf{I}_3 - \mathbf{A}) = \begin{vmatrix} \lambda & 0 & 0 \\ 0 & \lambda-5 & 0 \\ 0 & 0 & \lambda \end{vmatrix} = \lambda^2(\lambda - 5) = 0$$

and hence the eigenvalues are $\lambda = 5, 0, 0$. Thus the inertia of **A** is (1,0,2) and it follows from Table 7.2 that this quadric surface is a parabolic cylinder.

9. For the quadric surface

$$4x^2 + 9y^2 + z^2 + 8x - 18y - 4z - 19 = 0$$

$\mathbf{A} = \begin{bmatrix} 4 & 0 & 0 \\ 0 & 9 & 0 \\ 0 & 0 & 1 \end{bmatrix}$. Since **A** is diagonal, its eigenvalues are $\lambda = 9, 4, 1$. Thus the inertia of **A** is (3,0,0) and it follows from Table 7.2 that this quadric surface is an ellipsoid.

11. For the quadric surface

$$x^2 + 4y^2 + 4x + 16y - 16z - 4 = 0$$

$\mathbf{A} = \begin{bmatrix} 1 & 0 & 0 \\ 0 & 4 & 0 \\ 0 & 0 & 0 \end{bmatrix}$. Since **A** is diagonal, its eigenvalues are $\lambda = 4, 1, 0$. Thus the inertia of **A** is (2,0,1) and it follows from Table 7.2 that this quadric surface is an elliptic paraboloid.

13. For the quadric surface

$$x^2 - 4z^2 - 4x + 8z = 0$$

$$\mathbf{A} = \begin{bmatrix} 1 & 0 & 0 \\ 0 & 0 & 0 \\ 0 & 0 & -4 \end{bmatrix}.$$ Since **A** is diagonal, its eigenvalues are $\lambda = 1, -4, 0$. Thus the inertia of **A** is (1,1,1) and it follows from Table 7.2 that this quadric surface is a hyperbolic paraboloid.

<<**Strategy:** In Exercises 15-28 we proceed as above and then perform any rotations or translations that are required to obtain the standard form. If a rotation is required then we must find the eigenvectors of **A**.>>

15. For the quadric surface

$$x^2 + 2y^2 + 2z^2 + 2yz = 1$$

$$\mathbf{A} = \begin{bmatrix} 1 & 0 & 0 \\ 0 & 2 & 1 \\ 0 & 1 & 2 \end{bmatrix}.$$ The characteristic equation of **A** is

$$\det(\lambda \mathbf{I}_3 - \mathbf{A}) = \begin{vmatrix} \lambda-1 & 0 & 0 \\ 0 & \lambda-2 & -1 \\ 0 & -1 & \lambda-2 \end{vmatrix} = (\lambda - 1)(\lambda^2 - 4\lambda + 3)$$
$$= (\lambda - 1)(\lambda - 1)(\lambda - 3) = 0$$

and hence the eigenvalues are $\lambda = 3, 1, 1$. Thus the inertia of **A** is (3,0,0) and it follows from Table 7.2 that this quadric surface is an ellipsoid. Since there is a cross product term we must perform a rotation (and possibly a translation) to obtain the standard form. Hence we require the eigenvectors. Solving the appropriate homogeneous systems we have eigenvalue and eigenvector pairs

$$\lambda_1 = 3,\ \mathbf{X}_1 = \begin{bmatrix} 0 \\ 1 \\ 1 \end{bmatrix};\ \lambda_2 = 1,\ \mathbf{X}_2 = \begin{bmatrix} 1 \\ 0 \\ 0 \end{bmatrix};\ \lambda_3 = 1,\ \mathbf{X}_3 = \begin{bmatrix} 0 \\ -1 \\ 1 \end{bmatrix}.$$

Normalizing the eigenvectors we have

$$\mathbf{U}_1 = \begin{bmatrix} 0 \\ 1/\sqrt{2} \\ 1/\sqrt{2} \end{bmatrix},\ \mathbf{U}_2 = \mathbf{X}_2,\ \mathbf{U}_3 = \begin{bmatrix} 0 \\ -1/\sqrt{2} \\ 1/\sqrt{2} \end{bmatrix}$$

Let $\mathbf{P} = [\mathbf{U}_1\ \mathbf{U}_2\ \mathbf{U}_3]$. Then $\det(\mathbf{P}) = -1$, so to have a counter-clockwise rotation we will redefine the eigenvector $\mathbf{U}_2 = -\mathbf{X}_2$. (This is valid since any nonzero multiple of an eigenvector is another eigenvector.) Thus

$$\mathbf{P} = \begin{bmatrix} 0 & -1 & 0 \\ 1/\sqrt{2} & 0 & -1/\sqrt{2} \\ 1/\sqrt{2} & 0 & 1/\sqrt{2} \end{bmatrix}$$

Let $\mathbf{X} = \mathbf{PY}$ where $\mathbf{Y} = [x' \quad y' \quad z']^T$ then the quadric surface can be written as

$$(\mathbf{PY})^T\mathbf{A}(\mathbf{PY}) = x'^2 + y'^2 + 3z'^2 = 1$$

Hence the standard form is

$$\frac{x'^2}{1} + \frac{y'^2}{1} + \frac{z'^2}{(1/3)} = 1$$

Note for this problem no translations were required. The expression for the standard form is not unique. The roles of x', y', and z' may be interchanged depending upon the order of the eigenvectors used as columns in **P**.

17. For the quadric surface

$$2xz - 2x - 4y - 4z + 8 = 0$$

$\mathbf{A} = \begin{bmatrix} 0 & 0 & 1 \\ 0 & 0 & 0 \\ 1 & 0 & 0 \end{bmatrix}$. The characteristic equation of **A** is

$$\det(\lambda\mathbf{I}_3 - \mathbf{A}) = \begin{vmatrix} \lambda & 0 & -1 \\ 0 & \lambda & 0 \\ -1 & 0 & \lambda \end{vmatrix} = \lambda^3 - \lambda = \lambda(\lambda^2 - 1) = 0$$

and hence the eigenvalues are $\lambda = 1, -1, 0$. Thus the inertia of **A** is (1,1,1) and it follows from Table 7.2 that this quadric surface is a hyperbolic paraboloid. Since there is a cross product term we must perform a rotation (and possibly a translation) to obtain the standard form. Hence we require the eigenvectors. Solving the appropriate homogeneous systems we have eigenvalue and eigenvector pairs

$$\lambda_1 = 1,\ \mathbf{X}_1 = \begin{bmatrix} 1 \\ 0 \\ 1 \end{bmatrix};\ \lambda_2 = -1,\ \mathbf{X}_2 = \begin{bmatrix} 1 \\ 0 \\ -1 \end{bmatrix};\ \lambda_3 = 0,\ \mathbf{X}_3 = \begin{bmatrix} 0 \\ 1 \\ 0 \end{bmatrix}$$

Normalizing the eigenvectors we have

$$\mathbf{U}_1 = \begin{bmatrix} 1/\sqrt{2} \\ 0 \\ 1/\sqrt{2} \end{bmatrix}, \quad \mathbf{U}_2 = \begin{bmatrix} 1/\sqrt{2} \\ 0 \\ -1/\sqrt{2} \end{bmatrix}, \quad \mathbf{U}_3 = \begin{bmatrix} 0 \\ 1 \\ 0 \end{bmatrix}$$

Let $\mathbf{P} = [\mathbf{U}_1\ \mathbf{U}_2\ \mathbf{U}_3]$. Then $\det(\mathbf{P}) = 1$ so we have a counterclockwise rotation. Let $\mathbf{X} = \mathbf{PY}$ where $\mathbf{Y} = [x' \quad y' \quad z']^T$ then the quadric surface can be written as

$$(\mathbf{PY})^T\mathbf{A}(\mathbf{PY}) + [-2 \quad -4 \quad -4]\mathbf{PY} + 8 = 0$$

which simplifies to

$$x'^2 - y'^2 - 6/\sqrt{2}\ x' + 2/\sqrt{2}\ y' - 4\ z' + 8 = 0$$

Completing the square in x' and y' we have

$$(x'^2 - 6/\sqrt{2}\ x' + 9/2) - (y'^2 - 2/\sqrt{2}\ y' + 1/2) - 4z' + 8 = 9/2 - 1$$

$$(x' - 3/\sqrt{2}\ )^2 - (y' - 1/\sqrt{2}\ )^2 - 4z' + 8 = 8$$

Let $x'' = x' - 3/\sqrt{2}$, $y'' = y' - 1/\sqrt{2}$, and $z'' = z'$; then we have

$$x''^2 - y''^2 - 4z'' = 0$$

and the standard form is

$$\frac{x''^2}{4} - \frac{y''^2}{4} = z''$$

The expression for the standard form is not unique. The roles of x'', y'', and z'' may be interchanged depending upon the order of the eigenvectors used as columns in $\mathbf{P}$.

19. For the quadric surface

$$x^2 + y^2 + z^2 + 2xy = 8$$

$\mathbf{A} = \begin{bmatrix} 1 & 1 & 0 \\ 1 & 1 & 0 \\ 0 & 0 & 1 \end{bmatrix}$. The characteristic equation of $\mathbf{A}$ is

$$\det(\lambda\mathbf{I}_3 - \mathbf{A}) = \begin{vmatrix} \lambda-1 & -1 & 0 \\ -1 & \lambda-1 & 0 \\ 0 & 0 & \lambda-1 \end{vmatrix} = (\lambda - 1)^3 - (\lambda - 1)$$

$$= (\lambda - 1)(\lambda^2 - 2\lambda) = (\lambda - 1)(\lambda - 2)\lambda = 0$$

and hence the eigenvalues are $\lambda = 2, 1, 0$. Thus the inertia of **A** is (2,0,1) and it follows from Table 7.2 that this quadric surface is an elliptic paraboloid. Since there is a cross product term we must perform a rotation to obtain the standard form. Hence we require the eigenvectors. Solving the appropriate homogeneous systems we have eigenvalue and eigenvector pairs

$$\lambda_1 = 2,\ \mathbf{X}_1 = \begin{bmatrix} 1 \\ 1 \\ 0 \end{bmatrix};\ \lambda_2 = 1,\ \mathbf{X}_2 = \begin{bmatrix} 0 \\ 0 \\ 1 \end{bmatrix};\ \lambda_3 = 0,\ \mathbf{X}_3 = \begin{bmatrix} -1 \\ 1 \\ 0 \end{bmatrix}.$$

Normalizing the eigenvectors we have

$$\mathbf{U}_1 = \begin{bmatrix} 1/\sqrt{2} \\ 1/\sqrt{2} \\ 0 \end{bmatrix},\ \mathbf{U}_2 = \mathbf{X}_2,\ \mathbf{U}_3 = \begin{bmatrix} -1/\sqrt{2} \\ 1/\sqrt{2} \\ 0 \end{bmatrix}$$

Let $\mathbf{P} = [\mathbf{U}_1\ \mathbf{U}_2\ \mathbf{U}_3]$. Then $\det(\mathbf{P}) = -1$, so to have a counter-clockwise rotation we will redefine the eigenvector $\mathbf{U}_2 = -\mathbf{X}_2$. (This is valid since any nonzero multiple of an eigenvector is another eigenvector.) Thus

$$\mathbf{P} = \begin{bmatrix} 1/\sqrt{2} & 0 & -1/\sqrt{2} \\ 1/\sqrt{2} & 0 & 1/\sqrt{2} \\ 0 & -1 & 0 \end{bmatrix}$$

Let $\mathbf{X} = \mathbf{PY}$ where $\mathbf{Y} = [x'\ \ y'\ \ z']^T$; then the quadric surface can be written as

$$(\mathbf{PY})^T\mathbf{A}(\mathbf{PY}) = 2x'^2 + y'^2 = 8$$

Hence the standard form is

$$\frac{x'^2}{4} + \frac{y'^2}{8} = 1$$

Note that for this problem no translations were required. The expression for the standard form is not unique. The roles of x', y', and z' may interchanged depending upon the order of the eigenvectors used as columns in **P**.

21. For the quadric surface

$$2x^2 + 2y^2 + 4z^2 - 4xy - 8xz - 8yz + 8x = 15$$

$\mathbf{A} = \begin{bmatrix} 2 & -2 & -4 \\ -2 & 2 & -4 \\ -4 & -4 & 4 \end{bmatrix}$. The characteristic equation of **A** is

$$\det(\lambda \mathbf{I}_3 - \mathbf{A}) = \begin{vmatrix} \lambda-2 & 2 & 4 \\ 2 & \lambda-2 & 4 \\ 4 & 4 & \lambda-4 \end{vmatrix} = \lambda^3 - 8\lambda^2 - 16\lambda + 128$$
$$= \lambda^2(\lambda - 8) - 16(\lambda - 8) = (\lambda - 8)(\lambda^2 - 16)$$
$$= (\lambda - 8)(\lambda - 4)(\lambda + 4) = 0$$

and hence the eigenvalues are $\lambda = 8, 4, -4$. Thus the inertia of $\mathbf{A}$ is (2,1,0) and it follows from Table 7.2 that this quadric surface is a hyperboloid of one sheet. From the terms present we see that we must perform a rotation and a translation to obtain the standard form. Hence we require the eigenvectors. Solving the appropriate homogeneous systems we have eigenvalue and eigenvector pairs

$$\lambda_1 = 8,\ \mathbf{X}_1 = \begin{bmatrix} -1/2 \\ -1/2 \\ 1 \end{bmatrix};\ \lambda_2 = 4,\ \mathbf{X}_2 = \begin{bmatrix} -1 \\ 1 \\ 0 \end{bmatrix};\ \lambda_3 = -4,\ \mathbf{X}_3 = \begin{bmatrix} 1 \\ 1 \\ 1 \end{bmatrix}$$

Normalizing the eigenvectors we have

$$\mathbf{U}_1 = \begin{bmatrix} -1/\sqrt{6} \\ -1/\sqrt{6} \\ 1/\sqrt{3/2} \end{bmatrix},\ \mathbf{U}_2 = \begin{bmatrix} -1/\sqrt{2} \\ 1/\sqrt{2} \\ 0 \end{bmatrix},\ \mathbf{U}_3 = \begin{bmatrix} 1/\sqrt{3} \\ 1/\sqrt{3} \\ 1/\sqrt{3} \end{bmatrix}$$

Let $\mathbf{P} = [\mathbf{U}_1\ \mathbf{U}_2\ \mathbf{U}_3]$. Then $\det(\mathbf{P}) = -1$ so we do not have a counterclockwise rotation. Here we replace $\mathbf{U}_1$ by $-\mathbf{U}_1$ and redefine $\mathbf{P}$ to be

$$\mathbf{P} = \begin{bmatrix} 1/\sqrt{6} & -1/\sqrt{2} & 1/\sqrt{3} \\ 1/\sqrt{6} & 1/\sqrt{2} & 1/\sqrt{3} \\ 1/\sqrt{3/2} & 0 & 1/\sqrt{3} \end{bmatrix}$$

Then $\det(\mathbf{P}) = 1$ (verify). Let $\mathbf{X} = \mathbf{PY}$ where $\mathbf{Y} = [x' \quad y' \quad z']^T$; then the quadric surface can be written as

$$(\mathbf{PY})^T\mathbf{A}(\mathbf{PY}) + [8 \quad 0 \quad 0]\mathbf{PY} = 15$$

which simplifies to

$$8x'^2 + 4y'^2 - 4z'^2 - 8/\sqrt{6}\ x' - 8/\sqrt{2}\ y' + 8/\sqrt{3}\ z' = 15$$

Completing the square in x', y', and z' we have

$$8(x'^2 - 1/\sqrt{6}\ x' + 1/24) + 4(y'^2 - 2/\sqrt{2}\ y' + 1/2)$$
$$- 4(z'^2 - 2/\sqrt{3}\ z' + 1/3) = 8/24 + 2 - 4/3 + 15$$

An equivalent expression is

$$8(x' - 1/\sqrt{24})^2 + 4(y' - 1/\sqrt{2})^2 - 4(z' - 1/\sqrt{3})^2 = 16$$

Let $x'' = x' - 1/\sqrt{24}$, $y'' = y' - 1/\sqrt{2}$, and $z'' = z' - 1/\sqrt{3}$; then we have

$$8x''^2 + 4y''^2 - 4z''^2 = 16$$

and the standard form is

$$\frac{x''^2}{2} + \frac{y''^2}{4} - \frac{z''^2}{4} = 1$$

The expression for the standard form is not unique. The roles of x", y", and z" may interchanged depending upon the order of the eigenvectors used as columns in **P**.

23. For the quadric surface

$$2y^2 + 2z^2 + 4yz + 16/\sqrt{2}\, x + 4 = 0$$

$\mathbf{A} = \begin{bmatrix} 0 & 0 & 0 \\ 0 & 2 & 2 \\ 0 & 2 & 2 \end{bmatrix}$. The characteristic equation of **A** is

$$\det(\lambda \mathbf{I}_3 - \mathbf{A}) = \begin{vmatrix} \lambda & 0 & 0 \\ 0 & \lambda-2 & -2 \\ 0 & -2 & \lambda-2 \end{vmatrix} = \lambda((\lambda - 2)^2 - 4)$$
$$= \lambda^2(\lambda - 4) = 0$$

and thus the eigenvalues are $\lambda = 4, 0, 0$. Thus the inertia of **A** is (1,0,2) and it follows from Table 7.2 that this quadric surface is an parabolic cylinder. Since there is a cross product term we must perform a rotation to obtain the standard form. Hence we require the eigenvectors. Solving the appropriate homogeneous systems we have eigenvalue and eigenvector pairs

$$\lambda_1 = 4,\ \mathbf{X}_1 = \begin{bmatrix} 0 \\ 1 \\ 1 \end{bmatrix};\ \lambda_2 = 0,\ \mathbf{X}_2 = \begin{bmatrix} 1 \\ 0 \\ 0 \end{bmatrix};\ \lambda_3 = 0,\ \mathbf{X}_3 = \begin{bmatrix} 0 \\ -1 \\ 1 \end{bmatrix}.$$

Normalizing the eigenvectors we have

$$\mathbf{U}_1 = \begin{bmatrix} 0 \\ 1/\sqrt{2} \\ 1/\sqrt{2} \end{bmatrix}, \quad \mathbf{U}_2 = \mathbf{X}_2, \quad \mathbf{U}_3 = \begin{bmatrix} 0 \\ -1/\sqrt{2} \\ 1/\sqrt{2} \end{bmatrix}$$

Let $\mathbf{P} = [\mathbf{U}_1\ \mathbf{U}_2\ \mathbf{U}_3]$. Then $\det(\mathbf{P}) = -1$, so to have a counter-clockwise rotation we will redefine the eigenvector $\mathbf{U}_2 = -\mathbf{X}_2$. (This is valid since any nonzero multiple of an eigenvector is another eigenvector.) Thus

$$\mathbf{P} = \begin{bmatrix} 0 & -1 & 0 \\ 1/\sqrt{2} & 0 & -1/\sqrt{2} \\ 1/\sqrt{2} & 0 & 1/\sqrt{2} \end{bmatrix}$$

Let $\mathbf{X} = \mathbf{PY}$ where $\mathbf{Y} = [x'\ \ y'\ \ z']^T$; then the quadric surface can be written as

$$(\mathbf{PY})^T\mathbf{A}(\mathbf{PY}) + \begin{bmatrix} 16/\sqrt{2} & 0 & 0 \end{bmatrix}\mathbf{PY} + 4 = 0$$

or equivalently

$$4x'^2 - 16/\sqrt{2}\ y' + 4 = 0$$

Next we rearrange the equation to determine the translation required. We have

$$4x'^2 - 16/\sqrt{2}\ (y' - \sqrt{2}/4) = 0$$

Let $x'' = x'$ and $y'' = y' - \sqrt{2}/4$ and it follows that the equation is

$$4x''^2 - 16/\sqrt{2}\ y'' = 0$$

Hence the standard form is

$$x''^2 = 4/\sqrt{2}\ y''$$

The expression for the standard form is not unique. The roles of x", y", and z" may be interchanged depending upon the order of the eigenvectors used as columns in **P**.

25. For the quadric surface

$$-x^2 - y^2 - z^2 + 4xy + 4xz + 4yz + 3/\sqrt{2}\ x - 3/\sqrt{2}\ y = 6$$

Exercises 7.4

$$\mathbf{A} = \begin{bmatrix} -1 & 2 & 2 \\ 2 & -1 & 2 \\ 2 & 2 & -1 \end{bmatrix}.$$ The characteristic equation of **A** is

$$\det(\lambda \mathbf{I}_3 - \mathbf{A}) = \begin{vmatrix} \lambda+1 & -2 & -2 \\ -2 & \lambda+1 & -2 \\ -2 & -2 & \lambda+1 \end{vmatrix} = \lambda^3 + 3\lambda^2 - 9\lambda - 27$$
$$= (\lambda - 3)(\lambda + 3)^2 = 0$$

and hence the eigenvalues are $\lambda = 3, -3, -3$. Thus the inertia of **A** is (1,2,0) and it follows from Table 7.2 that this quadric surface is a hyperboloid of two sheets. From the terms present we see that we must perform a rotation and a translation to obtain the standard form. Hence we require the eigenvectors. Solving the appropriate homogeneous systems we have eigenvalue and eigenvector pairs

$$\lambda_1 = 3,\ \mathbf{X}_1 = \begin{bmatrix} 1 \\ 1 \\ 1 \end{bmatrix};\ \lambda_2 = -3,\ \mathbf{X}_2 = \begin{bmatrix} -1 \\ 1 \\ 0 \end{bmatrix};\ \lambda_3 = -3,\ \mathbf{X}_3 = \begin{bmatrix} -1 \\ 0 \\ 1 \end{bmatrix}$$

Unfortunately the eigenvectors corresponding to the eigenvalue $\lambda = -3$, which has multiplicity two, are not orthogonal. Thus we need to use the Gram-Schmidt process. For consistency of notation, let $\mathbf{Y}_1 = \mathbf{X}_1$ and proceed as follows:

let $\mathbf{Y}_2 = \mathbf{X}_2$, then define $\mathbf{Y}_3 = \mathbf{X}_3 - \dfrac{(\mathbf{X}_3, \mathbf{Y}_2)}{(\mathbf{Y}_2, \mathbf{Y}_2)} \mathbf{Y}_2 = \begin{bmatrix} -1/2 \\ -1/2 \\ 1 \end{bmatrix}$

Normalizing the eigenvectors $\mathbf{Y}_j$ we have

$$\mathbf{U}_1 = \begin{bmatrix} 1/\sqrt{3} \\ 1/\sqrt{3} \\ 1/\sqrt{3} \end{bmatrix},\ \mathbf{U}_2 = \begin{bmatrix} -1/\sqrt{2} \\ 1/\sqrt{2} \\ 0 \end{bmatrix},\ \mathbf{U}_3 = \begin{bmatrix} -1/\sqrt{6} \\ -1/\sqrt{6} \\ 1/\sqrt{3/2} \end{bmatrix}$$

Let $\mathbf{P} = [\mathbf{U}_1\ \mathbf{U}_2\ \mathbf{U}_3]$. Then $\det(\mathbf{P}) = 1$ so we do have a counter-clockwise rotation. Let $\mathbf{X} = \mathbf{PY}$ where $\mathbf{Y} = [x'\ \ y'\ \ z']^T$;then the quadric surface can be written as

$$(\mathbf{PY})^T \mathbf{A}(\mathbf{PY}) + \begin{bmatrix} 3/\sqrt{2} & -3/\sqrt{2} & 0 \end{bmatrix} \mathbf{PY} = 6$$

which simplifies to

$$3x'^2 - 3y'^2 - 3z'^2 - 3y' = 6$$

Completing the square in y' we have

$$3x'^2 - 3(y'^2 + y' + 1/4) - 3z'^2 = 6 - 3/4$$

$$3x'^2 - 3(y' + 1/2)^2 - 3z'^2 = 21/4$$

Let $x'' = x'$, $y'' = y' + 1/2$, and $z'' = z'$ then we have

$$3x''^2 - 3y''^2 - 3z''^2 = 21/4$$

and the standard form is

$$\frac{x''^2}{7/4} - \frac{y''^2}{7/4} - \frac{z''^2}{7/4} = 1$$

The expression for the standard form is not unique. The roles of x'', y'', and z'' may be interchanged depending upon the order of the eigenvectors used as columns in **P**.

27. For the quadric surface

$$x^2 + y^2 - z^2 - 2x - 4y - 4z + 1 = 0$$

$\mathbf{A} = \begin{bmatrix} 1 & 0 & 0 \\ 0 & 1 & 0 \\ 0 & 0 & -1 \end{bmatrix}$. The characteristic equation is

$$\det(\lambda \mathbf{I}_3 - \mathbf{A}) = \begin{vmatrix} \lambda-1 & 0 & 0 \\ 0 & \lambda-1 & 0 \\ 0 & 0 & \lambda+1 \end{vmatrix} = (\lambda - 1)^2(\lambda + 1) = 0$$

and hence the eigenvalues are $\lambda = 1, 1, -1$. Thus the inertia of **A** is (2,1,0) and it follows from Table 7.2 that this quadric surface is a hyperboloid of one sheet. Since there are no cross product terms we do not need the eigenvectors. We merely complete the square in each of the variables:

$$(x^2 - 2x + 1) + (y^2 - 4y + 4) - (z^2 + 4z + 4) = -1 + 1 + 4 - 4$$

$$(x - 1)^2 + (y - 2)^2 - (z + 2)^2 = 0$$

Let $x' = x - 1$, $y' = y - 2$, and $z' = z + 2$; then we have the standard form

$$x''^2 + y''^2 - z''^2 = 0$$

This is really a cone which is a special case of a hyperboloid of one sheet.

Exercises 7.5

1.

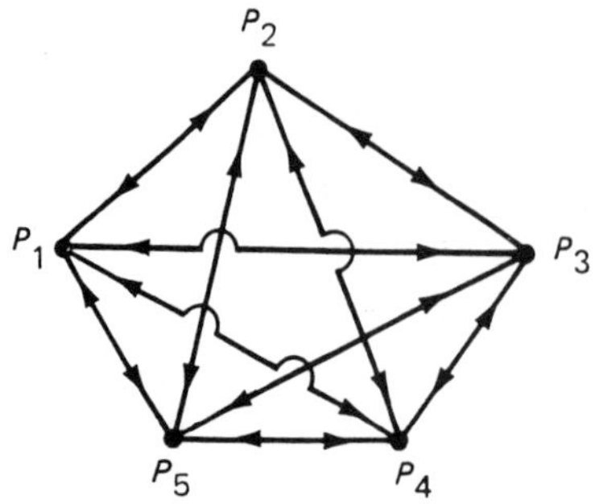

3. The statement in the hint is equivalent to the following:

> If the number of edges emanating from corresponding vertices is not the same, then the two digraphs are not equal.

The following table compares the number of edges emanating from the corresponding vertices in each of the digraphs.

vertex	P_1	P_2	P_3	P_4	P_5
edges in (3a)	2	0	2	3	1
edges in (3b)	2	0	3	2	0
edges in (3c)	2	0	2	3	1

Since the entries in the table are different for (3b) as compared to (3a) and (3c), digraph (3b) is not the same as digraphs (3a) or (3c). It follows that digraphs (3a) and (3c) are equal.

5. (a) The adjacency matrix for digraph (2a) is

$$\begin{array}{c} \\ P_1 \\ P_2 \\ P_3 \\ P_4 \\ P_5 \end{array}\begin{array}{c} \begin{array}{ccccc} P_1 & P_2 & P_3 & P_4 & P_5 \end{array} \\ \begin{bmatrix} 0 & 1 & 0 & 0 & 0 \\ 1 & 0 & 1 & 0 & 1 \\ 1 & 0 & 0 & 1 & 0 \\ 0 & 1 & 0 & 0 & 0 \\ 0 & 0 & 0 & 1 & 0 \end{bmatrix} \end{array}.$$

(b) The adjacency matrix for digraph (2b) is

$$\begin{array}{c} \\ P_1 \\ P_2 \\ P_3 \\ P_4 \\ P_5 \\ P_6 \end{array}\begin{array}{c} \begin{array}{cccccc} P_1 & P_2 & P_3 & P_4 & P_5 & P_6 \end{array} \\ \begin{bmatrix} 0 & 1 & 1 & 0 & 0 & 0 \\ 1 & 0 & 0 & 1 & 0 & 0 \\ 0 & 1 & 0 & 0 & 0 & 0 \\ 0 & 0 & 1 & 0 & 1 & 1 \\ 0 & 0 & 1 & 1 & 0 & 1 \\ 1 & 0 & 0 & 0 & 0 & 0 \end{bmatrix} \end{array}.$$

7. The adjacency matrix **A** of a dominance digraph is such that for $i \neq j$ either $a_{ij} = 1$ or $a_{ji} = 1$, but not both.

   (a) This adjacency matrix does not represent a dominance digraph since $a_{14} = a_{41} = 0$.

   (b) This adjaceny matrix represents a dominance digraph.

9. (a) P_2 has access only to P_4 and P_5. Similarly, P_4 does not have access to P_1, but P_5 does. Hence P_2 has access to P_1 through one individual in only one way: $P_2 \to P_5 \to P_1$.

   (b) From the row corresponding to P_2, we have that P_2 can access P_4 and P_5 directly. From the row corresponding to P_4, we have that P_4 can access only P_5: thus

$$P_2 \to P_4 \to P_5. \qquad (*)$$

Similarly, from the row corresponding to P_5 we have

$$P_2 \to P_5 \to P_1 \quad \text{and} \quad P_2 \to P_5 \to P_3. \qquad (**)$$

Next we have from (*)

$$P_2 \to P_4 \to P_5 \to P_1 \quad \text{and} \quad P_2 \to P_4 \to P_5 \to P_3.$$

Also from (**)

$$\begin{array}{l} P_2 \to P_5 \to P_1 \to P_2 \\ P_2 \to P_5 \to P_1 \to P_3 \\ P_2 \to P_5 \to P_1 \to P_4 \\ P_2 \to P_5 \to P_1 \to P_5 \end{array}$$

and

$$\begin{array}{l} P_2 \to P_5 \to P_3 \to P_2 \\ P_2 \to P_5 \to P_3 \to P_4. \end{array}$$

Thus we have that the smallest number of individuals through which P_2 can access himself is two:

Exercises 7.5

$$P_2 \to P_5 \to P_1 \to P_2$$
$$P_2 \to P_5 \to P_3 \to P_2.$$

<<**Strategy:** For Exercises 11 and 13, we follow the procedure for determining a clique in a digraph that is discussed following Theorem 7.5 and used in Examples 17 and 18.>>

11. Let $\mathbf{A} = \begin{bmatrix} 0 & 0 & 0 & 0 & 0 \\ 1 & 0 & 1 & 1 & 1 \\ 0 & 1 & 0 & 1 & 0 \\ 1 & 1 & 1 & 0 & 0 \\ 0 & 0 & 1 & 1 & 0 \end{bmatrix}$. Then the symmetric matrix $\mathbf{S}$ associated with $\mathbf{A}$ is

$$\mathbf{S} = \begin{bmatrix} 0 & 0 & 0 & 0 & 0 \\ 0 & 0 & 1 & 1 & 0 \\ 0 & 1 & 0 & 1 & 0 \\ 0 & 1 & 1 & 0 & 0 \\ 0 & 0 & 0 & 0 & 0 \end{bmatrix} \text{ and } \mathbf{S}^3 = \begin{bmatrix} 0 & 0 & 0 & 0 & 0 \\ 0 & 2 & 3 & 3 & 0 \\ 0 & 3 & 2 & 3 & 0 \\ 0 & 3 & 3 & 2 & 0 \\ 0 & 0 & 0 & 0 & 0 \end{bmatrix}.$$

Since $s_{22}^{(3)} = 2$, $s_{33}^{(3)} = 2$, $s_{44}^{(3)} = 2$, it follows that P_2, P_3, and P_4 form a clique.

13. Let $\mathbf{A} = \begin{bmatrix} 0 & 1 & 1 & 0 & 1 \\ 1 & 0 & 1 & 1 & 1 \\ 0 & 0 & 0 & 1 & 0 \\ 1 & 0 & 0 & 0 & 1 \\ 0 & 1 & 0 & 1 & 0 \end{bmatrix}$. Then the symmetric matrix $\mathbf{S}$ associated with $\mathbf{A}$ is

$$\mathbf{S} = \begin{bmatrix} 0 & 1 & 0 & 0 & 0 \\ 1 & 0 & 0 & 0 & 1 \\ 0 & 0 & 0 & 0 & 0 \\ 0 & 0 & 0 & 0 & 1 \\ 0 & 1 & 0 & 1 & 0 \end{bmatrix} \text{ and } \mathbf{S}^3 = \begin{bmatrix} 0 & 2 & 0 & 1 & 0 \\ 2 & 0 & 0 & 0 & 3 \\ 0 & 0 & 0 & 0 & 0 \\ 0 & 1 & 0 & 1 & 0 \\ 0 & 3 & 0 & 2 & 0 \end{bmatrix}.$$

Since $s_{44}^{(3)} = 1$ and no other diagonal element is positive, this digraph has no clique.

15. Using Theorem 7.6, we compute $\mathbf{A} + \mathbf{A}^2 + \cdots + \mathbf{A}^{n-1} = \mathbf{E}$.

(a) $\mathbf{E} = \mathbf{A} + \mathbf{A}^2 + \mathbf{A}^3 = \begin{bmatrix} 3 & 2 & 6 & 6 \\ 2 & 1 & 4 & 4 \\ 3 & 1 & 4 & 4 \\ 1 & 1 & 3 & 1 \end{bmatrix}$. Since all the entries of $\mathbf{E}$ are nonzero, the digraph is strongly connected.

(b) $\mathbf{E} = \mathbf{A} + \mathbf{A}^2 + \mathbf{A}^3 + \mathbf{A}^4 = \begin{bmatrix} 1 & 2 & 1 & 0 & 0 \\ 1 & 1 & 2 & 0 & 0 \\ 2 & 1 & 1 & 0 & 0 \\ 1 & 1 & 2 & 2 & 2 \\ 2 & 1 & 3 & 2 & 2 \end{bmatrix}$. Since **E** has zero entries, the digraph is not strongly connected.

17. Let A_k be the submatrix of the adjacency matrix obtained by deleting the kth row and kth column. We determine if there is a path of some length between every distinct person. That is, determine if there is a positive integer m such that $\mathbf{A}_k + \mathbf{A}_k^2 + \cdots + \mathbf{A}_k^m$ has all its nondiagonal entries different than zero.

$\mathbf{A}_1 = \begin{bmatrix} 0 & 0 & 0 & 1 \\ 0 & 0 & 1 & 0 \\ 0 & 0 & 0 & 1 \\ 1 & 1 & 0 & 0 \end{bmatrix}$ and m = 3, thus P_1 can be left out.

$\mathbf{A}_2 = \begin{bmatrix} 0 & 1 & 1 & 0 \\ 1 & 0 & 1 & 0 \\ 1 & 0 & 0 & 1 \\ 0 & 1 & 0 & 0 \end{bmatrix}$ and m = 2, thus P_2 can be left out.

$\mathbf{A}_3 = \begin{bmatrix} 0 & 0 & 1 & 0 \\ 1 & 0 & 0 & 1 \\ 1 & 0 & 0 & 1 \\ 0 & 1 & 0 & 0 \end{bmatrix}$ and m = 3, thus P_3 can be left out.

$\mathbf{A}_4 = \begin{bmatrix} 0 & 0 & 1 & 0 \\ 1 & 0 & 0 & 1 \\ 1 & 0 & 0 & 0 \\ 0 & 1 & 1 & 0 \end{bmatrix}$ and there is no value of m that works, thus P_4 cannot be left out.

$\mathbf{A}_5 = \begin{bmatrix} 0 & 0 & 1 & 1 \\ 1 & 0 & 0 & 0 \\ 1 & 0 & 0 & 1 \\ 1 & 0 & 0 & 0 \end{bmatrix}$ and there is no value of m that works, thus P_5 cannot be left out.

T.1. In a dominance digraph, for each i and j, it is not the case that both P_i dominates P_j and P_j dominates P_i.

Exercises 7.5

T.3. The implication in one direction is proved in the discussion following the theorem. Next suppose P_i belongs to the clique $\{P_i, P_j, P_k, \cdots, P_m\}$. According to the definition of clique, it contains at least three vertices so we may assume P_i, P_j, and P_k all exist in the clique. Then $s_{ij} = s_{ji} = s_{jk} = s_{kj} = s_{ik} = s_{ki} = 1$ and $s_{ii}^{(3)}$ is a sum of nonnegative integer terms including the positive term which represents three stage access from P_i to P_j to P_k to P_i. Thus $s_{ii}^{(3)}$ is positive.

<<**Strategy:** For ML.1 and ML.2 we will use MATLAB code to compute the matrix S. After you enter the adjacency matrix in ML.1 then enter the following two lines into MATLAB.

```
S=zeros(A); [k,m]=size(A); for i=1:k, for j=1:k,
if A(i,j)==1 & A(j,i)==1 & j~=i, S(i,j)=1; S(j,i)=1;end,end,end,S
```

The preceding code can be recalled using the up-arrow when doing ML.2.

ML.1.
```
A=[0 0 0 0 0;1 0 1 1 1;0 1 0 1 0;1 1 1 0 0;0 0 1 1 0];
S=zeros(A);[k,m]=size(A);for i=1:k,for j=1:k,
if A(i,j)==1&A(j,i)==1 & j~=i,S(i,j)=1;S(j,i)=1;end,end,end,S

S =

     0     0     0     0     0
     0     0     1     1     0
     0     1     0     1     0
     0     1     1     0     0
     0     0     0     0     0
```

Next we compute S3 as follows.

```
S^3

ans =

     0     0     0     0     0
     0     2     3     3     0
     0     3     2     3     0
     0     3     3     2     0
     0     0     0     0     0
```

It follows that P2, P3, and P4 form a clique.

ML.2.
```
A=[0 1 1 0 1;1 0 0 1 0;0 1 0 0 1;0 1 1 0 1;1 0 0 1 0];
```

(Using the up-arrow, recall the following lines.)

```
S=zeros(A);[k,m]=size(A);for i=1:k,for j=1:k,
if A(i,j)==1&A(j,i)==1 & j~=i,S(i,j)=1;S(j,i)=1;end,end,end,S
```

```
S =

     0     1     0     0     1
     1     0     0     1     0
     0     0     0     0     0
     0     1     0     0     1
     1     0     0     1     0
```

Next we compute S3.

```
S^3

ans =

     0     4     0     0     4
     4     0     0     4     0
     0     0     0     0     0
     0     4     0     0     4
     4     0     0     4     0
```

Since no diagonal entry is different than zero, there are no cliques.

ML.3. We use Theorem 7.6.

(a) `A = [0 0 1 1 1;1 0 1 1 0;0 1 0 0 0;0 1 0 0 1;1 1 0 0 0];`

Here n = 5, so we form

```
E=A+A^2+A^3+A^4

E =

     7    13    10    10    11
    10    11    12    12     9
     5     7     4     4     3
     8    13     7     7     8
    11    11     9     9     6
```

Since E has no zero entries the digraph represented by A is strongly connected.

(b) `A=[0 0 0 0 1;0 0 1 1 0;0 1 0 0 1;1 0 0 0 0;0 0 0 1 0];`

Here n = 5, so we form

```
E=A+A^2+A^3+A^4

E =

     1     0     0     1     2
     3     2     2     4     3
     2     2     2     4     4
     2     0     0     1     1
     1     0     0     2     1
```

Exercises 7.5

Since E has zero entries the digraph represented by A is not strongly connected.

Exercises 7.6

1.

		C: 2 fingers shown	C: 3 fingers shown
R	2 fingers shown	-4	5
R	3 fingers shown	5	-6

3.

		B: Abington	B: Wyncote
A	Abington	50	60
A	Wyncote	25	50

5. As in Example 3, we determine the row minima and column maxima.

(a)

			Row minima
	5	4	4
	3	-2	-2
Column maxima	5	4	

Since 4 is both a row minimum and a column maximum, it is a saddle point.

(b)

				Row minima
	2	1	0	0
	3	1	-2	-2
	4	2	-4	-4
Column maxima	4	2	0	

Since 0 is both a row minimum and a column maximum, it is a saddle point.

(c)

				Row minima
	3	4	5	3
	-2	5	1	-2
	-1	0	1	-1
Column maxima	3	5	5	

Since 3 is both a row minimum and a column maximum, it is a saddle point.

(d)

					Row minima
	5	2	4	2	2
	0	-1	2	0	-1
	3	2	3	2	2
	1	0	-1	-1	-1
Column maxima	5	2	4	2	

Since the 2's (there are four of them) are both a row minimum and a column maximum, each is a saddle point.

7. For a strictly determined game the optimal strategy for each player is to choose the move represented by the saddle point. Then both players choose with probability 1 the row and column representing their respective moves.

(a) Determine the saddle point v.

				Row minima
	2	1	3	1
	-2	0	2	-2
Column maxima	2	1	3	

Since 1 is both a row minimum and a column maximum it is a saddle point. Thus R should make the first move and C the second move (v = 1 is the 1,2 element). It follows that the optimal strategy for R is $\mathbf{P} = \begin{bmatrix} 1 & 0 \end{bmatrix}$ and the optimal strategy for C is $\mathbf{Q} = \begin{bmatrix} 0 \\ 1 \\ 0 \end{bmatrix}$. The payoff for R is the value of the saddle point, v = 1.

(b) Determine the saddle point v.

					Row minima
	-2	-2	4	5	-2
	-2	-2	1	0	-2
	0	1	1	2	0
Column maxima	0	1	4	5	

Since the 0 in the 3,1 element is both a row minimum and a column maximum, it is a saddle point. Thus R should make the third move and C the first move. It follows that the optimal strategy for R is $\mathbf{P} = \begin{bmatrix} 0 & 0 & 1 \end{bmatrix}$ and the optimal strategy for C is $\mathbf{Q} = \begin{bmatrix} 1 \\ 0 \\ 0 \\ 0 \end{bmatrix}$. The payoff for R is the value of the saddle point, v = 0. This game is a fair game.

(c) Determine the saddle point.

$$\begin{array}{lcc|c} & & & \textbf{Row minima} \\ & \left[\begin{matrix} 6 \\ 7 \end{matrix}\right. & \left.\begin{matrix} 4 \\ 4 \end{matrix}\right] & \begin{matrix} 4 \\ 4 \end{matrix} \\ \textbf{Column maxima} & 7 & 4 & \end{array}$$

Since the 4 in the 1,2 element and the 4 in the 2,2 element are both a row minimum and a column maximum each is a saddle point. There are two optimal strategies:

(i) R should make the first move and C the second move.

Then $\mathbf{P} = \begin{bmatrix} 1 & 0 \end{bmatrix}$ and $\mathbf{Q} = \begin{bmatrix} 0 \\ 1 \end{bmatrix}$.

(ii) R should make the second move and C the second move.

Then $\mathbf{P} = \begin{bmatrix} 0 & 1 \end{bmatrix}$ and $\mathbf{Q} = \begin{bmatrix} 0 \\ 1 \end{bmatrix}$.

In either case the payoff is v = 4.

9. Let $\mathbf{A} = \begin{bmatrix} 3 & -3 \\ 2 & 5 \\ 1 & 0 \end{bmatrix}$.

(a) $E(\mathbf{P},\mathbf{Q}) = \mathbf{PAQ} = \begin{bmatrix} 1/2 & 1/3 & 1/6 \end{bmatrix} \begin{bmatrix} 3 & -3 \\ 2 & 5 \\ 1 & 0 \end{bmatrix} \begin{bmatrix} 1/6 \\ 5/6 \end{bmatrix} = 19/36$

(b) Let $\mathbf{P} = \begin{bmatrix} 0 & 0 & 1 \end{bmatrix}$ and $\mathbf{Q} = \begin{bmatrix} 1/7 \\ 6/7 \end{bmatrix}$. Then $E(\mathbf{P},\mathbf{Q}) = \mathbf{PAQ} = 1/7$.

11. Let $\mathbf{A} = \begin{bmatrix} -3 & 2 \\ 4 & -5 \end{bmatrix}$. Then the optimal strategy for R is obtained from Equation (8):

$$p_1 = \frac{-5-4}{-3+(-5)-2-4} = 9/14 \quad \text{and} \quad p_2 = \frac{-3-2}{-3+(-5)-2-4} = 5/14.$$

The value of the game is obtained from Equation (9):

$$v = \frac{(-3)(-5)-(2)(4)}{-3+(-5)-2-4} = -1/2.$$

The optimal strategy for C is obtained from Equation (13):

$$q_1 = \frac{-5-2}{-3+(-5)-2-4} = 1/2 \quad \text{and } q_2 = \frac{-3-4}{-3+(-5)-2-4} = 1/2.$$

13. Follow Example 10. Here we add 4 to each entry of the payoff matrix $\begin{bmatrix} 2 & -3 & 4 \\ 4 & 0 & 1 \\ 3 & 2 & -2 \end{bmatrix}$ to obtain $\mathbf{A} = \begin{bmatrix} 6 & 1 & 8 \\ 8 & 4 & 5 \\ 7 & 6 & 2 \end{bmatrix}$, which has positive entries. From (18), the linear programming problem for C is

$$\begin{aligned} &\text{Maximize} \quad x_1 + x_2 + x_3 \\ &\text{subject to} \\ &\quad 6x_1 + x_2 + 8x_3 \le 1 \\ &\quad 8x_1 + 4x_2 + 5x_3 \le 1 \\ &\quad 7x_1 + 6x_2 + 2x_3 \le 1 \\ &\quad x_1 \ge 0,\ x_2 \ge 0,\ x_3 \ge 0 \end{aligned}$$

where $x_i = \frac{q_i}{v}$ and $x_1 + x_2 + x_3 = \frac{1}{v}$. Reformulating the problem using slack variables we have

$$\begin{aligned} &\text{Maximize} \quad x_1 + x_2 + x_3 \\ &\text{subject to} \\ &\quad 6x_1 + x_2 + 8x_3 + x_4 = 1 \\ &\quad 8x_1 + 4x_2 + 5x_3 + x_5 = 1 \\ &\quad 7x_1 + 6x_2 + 2x_3 + x_6 = 1 \\ &\quad x_1 \ge 0,\ x_2 \ge 0,\ x_3 \ge 0,\ x_4 \ge 0,\ x_5 \ge 0,\ x_6 \ge 0. \end{aligned}$$

Form the initial tableau:

	x_1	x_2	x_3 ↓	x_4	x_5	x_6	z	
← x_4	6	1	8	1	0	0	0	1
x_5	8	4	5	0	1	0	0	1
x_6	7	6	2	0	0	1	0	1
	-1	-1	-1	0	0	0	1	0

Determine the pivotal column and pivotal row. The most negative entry in the objective row is -1. We choose the "x_3"-column as the pivotal column. Thus the θ-ratios are {1/8 1/5 1/2} and 1/8 is the smallest. Thus the pivotal row is the "x_4"-row. It follows that the pivot has value 8. This information is marked on the preceding tableau. Next perform pivotal elimination. Take $(1/8)\mathbf{r}_1$, $-5\mathbf{r}_1+\mathbf{r}_2$, $-2\mathbf{r}_1+\mathbf{r}_3$, $\mathbf{r}_1+\mathbf{r}_4$, and relabel the departing variable with the entering variable name. The resulting tableau is:

	x_1	↓ x_2	x_3	x_4	x_5	x_6	z	
x_3	3/4	1/8	1	1/8	0	0	0	1/8
← x_5	17/4	27/8	0	-5/8	1	0	0	3/8
x_6	11/2	23/4	0	-1/4	0	1	0	3/4
	-1/4	-7/8	0	1/8	0	0	1	1/8

Since there is a negative entry in the objective row, we do not have an optimal solution. We repeat the process.

Determine the pivotal column and pivotal row. The most negative entry in the objective row is -7/8, hence the "x_2"-column is the pivotal column. Thus the θ-ratios are {1, 1/9, 3/23} and 1/9 is the smallest. Hence the pivotal row is the "x_5"-row. It follows that the pivot has value 27/8. This information is marked on the preceding tableau. Next perform pivotal elimination. Take $(8/27)\mathbf{r}_2$, $(-1/8)\mathbf{r}_2+\mathbf{r}_1$, $(-23/4)\mathbf{r}_2+\mathbf{r}_3$, $(7/8)\mathbf{r}_2+\mathbf{r}_4$, and relabel the departing variable with the entering variable name. The resulting tableau is:

	x_1	x_2	x_3	↓ x_4	x_5	x_6	z	
x_3	16/27	0	1	4/27	-1/27	0	0	1/9
x_2	34/27	1	0	-5/27	8/27	0	0	1/9
← x_6	-47/27	0	0	22/27	-46/27	1	0	1/9
	23/27	0	0	-1/27	7/27	0	1	2/9

Since there is a negative entry in the objective row, we do not have the optimal solution. We repeat the process.

Determine the pivotal column and pivotal row. The most negative entry in the objective row is -1/27, hence the "x_4"-column as the pivotal column. Thus the θ-ratios are {3/4, 3/22} and 3/22 is the smallest. Hence the pivotal row is the "x_6"-row. It follows that the pivot has value 22/27. This information is marked on the preceding tableau. Next perform pivotal elimination. Take $(27/22)\mathbf{r}_3$, $(-4/27)\mathbf{r}_3+\mathbf{r}_1$, $(5/27)\mathbf{r}_3+\mathbf{r}_2$, $(1/27)\mathbf{r}_3+\mathbf{r}_4$, and relabel the departing variable with the entering variable name. The resulting tableau is:

	x_1	x_2	x_3	x_4	x_5	x_6	z	
x_3	82/297	0	1	0	3/11	-2/11	0	1/11
x_2	19/22	1	0	0	-1/11	5/22	0	3/22
x_4	-47/22	0	0	1	-23/11	27/22	0	3/22
	17/22	0	0	0	2/11	1/22	1	5/22

Since all the entries of the objective row are nonnegative, we have the optimal solution,

$$x_1 = 0, \quad x_2 = 3/22, \quad x_3 = 1/11.$$

The maximum value of $v = 1/(5/22) = 22/5$. Hence

$$q_1 = x_1 \cdot v = 0,$$
$$q_2 = x_2 \cdot v = (3/22)(22/5) = 3/5,$$
$$q_3 = x_3 \cdot v = (1/11)(22/5) = 2/5.$$

Thus an optimal strategy for C is $\mathbf{Q} = \begin{bmatrix} 0 \\ 3/5 \\ 2/5 \end{bmatrix}$. An optimal strategy for R is found in the objective row under the columns of the slack variables. Thus,

$$y_1 = 0, \; y_2 = 2/11, \; y_3 = 1/22$$

and

$$p_1 = y_1 \cdot v = 0,$$
$$p_2 = y_2 \cdot v = 4/5,$$
$$p_3 = y_3 \cdot v = 1/5.$$

Hence an optimal strategy for R is $\mathbf{P} = \begin{bmatrix} 0 & 4/5 & 1/5 \end{bmatrix}$.

15. Let $\mathbf{A} = \begin{bmatrix} 0 & -4 & 3 & 0 \\ 2 & -3 & 4 & 1 \\ -1 & 2 & 2 & 2 \\ 1 & -4 & 3 & 0 \end{bmatrix}$. Examine **A** for a recessive row or a recessive column. Each element of row 1 is less than or equal to the corresponding element of row 2. Thus we can drop row 1. The new matrix is

$$\mathbf{A}_1 = \begin{bmatrix} 2 & -3 & 4 & 1 \\ -1 & 2 & 2 & 2 \\ 1 & -4 & 3 & 0 \end{bmatrix}.$$

Each element of column 3 is greater than or equal to the corresponding element in column 4. Thus we can drop column 3. The new matrix is

$$\mathbf{A}_2 = \begin{bmatrix} 2 & -3 & 1 \\ -1 & 2 & 2 \\ 1 & -4 & 0 \end{bmatrix}.$$

Each element of column 3 is greater than or equal to the corresponding element in column 2. Thus we can drop column 3. The new matrix is

$$\mathbf{A}_3 = \begin{bmatrix} 2 & -3 \\ -1 & 2 \\ 1 & -4 \end{bmatrix}.$$

Each element of row 3 is less than the corresponding element in row 1. Thus we can drop row 3. The new matrix is

$$\mathbf{A}_4 = \begin{bmatrix} 2 & -3 \\ -1 & 2 \end{bmatrix}.$$

We inspect $\mathbf{A}_4$ for a saddle point:

			Row minima
	2	-3	-3
	-1	2	-1
Column maxima	2	2	

$\mathbf{A}_4$ has no saddle point, but we can use Equations (8), (9), and (13). We obtain

$$p_1 = 3/8, \; p_2 = 5/8,$$

thus the strategy for R is $\mathbf{P} = \begin{bmatrix} 3/8 & 5/8 \end{bmatrix}$. Also,

$$q_1 = 5/8, \; q_2 = 3/8,$$

thus the strategy for C is $\mathbf{Q} = \begin{bmatrix} 5/8 \\ 3/8 \end{bmatrix}$. The value of the game is $v = 1/8$. To form $\mathbf{A}_4$ we dropped rows 1 and 4, hence for the original game $\mathbf{P} = \begin{bmatrix} 0 & 3/8 & 5/8 & 0 \end{bmatrix}$. Similarly, since columns 3 and 4 were dropped, $\mathbf{Q} = \begin{bmatrix} 5/8 \\ 3/8 \\ 0 \\ 0 \end{bmatrix}$. The value of the original game is the same as the game using matrix $\mathbf{A}_4$.

17\. The payoff matrix from Exercise 2 is $\begin{bmatrix} 0 & 1 & -1 \\ -1 & 0 & 1 \\ 1 & -1 & 0 \end{bmatrix}$. Proceed as in Example 10. Add 2 to each entry so that all the entries are positive. We have

$$\mathbf{A} = \begin{bmatrix} 2 & 3 & 1 \\ 1 & 2 & 3 \\ 3 & 1 & 2 \end{bmatrix}.$$

The linear programming problem for C is

$$\begin{array}{ll} \text{Maximize} & x_1 + x_2 + x_3 \\ \text{subject to} & \\ & 2x_1 + 3x_2 + x_3 \le 1 \\ & x_1 + 2x_2 + 3x_3 \le 1 \\ & 3x_1 + x_2 + 2x_3 \le 1 \\ & x_1 \ge 0,\ x_2 \ge 0,\ x_3 \ge 0 \end{array}$$

where $x_i = \frac{q_i}{v}$ and $x_1 + x_2 + x_3 = \frac{1}{v}$. Reformulate the problem using slack variables and we have

$$\begin{array}{ll} \text{Maximize} & x_1 + x_2 + x_3 \\ \text{subject to} & \\ & 2x_1 + 3x_2 + x_3 + x_4 = 1 \\ & x_1 + 2x_2 + 3x_3 + x_5 = 1 \\ & 3x_1 + x_2 + 2x_3 + x_6 = 1 \\ & x_1 \ge 0,\ x_2 \ge 0,\ x_3 \ge 0,\ x_4 \ge 0,\ x_5 \ge 0,\ x_6 \ge 0. \end{array}$$

Form the initial tableau:

	x_1	x_2	↓ x_3	x_4	x_5	x_6	z	
x_4	2	3	1	1	0	0	0	1
← x_5	1	2	3	0	1	0	0	1
x_6	3	1	2	0	0	1	0	1
	-1	-1	-1	0	0	0	1	0

Determine the pivotal column and pivotal row. The most negative entry in the objective row is -1. We choose the "x_3"-column as the pivotal column. Thus the θ-ratios are {1 1/3 1/2} and 1/3 is the smallest. Hence the pivotal row is the "x_5"-row. It follows that the pivot has value 3. This information is marked on the preceding tableau. Next perform pivotal elimination. Take $(1/3)\mathbf{r}_2$, $-\mathbf{r}_2+\mathbf{r}_1$, $-2\mathbf{r}_2+\mathbf{r}_3$, $\mathbf{r}_2+\mathbf{r}_4$, and relabel the departing variable with the entering variable name. The resulting tableau is:

	↓ x_1	x_2	x_3	x_4	x_5	x_6	z	
x_4	5/3	7/3	0	1	-1/3	0	0	2/3
x_3	1/3	2/3	1	0	1/3	0	0	1/3
← x_6	7/3	-1/3	0	0	-2/3	1	0	1/3
	-2/3	-1/3	0	0	1/3	0	1	1/3

Since there is a negative entry in the objective row, we do not have the optimal solution. We repeat the process.

Determine the pivotal column and pivotal row. The most negative entry in the objective row is -2/3. We choose the "x_1"-column as the pivotal column. Thus the θ-ratios are {2/5, 1, 1/7} and 1/7 is the smallest. Hence the pivotal row is the "x_6"-row. It follows that the pivot has value 7/3. This information is marked on the preceding tableau. Next perform pivotal elimination. Take $(3/7)\mathbf{r}_3$, $(-5/3)\mathbf{r}_3+\mathbf{r}_1$, $(-1/3)\mathbf{r}_3+\mathbf{r}_2$, $(2/3)\mathbf{r}_3+\mathbf{r}_4$, and relabel the departing variable with the entering variable name. The resulting tableau is:

	x_1	↓ x_2	x_3	x_4	x_5	x_6	z	
← x_4	0	18/7	0	1	1/7	-5/7	0	3/7
x_3	0	5/7	1	0	3/7	-1/7	0	2/7
x_1	1	-1/7	0	0	-2/7	3/7	0	1/7
	0	-3/7	0	0	1/7	2/7	1	3/7

Since there is a negative entry in the objective row, we do not have the optimal solution. We repeat the process.

Determine the pivotal column and pivotal row. The most negative entry in the objective row is -3/7. We choose the "x_2"-column as the pivotal column. Thus the θ-ratios are {1/6, 2/5} and 1/6 is the smallest. Hence the pivotal row is the "x_4"-row. It follows that the pivot has value 18/7. This information is marked on the preceding tableau. Next perform pivotal elimination. Take $(7/18)\mathbf{r}_1$, $(-5/7)\mathbf{r}_1+\mathbf{r}_2$, $(1/7)\mathbf{r}_1+\mathbf{r}_3$, $(3/7)\mathbf{r}_1+\mathbf{r}_4$, and relabel the departing variable with the entering variable name. The resulting tableau is:

	x_1	x_2	x_3	x_4	x_5	x_6	z	
x_2	0	1	0	7/18	1/18	-5/18	0	1/6
x_3	0	0	1	-5/18	7/18	1/18	0	1/6
x_1	1	0	0	1/18	37/126	7/18	0	1/6
	0	0	0	1/6	1/6	1/6	1	1/2

We have the optimal solution to the linear programming problem:

$$x_1 = 1/6, \; x_2 = 1/6, \; x_3 = 1/6$$

and

$$\text{the maximum value of } x_1 + x_2 + x_3 = 1/2.$$

It follows that $v = 2$ and

$$q_1 = x_1 \cdot v = 1/3,$$
$$q_2 = x_2 \cdot v = 1/3,$$
$$q_3 = x_3 \cdot v = 1/3.$$

Hence an optimal strategy for C is $\mathbf{Q} = \begin{bmatrix} 1/3 \\ 1/3 \\ 1/3 \end{bmatrix}$. An optimal strategy for R is found in the objective row under the columns corresponding to the slack variables. We have

$$y_1 = 1/6, \; y_2 = 1/6, \; y_3 = 1/6,$$

hence

$$p_1 = y_1 \cdot v = 1/3,$$
$$p_2 = y_2 \cdot v = 1/3,$$
$$p_3 = y_3 \cdot v = 1/3.$$

An optimal strategy for R is $\mathbf{P} = \begin{bmatrix} 1/3 & 1/3 & 1/3 \end{bmatrix}$. The value of the original game is $v - 2 = 0$.

19. The payoff matrix for Exercise 4 is $\begin{bmatrix} -5 & 5 \\ 10 & -10 \end{bmatrix}$. We attempt to find a saddle point:

			Row minima
	-5	5	-5
	10	-10	-10
Column maxima	10	5	

Hence there is no saddle point, so we use Equations (8), (9), and (13). We obtain

$$p_1 = 2/3 \quad \text{and} \quad p_2 = 1/3.$$

Thus the optimal strategy for R is $\mathbf{P} = \begin{bmatrix} 2/3 & 1/3 \end{bmatrix}$.

Similarly,

$$q_1 = 1/2 \quad \text{and} \quad q_2 = 1/2$$

which implies that the optimal strategy for C is $\mathbf{Q} = \begin{bmatrix} 1/2 \\ 1/2 \end{bmatrix}$.

The value of the game is $v = 0$.

T.1. The expected payoff to R is the sum of the terms of the form

(probability that R plays row i and C plays row j)

$\times$ (payoff to R when R plays i and C plays j)

$= (p_i q_j)(a_{ij})$.

Summing over all $1 \le i \le m$ and $1 \le j \le n$, we get

$$(\text{expected payoff to R}) = \sum_{i=1}^{m} \sum_{j=1}^{n} p_i a_{ij} q_j = \mathbf{PAQ}.$$

Exercises 7.7

1. Let $(x_1,y_1) = (2,1)$, $(x_2,y_2) = (3,2)$, $(x_3,y_3) = (4,3)$, $(x_4,y_4) = (5,2)$. Then

$$\sum_{i=1}^{4} x_i = 14,\quad \sum_{i=1}^{4} x_i^2 = 54,\quad \sum_{i=1}^{4} y_i = 8, \text{ and } \sum_{i=1}^{4} x_i y_i = 30.$$

From Equation (5) we have the linear system

$$\begin{aligned} 4b_0 + 14b_1 &= 8 \\ 14b_0 + 54b_1 &= 30. \end{aligned}$$

Form the augmented matrix and row reduce it to obtain

$$\left[\begin{array}{cc|c} 1 & 0 & 3/5 \\ 0 & 1 & 2/5 \end{array}\right].$$

Then the line of best fit is $y = b_0 + b_1x = 3/5 + 2/5\ x$.

3. Let $(x_1,y_1) = (2,3)$, $(x_2,y_2) = (3,4)$, $(x_3,y_3) = (4,3)$, $(x_4,y_4) = (5,4)$, $(x_5,y_5) = (6,3)$, $(x_6,y_6) = (7,4)$.

$$\sum_{i=1}^{6} x_i = 27,\quad \sum_{i=1}^{6} x_i^2 = 139,\quad \sum_{i=1}^{6} y_i = 21, \text{ and } \sum_{i=1}^{6} x_i y_i = 96.$$

From Equation (5) we have the linear system

$$\begin{aligned} 6b_0 + 27b_1 &= 21 \\ 27b_0 + 139b_1 &= 96. \end{aligned}$$

Form the augmented matrix and row reduce it to obtain

$$\left[\begin{array}{cc|c} 1 & 0 & 109/35 \\ 0 & 1 & 3/35 \end{array}\right].$$

Then the line of best fit is $y = b_0 + b_1x = 109/35 + 3/35\ x$.

5. Let x_i be the number of hours in the room and y_i be the corresponding number of minutes to find a way out of the maze. The data is

$$\{(x_i,y_i) \mid i = 1,2,\ldots,6\} = \{(1,.8),(2,2.1),(3,2.6),(4,2),(5,3.1),(6,3.3)\}.$$

(a) Following Example 4, we let

$$\mathbf{Y} = \begin{bmatrix} .8 \\ 2.1 \\ 2.6 \\ 2.0 \\ 3.1 \\ 3.3 \end{bmatrix}, \quad \mathbf{A} = \begin{bmatrix} 1 & 1 \\ 1 & 2 \\ 1 & 3 \\ 1 & 4 \\ 1 & 5 \\ 1 & 6 \end{bmatrix}, \quad \text{and } \mathbf{X} = \begin{bmatrix} b_0 \\ b_1 \end{bmatrix}.$$

Then

$$\mathbf{A}^T\mathbf{A} = \begin{bmatrix} 6 & 21 \\ 21 & 91 \end{bmatrix} \quad \text{and } \mathbf{A}^T\mathbf{Y} = \begin{bmatrix} 13.9 \\ 56.1 \end{bmatrix}.$$

Compute $(\mathbf{A}^T\mathbf{A})^{-1}$, then use Equation (8):

$$\mathbf{X} = (\mathbf{A}^T\mathbf{A})^{-1}\mathbf{A}^T\mathbf{Y} = \begin{bmatrix} 13/15 & -1/5 \\ -1/5 & 2/35 \end{bmatrix} \begin{bmatrix} 13.9 \\ 56.1 \end{bmatrix} = \begin{bmatrix} 0.827 \\ 0.426 \end{bmatrix} = \begin{bmatrix} b_0 \\ b_1 \end{bmatrix}.$$

Then the line of best fit is $y = 0.827 + 0.426\,x$.

(b) Set $x = 10$ and compute
$y = 0.827 + 0.426(10) = 5.087$ (minutes).

7. Let x_i be the number of salespersons and y_i be the corresponding annual sales (in millions). The data is
$\{(x_i, y_i) \mid i = 1,2,\ldots,6\}$
$=\{(5,2.3),(6,3.2),(7,4.1),(8,5.0),(9,6.1),(10,7.2)\}$.

(a) Following Example 4, we let

$$\mathbf{Y} = \begin{bmatrix} 2.3 \\ 3.2 \\ 4.1 \\ 5.0 \\ 6.1 \\ 7.2 \end{bmatrix}, \quad \mathbf{A} = \begin{bmatrix} 1 & 5 \\ 1 & 6 \\ 1 & 7 \\ 1 & 8 \\ 1 & 9 \\ 1 & 10 \end{bmatrix}, \quad \text{and } \mathbf{X} = \begin{bmatrix} b_0 \\ b_1 \end{bmatrix}.$$

Then

$$\mathbf{A}^T\mathbf{A} = \begin{bmatrix} 6 & 45 \\ 45 & 355 \end{bmatrix} \quad \text{and } \mathbf{A}^T\mathbf{Y} = \begin{bmatrix} 27.9 \\ 226.3 \end{bmatrix}.$$

Compute $(\mathbf{A}^T\mathbf{A})^{-1}$, then use Equation (8):

$$\mathbf{X} = (\mathbf{A}^T\mathbf{A})^{-1}\mathbf{A}^T\mathbf{Y} = \begin{bmatrix} 71/21 & -3/7 \\ -3/7 & 2/35 \end{bmatrix} \begin{bmatrix} 27.9 \\ 226.3 \end{bmatrix} = \begin{bmatrix} -2.657 \\ 0.974 \end{bmatrix} = \begin{bmatrix} b_0 \\ b_1 \end{bmatrix}.$$

Then the line of best fit is $y = -2.657 + 0.974\,x$.

(b) Set $x = 14$ and compute
$y = -2.657 + 0.974(14) = 10.979$ millions of dollars.

Exercises 7.7

T.1. $$\mathbf{A}^T\mathbf{A} = \begin{bmatrix} 1 & 1 & \cdot & \cdot & \cdot & 1 \\ x_1 & x_2 & \cdot & \cdot & \cdot & x_n \end{bmatrix} \begin{bmatrix} 1 & x_1 \\ 1 & x_2 \\ \cdot & \cdot \\ \cdot & \cdot \\ \cdot & \cdot \\ 1 & x_n \end{bmatrix} = \begin{bmatrix} n & \Sigma x_i \\ \Sigma x_i & \Sigma x_i^2 \end{bmatrix},$$

$$\mathbf{A}^T\mathbf{Y} = \begin{bmatrix} 1 & 1 & \cdot & \cdot & \cdot & 1 \\ x_1 & x_2 & \cdot & \cdot & \cdot & x_n \end{bmatrix} \begin{bmatrix} y_1 \\ y_2 \\ \cdot \\ \cdot \\ \cdot \\ y_n \end{bmatrix} = \begin{bmatrix} \Sigma y_i \\ \Sigma x_i y_i \end{bmatrix}.$$

Thus Equation (7) is $\begin{bmatrix} n & \Sigma x_i \\ \Sigma x_i & \Sigma x_i^2 \end{bmatrix} \begin{bmatrix} b_0 \\ b_1 \end{bmatrix} = \begin{bmatrix} \Sigma y_i \\ \Sigma x_i y_i \end{bmatrix}$, which gives Equation (5).

ML.1. Enter the data into MATLAB.

```
x = [2 3 4 5 6 7];y = [3 4 3 4 3 4];
c=lsqline(x,y)
```

We find that the least squares model is:

```
y =  0.08571*x +  3.114.
```

ML.2. Enter the data into MATLAB.

```
x=[1 2 3 4 5 6];y=[.8 2.1 2.6 2.0 3.1 3.3];
c=lsqline(x,y)
```

We find that the least squares model is:

```
y =  0.4257*x +  0.8267.
```

Using the option to evaluate the model, we find that x = 7 gives 3.8067, x = 8 gives 4.2324, and x = 9 gives 4.6581.

ML.3. Enter the data into MATLAB.

```
x=[0 2 3 5 9];y=[185 170 166 152 110];
```

a) Using command c=lsqline(x,y) we find that the least squares model is:

```
y =  -8.278*x +  188.1.
```

b) Using the option to evaluate the model, we that $x = 1$ gives 179.7778, $x = 6$ gives 138.3889, and $x = 8$ gives 121.8333.

c) In the equation for the least squares line set $y = 160$ and solve for x. We find $x = 3.3893$ min.

Exercises 7.8

1. Apply the definition of an exchange matrix.

   (a) This is not an exchange matrix since it has a negative entry: $a_{32} = -1/2$.

   (b) This is an exchange matrix: all entries are nonnegative and the sum of the entries in each column is 1.

   (c) This is not an exchange matrix since it has a negative entry: $a_{12} = -2/3$.

   (d) This is an exchange matrix: all entries are nonnegative and the sum of the entries in each column is 1.

3. Following Example 1, we find a solution **P** of the homogeneous system $(\mathbf{I}_3 - \mathbf{A})\mathbf{P} = \mathbf{0}$ such that $\mathbf{P} \geq \mathbf{0}$ with at least one positive component. Let

$$\mathbf{A} = \begin{bmatrix} 1/2 & 1 & 2/3 \\ 0 & 0 & 0 \\ 1/2 & 0 & 1/3 \end{bmatrix}.$$

Then,

$$(\mathbf{I}_3 - \mathbf{A})\mathbf{P} = \begin{bmatrix} 1/2 & -1 & -2/3 \\ 0 & 1 & 0 \\ -1/2 & 0 & 2/3 \end{bmatrix} \begin{bmatrix} p_1 \\ p_2 \\ p_3 \end{bmatrix} = \begin{bmatrix} 0 \\ 0 \\ 0 \end{bmatrix}.$$

Row reduce the coefficient matrix. Use row operations $2\mathbf{r}_1$, $(1/2)\mathbf{r}_1+\mathbf{r}_3$, $2\mathbf{r}_2+\mathbf{r}_1$, $\mathbf{r}_2+\mathbf{r}_3$ to obtain

$$\begin{bmatrix} 1 & 0 & -4/3 \\ 0 & 1 & 0 \\ 0 & 0 & 0 \end{bmatrix}.$$

The solution of the homogeneous system is $p_1 = (4/3)r$, $p_2 = 0$, $p_3 = r$, where r is any real number. Set r = 3, then one vector satisfying the requirements is $\mathbf{P} = \begin{bmatrix} 4 \\ 0 \\ 3 \end{bmatrix}$.

5. Follow the method of Example 1. Let p_1 be the price per unit of food, p_2 the price per unit of housing, and p_3 the price per unit of clothes. we have the following relations:

$$\begin{array}{rl} \text{farmer} & 2/5p_1 + 1/3p_2 + 1/2p_3 = p_1 \\ \text{carpenter} & 2/5p_1 + 1/3p_2 + 1/2p_3 = p_2 \\ \text{tailor} & 1/5p_1 + 1/3p_2 + \quad 0p_3 = p_3 \end{array}$$

Then the exchange matrix is $\mathbf{A} = \begin{bmatrix} 2/5 & 1/3 & 1/2 \\ 2/5 & 1/3 & 1/2 \\ 1/5 & 1/3 & 0 \end{bmatrix}$. We solve

$$(\mathbf{I}_3 - \mathbf{A})\mathbf{P} = \begin{bmatrix} 3/5 & -1/3 & -1/2 \\ -2/5 & 2/3 & -1/2 \\ -1/5 & -1/3 & 1 \end{bmatrix} \begin{bmatrix} p_1 \\ p_2 \\ p_3 \end{bmatrix} = \begin{bmatrix} 0 \\ 0 \\ 0 \end{bmatrix}.$$

Row reduce the coefficient matrix by applying row operations $(5/3)\mathbf{r}_1$, $(2/5)\mathbf{r}_1+\mathbf{r}_2$, $(1/5)\mathbf{r}_1+\mathbf{r}_3$, $(3/4)\mathbf{r}_2$, $(-5/3)\mathbf{r}_2+\mathbf{r}_1$, $(4/3)\mathbf{r}_2+\mathbf{r}_3$ to obtain

$$\begin{bmatrix} 1 & 0 & -15/8 \\ 0 & 1 & -15/8 \\ 0 & 0 & 0 \end{bmatrix}.$$

The general solution of the homogeneous system is $p_1 = (15/8)r$, $p_2 = (15/8)r$, $p_3 = r$ where r is any real number. Set $r = 40$, then one vector satisfying the requirements is

$$\mathbf{P} = \begin{bmatrix} 75 \\ 75 \\ 40 \end{bmatrix}.$$

<<**Strategy:** In Exercises 7 and 9, we determine if $(\mathbf{I}_n - \mathbf{C})^{-1}$ exists, and if it does whether all the entries are nonnegative. If both conditions are satisfied **C** is called productive.>>

7. Let $\mathbf{C} = \begin{bmatrix} 1/2 & 1/3 & 0 \\ 0 & 2/3 & 0 \\ 1 & 0 & 2 \end{bmatrix}$. Then $\mathbf{I}_3 - \mathbf{C} = \begin{bmatrix} 1/2 & -1/3 & 0 \\ 0 & 1/3 & 0 \\ -1 & 0 & -1 \end{bmatrix}$. Compute the determinant: $|\mathbf{I}_3 - \mathbf{C}| = -1/6 \neq 0$. Thus $\mathbf{I}_3 - \mathbf{C}$ is nonsingular. To compute its inverse we proceed as follows. Form the matrix $[\mathbf{I}_3 - \mathbf{C} \mid \mathbf{I}_3]$ and row reduce it. The result is

$$\left[\begin{array}{ccc|ccc} 1 & 0 & 0 & 2 & 2 & 0 \\ 0 & 1 & 0 & 0 & 3 & 0 \\ 0 & 0 & 1 & -2 & -2 & -1 \end{array}\right].$$

Thus $(\mathbf{I}_3 - \mathbf{C})^{-1} = \begin{bmatrix} 2 & 2 & 0 \\ 0 & 3 & 0 \\ -2 & -2 & -1 \end{bmatrix}$. Since there are negative entries **C** is not productive.

Exercises 7.8

9. Let $\mathbf{C} = \begin{bmatrix} 0 & 1/3 & 1/2 \\ 1/2 & 0 & 1/4 \\ 1/4 & 1/3 & 0 \end{bmatrix}$. Then $\mathbf{I}_3 - \mathbf{C} = \begin{bmatrix} 1 & -1/3 & -1/2 \\ -1/2 & 1 & -1/4 \\ -1/4 & -1/3 & 1 \end{bmatrix}$.

Compute the determinant: $|\mathbf{I}_3 - \mathbf{C}| = 25/48 \neq 0$. Thus $\mathbf{I}_3 - \mathbf{C}$ is nonsingular. To compute its inverse we proceed as follows. Form the matrix $[\mathbf{I}_3-\mathbf{C}|\mathbf{I}_3]$ and row reduce it. The result is

$$\left[\begin{array}{ccc|ccc} 1 & 0 & 0 & 44/25 & 24/25 & 28/25 \\ 0 & 1 & 0 & 27/25 & 42/25 & 24/25 \\ 0 & 0 & 1 & 4/5 & 4/5 & 8/5 \end{array}\right].$$

Thus $(\mathbf{I}_3 - \mathbf{C})^{-1} = \begin{bmatrix} 44/25 & 24/25 & 28/25 \\ 27/25 & 42/25 & 24/25 \\ 4/5 & 4/5 & 8/5 \end{bmatrix}$. Since all the entries are nonnegative, $\mathbf{C}$ is productive.

11. Let $\mathbf{C} = \begin{bmatrix} 1/2 & 1/2 \\ 1/2 & 1/4 \end{bmatrix}$. Following Example 7, we find $(\mathbf{I}_2 - \mathbf{C})^{-1}$.

Form the matrix $[\mathbf{I}_2-\mathbf{C}|\mathbf{I}_2]$ and row reduce it. We obtain

$$\left[\begin{array}{cc|cc} 1 & 0 & 6 & 4 \\ 0 & 1 & 4 & 4 \end{array}\right].$$

Thus $(\mathbf{I}_2 - \mathbf{C})^{-1} = \begin{bmatrix} 6 & 4 \\ 4 & 4 \end{bmatrix}$.

(a) Let $\mathbf{D} = \begin{bmatrix} 1 \\ 3 \end{bmatrix}$. Then the production vector is

$$\mathbf{X} = (\mathbf{I}_3 - \mathbf{C})^{-1}\mathbf{D} = \begin{bmatrix} 18 \\ 16 \end{bmatrix}.$$

(b) Let $\mathbf{D} = \begin{bmatrix} 2 \\ 0 \end{bmatrix}$. Then the production vector is

$$\mathbf{X} = (\mathbf{I}_3 - \mathbf{C})^{-1}\mathbf{D} = \begin{bmatrix} 12 \\ 8 \end{bmatrix}.$$

T.1. Let **A** = $[a_{ij}]$, **P** = $[p_j]$ and **AP** = **Y** = $[y_j]$. Then

$$\sum_{j=1}^{n} y_j = \sum_{j=1}^{n} \sum_{k=1}^{n} a_{jk} p_k = \sum_{k=1}^{n} \left[\sum_{j=1}^{n} a_{jk} \right] p_k = \sum_{k=1}^{n} p_k$$

since the sum of the entries in the kth column of **A** is 1.

Since **AP** $\le$ **P**, $y_j \le p_j$ for $j = 1, \ldots, n$. However, $\Sigma y_j = \Sigma p_j$, then the respective entries must be equal: $y_j = p_j$ for $j = 1, \ldots, n$. Thus **AP** = **P**.

Exercises 7.9

1. **T** is a transition matrix if and only if $0 \le t_{ij} \le 1$ for $1 \le i,j \le n$ and the sum of the entries in each column is 1.

   (a) Let $\mathbf{T} = \begin{bmatrix} .3 & .7 \\ .4 & .6 \end{bmatrix}$. **T** is not a transition matrix because the sum of the entries in each column is not 1.

   (b) Let $\mathbf{T} = \begin{bmatrix} .2 & .3 & .1 \\ .8 & .5 & .7 \\ 0 & .2 & .2 \end{bmatrix}$. **T** is a transition matrix.

   (c) Let $\mathbf{T} = \begin{bmatrix} .55 & .33 \\ .45 & .67 \end{bmatrix}$. **T** is a transition matrix.

   (d) Let $\mathbf{T} = \begin{bmatrix} .3 & .4 & .2 \\ .2 & 0 & .8 \\ .1 & .3 & .6 \end{bmatrix}$. **T** is not a transition matrix because the sum of the entries in each column is not 1.

3. Let $\mathbf{T} = \begin{bmatrix} .7 & .4 \\ .3 & .6 \end{bmatrix}$.

   (a) For $\mathbf{P} = \begin{bmatrix} 1 \\ 0 \end{bmatrix}$, $\mathbf{P}^{(1)} = \mathbf{TP} = \begin{bmatrix} .700 \\ .300 \end{bmatrix}$, $\mathbf{P}^{(2)} = \mathbf{TP}^{(1)} = \begin{bmatrix} .610 \\ .390 \end{bmatrix}$,

   $\mathbf{P}^{(3)} = \mathbf{TP}^{(2)} = \begin{bmatrix} .583 \\ .417 \end{bmatrix}$.

   (b) All the entries of **T** are positive thus **T** is regular. Following the method in Example 8, we solve for a vector **U** such that **TU** = **U**. Hence solve the homogeneous linear system

   $$(\mathbf{I}_3 - \mathbf{T})\mathbf{U} = \begin{bmatrix} .3 & -.4 \\ -.3 & .4 \end{bmatrix}\begin{bmatrix} u_1 \\ u_2 \end{bmatrix} = \mathbf{0}.$$

   Row reduce the coefficient matrix to obtain $\begin{bmatrix} 1 & -4/3 \\ 0 & 0 \end{bmatrix}$.

   The solution is $u_1 = (4/3)r$, $u_2 = r$, where r is any real number. Since **U** is to be a probability vector we require that $u_1 + u_2 = (4/3)r + r = (7/3)r = 1$. Set $r = 3/7$. Then the steady-state vector is

$$\mathbf{U} = \begin{bmatrix} 4/7 \\ 3/7 \end{bmatrix}.$$

5. (a) $\mathbf{T} = \begin{bmatrix} 0 & 1/2 \\ 1 & 1/2 \end{bmatrix}$ is regular since $\mathbf{T}^2 = \begin{bmatrix} 1/2 & 1/4 \\ 1/2 & 3/4 \end{bmatrix} > \mathbf{0}$.

(b) $\mathbf{T} = \begin{bmatrix} 1/2 & 0 & 0 \\ 0 & 1 & 1/2 \\ 1/2 & 0 & 1/2 \end{bmatrix}$ is not regular since $\mathbf{T}^k$ will always have $\begin{bmatrix} 0 \\ 1 \\ 0 \end{bmatrix}$ as its second column.

(c) $\mathbf{T} = \begin{bmatrix} 1 & 1/3 & 0 \\ 0 & 1/3 & 1 \\ 0 & 1/3 & 0 \end{bmatrix}$ is not regular since $\mathbf{T}^k$ will always have $\begin{bmatrix} 1 \\ 0 \\ 0 \end{bmatrix}$ as its first column.

(d) $\mathbf{T} = \begin{bmatrix} 1/4 & 3/5 & 1/2 \\ 1/2 & 0 & 0 \\ 1/4 & 2/5 & 1/2 \end{bmatrix}$ is regular since

$$\mathbf{T}^2 = \begin{bmatrix} .4875 & .35 & .375 \\ .1250 & .30 & .250 \\ .3875 & .35 & .375 \end{bmatrix} > \mathbf{0}.$$

7. Let $\mathbf{T} = \begin{bmatrix} 1/2 & 0 \\ 1/2 & 1 \end{bmatrix}$.

(a) $\mathbf{T}^k$ will have as its second column $\begin{bmatrix} 0 \\ 1 \end{bmatrix}$ for all k . Thus $\mathbf{T}$ is not regular.

(b) We have $\mathbf{T}^2 = \begin{bmatrix} \frac{1}{2^2} & 0 \\ \frac{1}{2^2} + \frac{1}{2} & 1 \end{bmatrix}$,

$$\mathbf{T}^3 = \begin{bmatrix} \frac{1}{2^3} & 0 \\ \frac{1}{2^3} + \frac{1}{2^2} + \frac{1}{2} & 1 \end{bmatrix}, \ldots,$$

$$\mathbf{T}^n = \begin{bmatrix} \frac{1}{2^n} & 0 \\ \frac{1}{2^n} + \frac{1}{2^{n-1}} + \cdots + \frac{1}{2} & 1 \end{bmatrix}$$

$$= \begin{bmatrix} \frac{1}{2^n} & 0 \\ \frac{(1/2)\cdot(1-(1/2^n))}{(1-1/2)} & 1 \end{bmatrix}.$$

This follows since $\frac{1}{2^n} + \frac{1}{2^{n-1}} + \cdots + \frac{1}{2}$ is a geometric progression. Thus for $\mathbf{X} = \begin{bmatrix} x_1 \\ x_2 \end{bmatrix}$ with $x_1 + x_2 = 1$ we have

$$\mathbf{T}^n\mathbf{X} = \begin{bmatrix} (1/2^n)\cdot x_1 \\ \frac{(1/2)\cdot(1 - (1/2^n))}{(1 - 1/2)} x_1 + x_2 \end{bmatrix} \to \begin{bmatrix} 0 \\ x_1 + x_2 \end{bmatrix} = \begin{bmatrix} 0 \\ 1 \end{bmatrix}$$

as $n \to \infty$. Thus $\begin{bmatrix} 0 \\ 1 \end{bmatrix}$ is a unique steady-state vector.

9\. (a) $\mathbf{T} = \begin{array}{c} \begin{matrix} A & B \end{matrix} \\ \begin{bmatrix} .3 & .4 \\ .7 & .6 \end{bmatrix} \end{array} \begin{matrix} A \\ B \end{matrix}$

(b) Compute $\mathbf{TP}^{(2)}$, where $\mathbf{P}^{(0)} = \begin{bmatrix} 1/2 \\ 1/2 \end{bmatrix}$:

$$\mathbf{TP}^{(0)} = \mathbf{P}^{(1)} = \begin{bmatrix} .35 \\ .65 \end{bmatrix}, \quad \mathbf{TP}^{(1)} = \mathbf{P}^{(2)} = \begin{bmatrix} .365 \\ .635 \end{bmatrix},$$

$$\mathbf{TP}^{(2)} = \mathbf{P}^{(3)} = \begin{bmatrix} .364 \\ .636 \end{bmatrix}.$$

The probability of the rat going through door A on the third day is $p_1^{(3)} = .364$.

(c) Solve $(\mathbf{I}_2 - \mathbf{T})\mathbf{U} = \begin{bmatrix} .7 & -.4 \\ -.7 & .4 \end{bmatrix}\begin{bmatrix} u_1 \\ u_2 \end{bmatrix} = \mathbf{0}$. Row reduce the coefficient matrix to obtain $\begin{bmatrix} 1 & -4/7 \\ 0 & 0 \end{bmatrix}$. Thus the solution is $u_1 = (4/7)r$, $u_2 = r$ where r is any real number. Since $\mathbf{U}$ is a probability vector, $u_1 + u_2 = (4/7)r + r = 1$. Hence $r = 7/11$ and $\mathbf{U} = \begin{bmatrix} 4/11 \\ 7/11 \end{bmatrix}$ which is approximately $\begin{bmatrix} .364 \\ .636 \end{bmatrix}$.

11. Let $\mathbf{T} = \begin{bmatrix} .8 & .3 & .2 \\ .1 & .5 & .2 \\ .1 & .2 & .6 \end{bmatrix}$.

(a) We have that an individual is a professional. Thus the current state is $\mathbf{P}^{(0)} = \begin{bmatrix} 1 \\ 0 \\ 0 \end{bmatrix}$. We are asked to determine the probability that a grandson will be a professional. This represents a transition through two stages:

$$\mathbf{P}^{(2)} = \mathbf{TP}^{(1)} = \mathbf{T}(\mathbf{TP}^{(0)}) = \mathbf{T}*\begin{bmatrix} .8 \\ .1 \\ .1 \end{bmatrix} = \begin{bmatrix} .69 \\ .15 \\ .16 \end{bmatrix}.$$

Thus the probability that a grandson of a professional will be a professional is .69.

(b) The "long-run" distribution is the steady-state vector.

Thus solve $(\mathbf{I}_3 - \mathbf{T})\mathbf{U} = \begin{bmatrix} .2 & -.3 & -.2 \\ -.1 & .5 & -.2 \\ -.1 & -.2 & .4 \end{bmatrix}\begin{bmatrix} u_1 \\ u_2 \\ u_3 \end{bmatrix} = \mathbf{0}$. Row reduce the coefficient matrix to obtain $\begin{bmatrix} 1 & 0 & -16/7 \\ 0 & 1 & -6/7 \\ 0 & 0 & 0 \end{bmatrix}$. The solution is $u_1 = (16/7)r$, $u_2 = (6/7)r$, $u_3 = r$, where r is any real number. Since the steady-state vector is a probability vector we have $u_1 + u_2 + u_3 = (16/7)r + (6/7)r + r = 1$. Thus $r = 7/29$ and $\mathbf{U} = \begin{bmatrix} 16/29 \\ 6/29 \\ 7/29 \end{bmatrix}$ which is approximately $\begin{bmatrix} .552 \\ .207 \\ .241 \end{bmatrix}$. The proportion of the population that will be farmers is .207 or 20.7%.

13. Let $\mathbf{T} = \begin{bmatrix} .7 & .2 \\ .3 & .8 \end{bmatrix}$. The initial state is that 30% of commuters use mass transit and 70% use their automobiles. Hence the initial state is $\mathbf{P}^{(0)} = \begin{bmatrix} .3 \\ .7 \end{bmatrix}$.

(a) The state 1 year from now is calculated as $\mathbf{P}^{(1)} = \mathbf{TP}^{(0)} = \begin{bmatrix} .35 \\ .65 \end{bmatrix}$. The state 2 years from now is $\mathbf{P}^{(2)} = \mathbf{TP}^{(1)} = \begin{bmatrix} .375 \\ .625 \end{bmatrix}$. Hence after one year 35% of the commuters will be using mass transit and at the end of two years 37.5%

(b) Find the steady-state vector. Solve the homogeneous linear system $(\mathbf{I}_2 - \mathbf{T})\mathbf{U} = \begin{bmatrix} .3 & -.2 \\ -.3 & .2 \end{bmatrix}\begin{bmatrix} u_1 \\ u_2 \end{bmatrix} = \mathbf{0}$. Row reduce the coefficient matrix to obtain $\begin{bmatrix} 1 & -2/3 \\ 0 & 0 \end{bmatrix}$. The solution is $u_1 = (2/3)r$, $u_2 = r$, where r is any real number. Since the steady-state vector is a probability vector $u_1 + u_2 = (5/3)r = 1$. Thus $r = 3/5$ and $\mathbf{U} = \begin{bmatrix} 2/5 \\ 3/5 \end{bmatrix}$.

In the long run 40% of commuters will be using mass transit.

ML.2. Enter the matrix T and initial state vector P into MATLAB.

```
T=[.5 .6 .4;.25 .3 .3;.25 .1 .3];
P=[.1 .3 .6]';
```

State vector P(5) is given by

```
P5=T^5*P

P5 =

    0.5055
    0.2747
    0.2198
```

ML.3. The command **sum** operating on a matrix computes the sum of the entries in each column and displays these totals as a row vector. If the output from the **sum** command is a row of ones, then the matrix is a Markov matrix.

(a) `A=[2/3 1/3 1/2;1/3 1/3 1/4;0 1/3 1/4]; sum(A)`

```
ans =

     1     1     1
```

Hence A is a Markov matrix.

(b) `A=[.5 .6 .7;.3 .2 .3;.1 .2 0]; sum(A)`

```
ans =

    0.9000    1.0000    1.0000
```

A is not a Markov matrix.

(c) `A=[.66 .25 .125;.33 .25 .625;0 .5 .25]; sum(A)`

```
ans =

    0.9900    1.0000    1.0000
```

A is not a Markov matrix.

Exercises 7.10

1. Let $\mathbf{A} = \begin{bmatrix} 1 & 1 \\ 1 & 0 \end{bmatrix}$. The characteristic polynomial is

$$|\lambda \mathbf{I}_2 - \mathbf{A}| = \begin{vmatrix} \lambda-1 & -1 \\ -1 & \lambda \end{vmatrix} = \lambda^2 - \lambda - 1.$$

Using the quadratic equation we find that the roots of $\lambda^2 - \lambda - 1 = 0$ are $\lambda_1 = \dfrac{1 + \sqrt{5}}{2}$ and $\lambda_2 = \dfrac{1 - \sqrt{5}}{2}$. We find the corresponding eigenvectors as follows:
Case $\lambda = \lambda_1$. Solve

$$(\lambda_1 \mathbf{I}_2 - \mathbf{A})\mathbf{X} = \begin{bmatrix} \dfrac{-1+\sqrt{5}}{2} & -1 \\ -1 & \dfrac{1+\sqrt{5}}{2} \end{bmatrix} \begin{bmatrix} x_1 \\ x_2 \end{bmatrix} = \mathbf{0}.$$

Row reduce the coefficient matrix to obtain

$\begin{bmatrix} 1 & \dfrac{-(1+\sqrt{5})}{2} \\ 0 & 0 \end{bmatrix}$. The solution is $x_1 = \dfrac{1+\sqrt{5}}{2} r$, $x_2 = r$

where r is any real number. Let r = 1, then we have eigenvector

$$\mathbf{X}_1 = \begin{bmatrix} \dfrac{1+\sqrt{5}}{2} \\ 1 \end{bmatrix}.$$

Case $\lambda = \lambda_2$. Solve

$$(\lambda_2 \mathbf{I}_2 - \mathbf{A})\mathbf{X} = \begin{bmatrix} \dfrac{-1-\sqrt{5}}{2} & -1 \\ -1 & \dfrac{1-\sqrt{5}}{2} \end{bmatrix} \begin{bmatrix} x_1 \\ x_2 \end{bmatrix} = \mathbf{0}.$$

Row reduce the coefficient matrix to obtain

$$\begin{bmatrix} 1 & \frac{-(1-\sqrt{5})}{2} \\ 0 & 0 \end{bmatrix}.$$ The solution is $x_1 = \frac{1-\sqrt{5}}{2} r$, $x_2 = r$,

where r is any real number. Let r = 1, then we have eigenvector

$$\mathbf{x}_2 = \begin{bmatrix} \frac{1-\sqrt{5}}{2} \\ 1 \end{bmatrix}.$$

3. The Fibonacci sequence is computed using the recursion relation

$$u_n = u_{n-1} + u_{n-2},\ n \geq 2,\ u_0 = u_1 = 1.$$

From the discussion in the text, we have $u_4 = 5$ and $u_5 = 8$.

(a)

$$\begin{aligned} u_6 &= u_5 + u_4 = 13 \\ u_7 &= u_6 + u_5 = 21 \\ u_8 &= u_7 + u_6 = 34. \end{aligned}$$

(b)

$$\begin{aligned} u_9 &= u_8 + u_7 = 55 \\ u_{10} &= u_9 + u_8 = 89 \\ u_{11} &= u_{10} + u_9 = 144 \\ u_{12} &= u_{11} + u_{10} = 233 \end{aligned}$$

(c)

$$u_{13} = 377,\ u_{14} = 610,\ u_{15} = 987,\ u_{16} = 1{,}597,$$
$$u_{17} = 2{,}584,\ u_{18} = 4{,}181,\ u_{19} = 6{,}765,\ u_{20} = 10{,}946$$

T.1. Let us define u_{-1} to be 0. Then for n = 0,

$$\mathbf{A}^1 = \mathbf{A} = \begin{bmatrix} 1 & 1 \\ 1 & 0 \end{bmatrix} = \begin{bmatrix} u_1 & u_0 \\ u_0 & u_{-1} \end{bmatrix},$$

for n = 1, $\mathbf{A}^2 = \begin{bmatrix} 2 & 1 \\ 1 & 1 \end{bmatrix} = \begin{bmatrix} u_2 & u_1 \\ u_1 & u_0 \end{bmatrix}$. Suppose that the formula

$$\mathbf{A}^{n+1} = \begin{bmatrix} u_{n+1} & u_n \\ u_n & u_{n-1} \end{bmatrix} \quad (*)$$

holds for values up to and including n, $n \geq 1$. Then

$$\mathbf{A}^{n+2} = \mathbf{A}\cdot\mathbf{A}^{n+1} = \begin{bmatrix} 1 & 1 \\ 1 & 0 \end{bmatrix} \begin{bmatrix} u_{n+1} & u_n \\ u_n & u_{n-1} \end{bmatrix}$$

$$= \begin{bmatrix} u_{n+1} + u_n & u_n + u_{n-1} \\ u_{n+1} & u_n \end{bmatrix} = \begin{bmatrix} u_{n+2} & u_{n+1} \\ u_{n+1} & u_n \end{bmatrix}.$$

Thus the formula (*) also holds for n+1.

Using (*), we see that

$$u_{n+1}u_{n-1} - u_n^2 = \begin{vmatrix} u_{n+1} & u_n \\ u_n & u_{n-1} \end{vmatrix} = |\mathbf{A}^{n+1}| = |\mathbf{A}|^{n+1} = (-1)^{n+1}.$$

Exercises 7.11

1. Follow the steps in Example 2.

   (a) Since the system is diagonal

$$x_1 = b_1e^{-3t}, \quad x_2 = b_2e^{4t}, \quad x_3 = b_3e^{2t},$$

   where b_1, b_2, b_3 are arbitrary constants. Thus the solution is

$$\mathbf{X}(t) = \begin{bmatrix} b_1e^{-3t} \\ b_2e^{4t} \\ b_3e^{2t} \end{bmatrix} = b_1\begin{bmatrix}1\\0\\0\end{bmatrix}e^{-3t} + b_2\begin{bmatrix}0\\1\\0\end{bmatrix}e^{4t} + b_3\begin{bmatrix}0\\0\\1\end{bmatrix}e^{2t}.$$

   (b) For the initial conditions $x_1(0) = 3$, $x_2(0) = 4$, $x_3(0) = 5$ we have

$$b_1 = 3, \quad b_2 = 4, \quad b_3 = 5.$$

   (Substitute $t = 0$ into the expressions for x_1, x_2, x_3.) The solution of the initial value problem is

$$\mathbf{X}(t) = 3\begin{bmatrix}1\\0\\0\end{bmatrix}e^{-3t} + 4\begin{bmatrix}0\\1\\0\end{bmatrix}e^{4t} + 5\begin{bmatrix}0\\0\\1\end{bmatrix}e^{2t}.$$

3. Let $\mathbf{A} = \begin{bmatrix} 4 & 0 & 0 \\ 3 & -5 & 0 \\ 2 & 1 & 2 \end{bmatrix}$. Since the $\mathbf{A}$ is lower triangular its eigenvalues are the diagonal entries: $\lambda_1 = 4$, $\lambda_2 = -5$, $\lambda_3 = 2$. We find the corresponding eigenvectors by solving the linear systems $(\lambda_i\mathbf{I}_3 - \mathbf{A})\mathbf{X} = \mathbf{0}$ with $i = 1,2,3$. We have:

   For $\lambda_1 = 4$ the solution is $x_1 = (6/7)r$, $x_2 = (2/7)r$ and $x_3 = r$. Set $r = 7$ and we have eigenvector

$$\mathbf{X}_1 = \begin{bmatrix}6\\2\\7\end{bmatrix}.$$

   For $\lambda_2 = -5$ the solution is $x_1 = 0$, $x_2 = -7r$, and $x_3 = r$. Set $r = -1$ and we have eigenvector

$$\mathbf{X}_2 = \begin{bmatrix}0\\7\\-1\end{bmatrix}.$$

   For $\lambda_3 = 2$ the solution is $x_1 = 0$, $x_2 = 0$, and $x_3 = r$. Set $r = 1$ and we have eigenvector

$$\mathbf{X}_3 = \begin{bmatrix} 0 \\ 0 \\ 1 \end{bmatrix}.$$

From Equation (17), the solution is given by

$$\mathbf{X}(t) = b_1 \begin{bmatrix} 6 \\ 2 \\ 7 \end{bmatrix} e^{4t} + b_2 \begin{bmatrix} 0 \\ 7 \\ -1 \end{bmatrix} e^{-5t} + b_3 \begin{bmatrix} 0 \\ 0 \\ 1 \end{bmatrix} e^{2t}.$$

5. Let $\mathbf{A} = \begin{bmatrix} 5 & 0 & 0 \\ 0 & -4 & 3 \\ 0 & 3 & 4 \end{bmatrix}$. We find the eigenvalues and associated eigenvectors of **A**. The characteristic polynomial is

$$|\lambda \mathbf{I}_3 - \mathbf{A}| = (\lambda - 5)^2(\lambda + 5).$$

Hence the eigenvalues are $\lambda_1 = \lambda_2 = 5$, $\lambda_3 = -5$. Find the associated eigenvectors:

Case $\lambda_1 = 5$. Solve the linear system

$$(5\mathbf{I}_3 - \mathbf{A})\mathbf{X} = \begin{bmatrix} 0 & 0 & 0 \\ 0 & 9 & -3 \\ 0 & -3 & 1 \end{bmatrix} \begin{bmatrix} x_1 \\ x_2 \\ x_3 \end{bmatrix} = \mathbf{0}.$$

Row reduce the coefficient matrix to obtain

$\begin{bmatrix} 0 & 1 & -1/3 \\ 0 & 0 & 0 \\ 0 & 0 & 0 \end{bmatrix}$. The solution is $x_1 = r$, $x_2 = (1/3)s$, $x_3 = s$, where r and s are any real numbers. To find a pair of linearly independent eigenvectors, set r = 1 and s = 0; then set r = 0 and s = 3. This gives

$$\mathbf{X}_1 = \begin{bmatrix} 1 \\ 0 \\ 0 \end{bmatrix} \text{ and } \mathbf{X}_2 = \begin{bmatrix} 0 \\ 1 \\ 3 \end{bmatrix}.$$

Case $\lambda_3 = -5$. Solve the linear system

$$(-5\mathbf{I}_3 - \mathbf{A})\mathbf{X} = \begin{bmatrix} -10 & 0 & 0 \\ 0 & -1 & -3 \\ 0 & -3 & -9 \end{bmatrix} \begin{bmatrix} x_1 \\ x_2 \\ x_3 \end{bmatrix} = \mathbf{0}.$$

Row reduce the coefficient matrix to obtain $\begin{bmatrix} 1 & 0 & 0 \\ 0 & 1 & 3 \\ 0 & 0 & 0 \end{bmatrix}$. The solution is $x_1 = 0$, $x_2 = -3r$, $x_3 = r$, where r is any real number. Set r = 1 and we have eigenvector

$$\mathbf{X}_3 = \begin{bmatrix} 0 \\ -3 \\ 1 \end{bmatrix}.$$

The solution is

$$\mathbf{X}(t) = b_1\begin{bmatrix} 1 \\ 0 \\ 0 \end{bmatrix}e^{5t} + b_2\begin{bmatrix} 0 \\ 1 \\ 3 \end{bmatrix}e^{5t} + b_3\begin{bmatrix} 0 \\ -3 \\ 1 \end{bmatrix}e^{-5t}.$$

7. Let $\mathbf{A} = \begin{bmatrix} 1 & 2 & 3 \\ 0 & 1 & 0 \\ 2 & 1 & 2 \end{bmatrix}$. We find the eigenvalues and associated eigenvectors of **A**. The characteristic polynomial is

$$|\lambda\mathbf{I}_3 - \mathbf{A}| = (\lambda - 4)(\lambda + 1)(\lambda - 1).$$

Hence the eigenvalues are $\lambda_1 = 4$, $\lambda_2 = -1$, $\lambda_3 = 1$. Find the associated eigenvectors:

Case $\lambda_1 = 4$. Solve the linear system

$$(4\mathbf{I}_3 - \mathbf{A})\mathbf{X} = \begin{bmatrix} 3 & -2 & -3 \\ 0 & 3 & 0 \\ -2 & -1 & 2 \end{bmatrix}\begin{bmatrix} x_1 \\ x_2 \\ x_3 \end{bmatrix} = \mathbf{0}.$$

Row reduce the coefficient matrix to obtain $\begin{bmatrix} 1 & 0 & -1 \\ 0 & 1 & 0 \\ 0 & 0 & 0 \end{bmatrix}$. The solution is $x_1 = r$, $x_2 = 0$, $x_3 = r$, where r is any real number. Set r = 1. This gives eigenvector

$$\mathbf{X}_1 = \begin{bmatrix} 1 \\ 0 \\ 1 \end{bmatrix}.$$

Case $\lambda_2 = -1$. Solve the linear system

$$(-1\mathbf{I}_3 - \mathbf{A})\mathbf{X} = \begin{bmatrix} -2 & -2 & -3 \\ 0 & -2 & 0 \\ -2 & -1 & -3 \end{bmatrix}\begin{bmatrix} x_1 \\ x_2 \\ x_3 \end{bmatrix} = \mathbf{0}.$$

Row reduce the coefficient matrix to obtain

$\begin{bmatrix} 1 & 0 & 3/2 \\ 0 & 1 & 0 \\ 0 & 0 & 0 \end{bmatrix}$. The solution is $x_1 = (-3/2)r$, $x_2 = 0$, $x_3 = r$, where r is any real number. Set r = 2 and we have eigenvector

$$\mathbf{X}_2 = \begin{bmatrix} -3 \\ 0 \\ 2 \end{bmatrix}.$$

Case $\lambda_2 = 1$. Solve the linear system

$$(1\mathbf{I}_3 - \mathbf{A})\mathbf{X} = \begin{bmatrix} 0 & -2 & -3 \\ 0 & 0 & 0 \\ -2 & -1 & -1 \end{bmatrix}\begin{bmatrix} x_1 \\ x_2 \\ x_3 \end{bmatrix} = \mathbf{0}.$$

Row reduce the coefficient matrix to obtain

$\begin{bmatrix} 1 & 0 & -1/4 \\ 0 & 1 & 3/2 \\ 0 & 0 & 0 \end{bmatrix}$. The solution is $x_1 = (1/4)r$, $x_2 = (-3/2)r$, $x_3 = r$, where r is any real number. Set r = 4 and we have eigenvector

$$\mathbf{X}_3 = \begin{bmatrix} 1 \\ -6 \\ 4 \end{bmatrix}.$$

The solution is

$$\mathbf{X}(t) = b_1\begin{bmatrix} 1 \\ 0 \\ 1 \end{bmatrix}e^{4t} + b_2\begin{bmatrix} -3 \\ 0 \\ 2 \end{bmatrix}e^{-t} + b_3\begin{bmatrix} 1 \\ -6 \\ 4 \end{bmatrix}e^{t}.$$

9. Rewrite the system of differential equations in matrix form as

$$\begin{bmatrix} x_1'(t) \\ x_2'(t) \end{bmatrix} = \begin{bmatrix} -3 & 6 \\ 1 & -2 \end{bmatrix}\begin{bmatrix} x_1(t) \\ x_2(t) \end{bmatrix}.$$

Let $\mathbf{A} = \begin{bmatrix} -3 & 6 \\ 1 & -2 \end{bmatrix}$. Find the eigenvalues and associated eigenvectors of **A**. The characteristic polynomial is

$$|\lambda\mathbf{I}_2 - \mathbf{A}| = \lambda(\lambda + 5).$$

The eigenvalues are $\lambda_1 = 0$ and $\lambda_2 = -5$. Find the associated eigenvectors:

Case $\lambda_1 = 0$. Solve the linear system

$$(0\mathbf{I}_2 - \mathbf{A})\mathbf{X} = \begin{bmatrix} 3 & -6 \\ -1 & 2 \end{bmatrix}\begin{bmatrix} x_1 \\ x_2 \end{bmatrix} = \mathbf{0}.$$

Row reduce the coefficient matrix to obtain $\begin{bmatrix} 1 & -2 \\ 0 & 0 \end{bmatrix}$. The solution is $x_1 = 2r$, $x_2 = r$, where r is any real number. Set r = 1 and we have eigenvector

$$\mathbf{X}_1 = \begin{bmatrix} 2 \\ 1 \end{bmatrix}.$$

Case $\lambda_2 = -5$. Solve the linear system

$$(-5\mathbf{I}_2 - \mathbf{A})\mathbf{X} = \begin{bmatrix} -2 & -6 \\ -1 & -3 \end{bmatrix}\begin{bmatrix} x_1 \\ x_2 \end{bmatrix} = \mathbf{0}.$$

Row reduce the coefficient matrix to obtain $\begin{bmatrix} 1 & 3 \\ 0 & 0 \end{bmatrix}$. The solution is $x_1 = -3r$, $x_2 = r$, where r is any real number. Set r = -1 and we have eigenvector

$$\mathbf{X}_2 = \begin{bmatrix} 3 \\ -1 \end{bmatrix}.$$

The solution is

$$\mathbf{X}(t) = b_1\begin{bmatrix} 2 \\ 1 \end{bmatrix}e^{0t} + b_2\begin{bmatrix} 3 \\ -1 \end{bmatrix}e^{-5t} = b_1\begin{bmatrix} 2 \\ 1 \end{bmatrix} + b_2\begin{bmatrix} 3 \\ -1 \end{bmatrix}e^{-5t}.$$

Using the initial conditions $x_1(0) = 500$ and $x_2(0) = 200$ we have in matrix form, $\mathbf{X}(0) = \begin{bmatrix} 500 \\ 200 \end{bmatrix}$. Set t = 0 in the solution and solve for b_1 and b_2. We have,

$$\mathbf{X}(0) = \begin{bmatrix} 500 \\ 200 \end{bmatrix} = b_1\begin{bmatrix} 2 \\ 1 \end{bmatrix} + b_2\begin{bmatrix} 3 \\ -1 \end{bmatrix} = \begin{bmatrix} 2 & 3 \\ 1 & -1 \end{bmatrix}\begin{bmatrix} b_1 \\ b_2 \end{bmatrix} = \mathbf{A}\begin{bmatrix} b_1 \\ b_2 \end{bmatrix}.$$

Solving this linear system we get $b_1 = 220$ and $b_2 = 20$. The solution of the initial value problem and the populations at time t are given by

$$\mathbf{X}(t) = 220\begin{bmatrix} 2 \\ 1 \end{bmatrix} + 20\begin{bmatrix} 3 \\ -1 \end{bmatrix}e^{-5t}.$$

Exercises 7.11

T.1. In Exercise 25(e) of Section 3.4, we showed that the set of all differentiable functions is a subspace of $C(-\infty,\infty)$. It follows that the set of all n-tuples of differentable functions is a subspace W of the vector space V of n-tuples of elements from $C(-\infty,\infty)$. We show that the set S of all solutions of $\mathbf{X}' = \mathbf{A}\mathbf{X}$ is a subspace of W.

Let $\mathbf{X}_1$ and $\mathbf{X}_2$ be in S. Then $\mathbf{X}_1' = \mathbf{A}\mathbf{X}_1$ and $\mathbf{X}_2' = \mathbf{A}\mathbf{X}_2$. Since

$$(\mathbf{X}_1 + \mathbf{X}_2)' = \mathbf{X}_1' + \mathbf{X}_2' = \mathbf{A}\mathbf{X}_1 + \mathbf{A}\mathbf{X}_2 = \mathbf{A}(\mathbf{X}_1 + \mathbf{X}_2)$$

$\mathbf{X}_1 + \mathbf{X}_2$ is in S. Let k be any scalar, then since

$$(k\mathbf{X}_1)' = k\mathbf{X}_1' = k(\mathbf{A}\mathbf{X}_1) = \mathbf{A}(k\mathbf{X}_1)$$

$k\mathbf{X}_1$ is in S. Hence S is closed with respect to vector addition and scalar multiplication. Thus S is a subspace.

ML.1.
```
A=[0 1 0;0 0 1;8 -14 7];
[v,d]=eig(A)

v =

   -0.5774    0.2182    0.0605
   -0.5774    0.4364    0.2421
   -0.5774    0.8729    0.9684


d =

    1.0000         0         0
         0    2.0000         0
         0         0    4.0000
```

The general solution is given by

$$\mathbf{X}(t) = b_1\begin{bmatrix}-0.5774\\-0.5774\\-0.5774\end{bmatrix} e^{t} + b_2\begin{bmatrix}0.2182\\0.4364\\0.8729\end{bmatrix} e^{2t} + b_3\begin{bmatrix}0.0605\\0.2421\\0.9684\end{bmatrix} e^{4t}$$

ML.2.
```
A=[1 0 0;0 3 -2;0 -2 3];
[v,d]=eig(A)

v =

    1.0000         0         0
         0   -0.7071   -0.7071
         0   -0.7071    0.7071
```

```
d =

     1     0     0
     0     1     0
     0     0     5
```

The general solution is given by

$$\mathbf{X}(t) = b_1 \begin{bmatrix} 1.0000 \\ 0 \\ 0 \end{bmatrix} e^{t} + b_2 \begin{bmatrix} 0 \\ -0.7071 \\ -0.7071 \end{bmatrix} e^{t} + b_3 \begin{bmatrix} 0 \\ -0.7071 \\ 0.7071 \end{bmatrix} e^{5t}$$

ML.3.
```
A=[1 2 3;0 1 0;2 1 2];
[v,d]=eig(A)

v =

   -0.8321   -0.7071   -0.1374
         0         0    0.8242
    0.5547   -0.7071   -0.5494


d =

    -1     0     0
     0     4     0
     0     0     1
```

The general solution is given by

$$x(t) = b_1 \begin{bmatrix} -0.8321 \\ 0 \\ 0.5547 \end{bmatrix} e^{-t} + b_2 \begin{bmatrix} -0.7071 \\ 0 \\ -0.7071 \end{bmatrix} e^{4t} + b_3 \begin{bmatrix} -0.1374 \\ 0.8242 \\ -0.5494 \end{bmatrix} e^{t}$$

Chapter 7 Supplementary Exercises

1. Substitute the coordinates of each point into the symmetric form of the equation of the line. If the result is a string of equal values, the point is on the line.

   (a) (1,2,3) is not on the line since

   $$\frac{1-3}{2} = -1,\ \frac{2+3}{4} = 5/4,\ \frac{3+5}{-4} = -2.$$

   (b) (5,1,9) is on the line since

   $$\frac{5-3}{2} = 1,\ \frac{1+3}{4} = 1,\ \frac{-9+5}{-4} = 1.$$

   (c) (1,-7,-1) is on the line since

   $$\frac{1-3}{2} = -1,\ \frac{-7+3}{4} = -1,\ \frac{-1+5}{-4} = -1.$$

3. Follow the method of Example 7 of Section 7.2. The matrix of the quadratic form $\mathbf{Q} = x^2 + 2y^2 + z^2 - 2xy - 2yz$ is

   $$\mathbf{A} = \begin{bmatrix} 1 & -1 & 0 \\ -1 & 2 & -1 \\ 0 & -1 & 1 \end{bmatrix}.$$

   Find the eigenvalues of **A**. The characteristic polynomial is

   $$|\lambda\mathbf{I}_3 - \mathbf{A}| = (\lambda - 3)(\lambda - 1)\lambda.$$

   The eigenvalues are $\lambda_1 = 3$, $\lambda_2 = 1$, $\lambda_3 = 0$. Hence **A** is congruent to $\mathbf{D} = \begin{bmatrix} 3 & 0 & 0 \\ 0 & 1 & 0 \\ 0 & 0 & 0 \end{bmatrix}$. Let $\mathbf{H} = \begin{bmatrix} 1/\sqrt{3} & 0 & 0 \\ 0 & 1 & 0 \\ 0 & 0 & 1 \end{bmatrix}$. Then

   $\mathbf{D}_1 = \mathbf{H}^T\mathbf{D}\mathbf{H} = \begin{bmatrix} 1 & 0 & 0 \\ 0 & 1 & 0 \\ 0 & 0 & 0 \end{bmatrix}$. It follows that **A** is congruent to $\mathbf{D}_1$.

   Then **Q** is congruent to $\mathbf{Q}' = y_1{}^2 + y_2{}^2$.

5. Let $\mathbf{A} = \begin{bmatrix} 6 & 2 & 3 \\ 3 & 4 & 2 \\ 4 & 1 & 2 \end{bmatrix}$. We first inspect **A** for saddle points:

$$\begin{array}{r c c} & \begin{bmatrix} 6 & 2 & 3 \\ 3 & 4 & 2 \\ 4 & 1 & 2 \end{bmatrix} & \begin{array}{c} \textbf{Row minima} \\ 2 \\ 2 \\ 1 \end{array} \\ \textbf{Column maxima} & \begin{array}{ccc} 6 & 4 & 3 \end{array} & \end{array}$$

There is no saddle point. Next we look for recessive rows or columns. Each entry of row 3 is less than the corresponding entry in row 1, thus we can drop row 3. We have

$$\mathbf{A}_1 = \begin{bmatrix} 6 & 2 & 3 \\ 3 & 4 & 2 \end{bmatrix}.$$

Each entry in column 1 is greater than the corresponding entry in column 3, thus we can drop column 1. We have

$$\mathbf{A}_2 = \begin{bmatrix} 2 & 3 \\ 4 & 2 \end{bmatrix}.$$

The solution of the matrix game with payoff matrix $\mathbf{A}_2$ is obtained from Equations (8), (9), and (13) of Section 7.4. It follows that

$$p_1 = 2/3,\ p_2 = 1/3,\ v = 8/3,\ q_1 = 1/3,\ q_2 = 2/3.$$

Thus the optimal strategy for R is $\mathbf{P} = [2/3 \quad 1/3]$ and the optimal strategy for C is $\mathbf{Q} = \begin{bmatrix} 1/3 \\ 2/3 \end{bmatrix}$. Since row 3 and column 1 were dropped from the original game, the optimal strategies for the original game are $\mathbf{P} = [2/3 \quad 1/3 \quad 0]$ and $\mathbf{Q} = \begin{bmatrix} 0 \\ 1/3 \\ 2/3 \end{bmatrix}$. The value of the game is $v = 8/3$.

7. Let $\mathbf{A} = \begin{bmatrix} 1/2 & 3/8 & 1/3 \\ 1/4 & 1/4 & 1/3 \\ 1/4 & 3/8 & 1/3 \end{bmatrix}$ be the exchange matrix. We seek a vector $\mathbf{P}$ such that $\mathbf{AP} = \mathbf{P}$. To find $\mathbf{P}$, solve the linear system

$$(\mathbf{I}_3 - \mathbf{A})\mathbf{P} = \begin{bmatrix} 1/2 & -3/8 & -1/3 \\ -1/4 & 3/4 & -1/3 \\ -1/4 & -3/8 & 2/3 \end{bmatrix} \begin{bmatrix} p_1 \\ p_2 \\ p_3 \end{bmatrix} = \mathbf{0}.$$

Row reduce the coefficient matrix to obtain $\begin{bmatrix} 1 & 0 & -4/3 \\ 0 & 1 & -8/9 \\ 0 & 0 & 0 \end{bmatrix}$.

The solution is $p_1 = (4/3)s$, $p_2 = (8/9)s$, $p_3 = s$, where s is any real number. Let $s = 9r$, then $p_1 = 12r$, $p_2 = 8r$, $p_3 = 9r$, where r is any real number.

9. In Exercise 3(c) of Section 7.8 we computed $u_{20} = 10{,}946$ and $u_{19} = 6{,}765$. Using these values in the recursion relation for the Fibonacci sequence we get

$$u_{21} = 17{,}711,\ u_{22} = 28{,}657,\ u_{23} = 46{,}368,$$
$$u_{24} = 75{,}025,\ u_{25} = 121{,}393.$$

Exercises 8.1

1. 34.7213 is written as 0.3472×10^2.

3. -284 is written as -0.2840×10^3.

5. 1.230 rounded to four significant digits is 0.1230×10^1.
   1.230 truncated to four significant digits is 0.1230×10^1.

7. 17/3 as a decimal is 5.666... ; a repeating decimal.
   17/3 rounded to four significant digits is 0.5667×10^1.
   17/3 truncated to four significant digits is 0.5666×10^1.

9. Let N = 12.341 and N~ = 12.362. Then

$$\text{absolute error} = N\sim - N = 12.362 - 12.341 = 0.21 \times 10^{-1},$$

$$\text{relative error} = \frac{N\sim - N}{N} = \frac{0.21 \times 10^{-1}}{12.341} = 0.17 \times 10^{-2}.$$

11. Let N = 6482.0 and N~ = 6483.1. Then

$$\text{absolute error} = N\sim - N = 6483.1 - 6482.0 = 0.11 \times 10^{1},$$

$$\text{relative error} = \frac{N\sim - N}{N} = \frac{0.11 \times 10^{1}}{6482.0} = 0.17 \times 10^{-3}.$$

ML.1. The sequence of commands with the displayed results are shown below.

```
format short e
pi

ans =

    3.1416e+000

format long e
pi

ans =

    3.141592653589793e+000

format short
```

ML.2. The sequence of commands with the displayed results are shown below.

```
format long e
v(1)=0.1; for i=2:15,v(i)=v(i-1)+.1;end,v'

ans =

    1.000000000000000e-001
    2.000000000000000e-001
    3.000000000000000e-001
    4.000000000000000e-001
    5.000000000000000e-001
    6.000000000000000e-001
    7.000000000000000e-001
    7.999999999999999e-001
    8.999999999999999e-001
    9.999999999999999e-001
    1.100000000000000e+000
    1.200000000000000e+000
    1.300000000000000e+000
    1.400000000000000e+000
    1.500000000000000e+000
```

Note that there is an accumulation of roundoff error that shows itself in the 8th, 9th, and 10th terms.

```
v(1)=0.1; for i=2:35,v(i)=v(i-1)+.1;end,v'
```

(The following display is shown in two columns for convenience. A single column would be displayed in MATLAB.)

```
ans =

    1.000000000000000e-001     1.900000000000001e+000
    2.000000000000000e-001     2.000000000000000e+000
    3.000000000000000e-001     2.100000000000001e+000
    4.000000000000000e-001     2.200000000000001e+000
    5.000000000000000e-001     2.300000000000001e+000
    6.000000000000000e-001     2.400000000000001e+000
    7.000000000000000e-001     2.500000000000001e+000
    7.999999999999999e-001     2.600000000000001e+000
    8.999999999999999e-001     2.700000000000001e+000
    9.999999999999999e-001     2.800000000000001e+000
    1.100000000000000e+000     2.900000000000001e+000
    1.200000000000000e+000     3.000000000000001e+000
    1.300000000000000e+000     3.100000000000001e+000
    1.400000000000000e+000     3.200000000000002e+000
    1.500000000000000e+000     3.300000000000002e+000
    1.600000000000000e+000     3.400000000000002e+000
    1.700000000000000e+000     3.500000000000002e+000
    1.800000000000000e+000
```

Again notice the accumulation of roundoff error in the last digit.

Exercises 8.2

1. Let $\mathbf{A} = \begin{bmatrix} 1 & -2 & 0 \\ 2 & -3 & -1 \\ 1 & 3 & 2 \end{bmatrix}$. Identify the pivot as the (1,1) element.

Perform row operations to make all the entries below the pivot in the pivotal column zero: $-2\mathbf{r}_1+\mathbf{r}_2$, $-\mathbf{r}_1+\mathbf{r}_3$. We have

$$\mathbf{A}_1 = \begin{bmatrix} 1 & -2 & 0 \\ 0 & 1 & -1 \\ 0 & 5 & 2 \end{bmatrix}.$$

Identify $\mathbf{B}$ as the 2×3 submatrix obtained by deleting, but not erasing, the first row of $\mathbf{A}$:

$$\begin{matrix} 1 & -2 & 0 \end{matrix}$$
$$\mathbf{B} = \begin{bmatrix} 0 & 1 & -1 \\ 0 & 5 & 2 \end{bmatrix}.$$

Identify the pivot of $\mathbf{B}$ as the first element of the second column. Perform the operation $-5\mathbf{r}_1+\mathbf{r}_2$ which makes the entry below the pivot equal to zero. We have

$$\begin{matrix} 1 & -2 & 0 \end{matrix}$$
$$\mathbf{B}_1 = \begin{bmatrix} 0 & 1 & -1 \\ 0 & 0 & 7 \end{bmatrix}.$$

Divide the last row by 7 and we have the row echelon form

$$\begin{bmatrix} 1 & -2 & 0 \\ 0 & 1 & -1 \\ 0 & 0 & 1 \end{bmatrix}.$$

3. Follow the technique in Example 3. Form the augmented matrix and apply row operations to put the system into row echelon form. We have

$$\left[\begin{array}{rrr|r} 1 & 2 & 1 & 0 \\ -3 & 3 & 2 & -7 \\ 4 & -2 & -3 & 2 \end{array}\right] \begin{matrix} \\ 3\mathbf{r}_1+\mathbf{r}_2 \\ -4\mathbf{r}_1+\mathbf{r}_3 \end{matrix} \rightarrow \left[\begin{array}{rrr|r} 1 & 2 & 1 & 0 \\ 0 & 9 & 5 & -7 \\ 0 & -10 & -7 & 2 \end{array}\right] (1/9)\mathbf{r}_2 \rightarrow$$

$$\left[\begin{array}{rrr|r} 1 & 2 & 1 & 0 \\ 0 & 1 & 5/9 & -7/9 \\ 0 & -10 & -7 & 2 \end{array}\right] 10\mathbf{r}_2+\mathbf{r}_3 \rightarrow$$

$$\begin{bmatrix} 1 & 2 & 1 & | & 0 \\ 0 & 1 & 5/9 & | & -7/9 \\ 0 & 0 & -13/9 & | & -52/9 \end{bmatrix} \begin{matrix} \\ \\ (-9/13)\mathbf{r}_3 \end{matrix} \rightarrow$$

$$\begin{bmatrix} 1 & 2 & 1 & | & 0 \\ 0 & 1 & 5/9 & | & -7/9 \\ 0 & 0 & 1 & | & 4 \end{bmatrix}.$$

Using back substitution on the final matrix we have:

$$x_3 = 4,\ x_2 = -7/9 - (5/9)x_3 = -3,\ x_1 = -x_3 - 2x_2 = 2.$$

The solution is $x_1 = 2$, $x_2 = -3$, $x_3 = 4$.

5. Follow the method in Example 4. Form the augmented matrix of the linear system and perform row operations with the partial pivoting strategy.

$$\begin{bmatrix} 3 & -2 & 3 & | & -8 \\ 6 & -4 & 5 & | & -14 \\ -12 & 6 & 7 & | & -8 \end{bmatrix}$$ Perform row operation $\mathbf{r}_1 \leftrightarrow \mathbf{r}_3$ to move the entry of largest magnitude in the first column to the (1,1) position.

The result is $\begin{bmatrix} -12 & 6 & 7 & | & -8 \\ 6 & -4 & 5 & | & -14 \\ 3 & -2 & 3 & | & -8 \end{bmatrix}$. Next perform row operation $(-1/12)\mathbf{r}_1$ to put a 1 in the pivot position. Do the calculations to three decimal places rounded. We get

$$\begin{bmatrix} 1.000 & -0.500 & -0.583 & | & 0.667 \\ 6.000 & -4.000 & 5.000 & | & -14.000 \\ 3.000 & -2.000 & 3.000 & | & -8.000 \end{bmatrix}$$

Apply row operations $-6\mathbf{r}_1+\mathbf{r}_2$ and $-3\mathbf{r}_1+\mathbf{r}_3$ to "zero-out" below the pivot in column 1:

$$\begin{bmatrix} 1.000 & -0.500 & -0.583 & | & 0.667 \\ 0.000 & -1.000 & 8.498 & | & -18.002 \\ 0.000 & -0.500 & 4.749 & | & -10.001 \end{bmatrix}.$$

To choose the second pivot, we inspect the (2,2) and (3,2) entries and use the one with largest absolute value. Clearly, here that is the (2,2) entry, -1.000. Since it is already in the (2,2) position, no row interchanges are needed. Next multiply row 2 by -1 to put a 1 into the second pivot position. We have

$$\left[\begin{array}{ccc|c} 1.000 & -0.500 & -0.583 & 0.667 \\ 0.000 & 1.000 & -8.498 & 18.002 \\ 0.000 & -0.500 & 4.749 & -10.001 \end{array}\right].$$

Use row operation $(1/2)\mathbf{r}_2+\mathbf{r}_3$ to "zero-out" below the pivot in column 2:

$$\left[\begin{array}{ccc|c} 1.000 & -0.500 & -0.583 & 0.667 \\ 0.000 & 1.000 & -8.498 & 18.002 \\ 0.000 & 0.000 & 0.500 & -1.000 \end{array}\right].$$

To get the row echelon form, use row operation $2\mathbf{r}_3$:

$$\left[\begin{array}{ccc|c} 1.000 & -0.500 & -0.583 & 0.667 \\ 0.000 & 1.000 & -8.498 & 18.002 \\ 0.000 & 0.000 & 1.000 & -2.000 \end{array}\right].$$

Using back substitution with three digit arithmetic we get

$$x_1 \simeq 0.004,\ x_2 \simeq 1.006,\ x_3 \simeq -2.000.$$

(We use the approximately equals symbol $\simeq$ because there may be a loss of accuracy since the calculations were performed in three digit arithmetic.)

7. Follow the method in Example 4. Form the augmented matrix of the linear system and perform row operations with the partial pivoting strategy.

$$\left[\begin{array}{ccc|c} 2.5 & 3.5 & -4.25 & 37.3 \\ 3.4 & 2.5 & -2.01 & 26.8 \\ 5.3 & -2.4 & 6.21 & -20.68 \end{array}\right]$$

Perform row operation $\mathbf{r}_1 \leftrightarrow \mathbf{r}_3$ to move the entry of largest magnitude in the first column to the (1,1) position. The result is $\left[\begin{array}{ccc|c} 5.3 & -2.4 & 6.21 & -20.68 \\ 3.4 & 2.5 & -2.01 & 26.8 \\ 2.5 & 3.5 & -4.25 & 37.3 \end{array}\right]$. Next perform row operation $(1/5.3)\mathbf{r}_1$ to put a 1 in the pivot position. Do the calculations to three decimal places rounded. We get

$$\left[\begin{array}{ccc|c} 1.000 & -0.453 & 1.172 & -3.902 \\ 3.400 & 2.500 & -2.010 & 26.800 \\ 2.500 & 3.500 & -4.250 & 37.300 \end{array}\right].$$

Apply row operations $-3.4\mathbf{r}_1+\mathbf{r}_2$ and $-2.5\mathbf{r}_1+\mathbf{r}_3$ to "zero-out" below the pivot in column 1:

$$\begin{bmatrix} 1.000 & -0.453 & 1.172 & | & -3.902 \\ 0.000 & 4.040 & -5.995 & | & 40.067 \\ 0.000 & 4.633 & -7.180 & | & 47.055 \end{bmatrix}.$$

To choose the next pivot we look at the (2,2) and (3,2) entries. The largest in absolute value, 4.633, is used as next pivot. Thus interchange rows 2 and 3 to move the pivot to the (2,2) position:

$$\begin{bmatrix} 1.000 & -0.453 & 1.172 & | & -3.902 \\ 0.000 & 4.633 & -7.180 & | & 47.055 \\ 0.000 & 4.040 & -5.995 & | & 40.067 \end{bmatrix}.$$

Next multiply row 2 by (1/4.633) to put a 1 into the second pivot position. We have

$$\begin{bmatrix} 1.000 & -0.453 & 1.172 & | & -3.902 \\ 0.000 & 1.000 & -1.550 & | & 10.156 \\ 0.000 & 4.040 & -5.995 & | & 40.067 \end{bmatrix}.$$

Use row operation $-4.040\mathbf{r}_2+\mathbf{r}_3$ to "zero-out" below the pivot in the second column:

$$\begin{bmatrix} 1.000 & -0.453 & 1.172 & | & -3.902 \\ 0.000 & 1.000 & -1.550 & | & 10.156 \\ 0.000 & 0.000 & 0.267 & | & -0.963 \end{bmatrix}.$$

To get the row echelon form, use row operation $(1/0.267)\mathbf{r}_3$:

$$\begin{bmatrix} 1.000 & -0.453 & 1.172 & | & -3.902 \\ 0.000 & 1.000 & -1.550 & | & 10.156 \\ 0.000 & 0.000 & 1.000 & | & -3.607 \end{bmatrix}.$$

Using back substitution with three digit arithmetic we get

$$x_1 \simeq 2.393,\ x_2 \simeq 4.565,\ x_3 \simeq -3.607.$$

(We use the approximately equals symbol $\simeq$ because there may be a loss of accuracy since the calculations were performed in three digit arithmetic.)

9. First solve the linear system

$$\begin{aligned} 2.121x + 3.421y &= 13.205 \\ 2.12x + \ \ 3.42y &= 13.200. \end{aligned}$$

Form the augmented matrix and use Gaussian elimination with partial pivoting. Round calculations to four decimal places.

$$\begin{bmatrix} 2.121 & 3.421 & | & 13.205 \\ 2.120 & 3.420 & | & 13.200 \end{bmatrix}$$

The first pivot is 2.121. Perform row operations $(1/2.121)\mathbf{r}_1$ and then $-2.120\mathbf{r}_1+\mathbf{r}_2$. The result is

$$\left[\begin{array}{cc|c} 1.0000 & 1.6129 & 6.2258 \\ 0.0000 & 0.0007 & 0.0013 \end{array}\right].$$

Row operation $(1/0.0007)\mathbf{r}_2$ gives the row echelon form

$$\left[\begin{array}{cc|c} 1.0000 & 1.6129 & 6.2258 \\ 0.0000 & 1.0000 & 1.8571 \end{array}\right].$$

Applying back substitution we have

$$x \simeq 3.2305,\ y \simeq 1.8571.$$

We solve the second system

$$\begin{aligned} 2.121x + 3.421y &= 13.205 \\ 2.12x + 3.42y &= 13.203. \end{aligned}$$

in a similar manner. Form the augmented matrix and use Gaussian elimination with partial pivoting. Round calculations to four decimal places.

$$\left[\begin{array}{cc|c} 2.121 & 3.421 & 13.205 \\ 2.120 & 3.420 & 13.203 \end{array}\right]$$

The first pivot is 2.121. Perform row operations $(1/2.121)\mathbf{r}_1$ and then $-2.120\mathbf{r}_1+\mathbf{r}_2$. The result is

$$\left[\begin{array}{cc|c} 1.0000 & 1.6129 & 6.2258 \\ 0.0000 & 0.0007 & 0.0043 \end{array}\right].$$

Row operation $(1/0.0007)\mathbf{r}_2$ gives the row echelon form

$$\left[\begin{array}{cc|c} 1.0000 & 1.6129 & 6.2258 \\ 0.0000 & 1.0000 & 6.1429 \end{array}\right].$$

Applying back substitution we have

$$x \simeq -3.6821,\ y \simeq 6.1429.$$

Thus a change of 0.003 in the right-hand side of the second equation has caused a much larger change in the solution of the system. In such cases we call the system ill-conditioned.

11. Use Jacobi's iteration method with three decimal calculations. Solve the first equation for x and the second for y:

$$\begin{aligned} x &= (-7/16) - (5/16)y = -0.438 - 0.313y \\ y &= (67/15) - (4/15)x = 4.467 - 0.267x. \end{aligned}$$

Let the initial guess be

$$x^{(0)} = 0 \text{ and } y^{(0)} = 0.$$

Compute

$$\begin{aligned} x^{(1)} &= -0.438 - 0.313y^{(0)} = -0.438 \\ y^{(1)} &= 4.467 - 0.267x^{(0)} = 4.467. \end{aligned}$$

Next compute

$$x^{(2)} = -0.438 - 0.313y^{(1)} = -1.836$$
$$y^{(2)} = 4.467 - 0.267x^{(1)} = 4.584.$$

We continue this process through $x^{(5)}$ and $y^{(5)}$. The calculations are summarized in the following table.

Jacobi's method

k	$x^{(k)}$	$y^{(k)}$
0	0.000	0.000
1	-0.438	4.467
2	-1.836	4.584
3	-1.873	4.957
4	-1.990	4.967
5	-1.993	4.998

Use Gauss-Seidel with the same initial guess. Then

$$x^{(1)} = -0.438 - 0.313y^{(0)} = -0.438$$
$$y^{(1)} = 4.467 - 0.267x^{(1)} = 4.584.$$

We use the value $x^{(1)}$ to compute $y^{(1)}$. Next compute

$$x^{(2)} = -0.438 - 0.313y^{(1)} = -1.873$$
$$y^{(2)} = 4.467 - 0.267x^{(2)} = 4.967.$$

Similarly we use $x^{(2)}$ to compute $y^{(2)}$. We continue this process through $x^{(5)}$ and $y^{(5)}$. The calculations are summarized in the following table.

Gauss-Seidel method

k	$x^{(k)}$	$y^{(k)}$
0	0.000	0.000
1	-0.438	4.584
2	-1.873	4.967
3	-1.993	4.999
4	-2.003	5.002
5	-2.004	5.002

13. Use Jacobi's iteration method with three decimal calculations. Solve the first equation for x_1, the second for x_2, and the third for x_3:

$$x_1 = 1 + (2/9)x_2 - (2/3)x_3 = 1.000 + 0.222x_2 - 0.667x_3$$

$$x_2 = 5/2 - (1/2)x_1 + (1/3)x_3 = 2.500 - 0.500x_1 + 0.333x_3$$

$$x_3 = 9/16 - (3/4)x_1 + (1/8)x_2 = 0.563 - 0.750x_1 + 0.125x_2.$$

Let the initial guess be

$$x_1^{(0)} = 0,\ x_2^{(0)} = 0,\ x_3^{(0)} = 0.$$

Substitute the initial values into the right side of the expressions for x_1, x_2, and x_3. The first approximation is

$$x_1^{(1)} = 1.000,\ x_2^{(1)} = 2.500,\ x_3^{(1)} = 0.563.$$

Next we compute

$$x_1^{(2)} = 1.000 + 0.222x_2^{(1)} - 0.667x_3^{(1)} = 1.179$$

$$x_2^{(2)} = 2.500 - 0.500x_1^{(1)} + 0.333x_3^{(1)} = 2.187$$

$$x_3^{(2)} = 0.563 - 0.750x_1^{(1)} + 0.125x_2^{(1)} = 0.126.$$

In a similar fashion we compute

$$x_1^{(3)} = 1.000 + 0.222x_2^{(2)} - 0.667x_3^{(2)} = 1.402$$

$$x_2^{(3)} = 2.500 - 0.500x_1^{(2)} + 0.333x_3^{(2)} = 1.952$$

$$x_3^{(3)} = 0.563 - 0.750x_1^{(2)} + 0.125x_2^{(2)} = -0.048.$$

Continuing in this way we compute $x_i^{(4)}$, and $x_i^{(5)}$ for i = 1, 2,3. The calculations are summarized in the following table.

Jacobi's method	k	$x_1^{(k)}$	$x_2^{(k)}$	$x_3^{(k)}$
	0	0.000	0.000	0.000
	1	1.000	2.500	0.563
	2	1.179	2.187	0.126
	3	1.402	1.952	-0.048
	4	1.465	1.783	-0.245
	5	1.559	1.685	-0.313

Use Gauss-Seidel with the same initial guess. Substitute the initial values into the right side of the expressions for x_1, x_2, and x_3. The first approximation is

$$x_1^{(1)} = 1.000 + 0.222x_2^{(0)} - 0.667x_3^{(0)} = 1.000$$

$$x_2^{(1)} = 2.500 - 0.500x_1^{(1)} + 0.333x_3^{(0)}$$

$$= 2.500 - 0.500(1.000) + 0.333(0) = 2.000$$

$$x_3^{(1)} = 0.563 - 0.750x_1^{(1)} + 0.125x_2^{(1)} =$$

$$= 0.563 - 0.750(1.000) + 0.125(2.000) = 0.063.$$

(Note: we used the most recent approximations to x_1 and x_2 as soon as they became available.)

Next we compute

$$x_1^{(2)} = 1.000 + 0.222x_2^{(1)} - 0.667x_3^{(1)} = 1.402$$

$$x_2^{(2)} = 2.500 - 0.500x_1^{(2)} + 0.333x_3^{(1)} = 1.820$$

$$x_3^{(2)} = 0.563 - 0.750x_1^{(2)} + 0.125x_2^{(2)} = -0.261$$

In a similar fashion we compute

$$x_1^{(3)} = 1.000 + 0.222x_2^{(2)} - 0.667x_3^{(2)} = 1.578$$

$$x_2^{(3)} = 2.500 - 0.500x_1^{(3)} + 0.333x_3^{(2)} = 1.624$$

$$x_3^{(3)} = 0.563 - 0.750x_1^{(3)} + 0.125x_2^{(3)} = -0.418$$

Continuing in this way we compute $x_i^{(4)}$ and $x_i^{(5)}$ for $i = 1,2,3$. The calculations are summarized in the following table.

Gauss-Seidel method

k	$x_1^{(k)}$	$x_2^{(k)}$	$x_3^{(k)}$
0	0.000	0.000	0.000
1	1.000	2.000	0.063
2	1.402	1.820	-0.261
3	1.578	1.624	-0.418
4	1.640	1.541	-0.474
5	1.658	1.513	-0.492

15. (a) Solve the first equation for x and the second equation for y:

$$x = 4 - 2y$$
$$y = 5 + x.$$

Let the initial guess be $x^{(0)} = 0$ and $y^{(0)} = 0$. Then the Gauss-Seidel iterations are computed from

$$x^{(k)} = 4 - 2y^{(k-1)}$$

$$y^{(k)} = 5 + x^{(k)}.$$

The calculations are summarized in the following table.

k	$x^{(k)}$	$y^{(k)}$
0	0	0
1	4	9
2	-14	-9
3	22	27
4	-50	-45
5	94	99
6	-194	-189

The values for x and y seem to be oscillating and are becoming large in absolute value. This suggests that the iterations are diverging.

(b) Solve the first equation for x and the second equation for y:

$$x = -4 + (1/2)y$$
$$y = -1 - x.$$

Let the initial guess be $x^{(0)} = 0$ and $y^{(0)} = 0$. Then the Gauss-Seidel iterations are computed from

$$x^{(k)} = -4 + (1/2)y^{(k-1)}$$

$$y^{(k)} = -1 - x^{(k)}.$$

The calculations are summarized in the following table.

k	$x^{(k)}$	$y^{(k)}$
0	0	0
1	-4	3
2	-5/2	3/2
3	-13/4	9/4
4	-23/8	15/8
5	-49/16	33/16
6	-95/32	63/32

The last approximation is:

$$x^{(6)} = -95/32 \simeq -2.969 \quad \text{and} \quad y^{(6)} = 63/32 \simeq 1.969.$$

The exact solution of this system is

$$x = -3 \quad \text{and} \quad y = 2.$$

Thus the iterations from the Gauss-Seidel method seem to be converging.

Exercises 8.2

ML.1. A=[1 2 3;2 -1 1;3 0 -1];B=[9 8 3]';
Q=reduce([A B])

(The following is a log of the routine **reduce**. A line of plus signs separates successive screens.)

```
++++++++++++++++++++++++++++++++++++++++++++++++++++++++++++++++++++++
            ***** "REDUCE" a Matrix by Row Reduction *****

The current matrix is:

A =

     1     2     3     9
     2    -1     1     8
     3     0    -1     3

            OPTIONS
 <1>   Interchange two rows.
 <2>   Multiply a row by a nonzero scalar.
 <3>   Replace a row by linear combination of
                itself with another row.
<-1>   "Undo" previous row operation.
 <0>   Quit reduce!

       ENTER your choice ---> 3

 Enter multiplier. -2

 Enter first row number. 1

 Enter number of row that changes. 2

++++++++++++++++++++++++++++++++++++++++++++++++++++++++++++++++++++++

   Replacement by Linear Combination Complete: -2 * Row 1 + Row 2.

The current matrix is:

A =

     1     2     3     9
     0    -5    -5   -10
     3     0    -1     3

            OPTIONS
 <1>   Interchange two rows.
 <2>   Multiply a row by a nonzero scalar.
 <3>   Replace a row by linear combination of
                itself with another row.
<-1>   "Undo" previous row operation.
 <0>   Quit reduce!

       ENTER your choice ---> 3
```

```
 Enter multiplier. -3

 Enter first row number. 1

 Enter number of row that changes. 3

++++++++++++++++++++++++++++++++++++++++++++++++++++++++++++++++++++++

Replacement by Linear Combination Complete: -3 * Row 1 + Row 3.

The current matrix is:

A =

     1     2     3     9
     0    -5    -5   -10
     0    -6   -10   -24

             OPTIONS
 <1>   Interchange two rows.
 <2>   Multiply a row by a nonzero scalar.
 <3>   Replace a row by linear combination of
               itself with another row.
<-1>   "Undo" previous row operation.
 <0>   Quit reduce!

       ENTER your choice ---> 2

 Enter multiplier. -1/5

 Enter row number. 2

 ++++++++++++++++++++++++++++++++++++++++++++++++++++++++++++++++++++

 Row Multiplication Complete: -0.2 * Row 2.

 The current matrix is:

A =

     1     2     3     9
     0     1     1     2
     0    -6   -10   -24

             OPTIONS
 <1>   Interchange two rows.
 <2>   Multiply a row by a nonzero scalar.
 <3>   Replace a row by linear combination of
               itself with another row.
<-1>   "Undo" previous row operation.
 <0>   Quit reduce!

       ENTER your choice ---> 3
```

Exercises 8.2

```
 Enter multiplier. 6

 Enter first row number. 2

 Enter number of row that changes. 3

 ++++++++++++++++++++++++++++++++++++++++++++++++++++++++++++++++++++++

 Replacement by Linear Combination Complete: 6 * Row 2 + Row 3.

 The current matrix is:

A =

     1     2     3     9
     0     1     1     2
     0     0    -4   -12

            OPTIONS
 <1>   Interchange two rows.
 <2>   Multiply a row by a nonzero scalar.
 <3>   Replace a row by linear combination of
                itself with another row.
<-1>   "Undo" previous row operation.
 <0>   Quit reduce!

       ENTER your choice ---> 2

 Enter multiplier. -1/4

 Enter row number. 3

+++++++++++++++++++++++++++++++++++++++++++++++++++++++++++++++++++++++

 Row Multiplication Complete: -0.25 * Row 3.

 The current matrix is:

A =

     1     2     3     9
     0     1     1     2
     0     0     1     3

            OPTIONS
 <1>   Interchange two rows.
 <2>   Multiply a row by a nonzero scalar.
 <3>   Replace a row by linear combination of
                itself with another row.
<-1>   "Undo" previous row operation.
 <0>   Quit reduce!

       ENTER your choice ---> 0

+++++++++++++++++++++++++++++++++++++++++++++++++++++++++++++++++++++++
```

```
*****  -->  REDUCE is over.

Your final matrix is:


A =

     1     2     3     9
     0     1     1     2
     0     0     1     3


Q =

     1     2     3     9
     0     1     1     2
     0     0     1     3

 X=bksub(Q(:,1:3),Q(:,4))

X =

     2
    -1
     3
```

ML.2. (a)
```
A=[3 4 -1;2 -2 0;1 -3 4];B=[-2 8 8]';
Q=reduce([A B])
```

(Here we omit the log from routine **reduce** and just show the final matrix and the result of the back substitution.)

```
Q =

     1    -3     4     8
     0     1    -2    -2
     0     0     1     0

X=bksub(Q(:,1:3),Q(:,4))

X =

     2
    -2
     0
```

(b)
```
A=[2 3 -2 1;0 4 -2 3;1 1 -2 1;1 2 3 0];B=[3 1 6 -12]';
Q=reduce([A B])
```

(Here we omit the log from routine **reduce** and just show the final matrix and the result of the back substitution.)

```
Q =

     1     2     3     0   -12
     0     1     5    -1   -18
     0     0     1     0    -3
     0     0     0     1     1

X=bksub(Q(:,1:4),Q(:,5))

X =

     1
    -2
    -3
     1
```

ML.3.
```
A=[3 -2 3;6 -4 5;-12 6 7];B=[-8 -14 -8]';
Q=reduce([A B])
```

```
+++++++++++++++++++++++++++++++++++++++++++++++++++++++++++++++++++++

              ***** "REDUCE" a Matrix by Row Reduction *****

The current matrix is:

A =

     3    -2     3    -8
     6    -4     5   -14
   -12     6     7    -8

            OPTIONS
 <1>  Interchange two rows.
 <2>  Multiply a row by a nonzero scalar.
 <3>  Replace a row by linear combination of
               itself with another row.
<-1>  "Undo" previous row operation.
 <0>  Quit reduce!

      ENTER your choice ---> 1

 Enter first row number. 1

 Enter second row number. 3

+++++++++++++++++++++++++++++++++++++++++++++++++++++++++++++++++++++

  Interchange Complete:  Row 1 <> Row 3.

The current matrix is:
```

```
A =

   -12      6      7     -8
     6     -4      5    -14
     3     -2      3     -8

             OPTIONS
 <1>   Interchange two rows.
 <2>   Multiply a row by a nonzero scalar.
 <3>   Replace a row by linear combination of
                itself with another row.
<-1>   "Undo" previous row operation.
 <0>   Quit reduce!

       ENTER your choice ---> 2

 Enter multiplier. -1/12

 Enter row number.  1

 +++++++++++++++++++++++++++++++++++++++++++++++++++++++++++++++++++++

 Row Multiplication Complete: -0.08333 * Row 1.

 The current matrix is:

A =

     1.0000    -0.5000    -0.5833     0.6667
     6.0000    -4.0000     5.0000   -14.0000
     3.0000    -2.0000     3.0000    -8.0000

             OPTIONS
 <1>   Interchange two rows.
 <2>   Multiply a row by a nonzero scalar.
 <3>   Replace a row by linear combination of
                itself with another row.
<-1>   "Undo" previous row operation.
 <0>   Quit reduce!

       ENTER your choice ---> 3

 Enter multiplier. -6

 Enter first row number. 1

 Enter number of row that changes. 2

++++++++++++++++++++++++++++++++++++++++++++++++++++++++++++++++++++++

   Replacement by Linear Combination Complete: -6 * Row 1 + Row 2.

The current matrix is:
```

Exercises 8.2

```
A =

    1.0000   -0.5000   -0.5833    0.6667
         0   -1.0000    8.5000  -18.0000
    3.0000   -2.0000    3.0000   -8.0000

            OPTIONS
 <1>   Interchange two rows.
 <2>   Multiply a row by a nonzero scalar.
 <3>   Replace a row by linear combination of
               itself with another row.
<-1>   "Undo" previous row operation.
 <0>   Quit reduce!

       ENTER your choice ---> 3

 Enter multiplier. -3

 Enter first row number. 1

 Enter number of row that changes. 3

++++++++++++++++++++++++++++++++++++++++++++++++++++++++++++++++++++++

  Replacement by Linear Combination Complete: -3 * Row 1 + Row 3.

The current matrix is:

A =

    1.0000   -0.5000   -0.5833    0.6667
         0   -1.0000    8.5000  -18.0000
         0   -0.5000    4.7500  -10.0000

            OPTIONS
 <1>   Interchange two rows.
 <2>   Multiply a row by a nonzero scalar.
 <3>   Replace a row by linear combination of
               itself with another row.
<-1>   "Undo" previous row operation.
 <0>   Quit reduce!

       ENTER your choice ---> 2

 Enter multiplier. -1

 Enter row number.  2

++++++++++++++++++++++++++++++++++++++++++++++++++++++++++++++++++++++

  Row Multiplication Complete: -1 * Row 2.

The current matrix is:
```

```
A =

    1.0000   -0.5000   -0.5833    0.6667
         0    1.0000   -8.5000   18.0000
         0   -0.5000    4.7500  -10.0000

         OPTIONS
 <1>  Interchange two rows.
 <2>  Multiply a row by a nonzero scalar.
 <3>  Replace a row by linear combination of
               itself with another row.
<-1>  "Undo" previous row operation.
 <0>  Quit reduce!

      ENTER your choice --->3

 Enter multiplier. 1/2

 Enter first row number. 2

 Enter number of row that changes.  3

++++++++++++++++++++++++++++++++++++++++++++++++++++++++++++++++++++++

 Replacement by Linear Combination Complete: 0.5 * Row 2 + Row 3.
    1.0000   -0.5000   -0.5833    0.6667
         0    1.0000   -8.5000   18.0000
         0         0    0.5000   -1.0000

         OPTIONS
 <1>  Interchange two rows.
 <2>  Multiply a row by a nonzero scalar.
 <3>  Replace a row by linear combination of
               itself with another row.
<-1>  "Undo" previous row operation.
 <0>  Quit reduce!

      ENTER your choice ---> 2

 Enter multiplier. 2

 Enter row number. 3

++++++++++++++++++++++++++++++++++++++++++++++++++++++++++++++++++++++

  Row Multiplication Complete: 2 * Row 3.

The current matrix is:

A =

    1.0000   -0.5000   -0.5833    0.6667
         0    1.0000   -8.5000   18.0000
         0         0    1.0000   -2.0000
```

Exercises 8.2

```
            OPTIONS
 <1>   Interchange two rows.
 <2>   Multiply a row by a nonzero scalar.
 <3>   Replace a row by linear combination of
                itself with another row.
<-1>   "Undo" previous row operation.
 <0>   Quit reduce!

       ENTER your choice ---> 0

+++++++++++++++++++++++++++++++++++++++++++++++++++++++++++++++++++++

*****  -->  REDUCE is over.

Your final matrix is:

A =

    1.0000   -0.5000   -0.5833    0.6667
         0    1.0000   -8.5000   18.0000
         0         0    1.0000   -2.0000


Q =

    1.0000   -0.5000   -0.5833    0.6667
         0    1.0000   -8.5000   18.0000
         0         0    1.0000   -2.0000

X=bksub(Q(:,1:3),Q(:,4))

X =

    0.0000
    1.0000
   -2.0000
```

ML.4. X=A\B

```
      X =

           0
           1
          -2
```

Exercises 8.3

1. Solve $\mathbf{LZ} = \begin{bmatrix} 2 & 0 & 0 \\ 2 & -3 & 0 \\ 1 & -1 & 4 \end{bmatrix} \begin{bmatrix} z_1 \\ z_2 \\ z_3 \end{bmatrix} = \mathbf{B} = \begin{bmatrix} 18 \\ 3 \\ 12 \end{bmatrix}$ by forward substitution:

$$\begin{aligned} z_1 &= 18/2 = 9 \\ z_2 &= (3 - 2z_1)/(-3) = (-15)/(-3) = 5 \\ z_3 &= (12 + z_2 - z_1)/4 = 8/4 = 2. \end{aligned}$$

Solve $\mathbf{UX} = \begin{bmatrix} 1 & 4 & 0 \\ 0 & 2 & 1 \\ 0 & 0 & 2 \end{bmatrix} \begin{bmatrix} x_1 \\ x_2 \\ x_3 \end{bmatrix} = \mathbf{Z} = \begin{bmatrix} 9 \\ 5 \\ 2 \end{bmatrix}$ by back substitution:

$$\begin{aligned} x_3 &= 2/2 = 1 \\ x_2 &= (5 - x_3)/2 = 4/2 = 2 \\ x_1 &= (9 - 0x_3 - 4x_2)/1 = 1/1 = 1. \end{aligned}$$

Thus the solution is $\mathbf{X} = \begin{bmatrix} 1 \\ 2 \\ 1 \end{bmatrix}$.

3. Solve $\mathbf{LZ} = \begin{bmatrix} 1 & 0 & 0 & 0 \\ 2 & 1 & 0 & 0 \\ -1 & 3 & 1 & 0 \\ 4 & 3 & 2 & 1 \end{bmatrix} \begin{bmatrix} z_1 \\ z_2 \\ z_3 \\ z_4 \end{bmatrix} = \mathbf{B} = \begin{bmatrix} -2 \\ -2 \\ -16 \\ -66 \end{bmatrix}$ by forward substitution:

$$\begin{aligned} z_1 &= -2/1 = -2 \\ z_2 &= (-2 - 2z_1)/1 = 2/1 = 2 \\ z_3 &= (-16 - 3z_2 + z_1)/1 = -24/1 = -24 \\ z_4 &= (-66 - 2z_3 - 3z_2 - 4z_1)/1 = -16/1 = -16. \end{aligned}$$

Solve $\mathbf{UX} = \begin{bmatrix} 2 & 3 & 0 & 1 \\ 0 & -1 & 3 & 1 \\ 0 & 0 & -2 & 5 \\ 0 & 0 & 0 & 4 \end{bmatrix} \begin{bmatrix} x_1 \\ x_2 \\ x_3 \\ x_4 \end{bmatrix} = \mathbf{Z} = \begin{bmatrix} -2 \\ 2 \\ -24 \\ -16 \end{bmatrix}$ by back substitution:

$$\begin{aligned} x_4 &= -16/4 = -4 \\ x_3 &= (-24 - 5x_4)/(-2) = -4/(-2) = 2 \\ x_2 &= (2 - x_4 - 3x_3)/(-1) = 0/(-1) = 0 \\ x_1 &= (-2 - x_4 + 0x_3 - 3x_2)/2 = 2/2 = 1. \end{aligned}$$

Thus the solution is $\mathbf{X} = \begin{bmatrix} 1 \\ 0 \\ 2 \\ -4 \end{bmatrix}$.

Exercises 8.3

5. To find an LU-factorization of $\mathbf{A} = \begin{bmatrix} 2 & 3 & 4 \\ 4 & 5 & 10 \\ 4 & 8 & 2 \end{bmatrix}$ we follow the procedure used in Example 3.

Step 1. "Zero out" below the first diagonal entry of **A**. Add -2 times the first row of **A** to the second row of **A**. That is, $-2\mathbf{r}_1+\mathbf{r}_2$. Add -2 times the first row of **A** to the third row of **A**. That is, $-2\mathbf{r}_1+\mathbf{r}_3$. Call the new matrix $\mathbf{U}_1$.

$$\mathbf{U}_1 = \begin{bmatrix} 2 & 3 & 4 \\ 0 & -1 & 2 \\ 0 & 2 & -6 \end{bmatrix}$$

We begin building a lower triangular matrix, with 1's on the main diagonal, to record the row operations. Enter the negatives of the multipliers used in the row operations in the first column of $\mathbf{L}_1$, below the first diagonal entry of $\mathbf{L}_1$.

$$\mathbf{L}_1 = \begin{bmatrix} 1 & 0 & 0 \\ 2 & 1 & 0 \\ 2 & * & 1 \end{bmatrix}$$

Step 2. "Zero out" below the second diagonal entry of $\mathbf{U}_1$. Add 2 times the second row of $\mathbf{U}_1$ to the third row of $\mathbf{U}_1$. That is, $2\mathbf{r}_2+\mathbf{r}_3$. Call the new matrix $\mathbf{U}_2$.

$$\mathbf{U}_2 = \begin{bmatrix} 2 & 3 & 4 \\ 0 & -1 & 2 \\ 0 & 0 & -2 \end{bmatrix}$$

Enter the negatives of the multipliers from the row operations below the second diagonal entry of $\mathbf{L}_1$. Call the new matrix $\mathbf{L}_2$.

$$\mathbf{L}_2 = \begin{bmatrix} 1 & 0 & 0 \\ 2 & 1 & 0 \\ 2 & -2 & 1 \end{bmatrix}$$

Let $L = L_2$ and $U = U_2$. Solve $\mathbf{LZ} = \begin{bmatrix} 1 & 0 & 0 \\ 2 & 1 & 0 \\ 2 & -2 & 1 \end{bmatrix}\begin{bmatrix} z_1 \\ z_2 \\ z_3 \end{bmatrix} = \mathbf{B} = \begin{bmatrix} 6 \\ 16 \\ 2 \end{bmatrix}$ by forward substitution:

$$z_1 = 6$$
$$z_2 = 16 - 2z_1 = 4$$
$$z_3 = 2 + 2z_2 - 2z_1 = -2.$$

Solve $\mathbf{UX} = \begin{bmatrix} 2 & 3 & 4 \\ 0 & -1 & 2 \\ 0 & 0 & -2 \end{bmatrix} \begin{bmatrix} x_1 \\ x_2 \\ x_3 \end{bmatrix} = \mathbf{Z} = \begin{bmatrix} 6 \\ 4 \\ -2 \end{bmatrix}$ by back substitution:

$$x_3 = (-2)/(-2) = 1$$
$$x_2 = (4 - 2x_3)/(-1) = -2$$
$$x_1 = (6 - 4x_3 - 3x_2)/2 = 4.$$

Thus the solution is $\mathbf{X} = \begin{bmatrix} 4 \\ -2 \\ 1 \end{bmatrix}$.

7. To find an LU-factorization of $\mathbf{A} = \begin{bmatrix} 4 & 2 & 3 \\ 2 & 0 & 5 \\ 1 & 2 & 1 \end{bmatrix}$ we follow the procedure used in Example 3.

Step 1. "Zero out" below the first diagonal entry of **A**. Add -1/2 times the first row of **A** to the second row of **A**. That is, $(-1/2)\mathbf{r}_1+\mathbf{r}_2$. Add -1/4 times the first row of **A** to the third row of **A**. That is, $(-1/4)\mathbf{r}_1+\mathbf{r}_3$. Call the new matrix $\mathbf{U}_1$.

$$\mathbf{U}_1 = \begin{bmatrix} 4 & 2 & 3 \\ 0 & -1 & 7/2 \\ 0 & 3/2 & 1/4 \end{bmatrix}$$

We begin building a lower triangular matrix, with 1's on the main diagonal, to record the row operations. Enter the negatives of the multipliers used in the row operations in the first column of $\mathbf{L}_1$, below the first diagonal entry of $\mathbf{L}_1$.

$$\mathbf{L}_1 = \begin{bmatrix} 1 & 0 & 0 \\ 1/2 & 1 & 0 \\ 1/4 & * & 1 \end{bmatrix}$$

Step 2. "Zero out" below the second diagonal entry of $\mathbf{U}_1$. Add 3/2 times the second row of $\mathbf{U}_1$ to the third row of $\mathbf{U}_1$. That is, $(3/2)\mathbf{r}_2+\mathbf{r}_3$. Call the new matrix $\mathbf{U}_2$.

$$\mathbf{U}_2 = \begin{bmatrix} 4 & 2 & 3 \\ 0 & -1 & 7/2 \\ 0 & 0 & 11/2 \end{bmatrix}$$

Enter the negatives of the multipliers from the row operations below the second diagonal entry of $\mathbf{L}_1$. Call the new matrix $\mathbf{L}_2$.

$$\mathbf{L}_2 = \begin{bmatrix} 1 & 0 & 0 \\ 1/2 & 1 & 0 \\ 1/4 & -3/2 & 1 \end{bmatrix}$$

Let $L = L_2$ and $U = U_2$. Use forward substitution to solve

$$\mathbf{LZ} = \begin{bmatrix} 1 & 0 & 0 \\ 1/2 & 1 & 0 \\ 1/4 & -3/2 & 1 \end{bmatrix}\begin{bmatrix} z_1 \\ z_2 \\ z_3 \end{bmatrix} = \mathbf{B} = \begin{bmatrix} 1 \\ -1 \\ -3 \end{bmatrix} :$$

$$\begin{aligned} z_1 &= 1 \\ z_2 &= -1 - (1/2)z_1 = -3/2 \\ z_3 &= -3 + (3/2)z_2 - (1/4)z_1 = -11/2. \end{aligned}$$

Solve $\mathbf{UX} = \begin{bmatrix} 4 & 2 & 3 \\ 0 & -1 & 7/2 \\ 0 & 0 & 11/2 \end{bmatrix}\begin{bmatrix} x_1 \\ x_2 \\ x_3 \end{bmatrix} = \mathbf{Z} = \begin{bmatrix} 1 \\ -3/2 \\ -11/2 \end{bmatrix}$ by back substitution:

$$\begin{aligned} x_3 &= -1 \\ x_2 &= ((-3/2) - (7/2)x_3)/(-1) = -2 \\ x_1 &= (1 - 3x_3 - 2x_2)/4 = 2. \end{aligned}$$

Thus the solution is $\mathbf{X} = \begin{bmatrix} 2 \\ -2 \\ -1 \end{bmatrix}$.

9. To find an LU-factorization of $\mathbf{A} = \begin{bmatrix} 2 & 1 & 0 & -4 \\ 1 & 0 & .25 & -1 \\ -2 & -1.1 & .25 & 6.2 \\ 4 & 2.2 & .30 & -2.4 \end{bmatrix}$ we follow the procedure used in Example 3.

Step 1. "Zero out" below the first diagonal entry of **A**. Add -.5 times the first row of **A** to the second row of **A**. That is, $-.5\mathbf{r}_1+\mathbf{r}_2$. Add 1 times the first row of **A** to the third row of **A**. That is, $\mathbf{r}_1+\mathbf{r}_3$. Add -2 times the first row of **A** to the fourth row of **A**. That is, $-2\mathbf{r}_1+\mathbf{r}_4$. Call the new matrix $\mathbf{U}_1$.

$$\mathbf{U}_1 = \begin{bmatrix} 2 & 1 & 0 & -4 \\ 0 & -.5 & .25 & 1 \\ 0 & -.1 & .25 & 2.2 \\ 0 & .2 & .30 & 5.6 \end{bmatrix}$$

We begin building a lower triangular matrix, with 1's on the main diagonal, to record the row operations. Enter the negatives of the multipliers used in the row operations in the first column of $\mathbf{L}_1$, below the first diagonal entry of $\mathbf{L}_1$.

$$L_1 = \begin{bmatrix} 1 & 0 & 0 & 0 \\ .5 & 1 & 0 & 0 \\ -1 & * & 1 & 0 \\ 2 & * & * & 1 \end{bmatrix}$$

Step 2. "Zero out" below the second diagonal entry of $\mathbf{U}_1$. Add $-.2$ times the second row of $\mathbf{U}_1$ to the third row of $\mathbf{U}_1$. That is, $-.2\mathbf{r}_2+\mathbf{r}_3$. Add .4 times the second row of $\mathbf{U}_1$ to the fourth row of $\mathbf{U}_1$. That is, $.4\mathbf{r}_2+\mathbf{r}_4$. Call the new matrix $\mathbf{U}_2$.

$$\mathbf{U}_2 = \begin{bmatrix} 2 & 1 & 0 & -4 \\ 0 & -.5 & .25 & 1 \\ 0 & 0 & .20 & 2 \\ 0 & 0 & .40 & 6 \end{bmatrix}$$

Enter the negatives of the multipliers from the row operations below the second diagonal entry of $\mathbf{L}_1$. Call the new matrix $\mathbf{L}_2$.

$$\mathbf{L}_2 = \begin{bmatrix} 1 & 0 & 0 & 0 \\ .5 & 1 & 0 & 0 \\ -1 & .2 & 1 & 0 \\ 2 & -.4 & * & 1 \end{bmatrix}$$

Step 3. "Zero out" below the third diagonal entry of $\mathbf{U}_2$. Add -2 times the third row of $\mathbf{U}_2$ to the fourth row of $\mathbf{U}_2$. That is, $-2\mathbf{r}_3+\mathbf{r}_4$. Call the new matrix $\mathbf{U}_3$.

$$\mathbf{U}_3 = \begin{bmatrix} 2 & 1 & 0 & -4 \\ 0 & -.5 & .25 & 1 \\ 0 & 0 & .20 & 2 \\ 0 & 0 & 0 & 2 \end{bmatrix}$$

Enter the negatives of the multipliers from the row operations below the third diagonal entry of $\mathbf{L}_2$. Call the new matrix $\mathbf{L}_3$.

$$\mathbf{L}_3 = \begin{bmatrix} 1 & 0 & 0 & 0 \\ .5 & 1 & 0 & 0 \\ -1 & .2 & 1 & 0 \\ 2 & -.4 & 2 & 1 \end{bmatrix}$$

Let $L = L_3$ and $U = U_3$. Use forward substitution to solve

$$\mathbf{LZ} = \begin{bmatrix} 1 & 0 & 0 & 0 \\ .5 & 1 & 0 & 0 \\ -1 & .2 & 1 & 0 \\ 2 & -.4 & 2 & 1 \end{bmatrix} \begin{bmatrix} z_1 \\ z_2 \\ z_3 \\ z_4 \end{bmatrix} = \mathbf{B} = \begin{bmatrix} -3 \\ -1.5 \\ 5.6 \\ 2.2 \end{bmatrix}:$$

$$\begin{aligned} z_1 &= -3 \\ z_2 &= -1.5 - .5z_1 = 0 \\ z_3 &= 5.6 - .2z_2 + z_1 = 2.6 \\ z_4 &= 2.2 - 2z_3 + .4z_2 - 2z_1 = 3. \end{aligned}$$

Solve $\mathbf{UX} = \begin{bmatrix} 2 & 1 & 0 & -4 \\ 0 & -.5 & .25 & 1 \\ 0 & 0 & .20 & 2 \\ 0 & 0 & 0 & 2 \end{bmatrix}\begin{bmatrix} x_1 \\ x_2 \\ x_3 \\ x_4 \end{bmatrix} = \mathbf{Z} = \begin{bmatrix} -3 \\ 0 \\ 2.6 \\ 3 \end{bmatrix}$ by back substitution:

$$\begin{aligned} x_4 &= 1.5 \\ x_3 &= (2.6 - 2x_4)/.2 = -2 \\ x_2 &= (0 - x_4 - .25x_3)/(-.5) = 2 \\ x_1 &= (-3 + 4x_4 - 0x_3 - x_2)/2 = .5. \end{aligned}$$

Thus the solution is $\mathbf{X} = \begin{bmatrix} .5 \\ 2 \\ -2 \\ 1.5 \end{bmatrix}$.

ML.1. We show the first few steps of the LU-factorization using routine **lupr** and then display the matrices **L** and **U**.

```
     [L,U]=lupr(A)

++++++++++++++++++++++++++++++++++++++++++++++++++++++++++++++++++++
          ***** Find an LU-FACTORIZATION by Row Reduction *****

L =                       U =
       1    0    0                 2    8    0
       0    1    0                 2    2   -3
       0    0    1                 1    2    7

          OPTIONS
 <1>  Replace a row by linear combination of
               itself with another row.
<-1>  "Undo" previous row operation.
 <0>  Quit lupr!

      ENTER your choice ---> 1

 Enter multiplier. -1

 Enter first row number. 1

 Enter number of row that changes. 2
```

```
++++++++++++++++++++++++++++++++++++++++++++++++++++++++++++++++++++++
  Replacement by Linear Combination Complete.

L =                          U =
       1    0    0                  2     8     0
       0    1    0                  0    -6    -3
       0    0    1                  1     2     7

You just performed operation -1*Row(1) + Row(2)

            OPTIONS
 <1>   Construct an element of matrix L.
<-1>   "Undo" previous row operation.
 <0>   Quit lupr!

      ENTER your choice ---> 1

++++++++++++++++++++++++++++++++++++++++++++++++++++++++++++++++++++++
      Replacement by Linear Combination Complete.

L =                          U =
       1    0    0                  2     8     0
       0    1    0                  0    -6    -3
       0    0    1                  1     2     7

You just performed operation -1*Row(1) + Row(2)

Insert a value in L in the position you just eliminated in U.
Let the multiplier you just used be called num.
It has the value -1.

 Enter row number of L to change. 2

 Enter column number of L to change. 1

 Value of L(2,1) = -num
  Correct:  L(2,1) = 1

++++++++++++++++++++++++++++++++++++++++++++++++++++++++++++++++++++++

   Continuing the factorization gives


L =                                        U =
            1          0         0                    2     8     0
            1          1         0                    0    -6    -3
          0.5     0.3333         1                    0     0     8
```

ML.2. We show the first few steps of the LU-factorization using routine lupr and then display the matrices **L** and **U**.

```
[L,U]=lupr(A)
```

Exercises 8.3

```
++++++++++++++++++++++++++++++++++++++++++++++++++++++++++++++++++++++
          ***** Find an LU-FACTORIZATION by Row Reduction *****

L =                        U =
        1    0    0                    8     -1      2
        0    1    0                    3      7      2
        0    0    1                    1      1      5

            OPTIONS
 <1>   Replace a row by linear combination of
               itself with another row.
<-1>   "Undo" previous row operation.
 <0>   Quit lupr!

      ENTER your choice ---> 1

 Enter multiplier. -3/8

 Enter first row number. 1

 Enter number of row that changes. 2

++++++++++++++++++++++++++++++++++++++++++++++++++++++++++++++++++++++
 Replacement by Linear Combination Complete.

L =                           U =
        1    0    0                       8         -1         2
        0    1    0                       0      7.375      1.25
        0    0    1                       1          1         5

You just performed operation -0.375*Row(1) + Row(2)

            OPTIONS
 <1>   Construct an element of matrix L.
<-1>   "Undo" previous row operation.
 <0>   Quit lupr!

      ENTER your choice ---> 1

++++++++++++++++++++++++++++++++++++++++++++++++++++++++++++++++++++++
      Replacement by Linear Combination Complete.

L =                           U =
        1    0    0                       8         -1         2
        0    1    0                       0      7.375      1.25
        0    0    1                       1          1         5

You just performed operation -0.375*Row(1) + Row(2)

Insert a value in L in the position you just eliminated in U.
Let the multiplier you just used be called num.
It has the value -0.375.
```

```
Enter row number of L to change. 2

Enter column number of L to change. 1

Value of L(2,1) = -num
Correct:  L(2,1) = 0.375
+++++++++++++++++++++++++++++++++++++++++++++++++++++++++++++++++++++++
```

Continuing the factorization process we obtain

```
L =                                         U =
          1          0          0                  8        -1         2
      0.375          1          0                  0     7.375      1.25
      0.125     0.1525          1                  0         0     4.559
```

Warning: It is recommended that the multipliers be written in terms of the entries of matrix **U** when entries are decimal expressions. For example, -U(3,2)/U(2,2). This assures that the exact numerical values are used rather than the decimal approximations shown on the screen. The preceding display of **L** and **U** appears in the routine **lupr**, but the following displays, which are shown upon exit from the routine, more accurately show the decimal values in the entries.

```
L =                                    U =

    1.0000         0         0          8.0000   -1.0000    2.0000
    0.3750    1.0000         0               0    7.3750    1.2500
    0.1250    0.1525    1.0000               0         0    4.5593
```

ML.3. We first use **lupr** to find an LU-factorization of **A**. The matrices **L** and **U** that we will find are different from those stated in Example 2. There can be many LU-factorizations for a matrix. (We omit the details from **lupr**. It is assumed that **A** and **B** have been entered

```
L =

    1.0000         0         0         0
    0.5000    1.0000         0         0
   -2.0000   -2.0000    1.0000         0
   -1.0000    1.0000   -2.0000    1.0000


U =

     6    -2    -4     4
     0    -2    -4    -1
     0     0     5    -2
     0     0     0     8


Z=forsub(L,B)
```

Exercises 8.3

```
Z =

    2
   -5
    2
  -32

X=bksub(U,Z)

X =

    4.5000
    6.9000
   -1.2000
   -4.0000
```

ML.4. The detailed steps of the solution of Exercises 7 and 8 are omitted. The solution to Exercise 7 is $[2 \quad -2 \quad -1]^T$ and the solution of Exercise 8 is $[1 \quad -2 \quad 5 \quad -4]^T$.

Exercises 8.4

1. Let $\mathbf{A} = \begin{bmatrix} -4 & 2 \\ 3 & 1 \end{bmatrix}$. Let the initial approximation to the eigenvector associated with the eigenvalue of largest magnitude be

$$\mathbf{U}_0 = \begin{bmatrix} 1 \\ 1 \end{bmatrix}.$$

Compute $\mathbf{AU}_0 = \mathbf{U}_1 = \begin{bmatrix} -2.000 \\ 4.000 \end{bmatrix}$. Let $u_{12} = 4.000$, the component of $\mathbf{U}_1$ with largest magnitude. Define

$$\mathbf{V}_1 = \frac{1}{|u_{12}|}\,\mathbf{U}_1 = \begin{bmatrix} -0.500 \\ 1.000 \end{bmatrix}.$$

Next compute $\mathbf{AV}_1 = \begin{bmatrix} 4.000 \\ -0.500 \end{bmatrix}$ and using inner products compute

$$\lambda_1 \cong \frac{(\mathbf{AV}_1, \mathbf{V}_1)}{(\mathbf{V}_1, \mathbf{V}_1)} = \frac{-2.500}{1.250} = -2.000.$$

This is our first approximation to the dominant eigenvalue and $\mathbf{V}_1$ is an approximation to the associated eigenvector. We repeat the process using $\mathbf{V}_1$ as the approximation to the eigenvector associated with the dominant eigenvalue. (It is important from a numerical point of view to use $\mathbf{V}_1$, not $\mathbf{U}_1$.)

Define $\mathbf{U}_2 = \mathbf{AV}_1 = \begin{bmatrix} 4.000 \\ -0.500 \end{bmatrix}$. Let $u_{21} = 4.000$, the component of $\mathbf{U}_2$ with largest magnitude. Define

$$\mathbf{V}_2 = \frac{1}{|u_{21}|}\,\mathbf{U}_2 = \begin{bmatrix} 1.000 \\ -0.125 \end{bmatrix}.$$

Next compute $\mathbf{AV}_2 = \begin{bmatrix} -4.250 \\ 2.875 \end{bmatrix}$ and using inner products compute

$$\lambda_1 \cong \frac{(\mathbf{AV}_2, \mathbf{V}_2)}{(\mathbf{V}_2, \mathbf{V}_2)} = \frac{-4.609}{1.016} = -4.536.$$

This is our second approximation to the dominant eigenvalue and $\mathbf{V}_2$ is an approximation to the associated eigenvector. We repeat the process using $\mathbf{V}_2$ as the approximation to the eigenvector associated with the dominant eigenvalue. (It is important from a numerical point of view to use $\mathbf{V}_2$, not $\mathbf{U}_2$.)

There are three more iterations to be done in a similar manner. We summarize the calculations in the following table.

k	U_k	V_k	λ_1
0	$[1 \quad 1]^T$		
1	$[-2.000 \quad 4.000]^T$	$[-0.500 \quad 1.000]^T$	-2.000
2	$[\ 4.000 \quad -0.500]^T$	$[\ 1.000 \quad -0.125]^T$	-4.536
3	$[-4.250 \quad 2.875]^T$	$[-1.000 \quad 0.676]^T$	-4.752
4	$[\ 5.352 \quad -2.324]^T$	$[\ 1.000 \quad -0.434]^T$	-5.035
5	$[-4.868 \quad 2.566]^T$	$[-1.000 \quad 0.527]^T$	-4.974

3. Let $\mathbf{A} = \begin{bmatrix} 4 & 3 \\ 2 & 3 \end{bmatrix}$. Let the initial approximation to the eigenvector associated with the eigenvalue of largest magnitude be

$$\mathbf{U}_0 = \begin{bmatrix} 1 \\ 1 \end{bmatrix}.$$

Compute $\mathbf{AU}_0 = \mathbf{U}_1 = \begin{bmatrix} 7.000 \\ 5.000 \end{bmatrix}$. Let $u_{11} = 7.000$, the component of $\mathbf{U}_1$ with largest magnitude. Define

$$\mathbf{V}_1 = \frac{1}{|u_{11}|}\,\mathbf{U}_1 = \begin{bmatrix} 1.000 \\ 0.714 \end{bmatrix}.$$

Next compute $\mathbf{AV}_1 = \begin{bmatrix} 6.142 \\ 4.142 \end{bmatrix}$ and using inner products compute

$$\lambda_1 \cong \frac{(\mathbf{AV}_1, \mathbf{V}_1)}{(\mathbf{V}_1, \mathbf{V}_1)} = \frac{9.099}{1.510} = 6.026.$$

This is our first approximation to the dominant eigenvalue and $\mathbf{V}_1$ is an approximation to the associated eigenvector. We repeat the process using $\mathbf{V}_1$ as the approximation to the eigenvector associated with the dominant eigenvalue. (It is important from a numerical point of view to use $\mathbf{V}_1$, not $\mathbf{U}_1$.)

Define $\mathbf{U}_2 = \mathbf{AV}_1 = \begin{bmatrix} 6.142 \\ 4.142 \end{bmatrix}$. Let $u_{21} = 6.142$, the component of $\mathbf{U}_2$ with largest magnitude. Define

$$\mathbf{V}_2 = \frac{1}{|u_{21}|}\,\mathbf{U}_2 = \begin{bmatrix} 1.000 \\ 0.674 \end{bmatrix}.$$

Next compute $\mathbf{AV}_2 = \begin{bmatrix} 6.022 \\ 4.022 \end{bmatrix}$ and using inner products compute

$$\lambda_1 \cong \frac{(\mathbf{AV}_2, \mathbf{V}_2)}{(\mathbf{V}_2, \mathbf{V}_2)} = \frac{8.733}{1.454} = 6.006.$$

This is our second approximation to the dominant eigenvalue and $\mathbf{V}_2$ is an approximation to the associated eigenvector. We repeat the process using $\mathbf{V}_2$ as the approximation to the eigenvector associated with the dominant eigenvalue. (It is important from a numerical point of view to use $\mathbf{V}_2$, not $\mathbf{U}_2$.)

There are three more iterations to be done in a similar manner. We summarize the calculations in the following table.

k	$\mathbf{U}_k$	$\mathbf{V}_k$	λ_1
0	$[1 \quad 1]^T$		
1	$[\,7.000 \quad 5.000]^T$	$[\,1.000 \quad 0.714]^T$	6.026
2	$[\,6.142 \quad 4.142]^T$	$[\,1.000 \quad 0.674]^T$	6.006
3	$[\,6.022 \quad 4.022]^T$	$[\,1.000 \quad 0.668]^T$	6.002
4	$[\,6.004 \quad 4.004]^T$	$[\,1.000 \quad 0.667]^T$	6.000
5	$[\,6.001 \quad 4.001]^T$	$[\,1.000 \quad 0.667]^T$	6.000

5. Let $\mathbf{A} = \begin{bmatrix} 8 & 8 \\ 3 & -2 \end{bmatrix}$. Let the initial approximation to the eigenvector associated with the eigenvalue of largest magnitude be

$$\mathbf{U}_0 = \begin{bmatrix} 1 \\ 1 \end{bmatrix}.$$

Compute $\mathbf{AU}_0 = \mathbf{U}_1 = \begin{bmatrix} 16.000 \\ 1.000 \end{bmatrix}$. Let $u_{11} = 16.000$, the component of $\mathbf{U}_1$ with largest magnitude. Define

$$\mathbf{V}_1 = \frac{1}{|u_{11}|}\,\mathbf{U}_1 = \begin{bmatrix} 1.000 \\ 0.063 \end{bmatrix}.$$

Next compute $\mathbf{AV}_1 = \begin{bmatrix} 8.504 \\ 2.874 \end{bmatrix}$ and using inner products compute

$$\lambda_1 \cong \frac{(\mathbf{AV}_1, \mathbf{V}_1)}{(\mathbf{V}_1, \mathbf{V}_1)} = \frac{8.685}{1.004} = 8.650.$$

This is our first approximation to the dominant eigenvalue and $\mathbf{V}_1$ is an approximation to the associated eigenvector. We repeat the process using $\mathbf{V}_1$ as the approximation to the eigenvector associated with the dominant eigenvalue. (It is important from a numerical point of view to use $\mathbf{V}_1$, not $\mathbf{U}_1$.)

Define $\mathbf{U}_2 = \mathbf{AV}_1 = \begin{bmatrix} 8.504 \\ 2.874 \end{bmatrix}$. Let $u_{21} = 8.504$, the component of $\mathbf{U}_2$ with largest magnitude. Define

$$\mathbf{V}_2 = \frac{1}{|u_{21}|}\mathbf{U}_2 = \begin{bmatrix} 1.000 \\ 0.338 \end{bmatrix}.$$

Next compute $\mathbf{AV}_2 = \begin{bmatrix} 10.704 \\ 2.324 \end{bmatrix}$ and using inner products compute

$$\lambda_1 \cong \frac{(\mathbf{AV}_2, \mathbf{V}_2)}{(\mathbf{V}_2, \mathbf{V}_2)} = \frac{11.490}{1.114} = 10.314.$$

This is our second approximation to the dominant eigenvalue and $\mathbf{V}_2$ is an approximation to the associated eigenvector. We repeat the process using $\mathbf{V}_2$ as the approximation to the eigenvector associated with the dominant eigenvalue. (It is important from a numerical point of view to use $\mathbf{V}_2$, not $\mathbf{U}_2$.)

There are three more iterations to be done in a similar manner. We summarize the calculations in the following table.

k	$\mathbf{U}_k$	$\mathbf{V}_k$	λ_1
0	$[1 \quad 1]^T$		
1	$[16.000 \quad 1.000]^T$	$[\ 1.000 \quad 0.063]^T$	8.650
2	$[\ 8.504 \quad 2.874]^T$	$[\ 1.000 \quad 0.338]^T$	10.314
3	$[10.704 \quad 2.324]^T$	$[\ 1.000 \quad 0.217]^T$	9.831
4	$[\ 9.736 \quad 2.566]^T$	$[\ 1.000 \quad 0.264]^T$	10.061
5	$[10.112 \quad 2.472]^T$	$[\ 1.000 \quad 0.244]^T$	9.974

7. Let $\mathbf{A} = \begin{bmatrix} 2 & 2 \\ 2 & 2 \end{bmatrix}$. Identify the nonzero element of $\mathbf{A}$ that has largest absolute value and is not on the main diagonal. Both a_{12} and a_{21} qualify. Choose a_{12}. Then compute,

$$f = -a_{12} = -2, \; g = (1/2)(a_{11} - a_{22}) = 0,$$

$$h = \frac{\text{sign}(g)\cdot f}{\sqrt{f^2+g^2}} = -1.000, \; \sin\theta = \frac{h}{\sqrt{2(1+\sqrt{1-h^2})}} = -0.707,$$

$$\cos\theta = \sqrt{1-\sin^2\theta} = .707.$$

Construct the 2 × 2 matrix

$$\mathbf{E}_1 = \mathbf{E}_{12} = \begin{bmatrix} \cos\theta & \sin\theta \\ -\sin\theta & \cos\theta \end{bmatrix} = \begin{bmatrix} 0.707 & -0.707 \\ 0.707 & 0.707 \end{bmatrix}.$$

Compute

$$\mathbf{B}_1 = \mathbf{E}_1^T(\mathbf{A}\mathbf{E}_1) = \mathbf{E}_1^T \begin{bmatrix} 2.828 & 0.000 \\ 2.828 & 0.000 \end{bmatrix}$$

$$= \begin{bmatrix} 3.998 & 0.000 \\ 0.000 & 0.000 \end{bmatrix}.$$

$\mathbf{B}_1$ is a diagonal matrix in the three decimal arithmetic used in the computations, so no further steps are needed. In this case $\mathbf{B}_2 = \mathbf{B}_3 = \mathbf{B}_1$. It follows that

$$\lambda_1 \simeq 3.998 \quad \text{and} \quad \lambda_2 \simeq 0.000$$

with associated approximate eigenvectors

$$\mathbf{X}_1 \simeq \begin{bmatrix} 0.707 \\ 0.707 \end{bmatrix} \quad \text{and} \quad \mathbf{X}_2 \simeq \begin{bmatrix} -0.707 \\ 0.707 \end{bmatrix},$$

the columns of $\mathbf{E}_1$.

9. Let $\mathbf{A} = \begin{bmatrix} 8 & -2 & 0 \\ -2 & 9 & -2 \\ 0 & -2 & 10 \end{bmatrix}$. Identify the nonzero element of $\mathbf{A}$ that has largest absolute value and is not on the main diagonal. All of a_{12}, a_{21}, a_{23}, and a_{32} qualify. Choose a_{12}. Then compute,

$$f = -a_{12} = 2, \; g = (1/2)(a_{11} - a_{22}) = -0.500$$

$$h = \frac{\text{sign}(g)\cdot f}{\sqrt{f^2+g^2}} = -0.970, \; \sin\theta = \frac{h}{\sqrt{2(1+\sqrt{1-h^2})}} = -0.615,$$

$$\cos\theta = \sqrt{1-\sin^2\theta} = 0.789.$$

Construct the 3 × 3 matrix

$$\mathbf{E}_1 = \mathbf{E}_{12} = \begin{bmatrix} \cos\theta & \sin\theta & 0 \\ -\sin\theta & \cos\theta & 0 \\ 0 & 0 & 1 \end{bmatrix} = \begin{bmatrix} 0.789 & -0.615 & 0.000 \\ 0.615 & 0.789 & 0.000 \\ 0.000 & 0.000 & 1.000 \end{bmatrix}.$$

Exercises 8.4

Compute

$$\mathbf{B}_1 = \mathbf{E}_1^T(\mathbf{A}\mathbf{E}_1) = \mathbf{E}_1^T\begin{bmatrix} 5.082 & -6.498 & 0.000 \\ 3.957 & 8.331 & -2.000 \\ -1.230 & -1.578 & 10.000 \end{bmatrix}$$

$$= \begin{bmatrix} 6.444 & -0.003 & -1.230 \\ -0.003 & 10.569 & -1.578 \\ -1.230 & -1.578 & 10.000 \end{bmatrix}.$$

Repeat this process with **A** replaced by $\mathbf{B}_1$. Identify the nonzero element of $\mathbf{B}_1$ that has largest absolute value and is not on the main diagonal. Choose a_{23}. Then compute,

$$f = -a_{23} = 1.578, \quad g = (1/2)(a_{22} - a_{33}) = 0.285,$$

$$h = \frac{\text{sign}(g)\cdot f}{\sqrt{f^2+g^2}} = 0.984, \quad \sin\theta = \frac{h}{\sqrt{2(1+\sqrt{1-h^2})}} = 0.641,$$

$$\cos\theta = \sqrt{1-\sin^2\theta} = 0.767.$$

Construct the 3×3 matrix

$$\mathbf{E}_2 = \mathbf{E}_{23} = \begin{bmatrix} 1.000 & 0.000 & 0.000 \\ 0.000 & \cos\theta & \sin\theta \\ 0.000 & -\sin\theta & \cos\theta \end{bmatrix} = \begin{bmatrix} 1.000 & 0.000 & 0.000 \\ 0.000 & 0.767 & 0.641 \\ 0.000 & -0.641 & 0.767 \end{bmatrix}.$$

Compute

$$\mathbf{B}_2 = \mathbf{E}_2^T(\mathbf{B}_1\mathbf{E}_2) = \mathbf{E}_2^T\begin{bmatrix} 6.444 & 0.786 & -0.945 \\ -0.003 & 9.117 & 5.565 \\ -1.230 & -7.620 & 6.659 \end{bmatrix}$$

$$= \begin{bmatrix} 6.444 & 0.786 & -0.945 \\ 0.786 & 11.877 & 0.000 \\ -0.945 & -0.001 & 8.674 \end{bmatrix}.$$

Repeat the process with $\mathbf{B}_2$ replacing $\mathbf{B}_1$. There are three more steps of this type required to find $\mathbf{B}_5$. We summarize these steps as follows:

	f	g	h	sin θ	cos θ
step 3	0.945	-1.115	-0.646	-0.344	0.939
step 4	-0.738	-2.890	0.247	0.124	0.992
step 5	0.269	-1.472	-0.180	-0.090	0.996

$$\mathbf{E}_3 = \mathbf{E}_{13} = \begin{bmatrix} 0.939 & 0.000 & -0.344 \\ 0.000 & 1.000 & 0.000 \\ 0.344 & 0.000 & 0.939 \end{bmatrix}, \mathbf{B}_3 = \begin{bmatrix} 6.098 & 0.738 & -0.001 \\ 0.738 & 11.877 & -0.270 \\ -0.001 & -0.271 & 9.021 \end{bmatrix}$$

$$\mathbf{E}_4 = \mathbf{E}_{12} = \begin{bmatrix} 0.992 & 0.124 & 0.000 \\ -0.124 & 0.992 & 0.000 \\ 0.000 & 0.000 & 1.000 \end{bmatrix}, \mathbf{B}_4 = \begin{bmatrix} 6.001 & 0.004 & 0.032 \\ 0.004 & 11.964 & -0.268 \\ 0.033 & -0.269 & 9.021 \end{bmatrix}$$

$$\mathbf{E}_5 = \mathbf{E}_{32} = \begin{bmatrix} 1.000 & 0.000 & 0.000 \\ 0.000 & 0.996 & 0.090 \\ 0.000 & -0.090 & 0.996 \end{bmatrix}, \mathbf{B}_5 = \begin{bmatrix} 6.001 & 0.001 & 0.032 \\ 0.001 & 11.989 & 0.001 \\ 0.033 & -0.001 & 8.998 \end{bmatrix}.$$

Stopping at this stage, we have approximate eigenvalues

$$\lambda_1 \simeq 6.001, \quad \lambda_2 \simeq 11.989, \quad \lambda_3 \simeq 8.998.$$

To find the corresponding eigenvectors we compute

$$\mathbf{E}_1(\mathbf{E}_2(\mathbf{E}_3(\mathbf{E}_4\mathbf{E}_5))) = \mathbf{P} = \begin{bmatrix} 0.666 & -0.333 & -0.666 \\ 0.666 & 0.666 & 0.331 \\ 0.334 & -0.666 & 0.667 \end{bmatrix}.$$

The columns of $\mathbf{P}$ are approximate eigenvectors

$$\mathbf{X}_1 \simeq \begin{bmatrix} 0.666 \\ 0.666 \\ 0.334 \end{bmatrix}, \quad \mathbf{X}_2 \simeq \begin{bmatrix} -0.333 \\ 0.666 \\ -0.666 \end{bmatrix}, \quad \mathbf{X}_3 \simeq \begin{bmatrix} -0.666 \\ 0.331 \\ 0.667 \end{bmatrix}$$

which are associated with λ_1, λ_2, and λ_3 respectively.

T.1. Let
$$\mathbf{E}_{pq} = \begin{bmatrix} 1 & 0 & \cdot & \cdot & \cdot & \cdot & \cdot & \cdot & \cdot & 0 \\ 0 & 1 & 0 & \cdot & \cdot & \cdot & \cdot & \cdot & \cdot & 0 \\ \cdot & & \cdot & & & & \cdot & & & \cdot \\ \cdot & & & \cdot & & & \cdot & & & \cdot \\ \cdot & & & & c & \cdot & s & & & \cdot \\ \cdot & & & & \cdot & \cdot & \cdot & & & \cdot \\ \cdot & & & & -s & \cdot & c & & & \cdot \\ \cdot & & & & & & & \cdot & & \cdot \\ \cdot & & & & & & & & \cdot & 0 \\ 0 & \cdot & \cdot & \cdot & \cdot & \cdot & \cdot & \cdot & 0 & 1 \end{bmatrix},$$ where

$c = \cos\theta$ and $s = \sin\theta$. Then $\mathbf{E}_{pq}\mathbf{E}_{pq}^T$ is a diagonal matrix with diagonal

$$[1 \cdots 1 \ \cos^2\theta + \sin^2\theta \ 1 \cdots 1 \ \cos^2\theta + \sin^2\theta \ 1 \cdots 1].$$

Thus $\mathbf{E}_{pq}\mathbf{E}_{pq}^T = \mathbf{I}_n$.

Chapter 8 Supplementary Exercises

1. (a) Use Jacobi's iteration method with three decimal calculations. Solve the first equation for x and the second for y:

$$x = (-24/10) - (3/10)y = -2.400 - 0.300y$$
$$y = (16/11) - (2/11)x = 1.455 - 0.182x.$$

Let the initial guess be

$$x^{(0)} = 0 \text{ and } y^{(0)} = 0.$$

Compute

$$x^{(1)} = -2.400 - 0.300y^{(0)} = -2.400$$
$$y^{(1)} = 1.455 - 0.182x^{(0)} = 1.455.$$

Next compute

$$x^{(2)} = -2.400 - 0.300y^{(1)} = -2.837$$
$$y^{(2)} = 1.455 - 0.182x^{(1)} = 1.892.$$

We continue this process through $x^{(5)}$ and $y^{(5)}$. The calculations are summarized in the following table.

Jacobi's method	k	$x^{(k)}$	$y^{(k)}$
	0	0.000	0.000
	1	-2.400	1.455
	2	-2.837	1.892
	3	-2.968	1.971
	4	-2.991	1.995
	5	-2.999	1.999

(b) Use Gauss-Seidel with the same initial guess. Then

$$x^{(1)} = -2.400 - 0.300y^{(0)} = -2.400$$
$$y^{(1)} = 1.455 - 0.182x^{(1)} = 1.892.$$

We use the value $x^{(1)}$ to compute $y^{(1)}$. Next compute

$$x^{(2)} = -2.400 - 0.300y^{(1)} = -2.968$$
$$y^{(2)} = 1.455 - 0.182x^{(2)} = 1.995.$$

Similarly we use $x^{(2)}$ to compute $y^{(2)}$. We continue this process through $x^{(5)}$ and $y^{(5)}$. The calculations are summarized in the following table.

Gauss-Seidel method

k	$x^{(k)}$	$y^{(k)}$
0	0.000	0.000
1	-2.400	1.892
2	-2.968	1.995
3	-2.999	2.001
4	-3.000	2.001
5	-3.000	2.001

3. To find an LU-factorization of $\mathbf{A} = \begin{bmatrix} -2 & 1 & -2 \\ 6 & 1 & 9 \\ -4 & 18 & 5 \end{bmatrix}$ we follow the procedure used in Example 3 in Section 8.3.

Step 1. "Zero out" below the first diagonal entry of **A**. Add 3 times the first row of **A** to the second row of **A**. That is, $3\mathbf{r}_1+\mathbf{r}_2$. Add -2 times the first row of **A** to the third row of **A**. That is, $-2\mathbf{r}_1+\mathbf{r}_3$. Call the new matrix $\mathbf{U}_1$.

$$\mathbf{U}_1 = \begin{bmatrix} -2 & 1 & -2 \\ 0 & 4 & 3 \\ 0 & 16 & 9 \end{bmatrix}$$

We begin building a lower triangular matrix, with 1's on the main diagonal, to record the row operations. Enter the negatives of the multipliers used in the row operations in the first column of $\mathbf{L}_1$, below the first diagonal entry of $\mathbf{L}_1$.

$$\mathbf{L}_1 = \begin{bmatrix} 1 & 0 & 0 \\ -3 & 1 & 0 \\ 2 & * & 1 \end{bmatrix}$$

Step 2. "Zero out" below the second diagonal entry of $\mathbf{U}_1$. Add -4 times the second row of $\mathbf{U}_1$ to the third row of $\mathbf{U}_1$. That is, $-4\mathbf{r}_2+\mathbf{r}_3$. Call the new matrix $\mathbf{U}_2$.

$$\mathbf{U}_2 = \begin{bmatrix} -2 & 1 & -2 \\ 0 & 4 & 3 \\ 0 & 0 & -3 \end{bmatrix}$$

Enter the negatives of the multipliers from the row operations below the second diagonal entry of $\mathbf{L}_1$. Call the new matrix $\mathbf{L}_2$.

$$\mathbf{L}_2 = \begin{bmatrix} 1 & 0 & 0 \\ -3 & 1 & 0 \\ 2 & 4 & 1 \end{bmatrix}$$

Let L = L_2 and U = U_2. Solve $\mathbf{LZ} = \begin{bmatrix} 1 & 0 & 0 \\ -3 & 1 & 0 \\ 2 & 4 & 1 \end{bmatrix}\begin{bmatrix} z_1 \\ z_2 \\ z_3 \end{bmatrix} = \mathbf{B} = \begin{bmatrix} -6 \\ 19 \\ -17 \end{bmatrix}$

by forward substitution:

$$z_1 = -6$$
$$z_2 = 19 + 3z_1 = 1$$
$$z_3 = -17 - 4z_2 - 2z_1 = -9.$$

Solve $\mathbf{UX} = \begin{bmatrix} -2 & 1 & -2 \\ 0 & 4 & 3 \\ 0 & 0 & -3 \end{bmatrix}\begin{bmatrix} x_1 \\ x_2 \\ x_3 \end{bmatrix} = \mathbf{Z} = \begin{bmatrix} -6 \\ 1 \\ -9 \end{bmatrix}$ by back substitution:

$$x_3 = (-9)/(-3) = 3$$
$$x_2 = (1 - 3x_3)/4 = -2$$
$$x_1 = (-6 + 2x_3 - x_2)/(-2) = -1.$$

Thus the solution is $\mathbf{X} = \begin{bmatrix} -1 \\ -2 \\ 3 \end{bmatrix}$.

5. Let $\mathbf{A} = \begin{bmatrix} 1 & -1 & -1 \\ -1 & 1 & -1 \\ -1 & -1 & 1 \end{bmatrix}$. Identify the nonzero element of **A** that has largest absolute value and is not on the main diagonal. All of a_{12}, a_{21}, a_{13}, a_{31}, a_{23}, and a_{32} qualify. Choose a_{12}. Then compute,

$$f = -a_{12} = 1, \; g = (1/2)(a_{11} - a_{22}) = 0$$

$$h = \frac{\text{sign}(g)\cdot f}{\sqrt{f^2+g^2}} = 1.000, \; \sin\theta = \frac{h}{\sqrt{2(1+\sqrt{1-h^2})}} = 0.707,$$

$$\cos\theta = \sqrt{1-\sin^2\theta} = 0.707.$$

Construct the 3×3 matrix

$$\mathbf{E}_1 = \mathbf{E}_{12} = \begin{bmatrix} \cos\theta & \sin\theta & 0 \\ -\sin\theta & \cos\theta & 0 \\ 0 & 0 & 1 \end{bmatrix} = \begin{bmatrix} 0.707 & 0.707 & 0.000 \\ -0.707 & 0.707 & 0.000 \\ 0.000 & 0.000 & 1.000 \end{bmatrix}.$$

Compute

$$\mathbf{B}_1 = \mathbf{E}_1{}^T(\mathbf{A}\mathbf{E}_1) = \mathbf{E}_1{}^T \begin{bmatrix} 1.414 & 0.000 & -1.000 \\ -1.414 & 0.000 & -1.000 \\ 0.000 & -1.414 & 1.000 \end{bmatrix}$$

$$= \begin{bmatrix} 2.000 & 0.000 & 0.000 \\ 0.000 & 0.000 & -1.414 \\ 0.000 & -1.414 & 1.000 \end{bmatrix}.$$

Repeat this process with **A** replaced by $\mathbf{B}_1$. Identify the nonzero element of $\mathbf{B}_1$ that has largest absolute value and is not on the main diagonal. Choose a_{23}. Then compute,

$$f = -a_{23} = 1.414, \quad g = (1/2)(a_{22} - a_{33}) = -0.500,$$

$$h = \frac{\text{sign}(g)\cdot f}{\sqrt{f^2+g^2}} = -0.943, \quad \sin\theta = \frac{h}{\sqrt{2(1+\sqrt{1-h^2})}} = -0.577,$$

$$\cos\theta = \sqrt{1-\sin^2\theta} = 0.817.$$

Construct the 3 × 3 matrix

$$\mathbf{E}_2 = \mathbf{E}_{23} = \begin{bmatrix} 1.000 & 0.000 & 0.000 \\ 0.000 & \cos\theta & \sin\theta \\ 0.000 & -\sin\theta & \cos\theta \end{bmatrix} = \begin{bmatrix} 1.000 & 0.000 & 0.000 \\ 0.000 & 0.817 & -0.577 \\ 0.000 & 0.577 & 0.817 \end{bmatrix}.$$

Compute

$$\mathbf{B}_2 = \mathbf{E}_2{}^T(\mathbf{B}_1\mathbf{E}_2) = \mathbf{E}_2{}^T \begin{bmatrix} 2.000 & 0.000 & 0.000 \\ 0.000 & -0.816 & -1.155 \\ 0.000 & -0.578 & 1.633 \end{bmatrix}$$

$$= \begin{bmatrix} 2.000 & 0.000 & 0.000 \\ 0.000 & -1.001 & -0.002 \\ 0.000 & -0.001 & 2.000 \end{bmatrix}.$$

Stopping at this stage, we have approximate eigenvalues

$$\lambda_1 \simeq 2.000, \quad \lambda_2 \simeq -1.001, \quad \lambda_3 \simeq 2.000.$$

To find the corresponding eigenvectors we compute

$$\mathbf{E}_1\mathbf{E}_2 = \mathbf{P} = \begin{bmatrix} 0.707 & 0.578 & -0.408 \\ -0.707 & 0.578 & -0.408 \\ 0.000 & -0.577 & 0.817 \end{bmatrix}.$$

The columns of $\mathbf{P}$ are approximate eigenvectors

$$\mathbf{x}_1 \simeq \begin{bmatrix} 0.707 \\ -0.707 \\ 0.000 \end{bmatrix}, \quad \mathbf{x}_2 \simeq \begin{bmatrix} 0.578 \\ 0.578 \\ -0.577 \end{bmatrix}, \quad \mathbf{x}_3 \simeq \begin{bmatrix} -0.408 \\ -0.408 \\ 0.817 \end{bmatrix}$$

which are associated with λ_1, λ_2, and λ_3 respectively.

Exercises A.1

1. Let $c_1 = 3 + 4i$, $c_2 = 1 - 2i$, and $c_3 = -1 + i$.

(a) $c_1 + c_2 = (3 + 4i) + (1 - 2i) = (3 + 1) + (4 - 2)i$
$= 4 + 2i$

(b) $c_3 - c_1 = (-1 + i) - (3 + 4i) = (-1 - 3) + (1 - 4)i$
$= -4 - 3i$

(c) $c_1c_2 = (3 + 4i)(1 - 2i)$
$= ((3)(1) - (4)(-2)) + ((3)(-2) + (4)(1))i = 11 - 2i$

(d) $c_2\overline{c_3} = (1 - 2i)(\overline{-1 + i}) = (1 - 2i)(-1 - i) = -3 + i$

(e) $4c_3 + \overline{c_2} = (-4 + 4i) + (1 + 2i) = -3 + 6i$

(f) $(-i)c_2 = (-i)(1 - 2i) = -i + 2i^2 = -2 - i$

(g) $\overline{3c_1 - ic_2} = \overline{(9 + 12i) - (i - 2i^2)} = \overline{7 + 11i} = 7 - 11i$

(h) $c_1c_2c_3 = (c_1c_2)c_3 = (11 - 2i)(-1 + i) = -9 + 13i$
(See part (c).)

3.

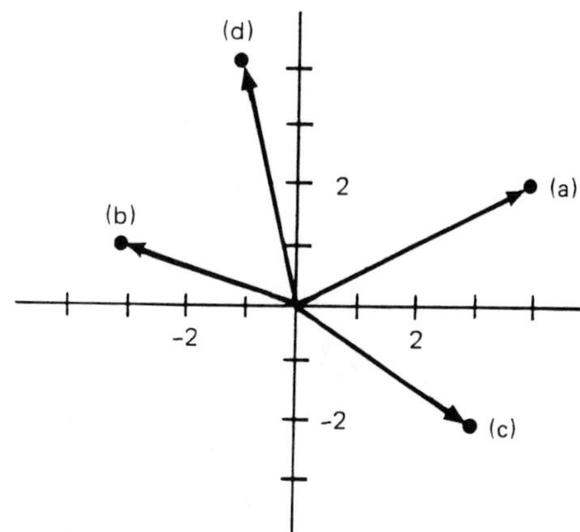

5.

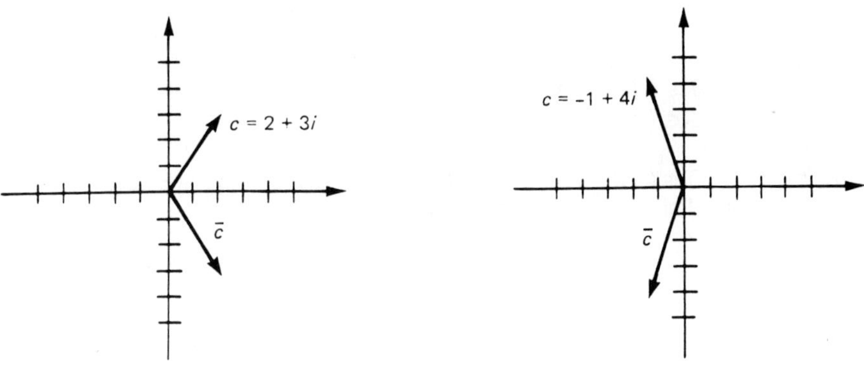

Exercises A.1

7. $\mathbf{A}^2 = \begin{bmatrix} -1 & 0 \\ 0 & -1 \end{bmatrix} = -\mathbf{I}_2$, $\mathbf{A}^3 = \mathbf{A}^2\mathbf{A} = -\mathbf{I}_2\mathbf{A} = -\mathbf{A}$, $\mathbf{A}^4 = \mathbf{A}^2\mathbf{A}^2 = \mathbf{I}_2$, and in general $\mathbf{A}^{4n} = \mathbf{I}_2$, $\mathbf{A}^{4n+1} = \mathbf{A}$, $\mathbf{A}^{4n+2} = \mathbf{A}^2 = -\mathbf{I}_2$, $\mathbf{A}^{4n+3} = -\mathbf{A}$.

9. (a) $\mathbf{A}^2 = \begin{bmatrix} 9 & 0 \\ 0 & 9 \end{bmatrix}$. Then

$$p(\mathbf{A}) = 2\begin{bmatrix} 9 & 0 \\ 0 & 9 \end{bmatrix} + 5\begin{bmatrix} -3 & 0 \\ 0 & -3 \end{bmatrix} - 3\begin{bmatrix} 1 & 0 \\ 0 & 1 \end{bmatrix} = \begin{bmatrix} 0 & 0 \\ 0 & 0 \end{bmatrix}.$$

(b) $\mathbf{A}^2 = \begin{bmatrix} 1 & 4 \\ 0 & 1 \end{bmatrix}$. Then

$$p(\mathbf{A}) = 2\begin{bmatrix} 1 & 4 \\ 0 & 1 \end{bmatrix} + 5\begin{bmatrix} 1 & 2 \\ 0 & 1 \end{bmatrix} - 3\begin{bmatrix} 1 & 0 \\ 0 & 1 \end{bmatrix} = \begin{bmatrix} 4 & 18 \\ 0 & 4 \end{bmatrix}.$$

(c) $\mathbf{A}^2 = \begin{bmatrix} -1 & 0 \\ 0 & -1 \end{bmatrix}$. Then

$$p(\mathbf{A}) = 2\begin{bmatrix} -1 & 0 \\ 0 & -1 \end{bmatrix} + 5\begin{bmatrix} 0 & i \\ i & 0 \end{bmatrix} - 3\begin{bmatrix} 1 & 0 \\ 0 & 1 \end{bmatrix} = \begin{bmatrix} -5 & 5i \\ 5i & -5 \end{bmatrix}.$$

(d) $\mathbf{A}^2 = \begin{bmatrix} 1 & i \\ 0 & 0 \end{bmatrix}$. Then

$$p(\mathbf{A}) = 2\begin{bmatrix} 1 & i \\ 0 & 0 \end{bmatrix} + 5\begin{bmatrix} 1 & i \\ 0 & 0 \end{bmatrix} - 3\begin{bmatrix} 1 & 0 \\ 0 & 1 \end{bmatrix} = \begin{bmatrix} 4 & 7i \\ 0 & -3 \end{bmatrix}.$$

11. $p(k\mathbf{I}_2) = (k\mathbf{I}_2)^2 - (k\mathbf{I}_2) - 2\mathbf{I}_2 = (k^2-k-2)\mathbf{I}_2 = \mathbf{0}$ if and only if $k^2-k-2 = (k-2)(k+1) = 0$. Thus $k = 2$ or $k = -1$. Hence $\begin{bmatrix} 2 & 0 \\ 0 & 2 \end{bmatrix} = 2\mathbf{I}_2$ and $\begin{bmatrix} -1 & 0 \\ 0 & -1 \end{bmatrix} = -1\mathbf{I}_2$ are the only solutions.

T.1. (a) $\text{Re}(c_1 + c_2) = \text{Re}((a_1+a_2) + (b_1+b_2)i) = a_1 + a_2 = \text{Re}(c_1) + \text{Re}(c_2)$

$\text{Im}(c_1 + c_2) = \text{Im}((a_1+a_2) + (b_1+b_2)i) = b_1 + b_2 = \text{Im}(c_1) + \text{Im}(c_2)$

(b) $\mathrm{Re}(kc) = \mathrm{Re}(ka + kbi) = ka = k\mathrm{Re}(c)$

$\mathrm{Im}(kc) = \mathrm{Im}(ka + kbi) = kb = k\mathrm{Im}(c)$

(c) No.

(d) $\mathrm{Re}(c_1c_2) = \mathrm{Re}((a_1+b_1i)(a_2+b_2i))$

$= \mathrm{Re}((a_1a_2-b_1b_2) + (a_1b_2+a_2b_1)i)$

$= a_1a_2-b_1b_2 \neq a_1a_2 = \mathrm{Re}(c_1)\ \mathrm{Re}(c_2)$

Exercises A.2

1. Form the augmented matrix and use row operations.

(a) $\left[\begin{array}{cc|c} 1+2i & -2+i & 1-3i \\ 2+i & -1+2i & -1-i \end{array}\right] \xrightarrow{1/(1+2i)\mathbf{r}_1}$

$\left[\begin{array}{cc|c} 1 & i & -1-i \\ 2+i & -1+2i & -1-i \end{array}\right] \xrightarrow{-(2+i)\mathbf{r}_1+\mathbf{r}_2} \left[\begin{array}{cc|c} 1 & i & -1-i \\ 0 & 0 & 2i \end{array}\right]$

The system is inconsistent; there is no solution.

(b) $\left[\begin{array}{cc|c} 2i & -(1-i) & 1+i \\ 1-i & 1 & 1-i \end{array}\right] \xrightarrow{1/(2i)\mathbf{r}_1}$

$\left[\begin{array}{cc|c} 1 & (1/2)+(1/2)i & (1/2)-(1/2)i \\ 1-i & 1 & 1-i \end{array}\right] \xrightarrow{-(1-i)\mathbf{r}_1+\mathbf{r}_2}$

$\left[\begin{array}{cc|c} 1 & (1/2)+(1/2)i & (1/2)-(1/2)i \\ 0 & 0 & 1 \end{array}\right]$

The system is inconsistent; there is no solution.

(c) $\left[\begin{array}{cc|c} 1+i & -1 & -2+i \\ 2i & 1-i & i \end{array}\right] \xrightarrow{1/(1+i)\mathbf{r}_1}$

$\left[\begin{array}{cc|c} 1 & (-1/2)+(1/2)i & (-1/2)+(3/2)i \\ 2i & 1-i & i \end{array}\right] \xrightarrow{-2i\mathbf{r}_1+\mathbf{r}_2}$

$\left[\begin{array}{cc|c} 1 & (-1/2)+(1/2)i & (-1/2)+(3/2)i \\ 0 & 2 & 3+2i \end{array}\right] \xrightarrow{(1/2)\mathbf{r}_2}$

$\left[\begin{array}{cc|c} 1 & (-1/2)+(1/2)i & (-1/2)+(3/2)i \\ 0 & 1 & (3/2)+i \end{array}\right] \xrightarrow{\frac{1-i}{2}\mathbf{r}_2+\mathbf{r}_1}$

$\left[\begin{array}{cc|c} 1 & 0 & (3/4)+(5/4)i \\ 0 & 1 & (3/2)+i \end{array}\right]$

Thus the solution is

$$x_1 = (3/4) + (5/4)i \text{ and } x_2 = (3/2) + i.$$

3. Form the augmented matrix and apply row operations to obtain row echelon form, then apply back substitution.

(a) $\left[\begin{array}{ccc|c} i & 1+i & 0 & i \\ 1-i & 1 & -i & 1 \\ 0 & i & 1 & 1 \end{array}\right] \begin{array}{l} (1/i)\mathbf{r}_1 \\ -(1-i)\mathbf{r}_1+\mathbf{r}_2 \end{array} \rightarrow$

$\left[\begin{array}{ccc|c} 1 & 1-i & 0 & 1 \\ 0 & 1+2i & -i & i \\ 0 & i & 1 & 1 \end{array}\right] \begin{array}{l} (1/(1+2i))\mathbf{r}_2 \\ -i\mathbf{r}_2+\mathbf{r}_3 \end{array} \rightarrow$

$\left[\begin{array}{ccc|c} 1 & 1-i & 0 & 1 \\ 0 & 1 & (-2/5)-(1/5)i & (2/5)+(1/5)i \\ 0 & 0 & (4/5)+(2/5)i & (6/5)-(2/5)i \end{array}\right]$

Next apply row operation $(1/((4/5)+(2/5)i))\mathbf{r}_3$ to obtain

$\left[\begin{array}{ccc|c} 1 & 1-i & 0 & 1 \\ 0 & 1 & (-2/5)-(1/5)i & (2/5)+(1/5)i \\ 0 & 0 & 1 & 1-i \end{array}\right]$

Use back substitution:

$x_3 = 1-i$
$x_2 = (2/5)+(1/5)i - ((-2/5)-(1/5)i)x_3 = 1$
$x_1 = 1-(1-i)x_2 = i$.

The solution is $\mathbf{X} = \begin{bmatrix} i \\ 1 \\ 1-i \end{bmatrix}$.

(b) $\left[\begin{array}{ccc|c} 1 & i & 1-i & 2+i \\ i & 0 & 1+i & -1+i \\ 0 & 2i & -1 & 2-i \end{array}\right] -i\mathbf{r}_1+\mathbf{r}_2 \rightarrow$

$\left[\begin{array}{ccc|c} 1 & i & 1-i & 2+i \\ 0 & 1 & 0 & -i \\ 0 & 2i & -1 & 2-i \end{array}\right] -2i\mathbf{r}_2+\mathbf{r}_3 \rightarrow \left[\begin{array}{ccc|c} 1 & i & 1-i & 2+i \\ 0 & 1 & 0 & -i \\ 0 & 0 & -1 & -i \end{array}\right] -1\mathbf{r}_3 \rightarrow$

$\left[\begin{array}{ccc|c} 1 & 1 & 1-i & 2+i \\ 0 & 1 & 0 & -i \\ 0 & 0 & 1 & i \end{array}\right]$ Use back substitution:

$x_3 = i$, $x_2 = -i$, $x_1 = (2+i) - (1-i)x_3 - x_2 = 0$. Thus the solution is $\mathbf{X} = \begin{bmatrix} 0 \\ -i \\ i \end{bmatrix}$.

5. (a) $\left[\begin{array}{cc|cc} i & 2 & 1 & 0 \\ 1+i & -i & 0 & 1 \end{array}\right] \xrightarrow[-(1+i)\mathbf{r}_1+\mathbf{r}_2]{(1/i)\mathbf{r}_1}$

$\left[\begin{array}{cc|cc} 1 & -2i & -i & 0 \\ 0 & -2+i & -1+i & 1 \end{array}\right] \xrightarrow[2i\mathbf{r}_2+\mathbf{r}_1]{1/(-2+i)\mathbf{r}_2}$

$\left[\begin{array}{cc|cc} 1 & 0 & (2/5)+(1/5)i & (2/5)-(4/5)i \\ 0 & 1 & (3/5)-(1/5)i & (-2/5)-(1/5)i \end{array}\right]$

Thus the inverse is $\begin{bmatrix} (2/5)+(1/5)i & (2/5)-(4/5)i \\ (3/5)-(1/5)i & (-2/5)-(1/5)i \end{bmatrix}$.

(b) $\left[\begin{array}{ccc|ccc} 2 & i & 3 & 1 & 0 & 0 \\ 1+i & 0 & 1-i & 0 & 1 & 0 \\ 2 & 1 & 2+i & 0 & 0 & 1 \end{array}\right] \xrightarrow[\substack{-(1+i)\mathbf{r}_1+\mathbf{r}_2 \\ -2\mathbf{r}_1+\mathbf{r}_3}]{(1/2)\mathbf{r}_1}$

$\left[\begin{array}{ccc|ccc} 1 & (1/2)i & 3/2 & 1/2 & 0 & 0 \\ 0 & (1/2)-(1/2)i & (-1/2)-(5/2)i & (-1/2)-(1/2)i & 1 & 0 \\ 0 & 1-i & -1+i & -1 & 0 & 1 \end{array}\right]$

Next apply row operations $(1/((1/2)-(1/2)i))\mathbf{r}_2$, $(-1/2)i\mathbf{r}_2+\mathbf{r}_1$, and $-(1-i)\mathbf{r}_2+\mathbf{r}_3$ to obtain

$\left[\begin{array}{ccc|ccc} 1 & 0 & -i & 0 & (1/2)-(1/2)i & 0 \\ 0 & 1 & 2-3i & -i & 1+i & 0 \\ 0 & 0 & 6i & i & -2 & 1 \end{array}\right] \xrightarrow[\substack{-(2-3i)\mathbf{r}_3+\mathbf{r}_2 \\ i\mathbf{r}_3+\mathbf{r}_1}]{(1/6i)\mathbf{r}_3}$

$\left[\begin{array}{ccc|ccc} 1 & 0 & 0 & (1/6)i & (1/6)-(1/2)i & (1/6) \\ 0 & 1 & 0 & (-1/3)-(1/2)i & (1/3)i & (1/2)+(1/3)i \\ 0 & 0 & 1 & 1/6 & (1/3)i & (-1/6)i \end{array}\right]$

Thus the inverse is

$$\begin{bmatrix} (1/6)i & (1/6)-(1/2)i & 1/6 \\ (-1/3)-(1/2)i & (1/3)i & (1/2)+(1/3)i \\ 1/6 & (1/3)i & (-1/6)i \end{bmatrix}.$$

7. $\mathbf{X}_1 = (-1+i,\ 2,\ 1)$, $\mathbf{X}_2 = (1, 1+i,\ i)$, $\mathbf{X}_3 = (-5+2i,\ -1-3i,\ 2-3i)$

(a) Let $\mathbf{X} = (i,\ 0,\ 0)$. Form the expression

$$c_1\mathbf{X}_1 + c_2\mathbf{X}_2 + c_3\mathbf{X}_3 = (c_1(-1+i)+c_2+c_3(-5+2i),\ c_1 2+c_2(1+i)+c_3(-1-3i),\ c_1+c_2 i+c_3(2-3i)) = \mathbf{X} = (i,\ 0\ ,0).$$

Equating corresponding components from both sides of the equation gives the linear system

$$\begin{aligned} (-1+i)c_1 + \quad c_2 + (-5+2i)c_3 &= i \\ 2c_1 + (1+i)c_2 + (-1-3i)c_3 &= 0 \\ c_1 + \quad ic_2 + \quad (2-3i)c_3 &= 0. \end{aligned}$$

Form the augmented matrix

$$\left[\begin{array}{ccc|c} -1+i & 1 & -5+2i & i \\ 2 & 1+i & -1-3i & 0 \\ 1 & i & 2-3i & 0 \end{array}\right]$$

and row reduce the matrix to obtain the reduced row echelon form

$$\left[\begin{array}{ccc|c} 1 & 0 & 0 & (-2/5)-(4/5)i \\ 0 & 1 & 0 & (11/10)+(7/10)i \\ 0 & 0 & 1 & (3/10)+(1/10)i \end{array}\right].$$

Thus the system is consistent which implies that $\mathbf{X}$ belongs to W.

(b) Form the expression

$$c_1\mathbf{X}_1 + c_2\mathbf{X}_2 + c_3\mathbf{X}_3 = \mathbf{0}.$$

Expand, add the vectors, and equate corresponding components from both sides of the equation. We obtain the homogeneous system

$$\begin{aligned} (-1+i)c_1 + \quad c_2 + (-5+2i)c_3 &= 0 \\ 2c_1 + (1+i)c_2 + (-1-3i)c_3 &= 0 \\ c_1 + \quad ic_2 + \quad (2-3i)c_3 &= 0. \end{aligned}$$

Form the coefficient matrix and apply row operations to obtain the reduced row echelon form which is $\mathbf{I}_3$. Hence the only solution to the linear system is the trivial solution. Thus $\{\mathbf{X}_1, \mathbf{X}_2, \mathbf{X}_3\}$ is linearly independent.

9. (a) $\mathbf{A} = \begin{bmatrix} 1 & 1 \\ -1 & 1 \end{bmatrix}$. Then $|\lambda\mathbf{I}_2 - \mathbf{A}| = \begin{vmatrix} \lambda-1 & -1 \\ 1 & \lambda-1 \end{vmatrix} = \lambda^2 - 2\lambda + 2$.

The eigenvalues are $\lambda = 1+i$ and $\lambda = 1-i$. To find the corresponding eigenvectors we proceed as follows:

Case $\lambda = 1+i$

$$(\lambda\mathbf{I}_2 - \mathbf{A})\mathbf{X} = \begin{bmatrix} i & -1 \\ 1 & i \end{bmatrix}\begin{bmatrix} x_1 \\ x_2 \end{bmatrix} = \begin{bmatrix} 0 \\ 0 \end{bmatrix}$$

The reduced row echelon form of the coefficient matrix is $\begin{bmatrix} 1 & i \\ 0 & 0 \end{bmatrix}$. Thus $x_1 = -ic$, $x_2 = c$, where c is any complex number. Let c = 1. Then $\mathbf{X} = \begin{bmatrix} -i \\ 1 \end{bmatrix}$ is an eigenvector corresponding to eigenvalue $\lambda = 1+i$.

Case $\lambda = 1-i$

$$(\lambda\mathbf{I}_2 - \mathbf{A})\ \mathbf{X} = \begin{bmatrix} -i & -1 \\ 1 & -i \end{bmatrix}\begin{bmatrix} x_1 \\ x_2 \end{bmatrix} = \begin{bmatrix} 0 \\ 0 \end{bmatrix}$$

The reduced row echelon form of the coefficient matrix is $\begin{bmatrix} 1 & -i \\ 0 & 0 \end{bmatrix}$. Thus $x_1 = ic$, $x_2 = c$, where c is any complex number. Let c = 1. Then $\mathbf{X} = \begin{bmatrix} i \\ 1 \end{bmatrix}$ is an eigenvector corresponding to eigenvalue $\lambda = 1-i$.

(b) $\mathbf{A} = \begin{bmatrix} 1 & i \\ -i & 1 \end{bmatrix}$. Then $|\lambda\mathbf{I}_2 - \mathbf{A}| = \begin{vmatrix} \lambda-1 & -i \\ i & \lambda-1 \end{vmatrix} = \lambda(\lambda - 2)$.

The eigenvalues are $\lambda = 0$ and $\lambda = 2$. To find the corresponding eigenvectors we proceed as follows:

Case $\lambda = 0$

$$(\lambda\mathbf{I}_2 - \mathbf{A})\ \mathbf{X} = \begin{bmatrix} -1 & -i \\ i & -1 \end{bmatrix}\begin{bmatrix} x_1 \\ x_2 \end{bmatrix} = \begin{bmatrix} 0 \\ 0 \end{bmatrix}$$

The reduced row echelon form of the coefficient matrix is $\begin{bmatrix} 1 & i \\ 0 & 0 \end{bmatrix}$. Thus $x_1 = -ic$, $x_2 = c$, where c is any complex number. Let c = 1. Then $\mathbf{X} = \begin{bmatrix} -i \\ 1 \end{bmatrix}$ is an eigenvector corresponding to eigenvalue $\lambda = 0$.

Case $\lambda = 2$

$$(\lambda \mathbf{I}_2 - \mathbf{A})\ \mathbf{X} = \begin{bmatrix} 1 & -i \\ i & 1 \end{bmatrix}\begin{bmatrix} x_1 \\ x_2 \end{bmatrix} = \begin{bmatrix} 0 \\ 0 \end{bmatrix}$$

The reduced row echelon form of the coefficient matrix is $\begin{bmatrix} 1 & -i \\ 0 & 0 \end{bmatrix}$. Thus $x_1 = ic$, $x_2 = c$, where c is any complex number. Let c = 1. Then $\mathbf{X} = \begin{bmatrix} i \\ 1 \end{bmatrix}$ is an eigenvector corresponding to eigenvalue $\lambda = 2$.

(c) $\mathbf{A} = \begin{bmatrix} 2 & 0 & 0 \\ 0 & 2 & i \\ 0 & -i & 2 \end{bmatrix}$. Then $|\lambda \mathbf{I}_3 - \mathbf{A}| = \begin{vmatrix} \lambda-2 & 0 & 0 \\ 0 & \lambda-2 & -i \\ 0 & i & \lambda-2 \end{vmatrix} =$ $(\lambda-1)(\lambda-2)(\lambda-3)$. The eigenvalues are $\lambda = 1, 2, 3$. To find the corresponding eigenvectors we proceed as follows:

Case $\lambda = 1$

$$(\lambda \mathbf{I}_3 - \mathbf{A})\ \mathbf{X} = \begin{bmatrix} -1 & 0 & 0 \\ 0 & -1 & -i \\ 0 & i & -1 \end{bmatrix}\begin{bmatrix} x_1 \\ x_2 \\ x_3 \end{bmatrix} = \begin{bmatrix} 0 \\ 0 \\ 0 \end{bmatrix}$$

The reduced row echelon form of the coefficient matrix is $\begin{bmatrix} 1 & 0 & 0 \\ 0 & 1 & i \\ 0 & 0 & 0 \end{bmatrix}$. Thus $x_1 = 0$, $x_2 = -ic$, $x_3 = c$, where c is any complex number. Let c = 1. Then $\mathbf{X} = \begin{bmatrix} 0 \\ -i \\ 1 \end{bmatrix}$ is an eigenvector corresponding to eigenvalue $\lambda = 1$.

Case $\lambda = 2$

$$(\lambda \mathbf{I}_3 - \mathbf{A})\,\mathbf{X} = \begin{bmatrix} 0 & 0 & 0 \\ 0 & 0 & -i \\ 0 & i & 0 \end{bmatrix} \begin{bmatrix} x_1 \\ x_2 \\ x_3 \end{bmatrix} = \begin{bmatrix} 0 \\ 0 \\ 0 \end{bmatrix}$$

The reduced row echelon form of the coefficient matrix is $\begin{bmatrix} 0 & 1 & 0 \\ 0 & 0 & 1 \\ 0 & 0 & 0 \end{bmatrix}$. Thus $x_1 = c$, $x_2 = 0$, $x_3 = 0$, where c is any complex number. Let c = 1. Then $\mathbf{X} = \begin{bmatrix} 1 \\ 0 \\ 0 \end{bmatrix}$ is an eigenvector corresponding to eigenvalue $\lambda = 2$.

Case $\lambda = 3$

$$(\lambda \mathbf{I}_3 - \mathbf{A})\,\mathbf{X} = \begin{bmatrix} 1 & 0 & 0 \\ 0 & 1 & -i \\ 0 & i & 1 \end{bmatrix} \begin{bmatrix} x_1 \\ x_2 \\ x_3 \end{bmatrix} = \begin{bmatrix} 0 \\ 0 \\ 0 \end{bmatrix}$$

The reduced row echelon form of the coefficient matrix is $\begin{bmatrix} 1 & 0 & 0 \\ 0 & 1 & -i \\ 0 & 0 & 0 \end{bmatrix}$. Thus $x_1 = 0$, $x_2 = ic$, $x_3 = c$, where c is any complex number. Let c = 1. Then $\mathbf{X} = \begin{bmatrix} 0 \\ i \\ 1 \end{bmatrix}$ is an eigenvector corresponding to eigenvalue $\lambda = 3$.

T.1. (a) $\overline{a_{ii}} = a_{ii}$, hence a_{ii} is real. See Property 4 in Appendix A.1.

(b) First, $\overline{\mathbf{A}}^T = \mathbf{A}$ implies that $\mathbf{A}^T = \overline{\mathbf{A}}$. Let $\mathbf{B} = \dfrac{\mathbf{A} + \overline{\mathbf{A}}}{2}$. Then

$$\overline{\mathbf{B}} = \overline{\left(\frac{\mathbf{A} + \overline{\mathbf{A}}}{2}\right)} = \frac{\overline{\mathbf{A}} + \overline{\overline{\mathbf{A}}}}{2} = \frac{\overline{\mathbf{A}} + \mathbf{A}}{2} = \frac{\mathbf{A} + \overline{\mathbf{A}}}{2} = \mathbf{B}$$

so $\mathbf{B}$ is a real matrix.

$$\text{Also, } \mathbf{B}^T = \left(\frac{\mathbf{A} + \overline{\mathbf{A}}}{2}\right)^T = \frac{\mathbf{A}^T + (\overline{\mathbf{A}})^T}{2} = \frac{\mathbf{A}^T + \overline{\mathbf{A}^T}}{2} = \frac{\overline{\mathbf{A}} + \overline{\overline{\mathbf{A}}}}{2}$$

$$= \frac{\overline{\mathbf{A}} + \mathbf{A}}{2} = \frac{\mathbf{A} + \overline{\mathbf{A}}}{2} = \mathbf{B} \quad \text{so } \mathbf{B} \text{ is symmetric.}$$

Next, let $\mathbf{C} = \dfrac{\mathbf{A} - \overline{\mathbf{A}}}{2i}$. Then $\overline{\mathbf{C}} = \overline{\left[\dfrac{\mathbf{A} - \overline{\mathbf{A}}}{2i}\right]} = \dfrac{\overline{\mathbf{A}} - \overline{\overline{\mathbf{A}}}}{-2i}$

$= \dfrac{\mathbf{A} - \overline{\mathbf{A}}}{2i} = \mathbf{C}$ so $\mathbf{C}$ is a real matrix.

Also, $\mathbf{C}^T = \left[\dfrac{\mathbf{A} - \overline{\mathbf{A}}}{2i}\right]^T = \dfrac{\mathbf{A}^T - (\overline{\mathbf{A}})^T}{2i} = \dfrac{\mathbf{A}^T - \overline{\mathbf{A}^T}}{2i} = \dfrac{\overline{\mathbf{A}} - \mathbf{A}}{2i}$

$$= -\left[\frac{\mathbf{A} - \overline{\mathbf{A}}}{2i}\right] = -\mathbf{C}$$

so $\mathbf{C}$ is skew symmetric. Moreover, $\mathbf{A} = \mathbf{B} + i\mathbf{C}$.

(c) Let $\mathbf{A}$ and $\mathbf{B}$ be Hermitian and let k be a complex scalar. Then $(\overline{\mathbf{A} + \mathbf{B}})^T = (\overline{\mathbf{A}} + \overline{\mathbf{B}})^T = \overline{\mathbf{A}}^T + \overline{\mathbf{B}}^T = \mathbf{A} + \mathbf{B}$, so the sum of Hermitian matrices is again Hermitian. Next, $(\overline{k\mathbf{A}})^T = (\overline{k}\ \overline{\mathbf{A}})^T = \overline{k}\ \overline{\mathbf{A}}^T = \overline{k}\ \mathbf{A} \neq k\mathbf{A}$, so the set of Hermitian matrices is not closed under scalar multiplication and hence is not a complex subspace of the complex vector space of $n \times n$ complex matrices.

(d) From (c), we have closure of addition and since the scalars are real here, $\overline{k} = k$, hence $(\overline{k\mathbf{A}})^T = k\mathbf{A}$. Thus W is a real subspace of the real vector space of all $n \times n$ complex matrices.

(e) If $\mathbf{A} = \mathbf{A}^T$ and $\mathbf{A} = \overline{\mathbf{A}}$ then $(\overline{\mathbf{A}}^T) = \mathbf{A}^T = \mathbf{A}$. Hence, $\mathbf{A}$ is Hermitian.

T.3. (a) Let $\mathbf{A}$ be Hermitian and suppose that $\mathbf{AX} = \lambda\mathbf{X}$, $\lambda \neq 0$. we show that $\lambda = \overline{\lambda}$. We have

$$(\overline{\mathbf{AX}})^T = (\overline{\mathbf{A}}\ \overline{\mathbf{X}})^T = \overline{\mathbf{X}}^T\overline{\mathbf{A}}^T = \overline{\mathbf{X}}^T\mathbf{A} = \overline{\lambda}\ \overline{\mathbf{X}}^T.$$

Multiplying both sides on the right by $\mathbf{X}$ we obtain

$$\overline{\mathbf{X}}^T\mathbf{AX} = \overline{\mathbf{X}}^T\lambda\mathbf{X} = \lambda\overline{\mathbf{X}}^T\mathbf{X} = \overline{\lambda}\ \overline{\mathbf{X}}^T\mathbf{X}.$$

Rearranging we have $(\lambda - \overline{\lambda})\overline{\mathbf{X}}^T\mathbf{X} = 0$ and since $\overline{\mathbf{X}}^T\mathbf{X} > 0$ we have $\lambda = \overline{\lambda}$. Thus the eigenvalues of a Hermitian matrix are real.

Exercises A.2

(b) $$\overline{\mathbf{A}}^T = \begin{bmatrix} 2 & 0 & 0 \\ 0 & 2 & \overline{-i} \\ 0 & \overline{i} & 2 \end{bmatrix} = \begin{bmatrix} 2 & 0 & 0 \\ 0 & 2 & i \\ 0 & -i & 2 \end{bmatrix} = \mathbf{A}$$

(c) No, see Exercise 9(b). An eigenvector $\mathbf{X}$ associated with a real eigenvalue λ of a complex matrix $\mathbf{A}$ is, in general, complex, because $\mathbf{AX}$ is, in general, complex. Thus $\lambda\mathbf{X}$ must also be complex.

T.5. (a) Let $\mathbf{B} = \dfrac{\mathbf{A} + \overline{\mathbf{A}^T}}{2}$ and $\mathbf{C} = \dfrac{\mathbf{A} - \overline{\mathbf{A}^T}}{2i}$. Then

$$\overline{\mathbf{B}^T} = \overline{\left[\frac{\mathbf{A} + \overline{\mathbf{A}^T}}{2}\right]^T} = \frac{\overline{\mathbf{A}^T} + \overline{(\overline{\mathbf{A}^T})^T}}{2} = \frac{\overline{\mathbf{A}^T} + \mathbf{A}}{2} = \frac{\mathbf{A} + \overline{\mathbf{A}^T}}{2} = \mathbf{B}$$

so $\mathbf{B}$ is Hermitian.

Also, $$\overline{\mathbf{C}^T} = \overline{\left[\frac{\mathbf{A} - \overline{\mathbf{A}^T}}{2i}\right]^T} = \frac{\overline{\mathbf{A}^T} - \overline{(\overline{\mathbf{A}^T})^T}}{-2i} = \frac{\mathbf{A} - \overline{\mathbf{A}^T}}{2i} = \mathbf{C}$$

hence $\mathbf{C}$ is Hermitian. Moreover, $\mathbf{A} = \mathbf{B} + i\mathbf{C}$.

(b) We have
$$\begin{aligned}\overline{\mathbf{A}^T}\mathbf{A} &= (\overline{\mathbf{B}^T + i\mathbf{C}^T})(\mathbf{B} + i\mathbf{C}) = (\overline{\mathbf{B}^T} + \overline{i\mathbf{C}^T})(\mathbf{B} + i\mathbf{C}) \\ &= (\mathbf{B} - i\mathbf{C})(\mathbf{B} + i\mathbf{C}) \\ &= \mathbf{B}^2 - i\mathbf{CB} + i\mathbf{BC} - i^2\mathbf{C}^2 \\ &= (\mathbf{B}^2 + \mathbf{C}^2) + i(\mathbf{BC} - \mathbf{CB}) .\end{aligned}$$

Similarly,
$$\begin{aligned}\mathbf{A}\overline{\mathbf{A}^T} &= (\mathbf{B} + i\mathbf{C})(\overline{\mathbf{B}^T + i\mathbf{C})^T} = (\mathbf{B} + i\mathbf{C})(\overline{\mathbf{B}^T} + \overline{i\mathbf{C}^T}) \\ &= (\mathbf{B} + i\mathbf{C})(\mathbf{B} - i\mathbf{C}) \\ &= \mathbf{B}^2 - i\mathbf{BC} + i\mathbf{CB} - i^2\mathbf{C}^2 \\ &= (\mathbf{B}^2 + \mathbf{C}^2) + i(\mathbf{CB} - \mathbf{BC}) .\end{aligned}$$

Since $\overline{\mathbf{A}^T}\mathbf{A} = \mathbf{A}\overline{\mathbf{A}^T}$, we equate imaginary parts obtaining

$$\mathbf{BC} - \mathbf{CB} = \mathbf{CB} - \mathbf{BC}$$

which implies that $\mathbf{BC} = \mathbf{CB}$. The steps are reversible, establishing the converse.